Wireless Java Programming for Enterprise Applications

Mobile Devices Go Corporate

Dan Harkey • Shan Appajodu • Mike Larkin

WILEY PUBLISHING, INC.

Publisher: Robert Ipsen
Editor: Theresa Hudson
Developmental Editor: Kathryn Malm
Managing Editor: Angela Smith
New Media Editor: Brian Snapp
Text Design & Composition: Dan Harkey, Shan Appajodu, and Mike Larkin
Graphic Art: Shweta Saokar

Library of Congress Cataloging-in-Publication Data:

ISBN 0-471-21878-2 (paper : alk. paper)

Wiley also publishes its books in a variety of electronic formats. Some content that appears in print may not be available in electronic versions.

For more information about Wiley products, visit our Web site at www.wiley.com.

Printed in the United States of America
10 9 8 7 6 5 4 3 2 1

Dedication

This book is dedicated to family and the ladies in my life, Tomomi Harkey, Wei Zhao, and Michiko Omatsu.

— ***Dan Harkey***

This book is dedicated to my family, especially my parents Rajyalakshmi and Chandramouli and my wife Deepika.

— ***Shan Appajodu***

This book is dedicated to my family and friends for all their support during the past six months.

— ***Mike Larkin***

Foreword

by Zog the Martian

Captain Zog

Greetings, Earthlings! I'm Zog, the captain of the Martian team. My team and I have visited Earth several times to understand what client/server and enterprise computing are all about. In our previous visits, we discovered the **Client/Server Survival Guide** and **The Essential Distributed Objects Survival Guide** and then we took a wild ride in **Client/Server Programming with Java and CORBA**. Now we are taking advantage of the new wireless technologies described in this book. It was just what we were looking for as it lets us watch what's happening on Earth from afar (even Mars).

So what did I like about this book? It felt like it was talking directly to me and to my crew in a friendly voice. That's very important when you're from a foreign planet. The artwork is absolutely wonderful. I like to see pictures of myself in books. It's not that inert boring stuff you read in the textbooks. It makes programming come to life.

We think you will find a lot of new opportunity with the wireless technologies that are presented in this book. Martians have always felt that they were untethered by technology and now we are glad that our Earthing friends can feel the same way. And, we're delighted to see how handheld devices, wireless technologies, and enterprise technologies are combined together to create some dynamite applications.

You might also be wondering what happened to Robert Orfali and Jeri Edwards. They decided to take the advice that was given to many readers in previous books to take a sabbatical by the beach. We wish them well and hope that we (and the new team of authors) can join them soon!

Zog

Preface

We're at the cusp of a new class of enterprise application, one that incorporates wireless devices as clients. These devices not only put a new face on our enterprise applications, but also open up new possibilities. A new class of application that makes *communication and access to personal information available anywhere and anytime* becomes possible. These applications not only take advantage of the convenience of a connected device that is always available, but also one that is connected to the power of enterprise applications.

The handheld device provides the convenience of an appliance that is always available, and that provides personal information and task management. It can also do much more. Wireless devices can provide a variety of communications types: from regional connections via a cellular network to Personal Area Networks, like those provided by Bluetooth. It can also provide services that are unique to handheld devices such as location-based services. Want to know where the closest Starbucks is located? Your handheld device will be able to tell you. As the features of handheld devices improve, there will also be many other possibilities—such as multimedia applications, voice recognition, and improved security and personalization via Java Card.

The handheld device as a standalone appliance is interesting, but limited to what can be accessed locally on the device. This is where the enterprise infrastructure steps in to help. The enterprise not only provides services for communication, collaboration, access to the Internet, and any of a number of business oriented tasks, it also complements the handheld device by providing a structure to compensate for some of the limitations of handheld devices, as well as a way to support a variety of handheld devices.

Java: From Micro to Enterprise

A few years back, the enterprise faced a challenge where there were many fragmented programming environments. There were many operating systems, each having its own programming environment. A variety of languages also flourished. A programmer was faced with writing programs for a limited market segment and not having skills that transferred to other programming environments.

These problems have not totally disappeared, but the introduction of *Java 2 Standard Edition (J2SE)* and *Java 2 Enterprise Edition (J2EE)* as a language and a programming platform has provided a common development environment for many, many enterprise programmers. A programmer's skills no longer need to be tied to a specific operating system. Java is also a rich platform that provides numerous services—including those for database access, communication, security, naming and directory services, and many more.

Wireless devices face the same fragmentation, but to a larger degree given the variety of handheld devices. *Java 2 Micro Edition (J2ME)* plays the same role for handheld devices as J2SE and J2EE play for enterprise applications. J2ME provides a programming platform with the familiar Java language that allows you to deploy the same application across a class of handheld devices with like capabilities. It also provides a set of APIs that are available for classes of devices. In addition, there is a strong affinity when connecting J2ME applications that run on a handheld device with J2EE applications that execute on enterprise hardware.

So, opportunity awaits you to make your applications available in a wireless enterprise setting.

What This Book Covers

This book explains the use of wireless handheld devices as clients for enterprise applications. Java 2 Micro Edition is used to provide a common programming platform for wireless devices. A number of complementary wireless technologies such as WAP, VoiceXML, and Java Card are also explored.

Using wireless devices as a client for enterprise applications presents a number of challenges. There is a large variety of wireless devices, they are very limited in their processing capability, they have inherently slow and unreliable network connections, and have limited user interface capabilities. Fortunately, the enterprise infrastructure can be leveraged to overcome some of these difficulties. This book explores using all of the available technologies, both handheld device and enterprise, to implement an architecture for enterprise applications that incorporates wireless device clients.

This book consists of four Parts that build on each other:

- **Part 1** starts with an overview of the technologies that influence enterprise application development for wireless devices. This includes enterprise architectures, device features, wireless networking protocols, wireless service technologies, middleware specific to wireless devices, and content formats for wireless devices. The part concludes with an overview of Java for wireless devices—

including *Java 2 Micro Edition (J2ME)* and its various profiles and configurations, and *Java Card*.

- **Part 2** starts our exploration of programming and content development models for wireless device application development. In this part we focus on developing client applications and don't worry about connecting them to the power the enterprise computing infrastructure provides. A user interface for a university registration example is developed using the *Wireless Markup Language (WML)*, J2ME's *Mobile Information Device Profile (MIDP)*, and *VoiceXML*. Some of the more advanced features of MIDP, such as networking (including Bluetooth), Record Store Management (RMS), and low-level GUI programming are also explored. Part 2 also includes a Java Card programming example.

- **Part 3** gives an introduction to the server-side technologies important for the development of enterprise applications that include wireless clients. It is server-side technologies that make it possible to support a variety of clients for each enterprise application. We introduce the concepts of *transcoding* and *personalization* (for individuals, devices, and applications) to avoid writing applications for each device supported and to enhance the user experience at the device. *Java 2 Enterprise Edition (J2EE)* is explored as a platform for developing the transcoding and personalization infrastructure for wireless devices and as a platform for writing applications. *Web Services* is also explored as a technology for extending the reach of our application and as an integration technology. Finally, we look at techniques for connecting our MIDP applications to the enterprise infrastructure using XML and give you some best practices for developing J2ME applications that connect to the enterprise.

- **Part 4** concludes—in grand finale style—with the *Campus Portal for Wireless Devices*. This Part brings it all together. The Campus Portal demonstrates providing services to clients that include browsers, handheld devices, and voice-only devices for students in a university setting. The architecture incorporates transcoding and personalization and delivers services to clients via Web Services. Small applications such as finding a room or a grade for a university course are demonstrated. The part concludes with a look ahead at the technologies for wireless enterprise application development.

A lot of this material covers previously uncharted territory. You may have read about some of the individual technologies but we think you will find that all of the technologies have been brought together like instruments into a fine symphonic orchestra.

What Are the Boxes For?

We use shaded boxes as a way to introduce concurrent threads in the presentation material. It's the book's version of multitasking. The *Soapboxes* introduce strong opinions or biases on some of the more controversial topics of distributed object computing. Because the discipline is so new and fuzzy, there's lots of room for interpretation and debate—so Soapboxes that are just another opinion (ours). The *Briefing* boxes give you background or tutorial type information. You can safely skip over them if you're already familiar with a topic. The *Details* boxes cover some esoteric area of technology that may not be of interest to the general readership. Typically, the same readers that skip over the briefings will find the details interesting (so you'll still get your money's worth). Lastly, we use *Warning* boxes to let you know where danger lies—this is, after all, previously uncharted territory.

How to Read This Book

As was recommended in previous books, it's best to ask your boss for a one-week, paid sabbatical to go sit on a beach and read this book. Tell him or her that it's the cheapest way to revitalize yourself technically and find out all there is to know about wireless Java technologies for the enterprise. Be sure to take a laptop with you and have a wireless connection so you explore the Web site for the book. You can even tell your boss that you will stay in touch via a wireless connection. Hopefully you will have a wireless device or two to test your programs. If not, there are emulators that can simulate your client devices on your laptop. So, once you sink into that comfortable chair overlooking the ocean, we think you'll find the book a lot of fun—maybe even downright relaxing.

Who Is This Book For?

If you are considering this book, you probably have some experience with either programming enterprise level applications or programming for handheld devices. This books takes you to the next level where you combine the handheld device skills with the enterprise skills to write new applications. This book is for anyone involved with delivering either enterprise or wireless applications. It's particularly important for the following readers:

- Wireless device programmers who are looking to add new value to their applications by extending them to the enterprise.
- Enterprise programmers that want to make their applications more accessible via wireless devices.

- MIS programmers who are evaluating the latest technologies.

- Architects who are thinking about combining wireless and enterprise technologies. This is the first book that exhaustively looks at Java as a technology from Micro to Enterprise. You'll get the good, the bad, and the ugly.

Eventually, every programmer will have to write a wireless client for an enterprise application. So you'll find that reading this book is time well spent.

The Book Web Site

The amount of code developed for this book was too voluminous to put every line in the book. So, be sure and visit the Web site at ***http://wirelessjava.engr.sjsu.edu***. Here are some of the items you'll find on the Web site:

- ✔ Source code for all the programs in this book
- ✔ Links to all of the programming tools used for developing the sample programs
- ✔ FAQs, code updates, and hints for program development

Because all this area is in flux, be sure to visit the Web site regularly as you read the book and develop new wireless enterprise applications.

Acknowledgments

A book like this does not get created without a lot of help. Many thanks to:

- The faculty and staff at San Jose State University for providing us with a world-class environment for a graduate research program.
- The many companies that have donated hardware and software to the lab at San Jose State University—including BEA Systems, Borland, Ericcson, Hewlett-Packard, IBM, Microsoft, Nokia, Oracle, Palm, Schlumberger, Sun Microsystems, Zucotto Wireless, and many more.
- The Campus Portal Team at San Jose State University (left to right: Mike Larkin, Mei Xue, Garrick Shin, Sirisha Bhusarapu, Hui (Harry) Tang, Tjahjadinata (TJ) Setiawan, Ravi Malhotra, Laxmi Nagarajan, Shan Appajodu, Vidya Shivkumar, Dan Harkey, Jahnavi Prasad, Malathi Gopalan, Yashwanth Rajaram, Ruby Yeung, Tommy So, Doris Hong, Prasanth Dhulipala, James Ye, Hans-Joachim Kirsch, John Gash, and Tom Chiang).

- A special thanks to Prasanth Dhulipala for his tireless help throughout the creation of this book.
- Shweta Saokar for creating the wonderful technical illustrations in this book .
- Our copyeditor, Dan MacWilliams.
- Our proofreader, Claudia Allen.
- The people at Wiley—especially Terri Hudson, Angela Smith, and Bob Ipsen.

Contents

Part 1
Technologies for Wireless Device Computing

An Introduction to Part 1

You're probably a user of handheld devices such as cellular phones, two-way pagers, and Personal Digital Assistants (PDAs). These devices provide very personal communication and productivity features. Many of these devices are limited to functions that are built into the device (especially cell phones) or that can be loaded onto the device from a personal computer using synchronization (as is commonly the case with PDAs).

A new class of connected and programmable device is becoming available in the market. These are cell phones and two-way pagers that allow new programs to be loaded onto the device, and PDAs that are connected to wireless networks. This means a handheld device can provide custom applications that connect to a network. The network can then make enterprise level services accessible via the handheld device. This allows users of handheld devices to have *personal communication and data access, anytime and anywhere* and opens up entirely new business opportunities for us as developers.

Is it a simple space to enter? Not at all. Handheld devices provide a very limited programming environment. There are also a variety of wireless networks to understand, plus choices between programming environments and content delivery systems. There are many different handheld devices (each with its own software and hardware characteristics), and special development techniques are required at the enterprise level to support handheld devices. And the list goes on. This book will introduce you to the technologies and concepts that you will need to understand before you can begin to write programs for handheld devices that effectively utilize enterprise services.

Handheld devices provide an explosion of choices when it comes to a programming environment. Fortunately, *Java 2 Micro Edition (J2ME)* provides a common software platform that can be used across a number of devices. This allows us to develop one program that can be used across a class of devices, all with similar capabilities. So, just as Java simplified our life at the enterprise level, it comes to the rescue again when writing applications for our handheld devices.

Here's a brief description of the chapters in Part 1:

- ***Chapter 1*** first gives you a bird's eye view of wireless and enterprise applications, and then how they might be combined together to deliver applications services to a handheld device. Different types of wireless networks are also described—from personal to regional—and an enterprise architecture for applications with wireless clients is proposed.

- ***Chapter 2*** categorizes the variety of handheld devices including two-way pagers, cell phones, Internet-ready phones, Java-enabled phones and PDAs, and high-end devices called communicators.

- *Chapter 3* introduces you to the terminology and standards for wireless networks.

- *Chapter 4* describes applications that use a wireless network to deliver data-driven functions that are built into a handheld device. This includes applications such as the Short Message Service (SMS), the Wireless Application Protocol (WAP) for delivering Wireless Markup Language (WML) content, and the i-mode service provided by NTT DoCoMo in Japan for wireless service delivery. We also give you a gentle introduction to VoiceXML for voice-driven applications.

- *Chapter 5* examines the programming environment for handheld devices. This includes middleware for handheld device operating systems and the special needs of wireless communication protocols, security, and micro-databases.

- *Chapter 6* looks at some of the content display formats that are used with handheld devices—including cHTML, XHTML, and WML. Formats for synchronizing data between a handheld device and a PC or other external systems, such as Web Clipping and SyncML, are also described.

- *Chapter 7* describes the Java programming environment and in particular how it applies to handheld and other limited devices. We also introduce the ultimate limited programming device, the Java Card, which implements a Java VM on a chip placed on a credit card.

We hope that Part 1 will give you the big picture overview, terminology, and confidence to tackle the programming examples in the remainder of the book.

Chapter 1

Introduction to Wireless Enterprise Applications

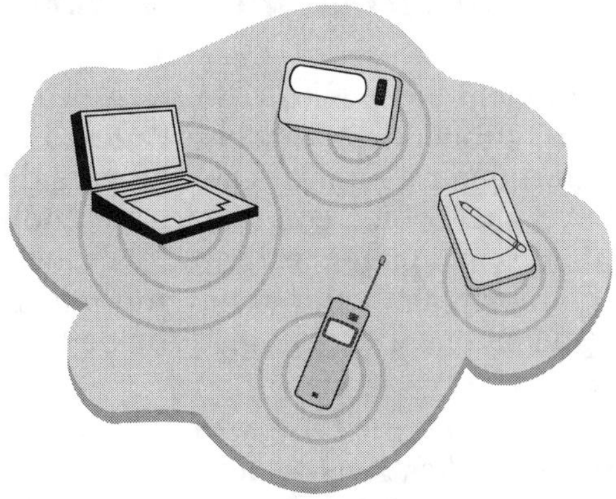

We begin our journey with an overview of applications for both wireless devices and enterprise computing, then a blend of both. Most likely you use wireless devices—such as pagers, cell phones, and perhaps a connected PDA—as simple communication devices. We will go beyond this limited functionality and characterize the usage of these devices for application delivery, first as micro-browser clients that can display Internet content, then as a programming platform where applications can be written, deployed, and executed. This initial discussion will set the stage for a more detailed discussion of the usage of wireless devices for application delivery in the remainder of Part 1.

Enterprise applications are probably more familiar to you. This is the world of Web servers, application servers, HTTP, HTML, XML, Java, Servlets, EJBs, and JSPs, just to name a few of the buzz-words that we have been chasing as part of the Internet revolution. We'll remind ourselves of these technologies, the software and hardware platforms they run on, and their predominant architectures.

We will then take a look at how wireless devices can be combined with enterprise applications to leverage the strengths of both to provide applications that offer personalized communication and information access, anytime and anywhere.

We will also begin the discussion of how to tailor the use of technologies often viewed as server technologies (such as Java and XML) on both handheld devices and servers, to efficiently deliver enterprise applications.

APPLICATIONS ON WIRELESS DEVICES

Handheld devices to date have primarily been personal communication and personal productivity devices. Pagers notify us when someone is trying to contact us, and the more sophisticated pagers deliver a simple text message. Cellular phones have made voice communication mobile. The most obvious advantage of both cell phones and pagers is they are always connected and always available. However, for the most part, cell phones and pagers do not provide a way to download and run applications, nor do they provide access to data or Internet content.

Personal Digital Assistants (PDAs), on the other hand, provide personal productivity applications—such as keeping our address books, calendars, memos, and to-do lists. PDAs also provide a programming platform, although, it's a limited platform with slow processor speed, limited memory, small persistent storage capacity, small screen size, and slow cumbersome data entry. Nevertheless, numerous applications exist for loading onto these devices, usually by synchronizing from a PC. Unlike with cell phones, some form of communication outside of a PDA device is not a given.

This characterization of wireless devices is quickly changing. Phones and pagers are becoming application clients—first as micro-browser clients that can display Internet content, then as clients that can download and run applications that take advantage of the ever-present communication network that cell phones and pagers rely on. PDAs, on the other hand, are becoming more connected. The connectivity may not actually be built into a PDA, but many add-on products exist that make PDAs as connected as cell phones. There is also a trend where cell phones and PDAs are combined into the same handheld device.

ENTERPRISE APPLICATIONS

Compared to applications for wireless devices, enterprise applications are behemoths. These are applications that depend on constant, reliable connectivity, powerful processors, unlimited memory, huge storage capacity, and sophisticated user interfaces.

Enterprise server hardware is gigantic. Server farms store corporate data and Internet content by the giga- and tera-byte. Multi-clustered processors access and process corporate data then return it to clients via very reliable high-speed networks.

Client hardware is no small item either. Processors in the gigahertz range, memory measured in the hundreds of megabytes, storage space measured in gigabytes, high-speed networking, and large true-color displays are quite commonplace.

In addition, the software infrastructure is just as large and sophisticated. On servers, relational databases navigate through tons of data at lightning speed, locating the few bytes of data important to a specific user's application. Data mining and data warehousing software chews through mountains of data, correlating a multitude of facts to reach a conclusion only a very powerful computer could discern. Web servers and application servers host enterprise-class applications and deliver Internet content and application information around the world. They also exchange information within a corporation using an intranet and with business partners over the Internet using a variety of communication mechanisms.

ENTERPRISE APPLICATIONS AND WIRELESS DEVICES COMBINED

So, we have the extremes of two worlds: that of handheld devices which are always available (but are very limited for application execution) and the enterprise computing infrastructure with its capacity to run enterprise-class applications (if you're willing to "go to the mountain"). It is the intersection of these two environments that allows a new class of application where personal communication and information access is available anytime and anywhere (see Figure 1-1).

So how does this intersection come about?

The first step is to have connectivity so that our handheld devices become wireless devices that can function as a client to our enterprise applications. We will explore the variety of connectivity and communication mechanisms, first at the wireless network level in Chapter 3, then at the communication middleware level in Chapter 5.

Connectivity allows the handheld device to act as a browser client (with the help of an embedded micro-browser) or as a voice client for a voice-driven application. Content browsing on handheld devices is an area whose standards are continually evolving. *WML (Wireless Markup Language)* is a dominant standard but devices also support other content types. The variety of content types supported by wireless devices as well as PC browsers and voice clients makes development of an application for these devices challenging. We'll explore the variety of content types in Chapter 6 and then go into the details of writing and deploying WML content in Part 2.

The next step in the evolution of a handheld device as a client for an enterprise application is to have a complete programming environment on the handheld device. Note that this is not the embedded programming environment used to

Figure 1-1. Personal Communication and Data Access, Anywhere and Anytime.

produce the device's firmware. The programming environment provides the framework for developing end-user applications that enhance the functionality of the device after the consumer purchases it. These applications should be easily downloadable from the device network connection at the user's whim. For our purposes, these applications also take advantage of an enterprise computing infrastructure.

A programming environment allows applications to run on a wireless device that is more sophisticated than those that can be provided by content browsers. It also allows some independence from the handheld device's connectivity, something that is not always available and is inherently unreliable when present. For example, how many of you have experienced going out of range of your cellular provider's service area or had a dropped call? We will explore handheld devices as programming platforms in more detail later, first as a hardware platform in Chapter 2 then revisit Chapter 5 for a discussion of operating systems and other middleware for handheld devices.

Programming environments exist on many devices today. Unfortunately, this is an area where there are a plethora of devices, each with its own proprietary programming environment. There is some hope though. In the same way Java has provided a more homogenous programming environment for enterprise computing, some headway is being made for a Java environment for smaller devices. Given the hardware and software constraints of consumer devices, this is no small feat.

We will explore Java as a programming environment for handheld devices in three broad categories:

- ***Connected, Limited Devices*** such as pagers, cell phones, and PDAs.
- ***Consumer Devices*** where a keyboard, a relatively large display, and a wireless connection exist.
- ***Java Cards*** or smart cards with Java programming capability.

We'll explore the details of Java support for these devices in Chapter 7. We will also explore the details of programming Java applications in Part 2.

An Enterprise Computing Architecture for Wireless Devices

We are now at a point where we can propose an architecture for using wireless devices for our enterprise applications (see Figure 1-2). We will spend a lot of time describing this architecture, particularly in Parts 3 and 4. At this point we will take some time to whet your appetite for what is to come by giving you the highlights. If you've read previous books by Orfali, Harkey, and Edwards, you will recognize the 3-tier client/server architecture in Figure 1-2. Starting from the left, there are wireless device clients, voice clients, and browser clients.

The middle tier is the bastion of *Java 2 Enterprise Edition (J2EE)* components: *Servlets*, *JavaServer Pages (JSPs)*, and *Enterprise JavaBeans (EJBs)*. These components are the building blocks for the server side of our enterprise applications. J2EE also incorporates a Web server for serving content to clients. There are also three new subsystems that are introduced in the middle tier to help us cope with the variety of wireless clients we wish to support:

- The ***Transcoding Engine*** treats the user interface as a virtual client and transforms the user interface to fit a specific client type. Client types include wireless devices that are Internet ready (such as WML/WAP phones), devices that support a programming environment like *Java 2 Micro Edition (J2ME)*, voice-driven clients (where the voice synthesis is controlled within the Wireless Portal using VoiceXML), and browsers. The transcoding engine uses *eXtensible Markup Language (XML)* and its related technologies to make the transformation to the markup required for a specific device.

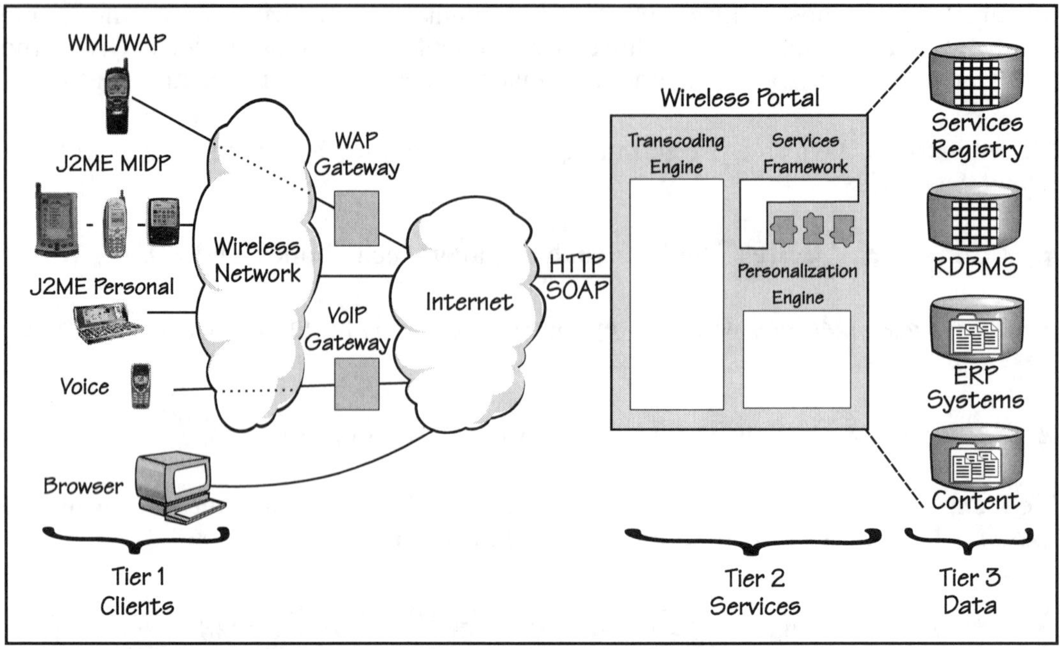

Figure 1-2. An Architecture for Enterprise Application Delivery to Wireless Devices.

- The ***Personalization Engine*** keeps information about each user's preferences and tailors the application interaction with a user and device based on the user's preferences. The personalization engine also records user input for application forms so that if the form is presented again, some of the entry fields are pre-filled to make entry easier for the user. This is important given the limited input capability of many wireless devices.

- The ***Services Framework*** allows enterprise applications to be added incrementally. This is important given the need to quickly add many small applications from a variety of sources. The services framework uses the developing Web Services standards such as the *Universal Description, Discovery, and Integration (UDDI)* registry (and its related Web Services technologies) for service publishing and discovery.

The third tier of the architecture is the data tier. This tier provides access to enterprise data of all types and is not specific to wireless devices. The Wireless Portal components will be described more extensively in Parts 3 and 4.

So, given all of this description, where does this leave us? It must be time for a prediction (see Figure 1-3). You will probably remember how books by Orfali, Harkey, and Edwards characterized the development of enterprise computing as a series of waves of increasing complexity in a wired, networked environment. In the

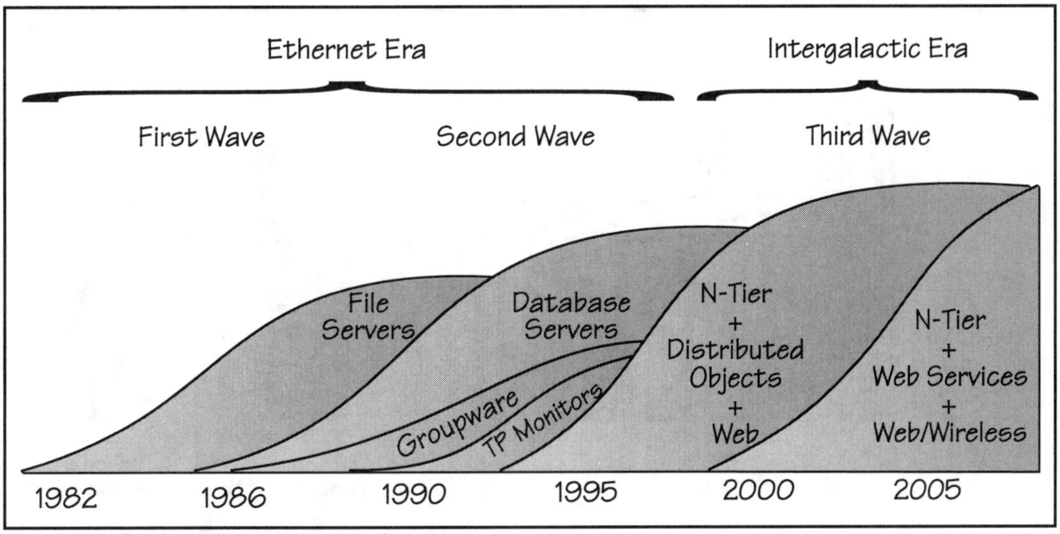

Figure 1-3. The Next Wave in Enterprise Computing.

last "waves" diagram you may have seen, it was purely a wired world and the software architecture featured n-tier client/server applications. The next wave is a combination of wired (the Web) and wireless delivery of applications. The n-tier architecture still exists but it is embellished with Web Services. This allows the incremental delivery of wireless applications and makes them available intergalactically.

Wireless Connectivity from Personal to Regional

This future world of wireless connectivity will also provide some truly exciting possibilities. Three classes of wireless connectivity for handheld devices are forming. They can be characterized as regional, local, and personal (see Figure 1-4).

- ***Regional*** wireless connectivity is best characterized by cellular phone networks. The coverage area is measured in miles or kilometers and gives access to phone and data services within a wide range. Today the data services are slow and are measured in the tens of Kb/sec (except for some specialized services), but you can't beat the coverage.

  The next generations of regional wireless coverage—called 3G (third generation), 4G (fourth generation), and so on—will substantially increase the data rates and add new communications services for handheld features like multimedia. Regional wireless connectivity also introduces technology for determining the location of a device and then providing services based on the location.

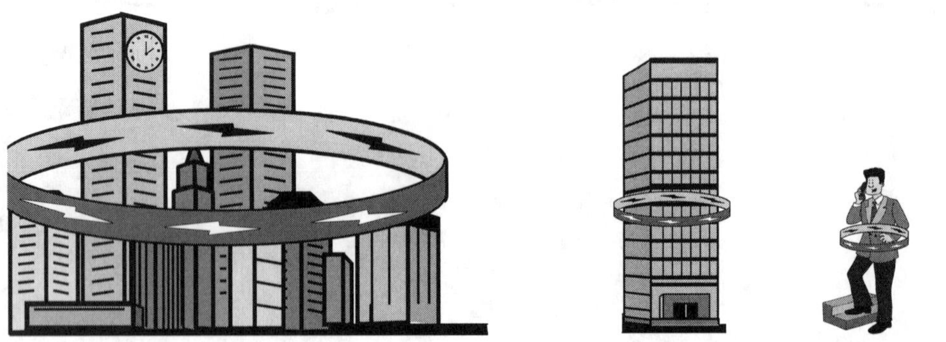

Figure 1-4. Wireless Connectivity from Personal to Regional.

- ***Local*** wireless connectivity, which is gaining momentum, is characterized by the 802.11 or the WiFi standard. With this technology there is a trade-off between distance (measured in a few hundred feet) and speed (up to 104 Mb/sec with 802.11a, more later). Coverage suffers because of the distance limitations, but if you identify where people congregate frequently, then you add a wireless access point, you're in business. At least that's what Starbucks, airports, and a few hotel chains think.

- ***Personal*** wireless connectivity is also gaining steam. This is characterized by a technology called Bluetooth that measures coverage in tens of feet and transfer rates of 1 Mb/sec. Personal connectivity will allow communication between multiple handheld devices and easily accessible devices such as printers.

So, wireless connectivity to our existing enterprise infrastructure is creating a new "world order" that will provide personal communication and data access, anytime and anywhere.

The handheld devices themselves are also on a fast evolutionary path with the addition of new multimedia features (such as cameras, MP3 players, imaging), and the integration of more telephony services. They are also growing as a platform with larger displays, more memory, faster processors, and higher data transfer rates. As more function and capability is added to these devices, they will be even better able to supply services to users everywhere.

SUMMARY

So, the stage is set, the actors have been identified, and the excitement level is high. We'll continue our journey by adding more detail as we progress through the book. But, before we conclude this chapter, let's send you away with a Soapbox about wireless, mobile, and pervasive computing.

Soapbox

Is It Wireless, Mobile, or Pervasive Computing?

Up to this point, we've made liberal use of the term *wireless*. So what about the term *mobile* you ask? Good question. You've probably noticed that there are numerous articles that use both terms in the title. There is actually a clear distinction between the terms. Wireless, as the name implies, means a connection without wires. Mobile, on the other hand, frequently refers to the techniques used for movement within a wireless network to create a seamless network from a collection of wireless transceivers. For instance, when a cell phone moves within range of one transceiver and out of range of another, mobility protocols define how the hand-off occurs.

So, why did we choose wireless and not mobile to describe the applications we will develop? This question was actually one of great debate among your authors and their publisher. In the end, we decided the term wireless conjures up the image of an untethered device and implies mobility at the same time.

So what about *m-commerce* you may ask? Doesn't the "m" stand for mobile? True. But, again we chose to stay with the term wireless, as it seems to be a more recognizable description of the overall use of handheld devices that act as clients to enterprise applications. We will use the term m-commerce to describe the business model where specific value is generated because a handheld device client can deliver a service that could not be done otherwise. For instance, if you're standing on a street corner wondering what today's specials are at the closest coffee shop, that's an application only a wireless device acting as a client to an enterprise application can provide.

So what's *pervasive* computing you ask (being the ever curious person you are). Another good question. Imagine that your refrigerator has a processor in it and your home has some sort of gateway (perhaps even wireless). The processor notices that your refrigerator is starting to malfunction so it notifies the service department of the refrigerator maker. Computing? Yes, but rather limited. Connected? Yes, again. Bring your toaster, microwave, coffee pot, stove, and dishwasher into the picture and you start to understand the concept of pervasive computing.

Some wireless devices can even be mobile. An automobile navigation system is wireless and mobile. Try taking it to your office though. Note that for the remainder of the book, we will focus on personal wireless devices that are always with you and provide applications that only have utility if they are always available. ❑

Chapter 2

Wireless Devices: The Hardware Platforms

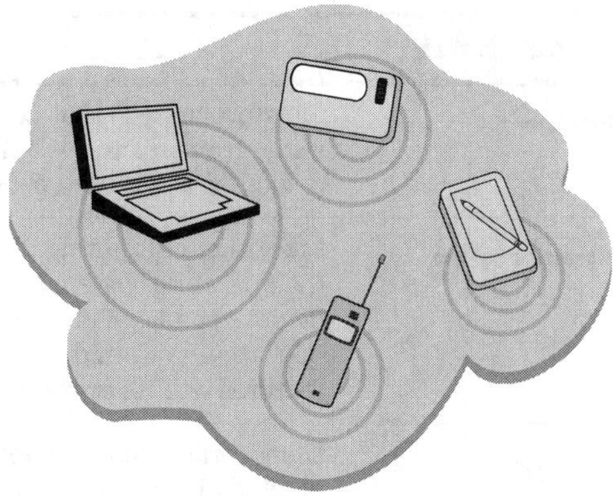

This chapter introduces you to handheld wireless devices as a hardware platform for client application delivery. Devices that act as clients fall into three categories:

- Voice devices that act as clients (for voice-driven applications)
- Devices that contain a micro-browser (also known as Internet-ready devices)
- Devices that provide a programming environment

As with PC platforms, executing client applications on wireless devices depends on such features as memory capacity, processor speed, display size, method of user input, and connectivity. However, wireless devices are much more limited and there is a larger variety of platforms. To add some sanity to our exploration, we will look at handheld devices that fall within a few categories:

- Simple cellular phones (for voice applications)
- Internet-ready cellular phones
- Mobile Information Devices with Java programming capability
  - Two-way pagers
  - Internet-ready phones
  - Connected PDAs (Personal Digital Assistants)

These are the device categories we will concentrate on for the remainder of the book. Table 2-1 gives a summary of the hardware features for these handheld devices.

Table 2-1. A Summary of Handheld Device Hardware Features.

Hardware Feature	Description
Processor	Slow processors that conserve battery power are the rule. Processors speeds range from around 16 – 233 MHz. Processor speed is primarily only an issue for devices that execute programs.
Operating System	Like other computing platforms, handheld device hardware is closely coupled to the operating system it supports. For instance, you are probably aware that Palm devices run the PalmOS and Pocket PC devices run Microsoft's WinCE. The operating system defines some basic device requirements. (Operating system features will be covered in more detail in Chapter 5.)
Display	Displays are small in keeping with the small size of a device. Displays are typically black and white or have limited grayscale capability. Some higher-end devices have color displays.
Memory	Memory comes in three types: ROM, RAM, and EEPROM and ranges in size from a few hundred kilobytes to a few megabytes. How memory is used depends on the device. For Internet-ready devices, memory to cache content pages is important. For devices that execute programs, the sizes of program storage, heap size, and persistent data storage are important.
Input	Providing a usable means of text entry is one of the challenges of designing an application capable of running on small devices. The types of input range from small alphanumeric keyboards on cell phones to handwriting recognition systems and mini-qwerty keyboards on more capable devices. Some devices also have special navigation features like scroll buttons and thumbwheels.
Connectivity	With handheld devices there are a number of ways to view connectivity. As a client for enterprise applications, connectivity to server-side applications is of primary concern.
Extensibility	Extensibility allows hardware expansion beyond the out-of-the-box configuration. This is a rare feature for handheld devices but some devices have expansion with "piggy-back" attachments and expansion slots. These expansion features allow the addition of memory, wireless modems, and even some more exotic attachments like cameras and Global Positioning System (GPS) receivers.
Price	Functionality usually comes at a price and can influence your choice of a device for an enterprise application.

So, with this brief introduction to handheld devices as a whole, we'll choose an example device in each category and explore it in more detail. If we don't choose your favorite device, we apologize.

SIMPLE CELL PHONES

The *Nokia 8260* (see Figure 2-1) is an example of a simple cell phone. Is it a programming platform? A definite no. It also is not Internet ready. However, it is a lightweight, easy-to-carry phone that makes a great handheld device for interaction with voice-driven applications. It also has features like *Short Message Service (SMS)* that make it very usable as the target (and source) of short text messages.

Figure 2-1. The Nokia 8260 Cellular Phone.

Table 2-2 shows some of the features of the Nokia 8260. It is one of many, many phones with similar features.

Table 2-2. Hardware Features of the Nokia 8260.

Hardware Feature	Description
Display	5 line LCD black-and-white display
Input	Telephone alphanumeric keypad
Connectivity	TDMA network (includes SMS text messaging)
Extensibility	None
Price	$179 + monthly wireless service charge

INTERNET-READY CELL PHONES

The *Nokia 7160* (see Figure 2-2) was one of the first Internet-ready cell phones. It does not provide a programming environment but it does provide a micro-browser that displays Wireless Markup Language (WML) and supports a connectivity protocol called *Wireless Application Protocol (WAP)*. We'll cover WAP in more detail in Chapter 4, WML in Chapter 6, then provide a sample application in Part 2.

For the first time, the size of the device's display becomes important to our client application. The larger the display, the more lines of content that can be viewed at once.

Internet-ready phones display *decks* of WML. These decks need to be downloaded to the phone via the relatively slow connection supported by the phone. The larger the onboard cache for the phone, the larger the number of WML decks that can be saved for future use. The users of your application perceive a faster connection as a result.

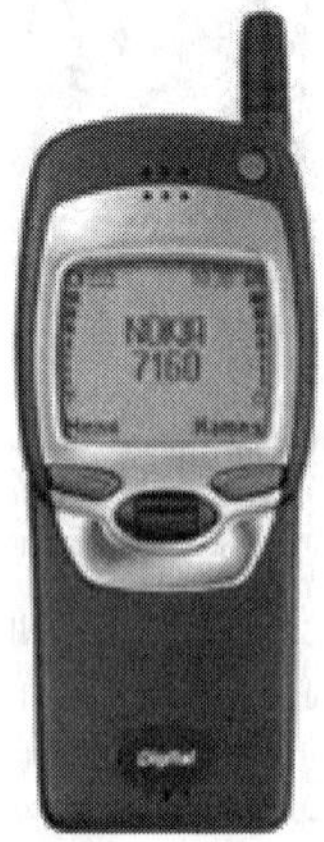

Figure 2-2. The Nokia 7160 WAP-enabled Cellular Phone.

Table 2-3 describes some of the hardware features of the Nokia 7160.

Table 2-3. Hardware Features of the Nokia 7160.

Hardware Feature	Description
Display	65 x 96 pixels, black and white
Memory	Max WML Deck Size: 1.3 KBytes WML Cache Size: 32.5 KBytes
Input	Telephone alphanumeric keypad
Connectivity	WAP 1.1
Price	$149 + monthly wireless service charge

MOBILE INFORMATION DEVICES

All of the devices in our next category have a Java programming environment. The programs that we write for these devices are written to Java classes defined by the *Mobile Information Device Profile (MIDP)* portion of *Java 2 Micro Edition (J2ME)* and run on a Java configuration called *Connected Limited Device Configuration (CLDC)*. Java will allow us to write a single program that can run on all of these devices without modification. We will explore the different versions of J2ME in great detail in Chapter 7 and provide programming examples in Part 2.

Java-enabled, Two-Way Pagers

The *Research in Motion (RIM) Blackberry 950* (see Figure 2-3) is an example of a two-way pager that supports the Java programming environment. The device has a small form-factor with a small keyboard for "two-thumb" text entry. Communication is via an embedded wireless modem. The Blackberry 950 represents the low-end of the J2ME devices. It has no voice capability and does not include a micro-browser.

Figure 2-3. The Research In Motion Blackberry 950.

Table 2-4 describes the hardware features of the RIM Blackberry 950.

Table 2-4. Hardware Features for the RIM Blackberry 950.

Hardware Feature	Description
Operating System	Proprietary
Display	User selectable 6- or 8-line display

Table 2-4. Hardware Features for the RIM Blackberry 950. (Continued)

Hardware Feature	Description
Memory	4-MB Flash Memory plus 0.5 MB SRAM
Input	Thumb-operated trackwheel plus 35-key Qwerty keyboard
Connectivity	900-MHz Mobitex Networks
Extensibility	Serial-port connectivity
Price	$399 + monthly wireless service charge

JAVA-ENABLED CELL PHONES

The Motorola i85s (see Figure 2-4) is one of the first J2ME (Java 2 Micro Edition) capable cellular phones. You can write your own Java programs or acquire programs from other sources and download them to your phone. Are your new Java applications going to calculate Pi to 100 decimal places? No, but your Java application might connect to a server application that can.

Of course, the phone is limited to only a small number of applications. However, you can choose your own suite of applications and, unlike a WAP application, your J2ME applications can also run independently of a network connection.

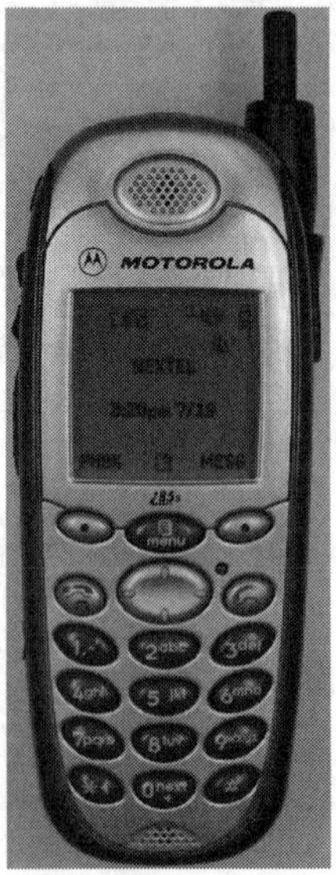

Figure 2-4. The Motorola i85s Java-enabled Cellular Phone.

Table 2-5 lists some of the hardware features important for Java programs running on the Motorola i85s phone. The phone also has a micro-browser for WML display.

Table 2-5. Hardware Features of the Motorola i85s.

Hardware Feature	Description
Processor	4-MHz proprietary processor
Operating System	Proprietary, T9 for keyboard input
Display	100 x 110 pixels, 4-level grayscale
Memory	256 KB for data, 384 KB for programs, and 256 KB to run applications (data memory shared with other non-Java applications like VoiceNotes)
Input	Telephone alphanumeric keypad
Connectivity	WAP 1.1, TCP/IP
Price	$199 + monthly wireless service charge

JAVA-ENABLED CONNECTED PDAS

PDAs, with their relatively large screens, powerful processors, and large amounts of memory, are the high end of handheld devices. PDAs are definitely a capable programming platform that can store and execute a number of applications. There is also a healthy programming community with many applications available.

Connected PDAs are not necessarily the norm, however. Unlike cellular phones and pagers which have a communication mechanism built in, for PDAs it is frequently an add-on option.

There are also two major camps in the PDA market: Palm devices and Pocket PC devices. At this point in time, Palm devices represent the PDA of choice for the consumer market with a lower cost, easy-to-use device. Pocket PCs, on the other hand, with their faster processors, large amounts of memory, and more expandability (with an associated heftier price), are the product of choice for corporations. We'll introduce you to both and let you be the judge of what is appropriate for your situation.

Palm Vx Plus Novatel Modem

Figure 2-5 shows the Palm Vx with an attached Novatel modem. With 8 MB of memory, the device can comfortably store several Java programs. The larger screen is also a welcome step up from the smaller screens most cell phones have. With the Novatel modem attached, the device is able to invoke server-side applications as well as provide email and Internet browsing capability.

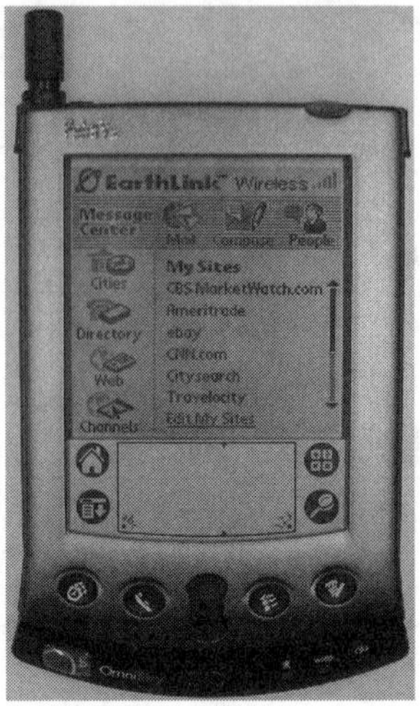

Figure 2-5. The Palm Vx with the Novatel Modem.

Table 2-6 summarizes the hardware features of the Palm Vx and Novatel modem combination.

Table 2-6. Hardware Features of the Palm Vx Plus the Novatel Modem.

Hardware Feature	Description	
Processor	20 MHz Motorola MC68EZ328 EZ (Dragonball EZ) chip	
Operating System	PalmOS	
Display	160 x 160 pixels, black and white	
Memory	8 MB RAM, 2 MB ROM, 256K Dynamic Heap	
Input	Graffiti	
Connectivity	TCP/IP (with Novatel Modem)	
Extensibility	Novatel Modem "back pack"	
Price	Palm Vx	$299
	Novatel Modem	$269
	plus monthly wireless service charge	

Compaq iPAQ plus Ricochet AirCard 300 Modem

Figure 2-6 shows the Compaq iPAQ 3135 with a Sierra AirCard 300 wireless modem PC card. This is a top-end configuration for a handheld device that does not have a full keyboard. The iPAQ has up to 64 MB of memory, and 240 x 320 color screens are available. There are also one or two slot expansion packs that can hold memory, network, modem, and micro hard disk drive PC and Compact Flash Cards.

In our sample configuration, we chose an expansion pack that allows the insertion of a PC card for our network connection. This allows a variety of wireless network and modem cards to use whatever network protocols are available.

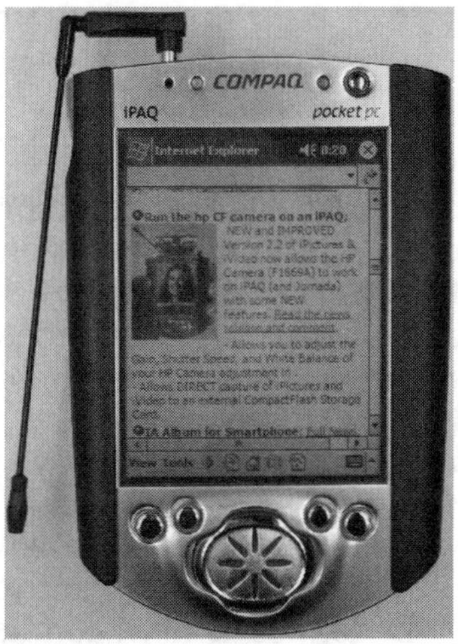

Figure 2-6. The Compaq iPAQ 3135 with the Sierra AirCard 300 Wireless Modem PC Card.

Table 2-7 summarizes the hardware features of the iPAQ 3135 with the Sierra AirCard 300 modem PC card.

Table 2-7. Hardware Features of the Compaq iPAQ 3135 with the Sierra AirCard 300 Modem PC Card.

Hardware Feature	Description
Processor	206 MHz Intel Strong ARM 32-bit RISC Processor
Operating System	WinCE
Display	240 x 320 pixels, 4048 Color TFT Display
Memory	32 MB RAM, 16 MB ROM

Table 2-7. Hardware Features of the Compaq iPAQ 3135 with the Sierra AirCard 300 Modem PC Card. (Continued)

Hardware Feature	Description	
Input	Handwriting recognition	
Connectivity	Via Wireless Network or Modem PCMCIA card	
Extensibility	One and two slot expansion packs for memory, modem, and networking cards are available.	
Price	iPAQ 3135	$499
	PC Card Expansion Pack	$150
	Sierra AirCard 300 Modem PC Card	$199
	Plus monthly wireless service charge	

COMMUNICATORS

There are also other Java-capable devices that are more capable than the Mobile Information Devices discussed in the previous sections. This includes handheld devices that have a full keyboard and a substantial display. These devices also include built-in communications or have expansion slots for adding wireless communication cards. We will label this category "communicators." These devices are also referred to as "handheld PCs." However, we want to emphasize that wireless connectivity is required for the class of applications we will develop.

Communicators are capable of executing the same Java programs as our MIDP devices; in fact, they are capable of doing much more. These devices have such high-end capabilities that J2ME defines a new configuration called *Connected Device Configuration (CDC)* and several *Profiles* specifically for these devices. That's the good news. The bad news is that the Java support for these devices is not complete as we go to press. So, we will only mention them briefly.

Nokia 9290 Communicator

The Nokia 9290 (see Figure 2-6) is representative of a Communicator device. It can function as a cell phone, is Internet ready, and provides a full programming environment. With a large keyboard and display for such a small device (158 x 56 x 27 mm), it is a device that is more powerful and easier to use than any of the other devices mentioned in this chapter.

Figure 2-7. The Nokia 9290 Communicator.

Table 2-8 summarizes the hardware features of the Nokia 9290 Communicator.

Table 2-8. Hardware Features of the Nokia 9290 Communicator.

Hardware Feature	Description	
Processor	32-bit ARM9-based RISC CPU	
Operating System	Symbian OS	
Display	640 x 200 pixels, 4048 Color Display	
Memory	Application memory: 14 MB User memory: 2 MB + 16 MB Application Card Execution memory: 8 MB (SD-RAM)	
Input	Full keyboard	
Connectivity	WAP 1.1, HSCSD high-speed data connection (up to 43.2 Kbps)	
Extensibility	Memory Card slot	
Price	Nokia 9290	$599

SUMMARY

This chapter has been a quick overview of a few devices and their hardware capabilities. As you can see, the range of capabilities is large. This allows us, as users of the devices, to choose a device that fits a specific need without depleting our bank account. On the other hand, this gives us some headaches as developers when developing for this variety of platforms. We will survive this by working with a few classes of devices and using standard platforms and middleware like Java and WAP to write applications that are portable across devices. Before we leave this chapter, let's take a peek at the future of handheld devices.

FYI Briefing

The Wireless Devices of the Future

In this chapter we've discussed some of the features of handheld devices that are necessary for programmers to develop enterprise applications. This includes processors, memory (both execution memory and persistent memory), displays, and connectivity. Of course we will continue to see and want more (bigger, faster, better) from these basic features.

But what about some features that can excite us as users *and* programmers of handheld devices? Consider some of the following:

- ***Multimedia features***, of course, will always be exciting. Audio and visual features such as music and video playback of your favorite songs and video clips will keep you entertained. These applications will grow to meet the available processing, storage capacity, and bandwidth available on devices. As the device hardware grows, middleware that facilitates the delivery of multimedia applications will also be required.
- ***Games*** are closely related to multimedia. As users, they will be another source of entertainment. For programmers, games can be both a creative outlet as well as a financial opportunity.
- ***Personal connectivity*** will create new application possibilities. As more and more devices embed the hardware needed for personal networks, our handheld devices will collaborate with each other as well as interact with other network nodes that are not as mobile—such as printers, televisions, and information kiosks—to provide a suite of personalized functions.

  Let's suppose that you are going to a movie. As you step in your car to drive to the movie, your handheld device transmits your personal settings to your car as you get in, adjusting your seat, mirror, temperature, and radio settings to your personal desires. Upon arrival at the theatre, your handheld device connects to a theatre electronic ticket dispenser and you choose, purchase your ticket, and enter the theatre without standing in line. As you pass by the concession stand, you choose your snacks via your handheld device and they are delivered to you. As you enter the theatre, your cell phone is forced into silent mode by the network in the theatre (no doubt a gratified smile just crossed your face). Let your imagination run wild and you can dream up many uses of this type of connectivity.

- ***Smart Cards*** will be used to enhance security for handheld devices. Security (as with all other aspects of computing) will always be of primary importance for handheld devices. Smart cards have embedded electronics that can be used to keep your digital certificates for authentication, your electronic dollars, and highly sensitive personalized information. With smart card readers incorporated into handheld devices, this highly personal information can be moved from device to device easily and securely.

So, new hardware features will drive new software requirements. This should indeed keep a smile on the faces of both users and developers of software for handheld devices. ❑

Chapter 3

Wireless Devices: Network Technologies

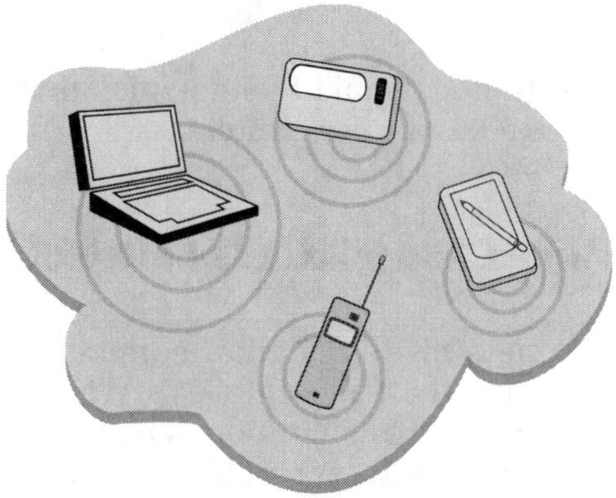

Network protocols and technologies form the infrastructure for both wired and wireless communications. As an application programmer, you might have little interest in these technologies, but it sometimes helps to understand the underlying network technology you are using. Your programming strategies might change based on whether you are using a circuit-switched network or a packet-switched network. If you use a cell phone, you might have already heard buzzwords such as GSM in Europe/Asia and CDMA in North America. We will briefly introduce you to the terminology and some useful information about network technologies. We categorize the supporting network technologies based on the following machine (or device) and communication characteristics:

- ***Fixed and Wired:*** This is typically a desktop PC connected via TCP/IP. We assume that you are already familiar with these networking technologies, at least from a programming point of view.[1]

- ***Mobile and Wired:*** This is categorized by laptops connected into a wired network. The laptop is mobile and it can connect into any available wired network.

---

[1] **Computer Networks** by Andrew S. Tanenbaum is an excellent source of information on wired networks and their protocols.

- ***Fixed and Wireless:*** In this scenario a desktop PC has a wireless connection to a localized wireless network.

- ***Mobile and Wireless:*** In this scenario, highly mobile and personal devices (such as two-way pagers, cell phones, and PDAs) connect via regional, local, or personal networks.

The protocols and technologies that form wireless networks are discussed in the remainder of this chapter.

MOBILE AND WIRELESS NETWORK PROTOCOLS

Wireless protocols allow communication between two devices without the use of wires (as the name implies). Wireless transmissions are typically limited to short distances. This limitation is overcome by having transmission points spaced sufficiently close to each other so a *service area* is covered. *Mobility protocols* define how a wireless device moves from one transmission point to another. The following sections give a bird's eye view of these mobile and wireless protocols.

Basic Terminology

Most likely you are a cell phone user. Let us take a look at the basic terminology of the cellular telecommunication networks and examine the way they work. Here are some definitions:

- ***Public Switched Telephone Network (PSTN):*** This is a wired telecommunications network. Wireless networks use PSTN to connect their various segments when a pure wireless link is not available.

- ***Mobile Terminal:*** This is the handset (your cell phone) that communicates with the wireless network. Capabilities of handsets include voice, data, and multimedia communications.

- ***Mobile Identity Number (MIN) and Electronic Serial Number (ESN):*** These are the mobile terminal identification numbers used by the wireless network for verification, authentication, and billing purposes. In some networks this number is stored in a SIM (Subscriber Identity Module) also know as a Smart Card, which is plugged into the device. In other cases, the manufacturer assigns this number.

- ***Cell Tower:*** Every cellular network provider installs towers (antennas) in overlapping geographic areas to provide cellular transmission. Theoretically a coverage area is divided into hexagonal boundaries; a cell tower serves that cell.

- ***Radio Frequency (RF) transceiver:*** An RF transceiver is a combination of a transmitter/receiver in a single package.
- ***Base Station Controller (BSC):*** A base station controls a cluster of cell towers. This provides the call setup with the mobile terminal and manages hand-off between cells.
- ***Mobile Switching Center (MSC):*** A MSC connects all the base stations to pass communication signals and messages to and from mobile terminals on the network. MSC is connected to a VLR, an HLR, an AuC, and an EIR, which are discussed below.
- ***Home Location Register (HLR):*** A HLR keeps track of information regarding the subscriber. Mobile terminals register with the wireless network every few seconds to identify their location. The HLR keeps a record of this information. So when a call needs to be set up, the BSCs can query the HLR to get this information. This speeds up the setup time.
- ***Visitor Location Register (VLR):*** A VLR tracks visitors and records information about the mobile terminals that have roamed into the local network from other networks.
- ***Equipment Identity Register (EIR):*** This is a database to store and check the status of MINs and ESNs.
- ***Authentication Center (AuC):*** An AuC is responsible for authentication and validation of services for every mobile terminal attempting to use the network.
- ***Operations and Maintenance Center (OMC):*** An OMC is connected to the network to provide functions such as billing, network management, service provisioning, and customer service.

A Cell Phone Operational Scenario

Let us look at a simple scenario (refer to Figure 3-1). When you buy a phone and sign up for a service plan, the OMC provisions the phone and makes an entry for the MIN and ESN into the EIR.

1. When you try to make a call, the mobile terminal's RF transceiver sends a message request to the nearest cell tower.
2. The cell tower then sends the request to the controlling base station.
3. The base station recognizes the call signal and routes it to the MSC.
4. The MSC queries the HLR or VLR (HLR if the call is local, VLR if the cell phone is in a roaming state) for location, EIR for service qualification and features, and AuC for authentication.

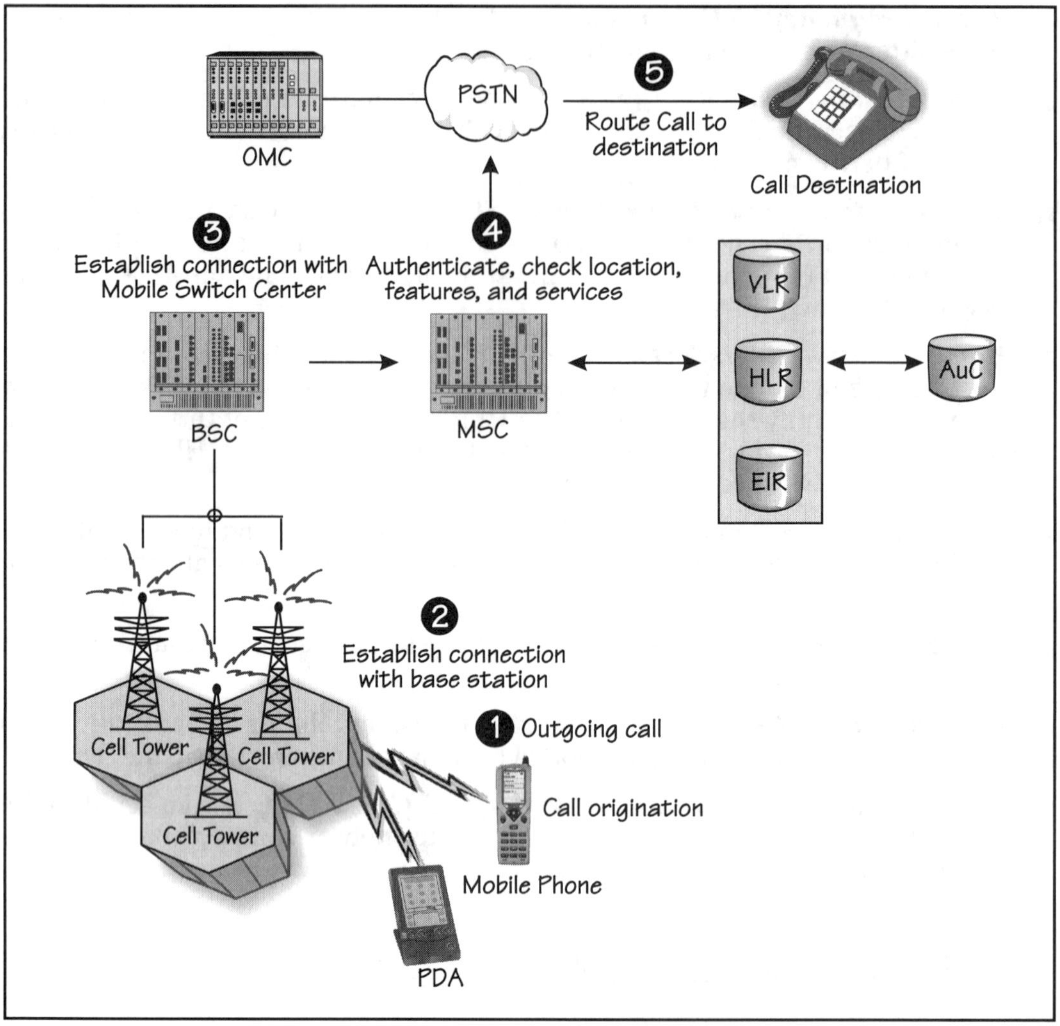

Figure 3-1. A Cell Phone Scenario from Call Origination to Call Destination.

5. Depending on the call destination, it is either routed to the same or another base station (depending on the area of the recipient) or to a PSTN (regular wired network) or to an Internet IP device.

Hand-off and Roaming

When a mobile terminal moves from one cell to the other, the control of the communication needs to transfer from one cell tower to the other without disruption. This mechanism is known as hand-off.

When a mobile terminal moves away from the coverage area of the servicing cellular network, there might be other networks covering that area. If there is compatibility between the networks, then that network can provide the service at an additional cost. This is called roaming. Figure 3-2 depicts the hand-off scenario.

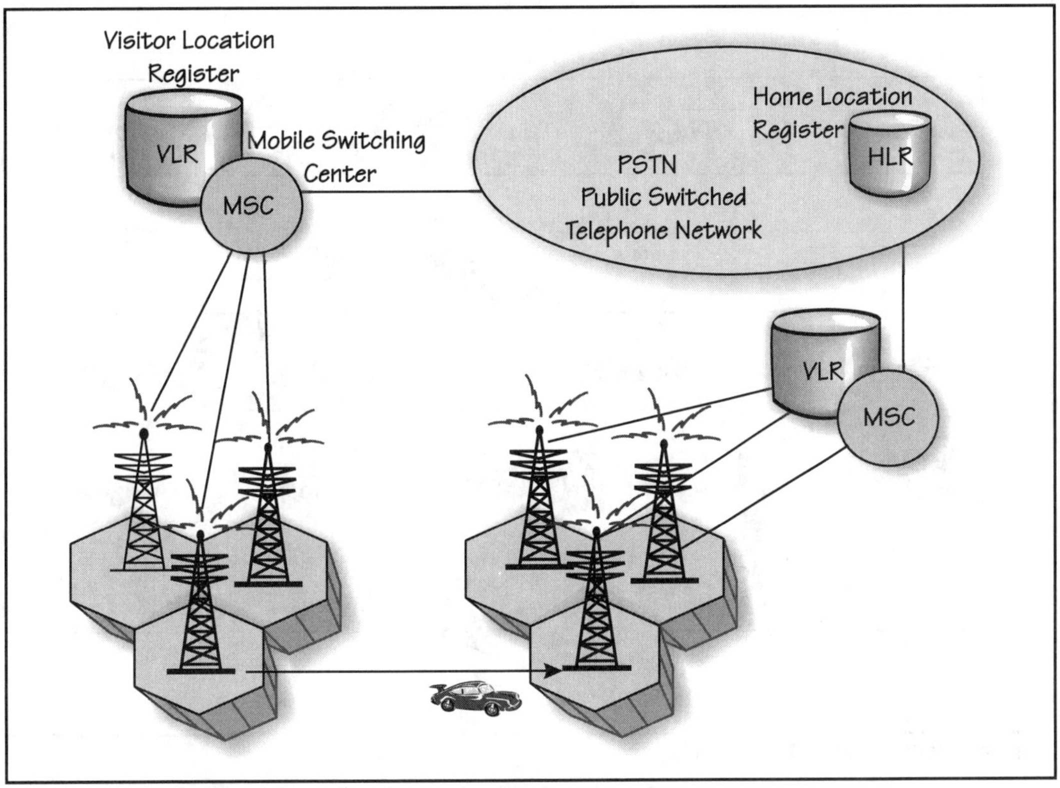

Figure 3-2. A Hand-off from One Cell to Another in a Wireless Network.

Wireless Network Technologies

With the basic terminology of a cellular network under your belt, we can begin to understand the basis of networking protocols and what they are trying to achieve.

Multiplexing

Wireless communication protocols mainly differ in the multiplexing mechanisms they follow. Multiplexing is the sharing of a medium by various users with minimal or no interference. This is analogous to many cars sharing the same highway with no interference (accidents) by using different lanes. In this case the space on the highway is multiplexed (space division multiplexing) and also different cars are sharing the same lane at different points of time (time division multiplexing).

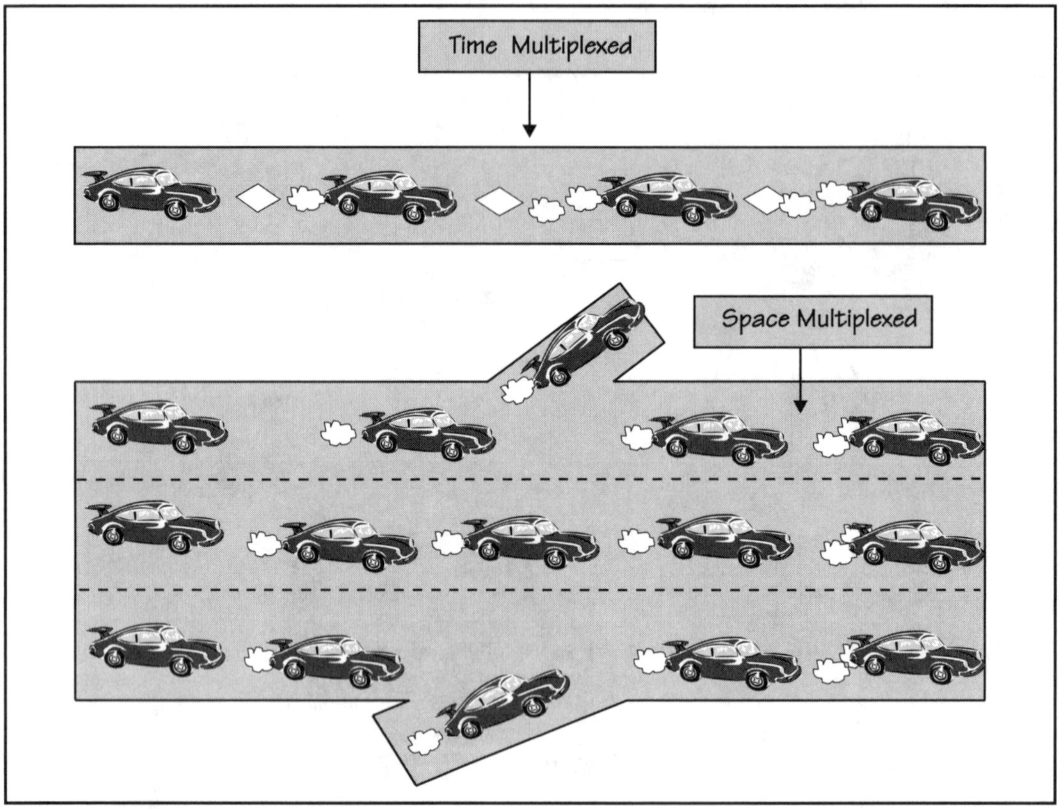

Figure 3-3. Multiplexing—Time and Space.

Frequency-division multiplexing (FDM) is a scheme where different data streams are placed on different frequencies. Each carrier frequency and associated data stream are then combined to be broadcast. The transmission media can be radio frequency or optical. The data streams are separated into their different frequencies at the receiver and handled as necessary. Cable television service often uses FDM as but one example.

Wireless communications also follow similar protocols to multiplex the medium they use. There are four dimensions that could be multiplexed: space, time, frequency and code. The first three schemes are used extensively and are easy to understand. *Code Division Multiplexing (CDM)* is relatively new and is being used extensively in North America. CDM uses a combination of other multiplexing schemes. Audio input is first digitized into binary elements. The frequency of the transmitted signal is then made to vary according to a defined pattern (code), so it can be intercepted only by a receiver whose frequency response is programmed with the same code and exactly follows the transmitter frequency.

Code is analogous to a spoken language. Imagine a room full of people, all trying to carry on one-on-one conversations. In *Time Division Multiplexing (TDM)*, each couple takes turns talking. They keep their turns short by saying only one sentence at a time. As there is never more than one person speaking in the room at any given moment, no one has to worry about being heard over the background noise. In CDM, each couple talks at the same time, but they all use a different language. Because none of the listeners understands any language other than that of the individual to whom he or she is listening, the background noise doesn't cause any real problems. In this case, the spectrum is divided into channels and each channel is assigned a code, so there is no interference. There are trillions of possible frequency-sequencing codes; this enhances privacy and makes cloning difficult.

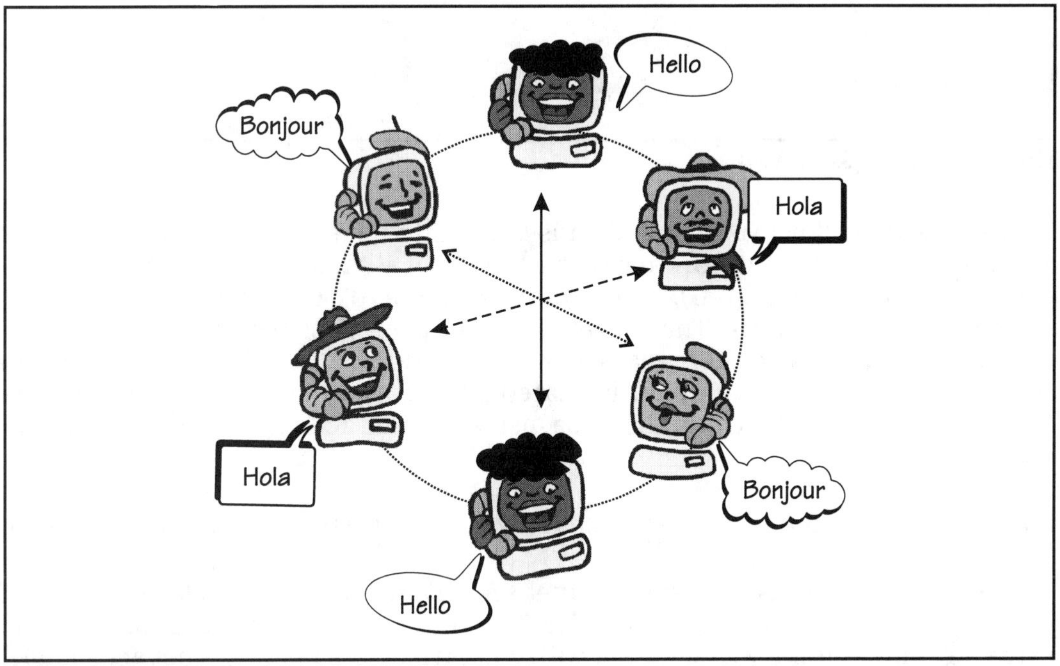

Figure 3-4. Code Division Multiplexing.

Data Transmission Mechanisms

Wireless data transmission can be either *circuit-switched* or *packet-switched*.

Circuit-switched networks use a physical path to transmit data between two end points (see Figure 3-5). This physical path is obtained and dedicated between two end points in the network for the duration of the connection. For example, the phone service is circuit-switched. So even with a wireless connection, the operator reserves a specific physical path to the number that is called for the duration of the call. This physical path cannot be used or shared by any other connection. In this mode of operation, users are generally charged for the duration of the connection time.

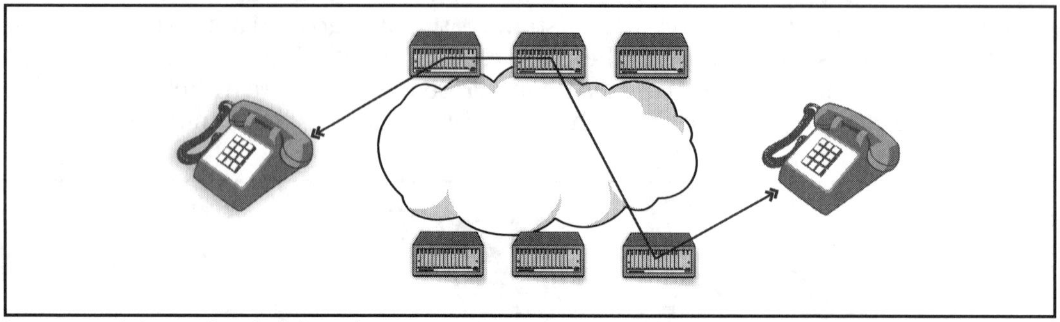

Figure 3-5. A Circuit-Switched Network.

In packet-switched networks, the data is transmitted in small units, called packets and routed through a network based on the destination address contained within each packet (see Figure 3-6). The data is broken into packets at the source and transmitted to the network. The packets may take many possible paths to reach their destination. These packets are reassembled at the destination. By breaking the data into packets, the same data path is shared among many users in a network. The Internet is packet-switched. In this case, users are charged based on the number of packets sent.

So now let us look at the key differences between the two types of data transmission.

Packet-switching works well with Internet scenarios, because the delay in arrival of packets might not affect the communication. For example, it is acceptable if email arrives a minute after it was sent. But with voice transmission, this is not acceptable. So if a packet-switched network has to be used for voice, it has to meet the Quality of Service (QoS) standards required for voice transmissions.

Circuit-switching, on the other hand, is expensive as resource sharing is not possible. As a programmer, if you open a connection using a circuit-switched network, you need to make sure you close it as soon as the work is done, because you will be charged for connection time. In the case of packet-switched networks you need not worry about this, as you would be only charged for the packets you transmit.

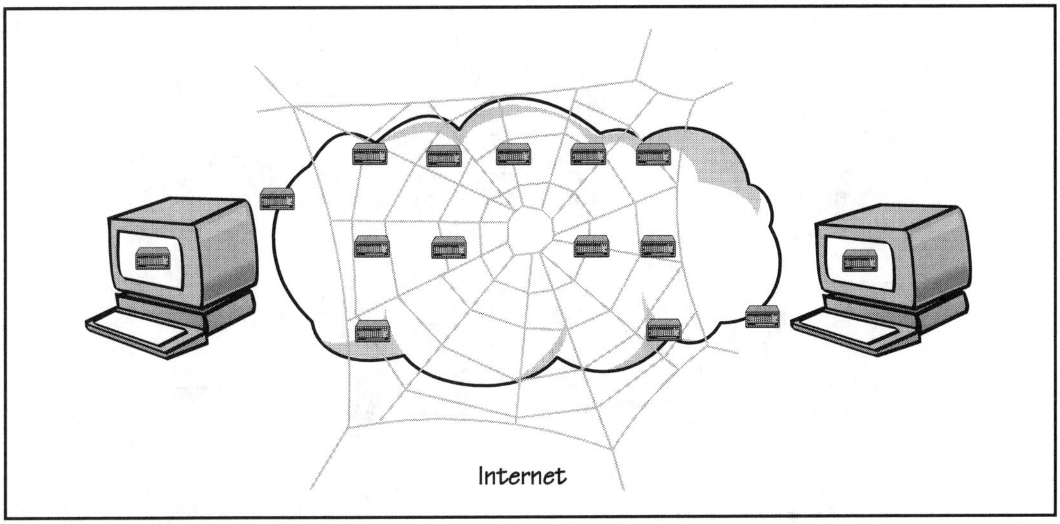

Figure 3-6. A Packet-Switched Network.

Table 3-1 lists the key differences between circuit-switched and packet-switched networks.

Table 3-1. Circuit-Switched vs. Packet-Switched Networks.

Circuit-Switched	Packet-Switched
Supporting always-on mode is expensive and not practical. So there is a delay in call setup and transmission	Can support Always-on mode. So call setup and termination is almost immediate
Logon for every call	One logon at power up
Good for voice communication	Good for data communications
Good for large transmissions	Good for short, bursty transmissions
Good for point-to-point connection	Good for broadcasting

WIRELESS NETWORK TECHNOLOGIES

The past twenty years marked the evolution and phenomenal growth in wireless network development. We will take a look at each of these technologies. Wireless network protocols are broadly categorized into generations based on their capabilities. Currently, the market is moving from 2G (2nd Generation) to 3G (3rd Generation) technologies. Figure 3-7 shows the wireless technologies transitioning from 2G to 3G. These technologies are discussed in the following sections.

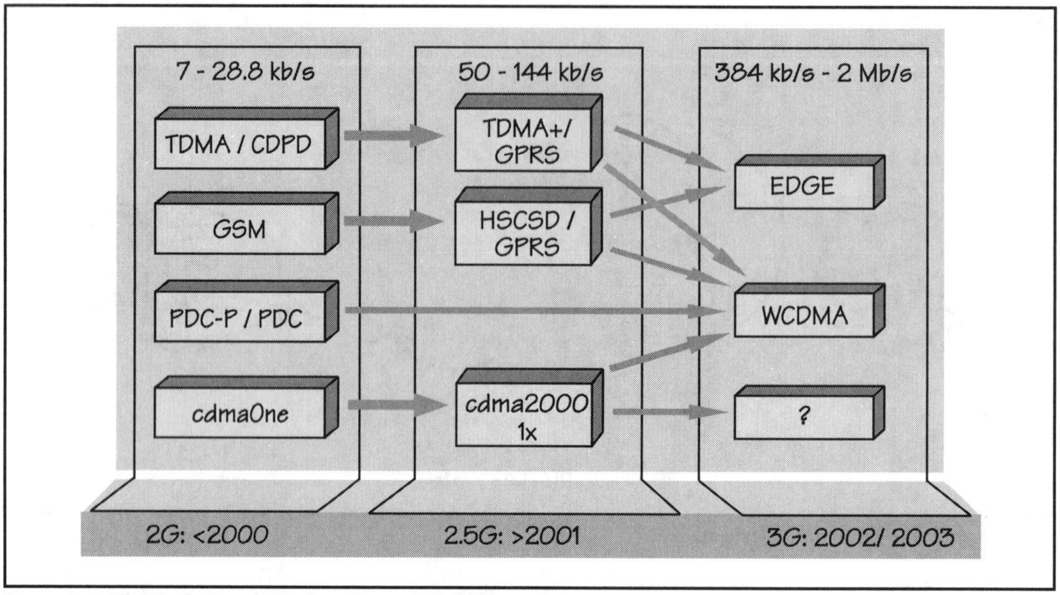

Figure 3-7. The Evolution of Wireless Network Technologies.

Advanced Mobile Phone Service (AMPS)

The analog cellular mobile phone system is used in North and South America and more than 35 other countries. This is the first cellular system, developed as a result of 10 years of research at Bell Labs. It uses Frequency Division Multiple Access (FDMA) transmission in the 800 MHz band. The first AMPS system in the U.S. was deployed in Chicago in 1983. This is the first generation (1G) of cellular transmission technologies.

Here are some terms that describe the evolution of AMPS.

- ***N-AMPS (Narrow-bandwidth AMPS)*** is a version of the analog cellular mobile phone system that uses a narrower bandwidth and provides more calling capacity.
- ***IS-54 (Interim Standard-54)*** is the first generation of the TDMA digital cellular system that was released in early 1991.
- **IS-136 (Interim Standard-136)** is the second generation of the TDMA digital cellular system. TDMA operates in North America in the 800 MHz band and 1.9 GHz Personal Communications Systems (PCS) band. First introduced in 1994, IS-136 is also known as "Digital AMPS" and "D-AMPS" (Digital-Advanced Mobile Phone Service).

- *NMT (Nordic Mobile Telephone)* is an analog cellular phone system deployed in more than 40 countries in Europe. Launched in the Scandinavian countries in 1979, NMT was the first analog cell phone system. Both 450 MHz and 900 MHz versions are available.

- *TACS (Total Access Communication System)* is an analog cellular phone system deployed mostly in Europe. It was modeled after the AMPS system in the U.S. In the U.K., ETACS (Extended TACS) transmits in the 871–904/916–949 MHz bands. International TACS (ITACS) and International ETACS (IETACS) are versions that operate outside the United Kingdom. Narrowband TACS (NTACS) operates in the 860–870/915–925 MHz band, and by using a narrower channel spacing, delivers more channels in the same amount of spectrum.

Global System for Mobile Communication (GSM)

GSM is a digital cellular system developed by Groupe Special Mobile of Conference Europeanee des Postes et Telecommunications (CEPT) and its successor European Telecommunications Standard Institute (ETSI). GSM operates in the 900-MHz and the 1800-MHz (1900 MHz in the US) frequency band and is the prevailing mobile standard in Europe and most of the Asia-Pacific region. GSM is used by more than 300 million people representing more than 50 percent of the world's mobile phone subscribers.[2] North America had only about 5 million GSM users in late 1999, while the majority of subscribers are using a variety of technologies for mobile communications, including pagers and a high percentage of analog devices. Additionally, the North American mobile market development is handicapped by the "Called Party Pays" principle, which has led to a low usage of mobile phones. In Europe, the common GSM standard provides the critical mass to make it economically feasible to develop a large variety of innovative applications and services. GSM combines both Time Division Multiple Access (TDMA) and FDMA. The other reason for GSM becoming popular is that the intellectual property rights of the GSM radio system were waived for all of the vendors. GSM represents the second generation (2G) of cellular transmission technologies.

High Speed Circuit-Switched Data (HSCSD)

High Speed Circuit Switched Data (HSCSD) is a new high-speed implementation of GSM data techniques. Here are the main features of HSCSD, with its advantages and disadvantages:

---

[2] As per EMC Research, by the end of 2004 there will be 733 million GSM subscribers globally.

- HSCSD is a circuit-switched wireless data transmission for mobile users at data rates up to 38.4 Kbps, which is faster than the standard data rates of the GSM (14.4 kbps). The advantage of this is HSCSD can be used for applications such as fax and file transfer.

- It gives a single terminal simultaneous access to multiple channels (up to 4). This translates to a higher transfer rate for the terminal. The disadvantage is the usage of more resources and the associated higher cost.

- HSCSD uses existing GSM infrastructure. This means that the base station operator can simply upgrade the base station software to provide HSCSD services. This allows network operators to provide more applications without a huge investment. This is important as the early migration to HSCSD will illustrate usage trends and end-user behavior. The provider can use this information when planning their migration to 3G or GPRS (discussed in coming sections).

- HSCSD requires a new type of handset that has supporting capabilities. This is a disadvantage.

HSCSD is especially well-suited for time-sensitive, real-time services. Examples could be transferring of large files with specified Quality of Service or video surveillance. This technology is deployed in a very limited way in some European countries.

General Packet Radio Service (GPRS)

GPRS is a packet-switched wireless protocol as defined in the GSM standard that offers instant access to data networks. It is a non-voice service that will permit bursty transmission speeds of up to 115 Kbit/s (or theoretically even 171 Kbit/s) when it is completely rolled out. The real advantage of GPRS is that it provides an "always on" connection (i.e., instant IP connectivity) between the mobile terminal and the network. Network capacity is only used when data is actually transmitted. The actual speed of GPRS will initially be much less than the theoretical maximums: 43.2 Kbit/s downstream and 14.4 Kbit/s upstream up to 56 Kbit/s bi-directional some time thereafter.

GPRS will be the first transport mode to allow full instant mobile Internet access and will become the enabler for a wide range of applications. In this sense it truly may pave the way for UMTS (explained later). Pilot GPRS networks are already in place today in many European countries. However, GPRS will require new handsets that support the higher data rates, and these seem to be the bottleneck to the early adoption of the technology. In order to push the development of applications for

GPRS, the GAA (GPRS Applications Alliance) was founded in October 1999 by Ericsson, Palm, IBM, Lotus, Oracle, and Symbian.

Availability of a large number of applications is critical for the adoption of GPRS as a bearer technology for Internet access. GPRS may be widely installed by operators as a step in the evolution towards the UMTS world. We will see how GPRS works later in this chapter.

Enhanced Data Rates for Global Evolution (EDGE)

Enhanced Data Rates for Global Evolution (EDGE) is a higher bandwidth version of GPRS permitting transmission speeds of up to 384 Kbit/s. It is a radio based high-speed mobile data standard. It allows data transmission speeds of up to 384 kbps to be achieved when all eight timeslots are used. This means a maximum bit rate of 48 kbps per timeslot, higher per timeslot than any other previously available mobile data services. It is also an evolution of the old GSM standard and will be available in the market for deployment by existing GSM operators during 2002.

Ericsson initially developed EDGE for GSM mobile network operators who fail to win a portion of the Third Generation (3G) spectrum. It was soon realized that EDGE would also work on TDMA networks. EDGE can also provide an evolutionary migration path to 3G by implementing now the changes in modulation that will be later necessary for implementing 3G. The idea behind EDGE is to provide even higher data rates on the current 200-KHz GSM radio carrier by changing the type of modulation used, while still working with current circuit (and packet) switches.

Given that a number of mobile operators are considering implementing EDGE as an interim data technology between GPRS and UMTS, the success of EDGE depends very much on the timely availability of the products and applications. The opportunity window for EDGE will be very short, unless major delays occur during UMTS deployment.

New operator licenses are not needed for EDGE. Hardware upgrades to the radio portion of the network will be needed to make the modulation upgrade. Each cell will need one new EDGE transceiver hardware unit. With most vendors, it is envisioned that software upgrades to the Base Station Controllers and Base Stations can be carried out remotely. The new EDGE capable transceiver can also handle standard GSM traffic and will automatically switch to EDGE mode when needed.

EDGE capable terminals will also be needed; existing GSM terminals do not support the new modulation techniques and will need to be upgraded to use EDGE network functionality. Some EDGE-capable terminals are expected to support high data rates in the downlink receiver only (i.e. high data rates can be received but not sent), whilst others will access EDGE in both uplink and downlinks (i.e., high data rates can be received and sent). The later device types will therefore need greater terminal modifications to both the receiver and the transmitter parts.

Deploying EDGE will allow mobile network operators to offer high-speed, mobile multimedia applications. It allows a migration path from GPRS to UMTS, because the modulation changes that will be necessary for UMTS at a later stage will already be implemented.

The research and development departments of infrastructure providers (such as Nokia and Ericsson) are committed to and preparing for EDGE, which is expected to be available commercially in 2001/2. Network operators are currently evaluating EDGE and determining whether or not to deploy it as an interim data technology while moving towards the Third Generation (3G). This decision will depend to a large part on their success in winning 3G licenses; those network operators that do not gain 3G licenses will likely adopt EDGE to maintain their competitive position in the mobile marketplace. Timescales are also an important decision factor.

Both High Speed Circuit Switched Data (HSCSD) and Third Generation (3G) are based on something called Gaussian Minimum Shift Keying (GMSK), which only yields a moderate increase in data bit rates per time slot. EDGE on the other hand is based on a new modulation scheme that allows a much higher bit rate across the air interface. This modulation technique is called eight-phase-shift keying (8 PSK). It automatically adapts to radio circumstances and thereby offers its highest rates in good propagation conditions close to the site of base stations. This shift in modulation from GMSK to 8 PSK is the central change with EDGE that prepares GSM and TDMA networks for 3G.

Bird's Eye View of GPRS

To better understand GPRS, we will take a quick tour beginning with the mobile PC and traversing through the network. First, we have a notebook computer connected to a GPRS-capable cell phone or modem, either through a serial cable or other type of connection such as Universal Serial Bus (USB) or local wireless link. The connection device could also take the form of a PC Card. The GPRS phone or modem communicates with GSM base stations, but unlike circuit-switched data calls, which are connected to voice networks by the mobile switching center, GPRS packets are sent from the base station to what is called a *Serving GPRS Support Node (SGSN)*.

The SGSN is the node within the GSM infrastructure that sends and receives data to and from the mobile stations. It also keeps track of the mobile devices within its service area. The SGSN communicates with what is called the *Gateway GPRS Support Node (GGSN)*, a system that maintains connections with other networks such as the Internet, X.25 networks, or private networks (see Figure 3-8). A GPRS

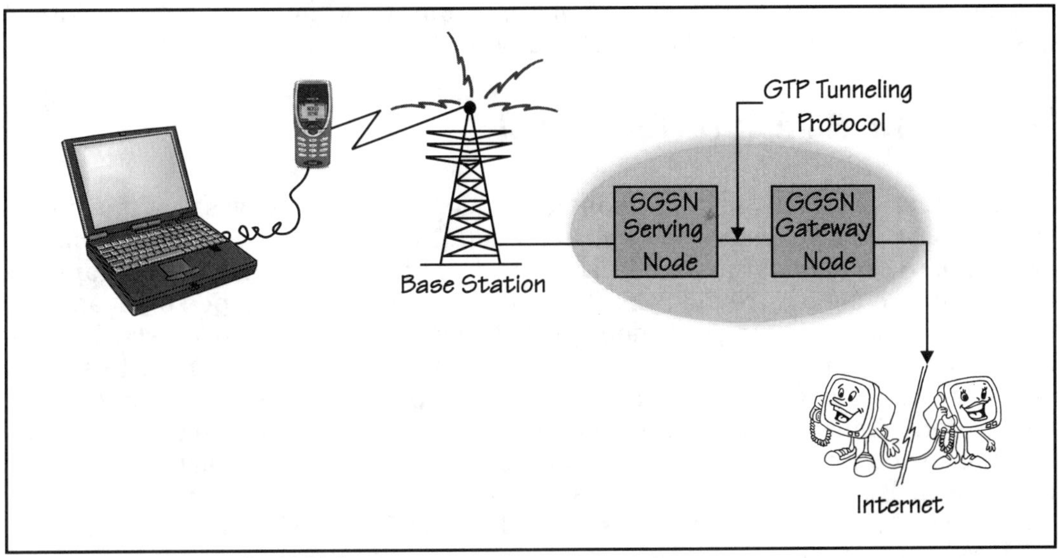

Figure 3-8. An Overview of a GPRS System.

network can use multiple serving nodes, but requires only one gateway node for connecting to an external network such as the Internet.

When the mobile station sends packets of data, it is via the SGSN to the GGSN, which converts the packets for transmission over the desired network. The transmission could be over the Internet, X.25 networks, or private networks. IP packets from the Internet addressed for the mobile station are received by the GGSN, forwarded to the SGSN, and then transmitted to the mobile station.

To forward IP or X.25 packets between each other, the SGSN and GGSN encapsulate these packets using a specialized protocol called the *GPRS tunnel protocol* (GTP), which operates on top of standard TCP/IP protocols. The details of the SGSN and GGSN are both invisible and irrelevant to the user who simply experiences a straightforward IP or X.25 connection that just happens to be wireless.

An interesting aspect of GPRS is how it achieves its high speeds of over 100 kbps when circuit-switched data today is limited to 9600 or 14.4 kbps. GPRS uses the same radio channel as voice calls, a channel that is 200 kHz wide. This radio channel carries a raw digital radio stream of 271 kbps, which for voice calls is divided into 8 separate data streams, each carrying about 34 kbps. After protocol and error correction overhead, 13 kbps is left for each voice connection or about 14 kbps for data. Circuit-switched data today uses only one voice channel. GPRS can combine up to 8 of these channels, and since each of these can deliver up to 14 kbps of data throughput, the net result is that users will be able to enjoy rates of over 100 Kbps. But not all eight-voice channels have to be used. In fact, the most economical phones will be the ones that are limited to 56 kbps. The GPRS standard defines a mechanism by which a mobile station can request the amount of bandwidth it desires at the time it establishes a data session.

Time Division Multiple Access (TDMA)

TDMA is a satellite and cellular phone technology that interleaves multiple digital signals onto a single high-speed channel. TDMA does this by chopping up the channel into sequential time slices. Each user of the channel takes turns transmitting and receiving in a round-robin fashion. In reality, only one person is actually using the channel at any given moment, but only uses it for short bursts. He then gives up the channel momentarily to allow the other users to have their turn. This is very similar to how a computer with just one processor can seem to run multiple applications simultaneously. For cellular networks, TDMA triples the capacity of the original analog Frequency Division Multiple Access (FDMA) modulation. It divides each channel into three sub-channels providing service to three users instead of one. The GSM cellular system is also based on TDMA, but GSM defines the entire network, not just the air interface.

TDMA is also known as D-AMPS (Digital Advanced Mobile Phone System) and is defined in the ANSI-136 standard. There are more than 30 million customers in 36 countries using TDMA digital services. It is claimed that TDMA already covers more than 85 percent of the ITU's IMT-2000 3G functionality and EDGE will deliver all the 3G capability in the existing 800- and 1900-MHz TDMA frequency bands.

TDMA supports the receipt of short messages—the Short Message Service (SMS) Mobile Terminate—but not the origination of short messages (SMS is discussed in detail in Chapter 4).

The Universal Wireless Communications Consortium (UWCC) introduced what it calls *IS136HS (High Speed)/EDGE Compact*.[3] This is a spectrum efficient version of EDGE that will support the 384 kbit/s mandated packet data rates, but will require only minimum spectral clearing and therefore could work for network operators with limited spectrum allocations. In fact, as a result of this, EDGE has been renamed Enhanced Data Rates for Global Evolution (replacing "GSM" with "Global").

Code Division Multiple Access (CDMA)

Code Division Multiple Access, a cellular technology also known as *IS-95*, competes with the GSM technology for dominance in the cellular world.

---

[3] The UWCC was disbanded in December 2001. For information on TDMA or EDGE technologies and their interoperability with GSM and UMTS, visit *3GAmericas.org*.

There are different variations; the original CDMA is now known as *cdmaOne*. Developed originally by QUALCOMM and enhanced by Ericsson, CDMA is characterized by high capacity and small cell radius, employing spread-spectrum technology and a special coding scheme. The Telecommunications Industry Association (TIA) adopted CDMA in 1993. By December 2000 there were 27 million subscribers on cdmaOne systems worldwide.

CDMA, unlike TDMA, lets everyone transmit at the same time. Using conventional modulation techniques, this would certainly be impossible. What makes CDMA work is a special type of digital modulation called "Spread Spectrum." This form of modulation takes the user's stream of bits and splatters them across a very wide channel in a pseudo-random fashion. The "pseudo" part is very important here, since the receiver must be able to undo the randomization in order to collect the bits together in a coherent order.

CDMA also supports "soft hand-off." A hand-off occurs in *any* cellular system when a call switches from one cell site to another as the caller travels. In all other technologies, this hand-off occurs when the network informs the caller's phone of the new channel to which it must switch. The phone then stops receiving and transmitting on the old channel, and commences transmitting and receiving on the new channel. This is known as a "hard hand-off." In CDMA however, every phone and every site are on the same frequency. In order to begin listening to a new site, the phone only needs to change the pseudo-random sequence it uses to decode the desired data from the jumble of bits sent. While a call is in progress, the network chooses two or more alternate sites that it feels are hand-off candidates. It simultaneously broadcasts a copy of the call on each of these sites. The caller's phone can then pick and choose between the different sources for the call, and move between them whenever it feels like it. It can even combine the data received from two different sites to ease the transition from one to the other.

This arrangement therefore puts the phone in almost complete control of the hand-off process. Such an arrangement should ensure that there is always a new site primed and ready to take over the call at a moment's notice. In theory, this should put an end to dropped calls and audio interruptions during the hand-off process. In practice it works quite well, but dropped calls are still a fact of life in a mobile environment. However, CDMA rarely drops a call due to a failed hand-off.

A major problem in CDMA is channel pollution, which occurs when signals from too many base stations are present at the mobile terminal, but none are dominant. When this situation occurs, audio quality degrades rapidly, even when signals seem otherwise very strong. Pollution occurs frequently in densely populated urban environments where service providers must build many sites in close proximity. Channel pollution can also result from massive multipath problems caused by many tall buildings. Taming pollution is a tuning and system design issue. It is up to the service provider to reduce this phenomenon as much as possible.

Integrated Digital Enhanced Network (iDEN)

iDEN, launched by Motorola in 1994, is a Private Mobile radio system from Motorola's Land Mobile Products Sector (LMPS). iDEN technology is currently available in the 800-MHz, 900-MHz, and 1.5-GHz bands. It utilizes a variety of advanced technologies, including state-of-the-art voice coders (vocoders), M16QAM modulation and TDMA (Time Division Multiple Access). It allows Commercial Mobile Radio Service (CMRS) operators to maximize the dispatch capacity and provides the flexibility to add optional services such as full-duplex telephone interconnect, alphanumeric paging and data/fax communication services.

I-mode

i-mode (the 'i' stands for information) is a service that was launched in February 1999 by NTT DoCoMo, Japan's leading cellular phone operator. DoCoMo in Japanese means 'anywhere.' Instant access to Web sites compatible with i-mode can be achieved at the touch of a button. You can receive email, exchange photographs, receive news and stock quotes, shop online, receive weather forecasts, play online games, and access music files online. Every subscriber is given an email ID, which is his or her cellular phone number concatenated with @docomo.ne.jp. Of course, amidst all this excitement of being able to access the Net on the go, one tends to forget that you can also make and receive phone calls from your i-mode (or WAP) enabled phone.

It is the packet-switched data network that gives i-mode its popular characteristics such as constant access, alerts, and instant messaging. WAP services in Japan use similar technology and enjoy similar popularity. The difference in the transmission technology between GSM and i-mode may be attributed to differences between circuit switching in GSM versus packet-switching in i-mode.

How Does It Work?

Components required for i-mode services are:

- An i-mode cellular phone (a phone capable of voice and packet transmission along with an installed browser)
- A packet network
- An i-mode server
- Content in c-HTML

Mode of Transmission

I-mode uses a *Personal Digital Cellular-Packet (PDC-P)* method of data transmission over the existing PDC network used for ordinary voice traffic. This is a packet-switched network that is well suited for wireless communication. In a packet-switched network, data is broken into small units called packets and routed over the net. This mode of transmission where communication is broken into packets allows the same data path to be shared among many users in the network.

Thus, users can stay online throughout and yet not be charged for the time spent online. Rather, they only pay for the amount of data that they retrieve. This is in contrast to the circuit-switched network like the regular voice telephone network where the communication path is dedicated to the callers thus blocking that path to other users for that period of time.

As of now, the rate at which data is being transmitted is 9.6 Kbps but with DoCoMo going all out to develop the *Wideband-Code Division Multiple Access (W-CDMA)* technology, a third generation mobile communication system, speeds of 384 Kbps or more could be in the offing. W-CDMA (discussed later) allows high-speed data transmission of video and large-volume data.

3G

3rd generation (3G) is the generic term for the next big step in mobile technology development. The formal standard for 3G is the IMT-2000 (International Mobile Telecommunications 2000). The different developer communities have pushed this standard: W-CDMA as backed by Ericsson, Nokia, and Japanese handset manufacturers and cdma2000 as backed by the U.S. vendors, QUALCOMM and Lucent. After long negotiations, the intellectual property rights were cross-licensed between Ericsson and QUALCOMM in June 1999, and it looks as though there will be a "peaceful co-existence" of standards.

The goal of being able to have one single network standard (CDMA) and use one handset throughout the world now seems to be attainable. But within the one standard there will be three optional, harmonized modes:

- W-CDMA for Europe and the Asian GSM countries
- Multi-carrier CDMA for North America
- TDD (Time Division Duplex)/CDMA for the Chinese

The first 3G network was deployed by NTT DoCoMo in Japan in the second quarter of 2002. UMTS (Universal Mobile Telephone System) is the third generation mobile phone system that will be commercially available from 2003 in Europe where the first licenses were granted in Finland. Although many people associate UMTS with a speed of 2 Bit/s, this will be reached only within a networked building and indeed only with some further development of the technology. Realistic expectations suggest a maximum capacity in metropolitan areas of 384 Bit/s, at least until 2005.

This is the same transmission rate that can be realized much earlier with EDGE. Some mobile operators are currently reconsidering their UMTS rollout plans due to concerns relating to the capacity differentials and the cost/benefit of migrating from one to the other.

W-CDMA

W-CDMA is a 3G technology that increases data transmission rates in GSM systems by using the CDMA air interface instead of TDMA. In the ITU's IMT-2000 3G specification, W-CDMA has become known as the Direct Sequence (DS) mode. Broadband CDMA is known as W-CDMA. This is one of the latest components of UMTS, along with TDMA and cdma2000. It has a 5-MHz air interface and is the basis of higher-bandwidth data rates.

UMTS

UMTS (Universal Mobile Telephone Service) is a 3rd Generation (3G) Mobile System being developed by ETS within the ITU's IMT-2000 framework.

It will provide data speeds of up to 2 Mbps, making portable videophones a reality.

UMTS has the support of many major telecommunications operators and manufacturers because it represents a unique opportunity to create a mass market for highly personalized and user-friendly mobile access to the Information Society.

UMTS seeks to build on and extend the capability of today's mobile, cordless, and satellite technologies by providing increased capacity, data capability, and a far greater range of services using an innovative radio access scheme and an enhanced, evolving core network.

The frequency bands 1885–2025 MHz and 2110–2200 MHz are reserved for future IMT-2000 systems, and the bands 1980–2010 MHz and 2170–2200 MHz are intended for the satellite part of these future systems.

Table 3-2 shows how the various cellular technologies are interdependent.

Table 3-2. Summary of Forthcoming Cellular-Data Services.

Core Technology	Service	Data Capability
GSM	Circuit-switched data based on the standard GSM 07.07	9.6 Kbps or 14.4 Kbps
	High-speed circuit-switched data (HSCSD)	28.8 to 56 Kbps service likely
	General Packet Radio Service (GPRS)	IP and X.25 communications over Kbps
	Enhanced Data Rates for GSM Evolution (EDGE)	IP communications to 384 Kbps. Roaming with IS-136 networks possible.
	Wideband CDMA (W-CDMA)	Similar to EDGE but adds 2 Mbps indoor capability. Increased capacity for voice.
IS-136	Circuit-switched data based on the standard IS-135	9.6 Kbps
	EDGE	IP communications to 384 Kbps. Roaming with GSM networks possible.
	W-CDMA or Wideband TDMA (WTDMA)	Similar to EDGE but adds 2 Mbps indoor capability
CDMA	Circuit-switched data based on the standard IS-707	9.6 Kbps or 14.4 Kbps
	IS-95B	IP communications to 64 Kbps
	cdma2000 - 1XRTT	IP communications to 144 Kbps
	cdma2000 - 3XRTT	IP communications to 384 Kbps outdoors and 2 Mbps indoors

Data Rates

Table 3-3 summarizes the data rates of wireless technologies.

Table 3-3. Data Rates of Wireless Technologies.

Technology Class	Technology	Data Rate
2G	GSM	~9.6 kbps
	CDMA	~14 kbps
	CDPD	~19.2 kbps

Table 3-3. Data Rates of Wireless Technologies. (Continued)

Technology Class	Technology	Data Rate
2.5G	GPRS	~115 kbps
	Ricochet	~128 kbps
3G	cdma2000, W-CDMA	~144 kbps ~384 kbps ~2 mbps

FYI Briefing

Why So Many Standards?

You must be wondering: "I could use my GSM phone in all the European and Asian countries by just changing the SIM card, but I could not use the same phone in the U.S." So you bought a new CDMA phone to use in U.S. Now you must be wondering: "I could use this phone in California, but why can't I use this phone in Texas?" I had similar doubts when I looked at a complex coverage map provided by my cell phone service vendor. Why don't these cellular technologies exist in harmony? This could mostly be attributed to the standards bodies.

International Telecommunications Union (ITU) located in Geneva (formerly Consultative Committee for International Radio Communications (CCIR)) is the international standards body for the radio and wireless communication sector. This body handles frequency planning and other standards definition tasks. But to reflect national interests, this standards body delegated some standards efforts to regional agencies. The European Conference for Posts and Telecommunications (CEPT), that further delegated some work to the European Telecommunications Standards Institute (ETSI), covers European countries and several other nations; the Federal Communications Commission (FCC) covers the U.S.

ETSI adopted and standardized GSM (at frequencies 900 MHz and 1800 MHz) to be used by all the European nations. In contrast, the FCC allowed several cellular technologies in the same frequency bands around 850 MHz. This led to hybrid solutions such as dual mode mobile phones supporting digital and analog systems. Later TDMA, CDMA, and variations in CDMA were also adopted. This led to islands of different systems. Thus, the FCC's decisions were led more by market forces than by the FCC leading the market. Hopefully 3G will bring much needed harmony by providing interoperability. ❑

FYI

4G - The Next Wireless Technology

Briefing

While we haven't seen the power of 3G yet, there has been talk of 4G already. This is not because of the accelerated pace of technology, it's because 3G has not delivered on its promise. The technology has not matured quickly enough and has not been able to provide the promised data speeds. The bandwidth is barely sufficient for email transfers, so video streaming and multimedia capabilities are far away.

Some market segments feel that 4G might be the solution for problems that 3G is not capable of answering. 4G is not just about wireless networking. 4G aims at using the ubiquitous IP network for both voice and data. IP networks are mature enough to carry voice and deliver real-time data. For more information about 4G, visit *http://www.4gmobile.com.* ❑

Mobile Position Location Technologies

The ability to determine the position of a mobile device opens a wide range of opportunities for application development in geography-aware information services. The FCC mandated that by October 2001 all the carriers must have the capability to identify the latitude and longitude of mobile devices making 911 calls within a radius of 125 meters in 67 percent of the cases. This level of precision might take more time. But when this happens even applications can take advantage of these capabilities. We envision that regardless of the underlying technology that is providing this service, there will be a programming abstraction, which application developers can use to determine the location of the mobile device. This enables the application to be location-sensitive. Some example applications where this information could be used are interactive map applications, location-based advertising and location-sensitive information services such as weather forecasts, and yellow pages.

There are two approaches: network-centric and handset-centric. The latter requires modification to the mobile terminal; the former requires changes to the network infrastructure while still using existing mobile terminals. In some network-centric cases, there might be changes required at both the terminal and the network. Let us take a look at these approaches.

There are four major network-centric techniques that could be used for location detection: Cell of Origin (COO), Time of Arrival (TOA), E-OTD (Enhanced Observed Time Difference), and Angle of Arrival (AOA); and one handset-centric location technique: *Global Positioning System (GPS)*. These technologies are described in the following sections.

Cell of Origin (COO)

Cell of Origin (COO) is the simplest and fastest of all the techniques, as this requires no upgrade to the network or to the handset. This can be used as a location-fixing scheme to locate a subscriber, but it is less accurate than the other methods. The accuracy varies on the cell size. In this system the base-station cell area is used as the location of the caller. Positioning accuracy generally depends upon the size of the cell but positioning down to 150 meters is possible in urban areas with the deployment of pico (smaller) cell sites. Although other schemes offer higher degrees of positioning accuracy than COO, its main advantages are that the speed of response in getting a location fix is fast (typically around three seconds) and that given no handset or network upgrade is required, it can be used to provide location specific services to existing customer bases.

Angle of Arrival (AOA)

Angle of Arrival (AOA) was first developed for military and government organizations since it operates with no modification to a mobile device, and was later applied to analog cellular systems. When deployed in digital systems, the brevity of the signals and the channel sharing means that AOA schemes are difficult to deploy successfully. Figure 3-9 illustrates the Angle of Arrival location technique.

The most common version of this technique is known as small aperture direction finding, which requires a complex 4–12 antenna array at each of several cell site locations. These antenna arrays can, in principle, work together to determine the angle (relative to the cell site) from which a cellular signal originated. When several cell sites can each determine their respective angles of arrival, cellular telephone location can be estimated from the point of intersection of projected lines drawn out from the cell site at the angle from which the signal originated.

Small aperture Angle of Arrival systems suffer from distortion of the wavefront of the cellular signal, caused by multipath and other environmental factors. The technology can perform acceptably when tracking a continuous transmission, such as a voice transmission, if longer integration times are used. As the caller moves from cell to cell, the system must follow each voice channel assignment as the call is handed off from channel to channel. This can be difficult if the Angle of Arrival antenna is not positioned to interpret the in-band voice channel signaling.

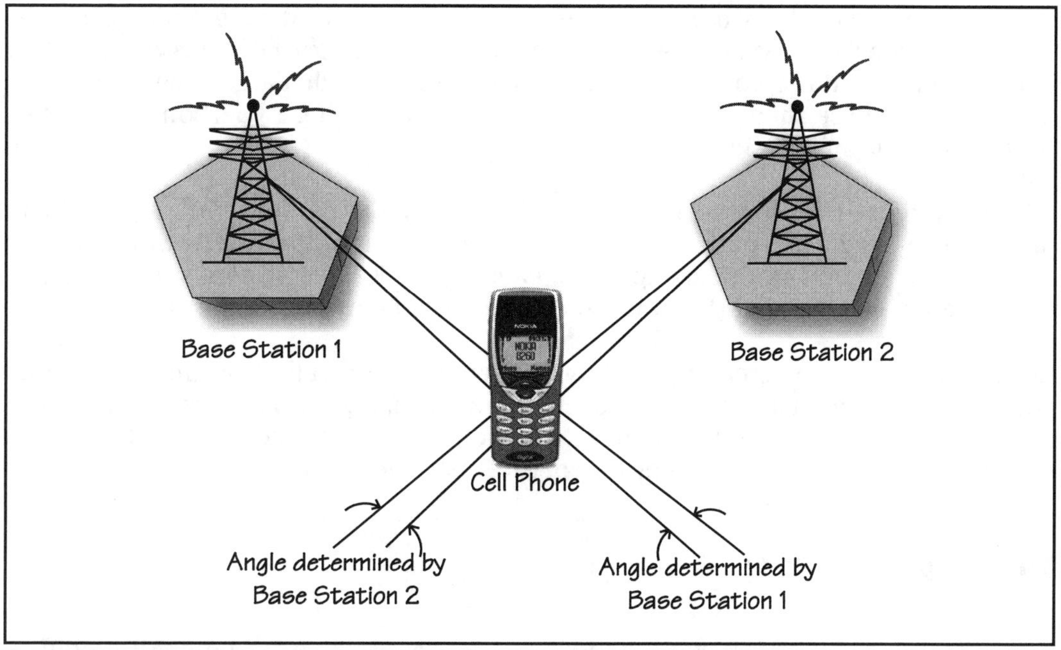

Figure 3-9. Position Location using Angle of Arrival.

A significant drawback to the Angle of Arrival systems is the logistical and aesthetic dilemma of adding antenna arrays to cell sites at a time when communities are enacting increasingly restrictive planning regulations. Existing cell site antennae are not suitable for angle of arrival. Another drawback to Angle of Arrival systems is that the angular error of the antenna array can translate into a significant error in lateral distance if the cellular telephone is far from the cell sites.

Enhanced Observed Time Difference (E-OTD)

The *Enhanced Observed Time Difference* system works by using the existing GSM infrastructure to determine the mobile phone's location. When a user calls selected service providers, E-OTD simultaneously sends data indicating the phone's position. It works by comparing the relative times of arrival, at the handset and at a nearby fixed receiver, of signals transmitted by the underlying mobile network base stations. The E-OTD system overlays the existing mobile network. Suppliers for E-OTD solutions include CPS, Ericsson, and BT Cellnet.

E-OTD systems operate by placing location receivers or reference beacons, overlaid on the cellular network as a Location Measurement Unit (LMU) at multiple sites geographically dispersed in a wide area. Each of these beacons has an accurate timing source, and when a signal from at least three base stations is received by an

E-OTD software-enabled mobile phone and the Location Measurement Unit, the time differences of arrival of the signal from each *Base Transceiver Station (BTS)* at the handset and the Location Measurement Unit are calculated. The differences in time stamps are then combined to produce intersecting hyperbolic lines from which the location is estimated.

E-OTD systems may however be subject to many of the same urban multipath problems as Angle of Arrival systems. In this case, multipath effects distort the shape of the signal and the group delay, causing the E-OTD system some difficulty in accurately determining the point in the signal to be measured by all receivers.

E-OTD schemes offer greater positioning accuracy than cell of origin, between 50 and 125 meters, but have a slower speed of response, typically around five seconds, and require software modified handsets, which means that they cannot be used to provide location-specific services to existing customer bases.

Time of Arrival (TOA)

Time Of Arrival technology (see Figure 3-10) requires larger network modifications and is therefore not very cost-effective. Rolling out TOA for an entire network is estimated to cost as much as 10 times the price of an E-OTD system.

In a similar manner to E-OTD, the difference in time of arrival of a signal from a mobile device to three base stations is used to calculate the location. In this scheme however, there may be no overlay network used as the Location Measurement Unit. Instead, this functionality is provided by synchronization of the cellular network, using GPS or atomic clocks at each BTS. This capability is found in cdmaOne networks in the USA which are synchronized, whereas it is not provided as a given in asynchronous GSM networks. Further problems associated with TOA schemes are that the accuracy offered might be little better than that of Cell of Origin in urban areas, but with a far longer speed of response (typically around 10 seconds) than Cell of Origin or E-OTD.

The cost of synchronizing a GSM network is therefore relatively high for little increase in performance offered over the use of cell of origin and the investment required for an operator may be prohibitive. However, no handset modification is required and services can therefore be offered to existing customers.

GPS

GPS (Global Positioning System) is a system that consists of 24 satellites circling the earth in a location relative to each other so that several satellites fall within line

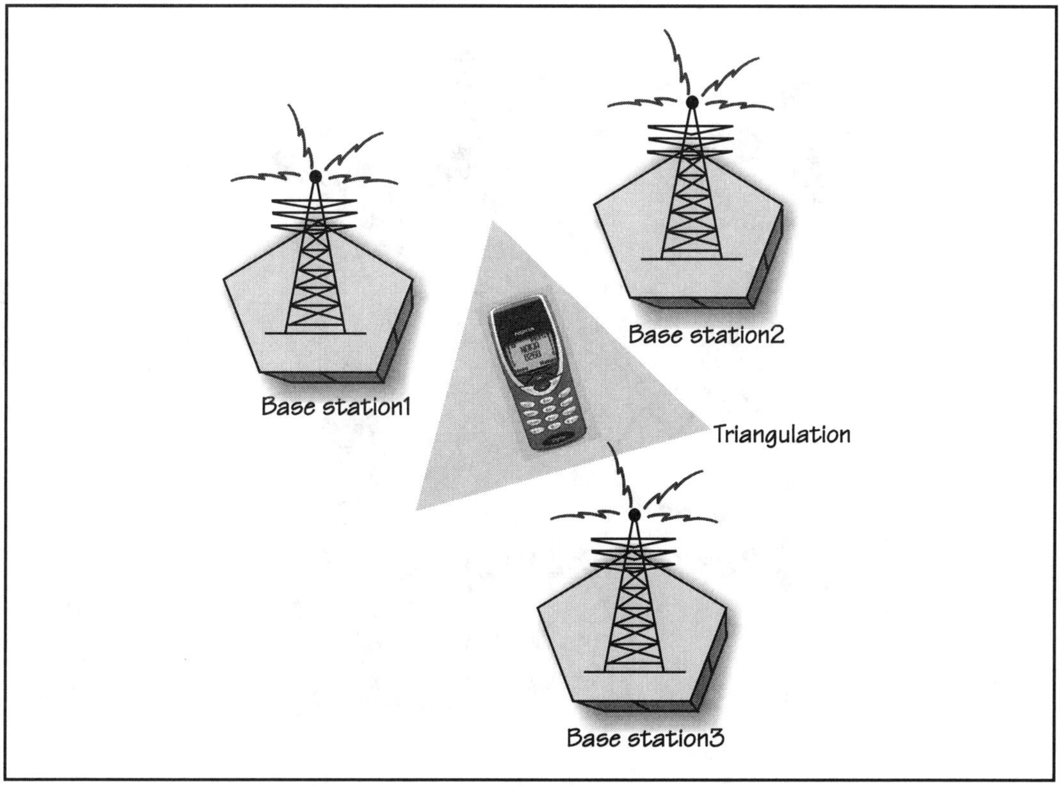

Figure 3-10. Position Location using Time of Arrival.

of sight for any GPS receiver on Earth (see Figure 3-11). Because the satellites are continuously broadcasting their own position and direction, the GPS receiver can calculate its position very exactly. Anybody can use the GPS system for free with an appropriate receiver. GPS has been developed in the U.S. for military use, but from the beginning of the '90s it has been usable (with lower resolution) for civilian purposes. GPS requires additional equipment or some modification in the mobile device, so that it can become a GPS receiver. This technology is used in car navigation systems.

WIRELESS LANS AND PERSONAL AREA NETWORKS (PANS)

Wireless LAN and PAN technologies provide short to medium-range wireless networking capabilities eliminating the need to use expensive cellular technologies. Note that the purpose of wireless LANs and PANs are different. Wireless LANs provide a way for you to connect to your network without using wires. Wireless PAN is used to connect between pervasive devices to achieve a specific job. We will discuss these in more detail in the next sections.

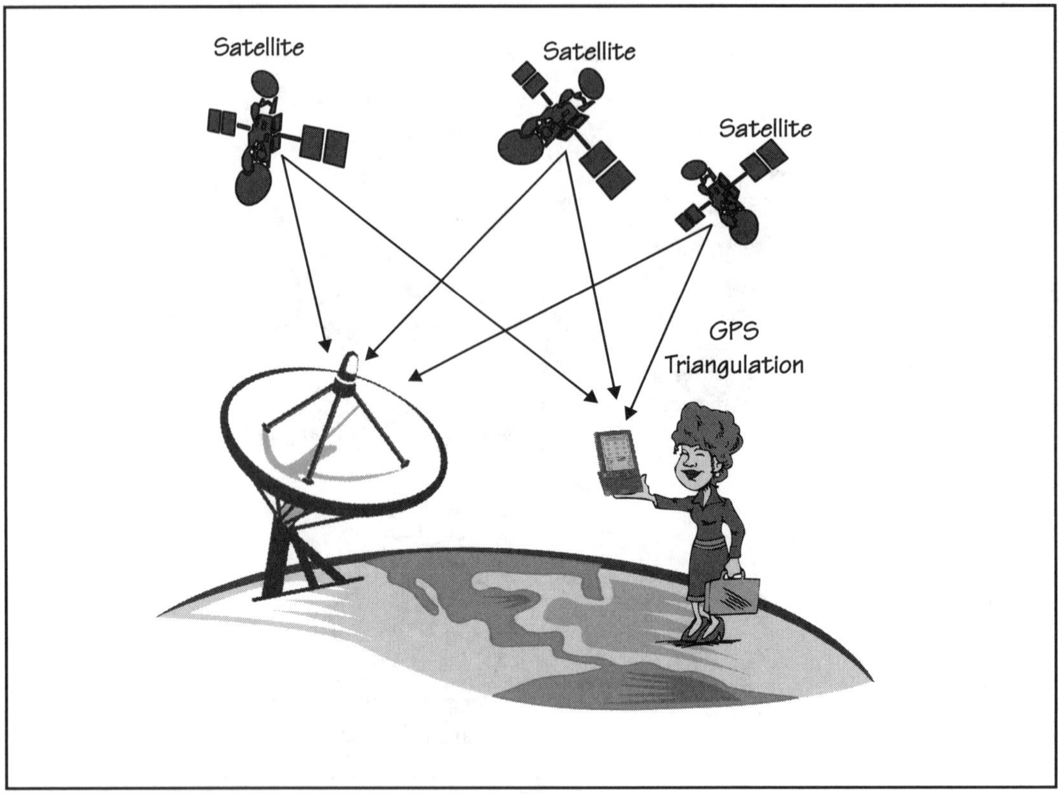

Figure 3-11. Position Location using Global Positioning Satellites.

IEEE 802.11 Wireless LAN

This is the Institute of Electrical and Electronics Engineers (IEEE) standard for medium-range, higher data rate wireless communications. This can typically replace a limited wired LAN. The data rates are up to 108 Mbps with 802.11a, higher later. Most computer manufacturers integrate 802.11 chips in their laptops.

Bluetooth—A Revolutionary Basis for Developing Wireless Applications

Can you imagine walking into a store and having all the sale items automatically available on your cell phone or PDA? This is a definite possibility with Bluetooth. We cover Bluetooth as an emerging standard with its own APIs that developers can use to build applications. There has been lot of activity in the Java Community Process (JCP) in the Java Specification Request (JSR) 82 to define a set of Java APIs to access the Bluetooth stack and services.[4] This section gives you a basic understanding of Bluetooth. Bluetooth, the new technology named after the 10th Century Danish King Harold Bluetooth, is a hot topic among wireless developers. In Chapter 10, we'll introduce Bluetooth programming, and examine the JSR-82 API set in detail. Chapter 10 will also walk you through creating your first Bluetooth application.

Bluetooth is being designed to allow low bandwidth wireless connections to become so simple to use that they seamlessly integrate into your daily life. A simple example of a Bluetooth application is updating the phone directory of your mobile phone. Today, you would have to either manually enter the names and phone numbers of all your contacts or use a cable or IR link between your phone and your PC and start an application to synchronize the contact information. With Bluetooth, this could all happen automatically and without any user involvement as soon as the phone comes within range of the PC! Of course, you can easily see this expanding to include your calendar, to do list, memos, email, etc. This is just one of many exciting applications for this new technology.

Bluetooth is a low power radio technology that is being developed to replace the cables and infrared links for distances up to ten meters. Devices such as PCs, printers, mobile phones, and PDAs can be linked together to communicate and exchange data via a wireless transceiver that fits on a single chip. We estimate that the unit cost will drop from $50 today to below $5 in a few years.

Using Bluetooth, it will be possible to separate the transceiver unit of a cell phone from the earpiece and the display. Thus, the transceiver unit could be in a belt buckle and the display in a watch with no wires needed.

Additionally, data exchange—for example with POS (Point Of Sale) terminals, ticketing, or e-wallet applications for mobile commerce—might also boost the success of Bluetooth. Since it has a throughput of about 1 Mbit/s, Bluetooth might also be used in wireless LAN applications.

We believe that although mobile commerce would be possible without Bluetooth, the technology is adding convenience for both mobile payment and security. Bluetooth provides some of the key functionality to change a mobile phone into a "lifetool."

The Bluetooth specification is an open specification that is governed by the Bluetooth Special Interest Group (SIG). The Bluetooth SIG is led by its five founding

---

[4] The Java Community Process is the organization that defines Java-related technologies. Under the guidance of Sun Microsystems, member companies and individuals form expert committees to define new technologies. Descriptions of proposed and final Java technology definitions are called Java Specification Requests or JSRs.

companies and four new member companies who were added in late 1999 (see Table 3-4). These nine companies form the *Promoter Group* of the Bluetooth SIG.

Table 3-4. The Bluetooth Special Interest Group (SIG) Founding Membership.

Founding Companies	New Members
Ericsson	3Com Corporation
IBM Corporation	Lucent Technologies
Intel Corporation	Microsoft Corporation
Nokia	Motorola Inc.
Toshiba Corporation	

More than 1200 additional companies are members of the Bluetooth SIG. The magnitude of industry involvement should ensure that Bluetooth becomes a widely adopted technology. The first Bluetooth products should begin to appear this year. The first Bluetooth product from Ericsson is a wireless cellular phone headset already available in the market.

Bluetooth System Architecture

Bluetooth communication occurs in the unlicensed ISM band at 2.4 GHz. The transceiver utilizes frequency hopping to reduce interference and fading. A typical Bluetooth device has a range of about 10 meters. The communication channel can support both data (asynchronous) and voice (synchronous) communications with a total bandwidth of 1 Mb/sec. Table 3-5 shows the supported channel configurations.

Table 3-5. Channel Configurations Supported by Bluetooth.

Configuration	Max. Data Rate Upstream	Max. Data Rate Downstream
3 Simultaneous Voice Channels	64 kb/sec X 3 channels	64 kb/sec X 3 channels
Symmetric Data	433.9 kb/sec	433.9 kb/sec
Asymmetric Data	723.2 kb/sec or 57.6 kb/sec	57.6 kb/sec or 723.2 kb/sec

The synchronous voice channels are provided using circuit switching with a slot reservation at fixed intervals. A synchronous link is referred to as an SCO (synchronous connection-oriented) link. The asynchronous data channels are provided using packet switching utilizing a polling access scheme. An asynchronous link is referred to as an ACL (asynchronous connection-less) link. A combined data-voice SCO packet is also defined. This can provide 64 kb/sec voice and 64 kb/sec data in each direction.

Bluetooth and Piconets

Bluetooth devices can interact with one or more other Bluetooth devices in several different ways. The simplest scheme is when only two devices are involved. This is referred to as point-to-point. One of the devices acts as the master and the other as a slave. This ad-hoc network is referred to as a *piconet* (see Figure 3-12). As a matter of fact, a piconet is any such Bluetooth network with one master and one or more slaves. In the case of multiple slaves, the communication topology is referred to as point-to-multipoint. In this case, the channel (and bandwidth) is shared among all the devices in the piconet.

There can be up to seven active slaves in a piconet. Each of the active slaves has an assigned 3-bit Active Member address (AM_ADDR). There can be additional slaves that remain synchronized to the master, but do not have an Active Member address. These slaves are not active and are referred to as parked. For the case of both active and parked units, all channel access is regulated by the master. A parked device has an 8-bit Parked Member Address (PM_ADDR), thus limiting the number of parked members to 256. A parked device remains synchronized to the master clock and can very quickly become active and begin communicating in the piconet.

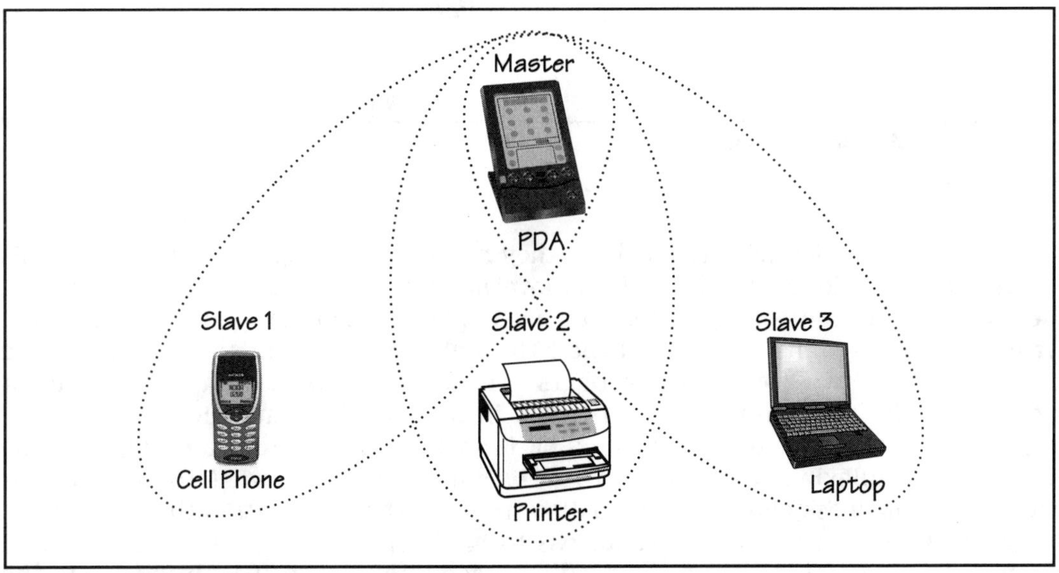

Figure 3-12. A Bluetooth Piconet.

Bluetooth and Scatternets

You may be wondering what would happen if two piconets are within the same coverage area. For example, you might have a piconet consisting of your cell phone and your PC, while the person in the neighboring cubicle has a piconet consisting of a cell phone, headset, and business card scanner. Figure 3-13 illustrates a Bluetooth scatternet.

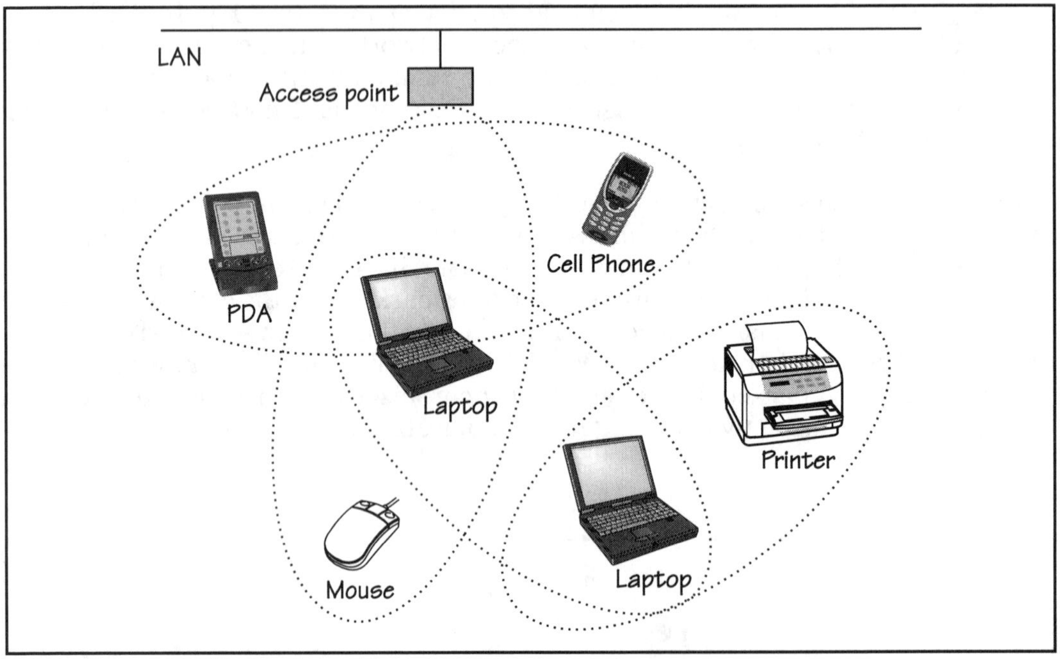

Figure 3-13. A Bluetooth Scatternet.

Because the two piconets are so close, they have overlapping coverage areas. This scenario is provided for in the Bluetooth specification and is referred to as a scatternet. Slaves in one piconet can participate in another piconet as either a master or slave. This is accomplished through time division multiplexing. In a scatternet, the two (or more) piconets are not synchronized in either time or frequency. Each of the piconets operates in its own frequency-hopping channel while any devices in multiple piconets participate at the appropriate time via time division multiplexing. Returning to the example, you may want to set up your neighbor's business card scanner to also transmit the information that is scanned to your PC so that you will have access to his business contacts information. Of course, this would have to be a mutually agreed upon usage. This brings us to the next topic, Bluetooth security.

Bluetooth Authentication and Privacy

While authentication and privacy could be handled at the software protocol layer, it is also provided in the Bluetooth physical layer. A particular connection can be specified to require either one-way, two-way, or no authentication. The authentication is provided using a challenge/response system. The system supports key lengths of 40 or 64 bits. The key management is left to software layers. These security mechanisms and the associated software allow a user to set up his or her devices to only communicate with each other. All Bluetooth devices implement this physical layer security in the same way. Of course, for highly sensitive applications, it is also recommended that you utilize more advanced algorithms in the network transport or application layer.

Bluetooth and Hardware Requirements

Now that you have a basic understanding of what the Bluetooth specification is and how Bluetooth devices can interact with each other, you are probably beginning to wonder what makes all this work in terms of real hardware and software. Let us first discuss the hardware modules before we plunge into details of the software.

Bluetooth hardware can be divided into two primary functions: the Radio Module and the Link Module. At this time, a complete Bluetooth hardware module including both Radio and Link subsystems costs between $25–30 in additional parts. However, within the next couple of years this is expected by most in the industry to approach $5.

Bluetooth: The Radio Module

As mentioned above, Bluetooth devices operate in the 2.4 GHz Industrial Scientific Medicine (ISM) band. This is an unlicensed band and, in most countries, includes the frequency range from 2400 to 2483.5 MHz. Of course, as always when dealing with international standards, there are a few exceptions. The primary geographies with exceptions are France (2446.5 to 2483.5 MHz) and Spain (2445 to 2475 MHz). At this time, Bluetooth products for these two markets are local versions that are not interoperable with the international versions that implement the full range. These localized versions have a reduced frequency band and a different hopping algorithm. However, the Bluetooth SIG is working with authorities in both countries to open the full range. For the sake of simplicity, we will only deal with the international frequency range implementation.

The RF channels used are from 2402 to 2480 MHz with a channel spacing of 1 MHz. Frequency hopping has been implemented to reduce interference and fading. This means that every 625 μsec the channel will hop to another frequency within the 2402 to 2480 MHz range. This translates to 1600 hops every second. Each piconet has a unique hopping sequence that is determined using an algorithm that uses the Bluetooth device address of the master device. All Bluetooth units in the piconet are then synchronized to this hopping sequence.

All packet transmissions are started at the beginning of one of the 625 μsec time slots. A packet may last up to 5 time slots. A time division duplex scheme is used to facilitate full duplex transmission. During even numbered slots, the master may begin a transmission. During odd numbered slots, a slave may begin a transmission. In addition, these time slots can be reserved for synchronous applications such as voice data.

Bluetooth devices are classified according to three different power classes, as shown Table 3-6.

Table 3-6. Bluetooth Devices by Power Class.

Power Class	Maximum Output	Power	Range
1	100.0 mW	20 dBm	100 M
2	2.5 mW	4 dBm	10 M
3	1.0 mW	0 dBm	2 M

Most portable Bluetooth devices will probably be in Power Class 1 or 2 (with a nominal output power of 0 dBm) due to cost and battery life issues. A Power Class 1 device requires that you utilize a power control to limit the transmitted power over 0 dBm. While a little more costly and power hungry, this will provide up to 100 M of range, which should be sufficient for home networking and other applications that require a greater range.

Bluetooth radio modules use Gaussian Frequency Shift Keying (GFSK) for modulation. A binary system is used where a one is signified by a positive frequency deviation and a zero is signified by a negative frequency deviation. The data is transmitted at a symbol rate of 1 Ms/sec. The radio module is covered in detail in Part A of the Specification of the Bluetooth System published by the Bluetooth SIG.

Bluetooth: The Link Module

The Link Module and the closely associated Link Manager software are responsible for the baseband protocols and some other low-level link functions. This includes sending/receiving data, setting up connections, error detection and correction, data whitening (simple scrambling scheme of both the header and payload—not the same as encryption), power management, and authentication.

The link module is responsible for deriving the hop sequence. This is accomplished using the Bluetooth Device Address (BD_ADDR) of the master device. All Bluetooth devices are assigned a 48-bit IEEE 802 address. This 48-bit master device address is used by each of the devices in the piconet to derive the hop sequence.

The Link Module is also responsible for performing the three error correction schemes that are defined for Bluetooth:

- 1/3 rate FEC
- 2/3 rate FEC
- ARQ scheme for the data

The purpose of the two FEC (forward error correction) schemes is to reduce the number of retransmissions. The ARQ scheme (automatic retransmission request) will cause the data to be retransmitted until an acknowledgement is received indicating a successful transmission (or until a pre-defined time-out occurs). A CRC (cyclic redundancy check) code is added to each packet and used by the receiver to decide whether or not the packet has arrived error free. Note that the ARQ scheme is only used for data packets, not synchronous payloads such as voice.

In order to reduce highly redundant data and minimize signal degradation, a data whitening scheme is used to randomize the data. The data is scrambled by a data whitening word and then unscrambled using the same word at the receiver. This descrambling is done after the error detection/correction process.

Bluetooth provides provisions for three low power modes to conserve battery life. These states, in decreasing order of power requirements are Sniff Mode, Hold Mode, and Park Mode. While in the Sniff mode, a device listens to the piconet at a reduced rate. The Sniff interval is programmable, providing flexibility for different applications. The Hold mode is similar to the Park mode, except that the Active Member address (AM_ADDR) is retained. In the Park mode, the device's clock continues to run and remains synchronized to the master, but the device does not participate at all in the piconet.

Bluetooth: Hardware Implementations

Several companies are working on single chip Bluetooth solutions with roadmaps toward the $5 cost target. In future designs the host controller software will be frozen and the flash memory will be integrated as mask ROM into the silicon, further reducing the size of the design.

SUMMARY

We have covered the basics of wireless communications, supporting technologies from 1st generation to 3rd generation. With the deployment of 3G networks by a majority of the operators, we will see a sudden spike in the connection speeds provided by the wireless networks. This is good news for the programming community as this means that we can build richer applications such as wireless video conferencing, multimedia applications, and games.

We also covered the basics of Bluetooth, an evolving standard in Personal Area Networks. This enables us to build applications, which use Bluetooth for network discovery and application communication.

Chapter 4

Wireless Devices: Service Technologies

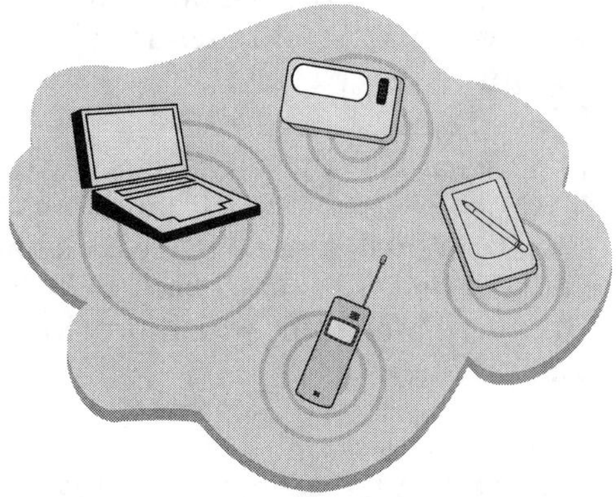

In the previous chapter, we described the various types of networks that support wireless transmission of voice and data. The business need for wireless data transmission is increasing every day. The Internet and the World Wide Web brought new business opportunities to our corporate infrastructure and the need is just as compelling to bring some of these services—such as messaging, email, and content browsing—to our handheld devices. So, wireless technologies are rapidly moving toward providing support for data transmission to complement wireless voice transmission.

This has not been an easy task. Voice transmission has a much lower bandwidth requirement. New technologies had to be developed, first with 2G then the 3G technologies (described in Chapter 3), to increase the bandwidth for data transmission. In addition, the connection model for voice transmission is a circuit-switched model and billing is based on the time the circuit is active. Data transmission, on the other hand, is better served by packet-based transmission.

In the packet-based model, the data to be transmitted is broken up into small pieces and routed over the network. These pieces are then reassembled at the receiving end. In this case a circuit does not have to be established, thus facilitating quicker data transmission and better sharing of the wireless medium. In addition, billing can now be charged based on the amount of data transmitted.

The ability to transmit data, however, is not sufficient. There must be applications that take advantage of the data transmission. The first type of these applications, which we will call *service technologies*, is an intermediary step towards providing a full programming platform on handheld devices. Service technologies are simple, providing either messaging or content browsing. For the most part, they are "hardwired" into the handheld device and do not support add-on programs. They also have a back-end infrastructure that is specific to the service technology they support.

In this chapter we will discuss two message-based service technologies: *Short Message Service (SMS)* and *Unstructured Supplementary Services Data (USSD)*. We will also cover two technologies for delivering HTML-like content to a handheld device: *i-mode* and *Wireless Application Protocol (WAP)*. Finally, we will cover *VoiceXML*, which provides for the delivery of a voice-driven application to any phone.

SHORT MESSAGING SERVICE (SMS)

The *Short Messaging Service (SMS)* allows a handheld device to generate and receive short text messages. It was introduced in Europe in 1991 and became an instant hit. The Global System for Mobile Communications (GSM)—formerly the European standard for digital wireless—has supported SMS from the beginning. In the USA, SMS was available initially on digital wireless networks built by BellSouth Mobility, PrimeCo, and Nextel, among others.

Short Messaging Service is an industry standard packet-based wireless service that enables the transmission of alphanumeric messages between mobile subscribers and systems such as electronic mail, paging, and voicemail systems.

SMS guarantees delivery of these notifications and alerts. It can support email generation and multicasting (i.e., sending messages to more than one end user). SMS supports out-of-band packet delivery and low-bandwidth message transfer. It is a highly efficient mechanism for transmitting short bursts of data. SMS was positioned initially as an attractive alternative to alphanumeric paging, since it permits two-way general-purpose messaging and notification services—primarily for notification of the arrival of voice mail. As technology and networks evolved, the power and efficiency of SMS was realized and a variety of services that use SMS were introduced. Typically service providers use SMS to notify users of significant events for stock quotes, email, voice mail, fax, reminder services, and airline schedules.

SMS is also employed in wireless data applications for Subscriber Identity Module (SIM) card activation, wireless Point Of Sale (POS), and field-service applications such as automatic meter reading, remote sensing, and location-based services.

SMS Under the Covers

Figure 4-1 illustrates the basic network architecture for SMS deployment. The key components of the SMS system along with the mobile communication components are depicted. Here's a description of the SMS system components:

- The ***External Short Messaging Entities (ESME)*** is a device that may receive or send short messages. The short message originator may be located in the fixed network, a mobile device, or a service center.
- The ***Short Messaging Service Center (SMSC)*** is a combination of hardware and software responsible for relaying, storing, and forwarding a short message between the SMS message source and destination. SMSC is a crucial component and so it must be reliable, scalable, flexible, and support high message throughput.

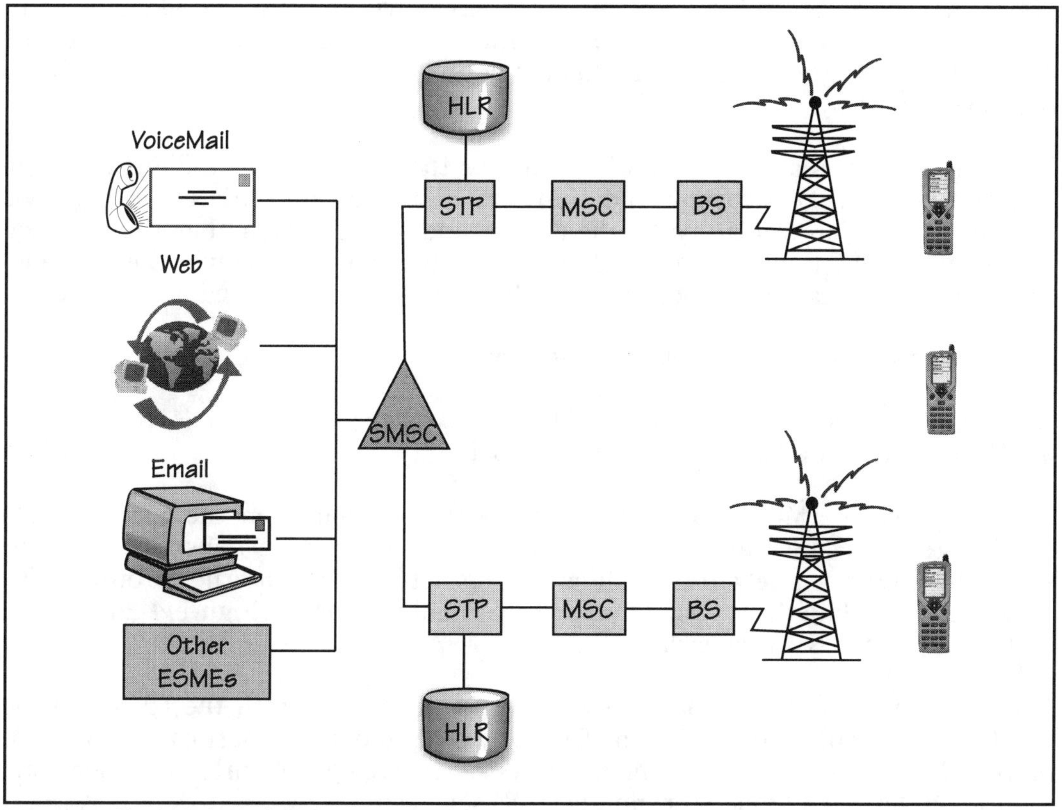

Figure 4-1. The SMS Network Architecture.

- The ***Signal Transfer Point (STP)*** is responsible for communication with the wireless network elements such as the HLR and mobile switching center (MSC).

This architecture supports multiple input sources, including a voice-mail system, Web-based messaging, email integration, and ESMEs that use an SMSC as a store-and-forward system for short messages. So, SMS interconnects different message sources and destinations. The Signal Transfer Point (STP) is responsible for communication with the wireless network elements such as the Home Location Register (HLR) and Mobile Switching Center (MSC).

The message can be originated from a mobile terminal or other input sources (ESMEs). To deliver the message, the SMSC sends a SMS request to the HLR to find if the customer is roaming. The HLR receives the request, and responds to the SMSC with the subscriber's status. The status indicates if the device is inactive, active, and if and where the subscriber is roaming.

If the device is in inactive status, the SMSC will store the message until the device is activated. Once the device is active, the HLR sends a SMS notification to the SMSC, and the SMSC will attempt delivery. SMS allows an active mobile handset to receive or submit a short message at anytime, independent of whether a voice or data call is in progress (in some implementations, this may depend on the MSC or SMSC capabilities).

When the device is active, the SMSC converts the message into a Short Message Delivery Point-to-Point format and transfers it to the appropriate serving base station system. This system pages the device, and if it responds, then the message is delivered. SMSC then receives verification that the end user received the message, categorizes the message as "sent," and will not attempt to send again.

SMS comprises two basic point-to-point services:

- Mobile-originated short message (MO–SM)
- Mobile-terminated short message (MT–SM)

Mobile-originated (MO) short messages are transported from the SMS–capable handset to the SMSC and can be destined to other mobile subscribers or for subscribers on fixed networks such as paging networks or Internet protocol (IP) networks (including the Internet and private email networks). Figure 4-2 depicts a mobile-originated SMS scenario.

Mobile-terminated (MT) short messages are transported from the SMSC to the handset and can be submitted to the SMSC by other mobile subscribers via MO–SM or by other sources such as voicemail systems, paging networks, or operators. Figure 4-3 depicts a Mobile-terminated SMS scenario.

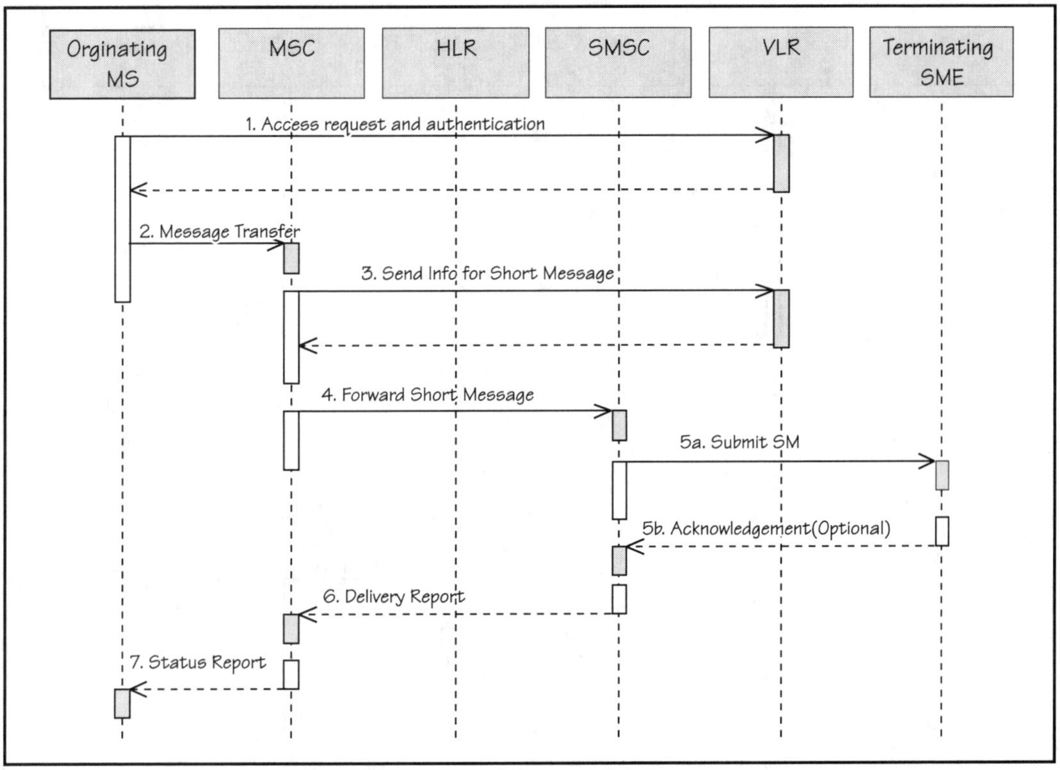

Figure 4-2. A Scenario for an SMS Message Originated from a Mobile Device.

For MT–SM, a report is always returned to the SMSC either confirming the short message delivery to the handset or informing the SMSC of the short message delivery failure and identifying the reason for failure (cause code). Similarly, for MO–SM, a report is always returned to the handset either confirming the short message delivery to the SMSC or informing of delivery failure and identifying the reason.

Service Elements

There are several service policies that affect the reception and submission of short messages:

- ***Message Expiration:*** The SMSC will store and reattempt delivery of messages for unavailable recipients until either the delivery is successful or the expiration time (set on a per-message or platform-wide basis) is reached.

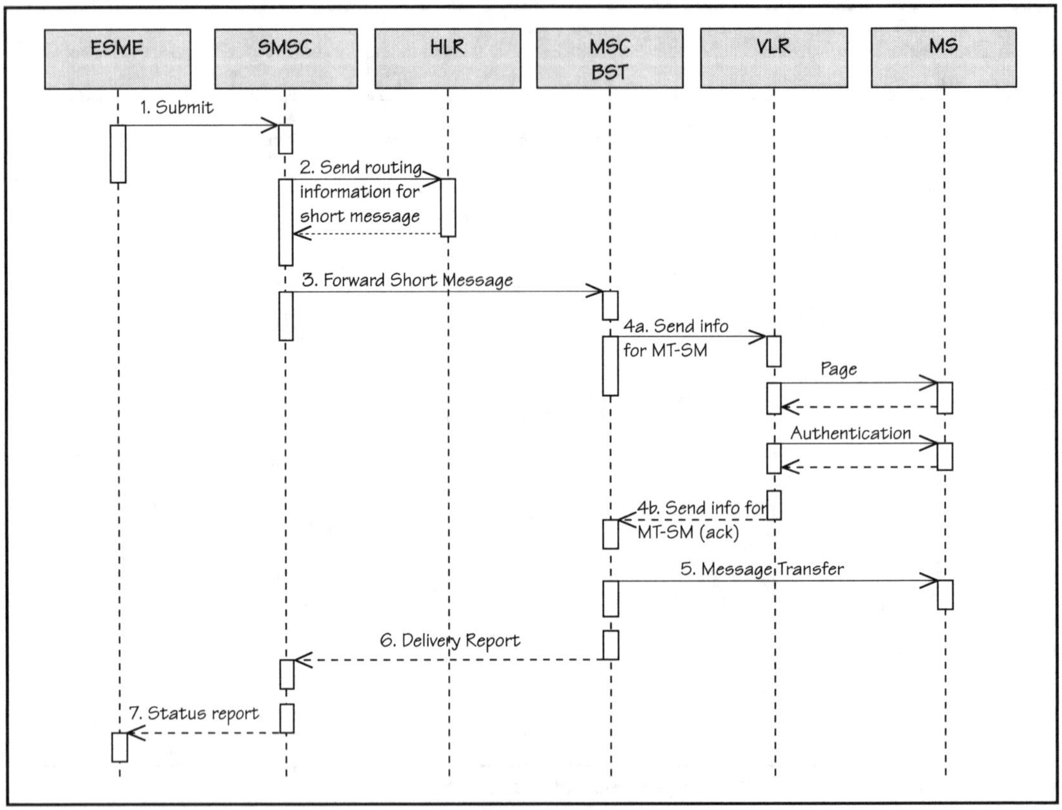

Figure 4-3. A Scenario for an SMS Message Delivered to a Mobile Device.

- ***Priority:*** This is the information element provided by an SME to indicate urgent messages and differentiate them from the normal priority messages. Urgent messages take priority over normal messages, regardless of the time of arrival at the SMSC platform.

- ***Message Escalation:*** After reaching an escalation time, a message will be sent to an alternate message system (such as a paging network or an email server) for delivery to the user. It is assumed that the escalation time is smaller than the expiration time associated with the message. The SMSC will store the message for a period no longer than the expiration time.

In addition, SMS provides a time stamp reporting the time of submission of the message to the SMSC and an indication to the handset of whether or not there are more messages to be sent or the number of additional messages to be sent.

Advantages of SMS

SMS, despite its limitations, is still a very popular service used by millions of people. Its success can be attributed to the simplicity and ease of use. These are the main advantages of the Short Messaging Service:

- Good for short, bursty transmissions.
- Good for point-to-point communication from mobile to mobile or from other external sources to mobile.
- Good alternative for alphanumeric paging.
- Very reliable for message delivery and can be used as an event delivery mechanism.
- Access from external sources such as email makes it convenient to generate messages.

Disadvantages of SMS

SMS, however, is not a solution for all problems. Here are some disadvantages:

- Not suitable for large data transmissions such as those required for Internet access.
- SMS messages must be no longer than 160 alphanumeric characters for GSM, 255 characters for CDMA, and 256 characters for TDMA, and cannot contain any images or graphics.
- Closely integrated with the network. Thus, there are a number of factors that are controlled by the system such as message priorities, message expirations, and message escalations.

The Future of SMS

Mobile messaging is no longer just text. It is evolving beyond text from Short Message Service (SMS) to *Enhanced Messaging Service (EMS)* to *Multimedia Messaging Service (MMS)*. EMS is the ability to send any combination of simple melodies, pictures, sounds, animations, modified text, and standard text as an integrated message for display on an EMS-compliant mobile phone. MMS is the ability to send messages comprising a combination of text, sounds, images, and video to MMS-capable handsets.

UNSTRUCTURED SUPPLEMENTARY SERVICES DATA (USSD)

Unstructured Supplementary Services Data, or USSD, is also a means of transmitting short information or instructions over a GSM network. It is defined within the GSM standards documents, GSM 02.90 (for USSD Stage 1) and GSM 03.90 (for USSD Stage 2). A USSD message can be up to 182 characters in length. Unlike SMS, which uses a store-and-forward mechanism, USSD is a session-based service. Typically, when a user accesses a USSD service, a session is established and the radio connection will remain open until the end of the session.

USSD, like SMS, can be used when making or receiving calls. However, USSD cannot be used when using SMS since both use the same channel.

USSD provides a simple TCP/IP interface to applications so that they do not have to know about network internals. The external applications can reside on any machine that is reachable via TCP/IP. Also, the service code in the USSD message is used to route the message to applications; the service code is interpreted based on the USSD Gateway configuration.

With this simple TCP/IP support, USSD can be used as a rudimentary service to access the Internet from the phone. Let us consider one example. Here is a USSD message submitted from the phone to retrieve the weather in Helsinki.

```
*14*WEATHER*HELSINSKI#
```

The device establishes a USSD connection to the network. The network uses service code 14 to route the message to a particular back-end service providing WEATHER information via TCP/IP. This back-end service could be a regular Internet service. In this scenario, weather information for the city Helsinki (which is passed as a parameter) is requested. The service responds with the requested information to the USSD connection that returns the information to the phone.

Advantages of USSD

Using USSD as a mechanism for providing network services will result in faster responses when compared with SMS because USSD is session-based. Also, since SMS stores and forwards messages, the turnaround response will generally be slower because it has to access data storage devices. Nokia believes that USSD could be up to seven times faster than SMS in carrying out the same two-way transaction.

Users do not have to navigate through phone menus to initiate a USSD message. They can enter messages right from the main phone menu. Also the USSD commands are always routed back to the mobile network's Home Location Register (HLR), so USSD services work the same way whether you are roaming or not.

USSD is available on all GSM phones. Recently, USSD stage 2 was added to the GSM standard. In the early stages, USSD was mostly used to send one-way commands. However, USSD stage 2 messages are more advanced and interactive. For example, a single USSD stage 2 command can fetch an information services menu to provide content browsing functions, similar to i-mode or WAP, on existing GSM phones. Another improvement in stage 2 is the ability for an application to initiate USSD-based transactions. In stage 1 only the phone could initiate interactions.

I-MODE

I-mode is the first successful implementation for providing Internet services on mobile phones. Mari Matsunaga of NTT DoCoMo in Japan designed i-mode and its associated mobile services technology. I-mode is a brand and a service for wireless Internet connection at the touch of a button on a cell phone. NTT DoCoMo started this service in February 1999 in Japan. The majority of i-mode customers use email and information services.

I-mode Technology

I-mode services require the following components (see Figure 4-4):

- An i-mode mobile phone. This is a phone capable of voice and packet transmission and includes a browser.
- A packet network.
- An i-mode server.
- Content in compact HTML (cHTML).

Mode of Transmission

I-mode uses the *PDC-P (Personal Digital Cellular-Packet)* method of data transmission over the existing CDMA network used for ordinary voice traffic. This is a packet-switched network that is well suited for wireless communication.

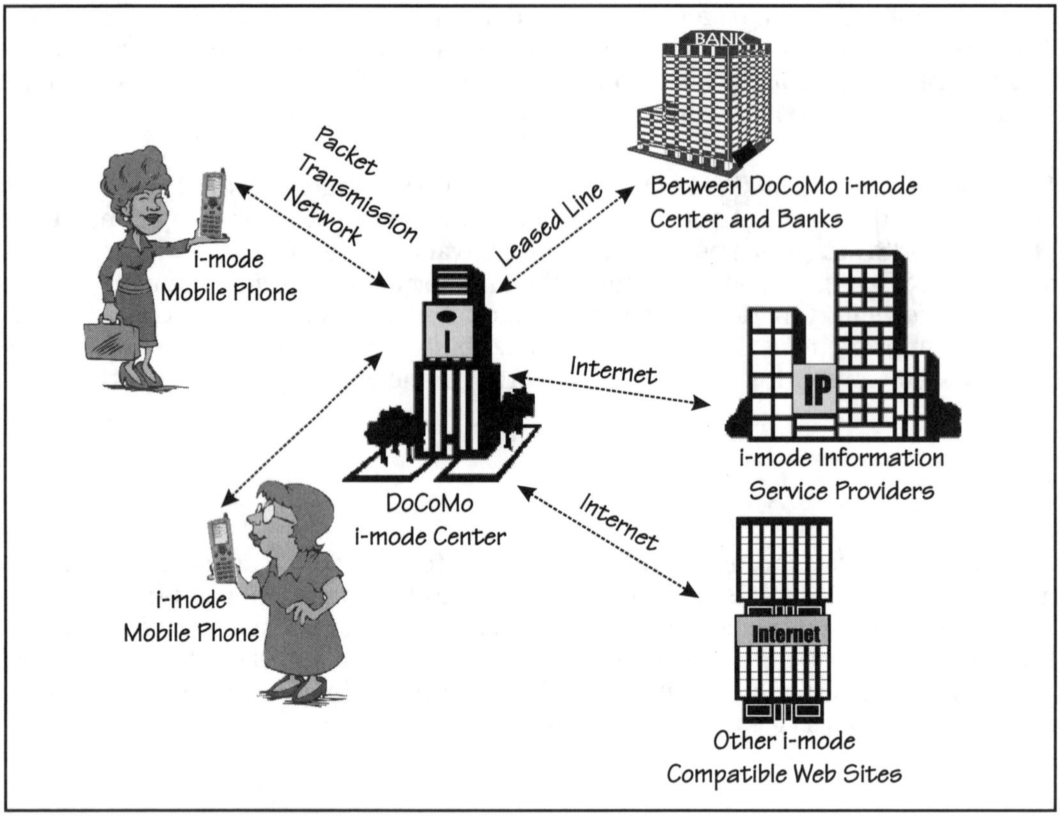

Figure 4-4. I-mode from NTT DoCoMo.

The Markup Language

The markup language used by i-mode is cHTML, or compact HTML, a subset of HTML. It is designed for devices with slower data transmission speeds. An i-mode Web site contains pages written in cHTML. Due to the similarity to HTML, it is easy for Web designers to create cHTML pages. One main difference between the two languages is that some of the more resource intensive areas of the markup language (such as tables and frames) have been removed from cHTML. This was done to reduce the download time to mobile devices. cHTML content needs to pass through a cHTML gateway before users can access it on their mobile phones using the i-mode browser. When an i-mode compatible wireless device makes a wireless request, the gateway translates the request and passes it to the server. The reply from the server is then translated before returning the reply to the handheld device.

The Future of I-Mode

Just as PC-based browsers are used to display an HTML page, the micro-browser in an i-mode phone is software designed to run on a handheld device and display a cHTML page. As of now, the rate at which data is being transmitted is 9.6 Kbps. NTT DoCoMo is testing W-CDMA (DoCoMo's 3G technology) which can achieve theoretical speeds of 384 Kbps. This will allow high-speed transmission of video and large-volume data.

Java on I-Mode

NTT DoCoMo released new handsets with Java capabilities in early 2001. Java applications developed for these devices are branded as "*iAPPLI*". We will discuss the Java capabilities of mobile handsets in Chapter 7.

WIRELESS APPLICATION PROTOCOL (WAP)

Wireless Application Protocol (WAP) is a lightweight protocol designed for providing Internet communications on resource-constrained devices such as mobile phones, pagers, PDAs, and other wireless terminals. These devices usually have a small screen area, minimal color or gray scale support for display, limited processing power, minimal network bandwidth, and low battery life. Unlike the proprietary i-mode, WAP was created by a consortium of leading companies including Openwave (formerly *Phone.com*), Ericsson, Motorola, and Nokia that founded the WAP Forum (*www.wapforum.org*).

The WAP Forum focuses on mobile technologies and devices. The WAP Forum publishes specifications, which are ratified by the consortium, as recommendations to standards bodies such as W3C, ETSI, and IETF. As of this writing, WAP specification 1.2 is in recommendation status (July 2000) and WAP specification 2.0 is in public review status. The next sections describe WAP based on the version 1.1 specification. We will look into some details of WAP1.2 and WAP 2.0 in later sections.

WAP Architecture

WAP is a combination of several layers of communication protocols working together to provide Internet access from mobile terminals. These protocols are implemented as two pieces of software: 1) The micro-browser and its associated protocol stacks that run on the mobile terminal, and 2) The WAP gateway server that provides a gateway between the wireless network and the Internet. The micro-browser handles user interactions and presentation of content. Under the covers, WAP protocols handle other tasks such as content encoding, encryption, and security. The gateway is installed on the wireless service provider's network. This gateway handles all the requests from the micro-browser and forwards requests to the Internet. Figure 4-5 shows the architecture of WAP. This architecture is very analogous to the Internet itself. The WAP wireless protocols are based on Internet standards such as XML, UDP, IP, HTTP, and TLS, but have been optimized for the unique constraints of the wireless environment.

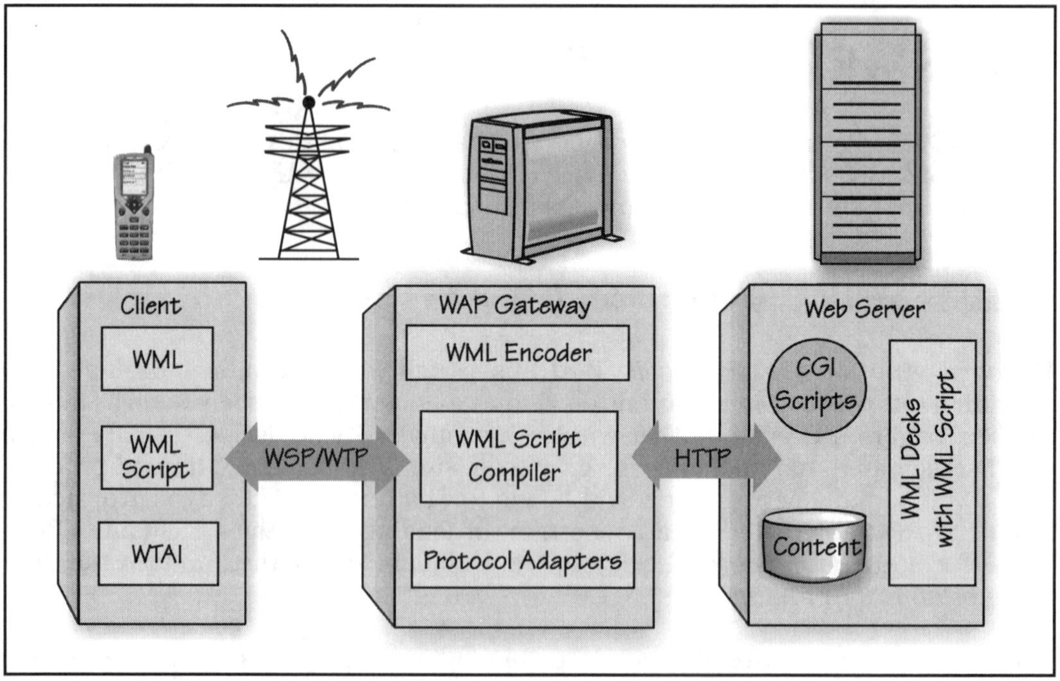

Figure 4-5. The WAP Architecture.

A WAP Scenario

The following is a simple scenario that depicts how WAP works under the covers when a mobile terminal tries to access the Internet.

- The micro-browser is invoked on the mobile terminal and a URL is requested by the user. The request is encoded in a compact binary format and transmitted to the wireless network.

- The requests are handed over to a WAP Gateway, installed by the service providers. On the WAP Gateway, the requests get converted to the World Wide

Web standards (HTML/HTTP) and are routed to the content server or a back-end service.

- The service will perform any required operation and return the response to the gateway, using HTTP in either standard HTML or in "wireless-ready" WML. Wireless Mark-up Language (WML) is a mark-up language based on XML. WML follows a stricter syntax than HTML and allows for the use of variables. If the information received by the gateway is not in WML, the gateway will convert it to WML. The gateway will parse and tokenize the response into a binary format, i.e., the WML elements will be replaced by a byte code format known as the WBXML format.

- This will then be sent (using WAP) to the device that requested the service. The device will decode the response and render it on the screen.

WAP Protocol Stack

To help explain the WAP protocol stack, we will compare WAP with the OSI ISO data communication model and the Internet HTTP protocol stack. We will look at the functionality of each layer in detail. Figure 4-6 shows the protocol stacks for OSI, HTTP, and WAP.

- ***Physical and Data link layers:*** Depends in part on the type of wireless network, but with WAP it will be Point-to-Point Protocol (PPP) over one or more over-the-air bearer protocols.

- ***Network layer:*** Internet Protocol (IP) is the network layer protocol of choice, although not all wireless networks are capable of transmitting IP. Thus, SMS or some other non-packet network protocol may be used.

- ***Transport layer:*** The transport layer protocol of choice is User Datagram Protocol (UDP). But UDP may not be possible over non-IP networks, so WAP defines an additional transport layer protocol, *Wireless Datagram Protocol (WDP)*, which can be used where UDP is not available.

- ***Session layer:*** In the wireless world some of the functionality of the session layer is incorporated into *Wireless Transport Protocol (WTP)*, while other aspects are included with *Wireless Session Protocol (WSP)*.

- ***Presentation layer:*** This functionality is included in WSP.

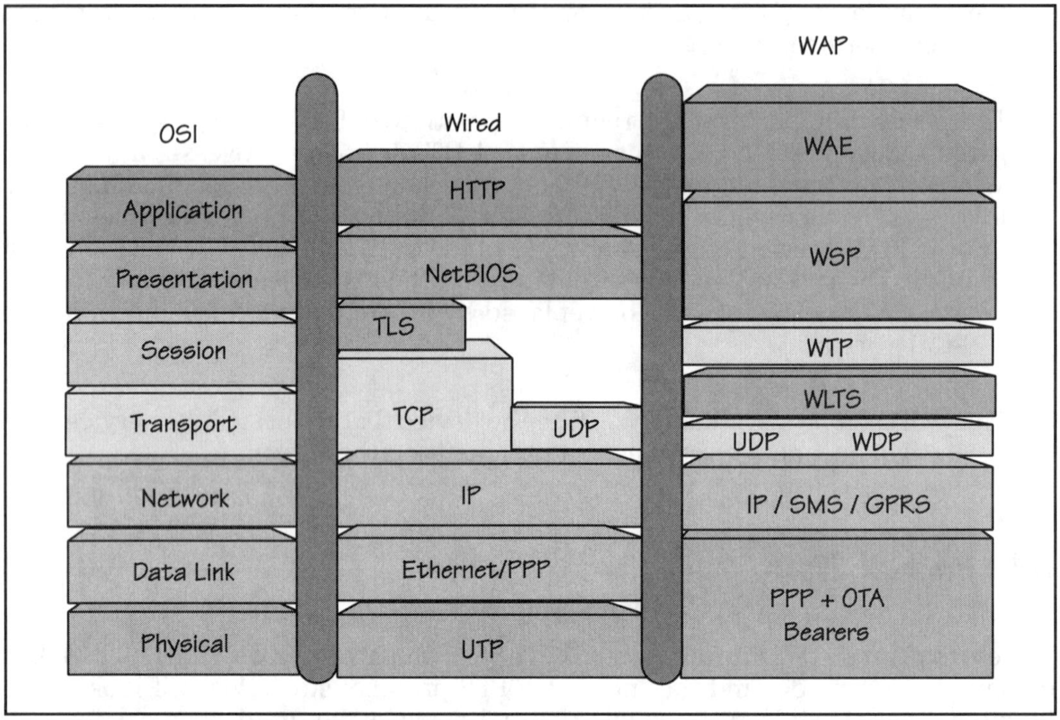

Figure 4-6. OSI, Wired, and WAP Protocols.

- ***Application layer:*** Some aspects of application layer functionality are taken care of by WSP, whereas others are implemented in the *Wireless Application Environment (WAE)*.

Figure 4-7 compares the WAP protocol stack with the Internet protocol stack.

Wireless Application Environment (WAE)

WAE is a general-purpose, interoperable application environment that allows service and application development. It is based on the World Wide Web and mobile telephony technologies. WAE consists of the following components:

- ***WML Micro-browser*** for interpreting Wireless Markup Language (WML).

- ***WMLScript Virtual Machine*** for providing support for WMLScript. The virtual machine and interpreter are optimized to better utilize the bandwidth constraints and memory constraints of mobile devices.

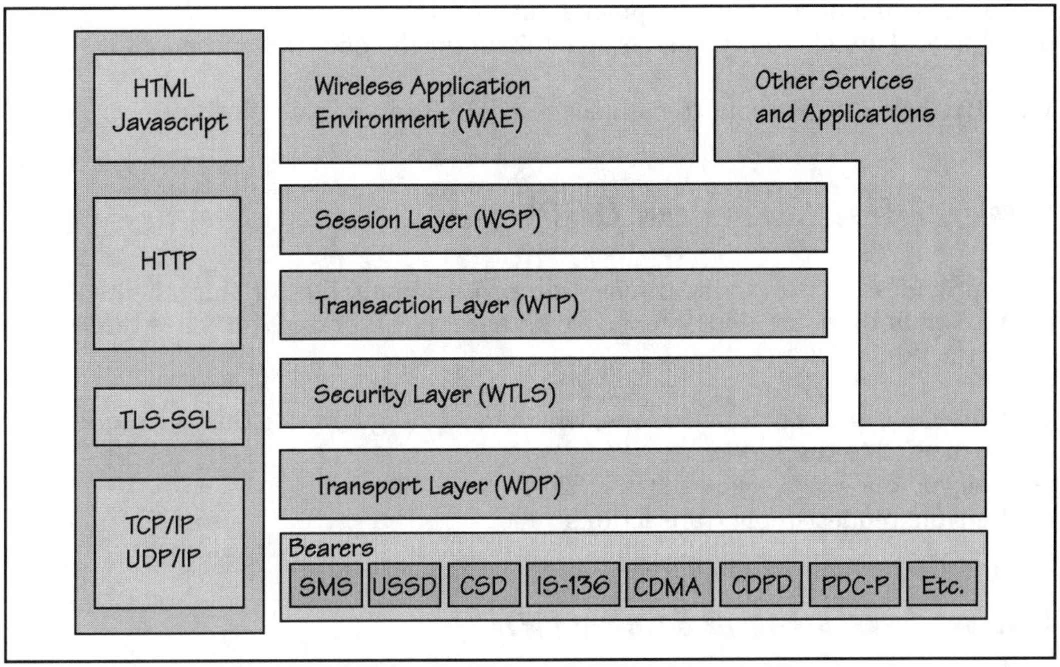

Figure 4-7. WAP Protocols vs. Internet Protocols.

- ***WMLScript Standard Library*** provides WMLScript libraries based on ECMAScript.
- ***Wireless Telephony Application Interface (WTAI)*** for accessing telephone functionality.
- ***WAP Content Types*** to define content formats for images, phone records, and calendars.

Wireless Session Protocol (WSP)

WSP provides consistent interfaces to connection-oriented and connectionless session services. Connection-oriented services operate above the transaction layer protocol (WTP) and connectionless services operate above the secure or non-secure datagram service (WDP). Protocols in both of these categories are optimized for low-bandwidth networks. Currently WSP provides services for browsing applications (referred to as WSP/B). It provides:

- Long-lived session state
- Session mitigation for suspending and resuming sessions

- Reliable and unreliable data push mechanism
- HTTP/1.1 functionality in a compact form for the over-the-air encoding

WSP/B allows a WAP client to connect to a HTTP server via a WAP proxy.

Wireless Transaction Protocol (WTP)

WTP provides a light-weight transaction protocol suitable for thin clients. It can operate on both secure and non-secure datagram networks. It provides three types of transaction services:

- Unreliable one-way transactions, which make no guarantee that the request will be handed to the server
- Reliable one-way requests transactions
- Reliable request-reply transactions

Wireless Transport Layer Security (WTLS)

WTLS is the transport-layer security protocol for WAP and is based on Secure Sockets Layer, which is now known as the Transport Layer Security (TLS) protocol. This is also optimized for use over narrow-band communication channels. WTLS ensures the following:

- ***Data Integrity:*** Any data exchanged will not be corrupted.

- ***Privacy:*** Any data exchanged will be unintelligible to any intruder.

- ***Authentication:*** Endorses the participants in an exchange.

- ***Denial-of-service protection:*** WTLS provides features that make it hard to perform a denial-of-service attack. It also detects and rejects unverified and replayed data.

It is also possible to selectively enable and disable WTLS programmatically.

Wireless Data Protocol (WDP)

Where UDP is not available, WDP provides the transport layer capabilities for WAP and provides a consistent interface to various bearers. WDP also provides data error correction.

Bearers

WAP supports a variety of bearers. IP-based wireless bearers like GPRS, CDPD (Cellular Digital Packet Data), CSD (Circuit Switch Data), and iDEN will use UDP, while non-IP-based wireless bearers like SMS and USSD will use WDP and thus provide seamless bridging via the transport layer to the upper layers of the WAP stack.

Other Services and Applications

WAP allows other services and applications to directly interact with session, transaction, security, and transport layers. This allows for development of a variety of applications.

The Web and WAP Content Presentation Models

The Internet World-Wide Web (WWW) architecture provides a very flexible and powerful content presentation model based on "request/response" exchanges. The WAP content presentation model is similar to the Web model. It has been developed to provide the same benefits to the application developer, including a familiar programming model, a proven architecture, and the ability to leverage existing tools like Web servers and XML tools. The protocol has been optimized and extended wherever needed to match the constraints and characteristics of the wireless environment. Let us now proceed to a comparison of the Web model and WAP model.

In the Web model, a browser is used to display the content. Similarly in case of WAP, a *micro-browser* is used. A micro-browser is a small piece of software that makes minimal demands on hardware, memory and CPU. It can display information written in the Wireless Markup Language. These micro-browsers are typically available out-of-the-box, on any WAP-enabled phone. For content development, micro-browser simulators are used to test applications. There are a number of WML micro-browser simulators and WAP development kits from vendors such as Nokia and Openwave (previously *Phone.com*).

World-Wide Web Content Presentation Model

The following are some of the standards used in the Web model:

- ***Standard naming model:*** All servers and content on the WWW are named with an Internet-standard *Uniform Resource Locator* (URL).
- ***Content typing:*** All content on the Web is given a specific type by assigning Multipurpose Internet Mail Extensions (MIME), thereby allowing Web browsers to correctly process the content based on its type.
- ***Standard content formats:*** All Web browsers support a set of standard content formats. These include the HyperText Markup Language (HTML), the JavaScript scripting language, and a large number of other formats.
- ***Standard Protocols:*** Standard networking protocols allow any Web browser to communicate with any Web server. The most commonly used protocol on the Web is the HyperText Transport Protocol (HTTP).

This infrastructure allows application developers to easily create applications and content services for a large community of clients.

The World Wide Web protocols define three classes of servers:

- ***Origin server:*** The server on which a given resource (content) resides or is to be created.
- ***Proxy server:*** An intermediary program that acts as both a server and a client for the purpose of making requests on behalf of other clients. The proxy typically resides between clients and servers that have no means of direct communication, for example, across a firewall. Requests are either serviced by the proxy program or passed on, with possible translation, to other servers. A proxy must implement both the client and server requirements of the WWW specifications.
- ***Gateway server:*** A server, which acts as an intermediary for some other server. Unlike a proxy, a gateway receives requests as if it were the originating server for the requested resource. The requesting client may not be aware that it is communicating with a gateway.

WAP Content Presentation Model

WAP content and applications are specified in a set of well-known content formats based on the familiar World Wide Web content formats. Content is transported using a set of standard communication protocols based on the World Wide Web communication protocols.

WAP defines a set of standard components that enable communication between mobile devices and network servers, including:

- ***Standard naming model:*** World Wide Web-standard URLs are used to identify WAP content on origin servers. World Wide Web-standard URIs are used to identify local resources (for example, call control functions).
- ***Content typing:*** All WAP content is given a specific type consistent with World Wide Web typing. This allows WAP user agents to correctly process the content based on its type.
- ***Standard content formats:*** WAP content formats are based on World Wide Web technology and include display markup, calendar information, electronic business card objects, images, and scripting language.
- ***Standard communication protocols:*** WAP communication protocols enable the communication of browser requests from the mobile device to the network Web server. The WAP content types and protocols have been optimized for mass market, hand-held wireless devices. WAP utilizes proxy technology to connect between the wireless domain and the World Wide Web. The WAP proxy typically is comprised of the following functionality:
    - ***WAP Gateway:*** The protocol gateway translates requests from the WAP protocol stack (WSP, WTP, WTLS, and WDP) to the World Wide Web protocol stack (HTTP and TCP/IP). This is an optional component and may facilitate full integration to a service provider's facilities.
    - ***Content Encoders and Decoders:*** The content encoders translate WAP content into compact encoded formats to reduce the size of data over the network. This infrastructure ensures that mobile device users can browse a wide variety of WAP content and applications, and that the application developer could build content services and applications that run on a large base of mobile devices. The WAP proxy allows content and applications to be hosted on standard WWW servers and to be developed using proven WWW technologies such as CGI scripting. While the nominal use of WAP will include a Web server, WAP proxy, and WAP client, the WAP architecture can quite easily support other configurations. It is possible to create an origin server that includes the WAP proxy functionality. Such a server might be used to facilitate end-to-end security solutions, or applications that require better access control or a guarantee of responsiveness, for example, the Wireless Telephony Application (WTA).

WAP Security and WTLS

Wireless Transport Layer Security (WTLS), as the name implies, is the security protocol that operates at, or more correctly, just above the transport layer in the OSI protocol stack. It is based on Transport Layer Security (TLS), which is the de facto security implementation on the Internet. It works by establishing a session between a client and a server (which, in the case of WTLS, is the WAP gateway), during which it negotiates security parameters to be used to protect the session. These include the encryption protocols to be used, signature algorithms, public keys, pre-master secrets, or the exchange of certificates, depending on the capabilities of both the client and the server and the required level of security. The process of establishing a session is called the *handshake*. Once a session has been established all communications between the mobile device and the WAP gateway are encrypted, and therefore should be unintelligible if they are intercepted.

WTLS includes support for both a full handshake, with negotiation of all security parameters, and for a "lightweight" handshake in which the security parameters of another session are reused. Support is also provided for a session to be suspended and resumed, which is useful in a wireless environment where reception quality is not always that good and where connections can easily be lost. The sessions can continue to exist despite a terminated connection and can be resumed on reconnection. Using this facility, it is possible to have sessions that last for days at a time.

The advantages of sessions that can continue to exist for days at a time must be weighed against the security implications of this feature. The longer the session remains valid, the longer the secret keys remain valid, and, presumably, the greater the number of messages exchanged that are encrypted using this key. This all provides material to someone wanting to crack the security protecting the session and compromise the messages. To guard against this, WTLS allows keys to be renegotiated periodically during a session. Renegotiating keys is not as computationally expensive as establishing the keys in the first place, so this is still more efficient that tearing down and re-establishing the session.

Another advantage of WTLS versus TLS is that it operates over UDP. TLS requires a reliable transport protocol, in particular TCP, so it cannot be used over UDP. WTLS addresses this shortcoming, and also functions over WDP in the absence of UDP.

Certificates, for all of their usefulness, were not designed with mobile devices in mind. WAP defines a new format of certificate that is optimized for storage on mobile devices and for transmission over constrained networks. These certificates still provide all of the functionality and security of their more heavyweight counterparts, but rely on the server to perform more of the processing under some circumstances.

WTLS thus provides a comprehensive, optimized solution for both client- and server-based authentication-using certificates, secure exchange of symmetric keys, anonymous and authenticated encryption of data, and support for digital signing of data.

WTLS is rich in features, making it possible, for example to:

- Transparently encrypt and decrypt information sent between a WAP client and a WAP gateway so that a third party cannot understand communication between the two.
- Authenticate the communicating parties so both parties can be sure that the other party is who it claims to be.
- Ensure the integrity of the communication so that the recipient of secure information can verify that the content of the data has not been changed since it was sent.

Specific Features and Benefits of WTLS

There are three classes of WTLS defined in the WAP specification used when negotiating the security parameters during session set up:

- ***Class 1 (No security):*** Anonymous key exchange with no authentication. This means that the client and the server involved in a transaction remain anonymous, that is, they are not verified as to who they really are by using certificates or by other means. They just exchange keys and proceed with the transaction.
- ***Class 2 (Shared secret-based authentication):*** Certificate-based server authentication. The server key is anonymous or authenticated and the client key is anonymous. Certificate-based server authentication means that the server will be validated as the one it claims to be, by using its certificate for verification.
- ***Class 3 (Public key-based authentication):*** Certificate-based client and server authentication. Both client and server keys are anonymous or authenticated. In this case, both the client and server will be validated based on their individual certificates.

No Security

The aptly named no-security protocol is used to establish a connection without using any cryptography algorithms. In other words, no data will be encrypted or decrypted.

Shared Secret-based Authentication.

In this version, both parties share a secret key (pre-master secret) that is used to calculate the master secret. The master secret is a symmetric key that is used for encrypting and decrypting messages.

The main purpose of the shared secret protocol is the distribution of the keys. The shared secret cannot be transferred over the air since it reveals the master secret. In practice this means that the shared secret should be entered in the phone (or SIM) by the operator or the creator of the pre-master secret. The user should also be able to manually enter the pre-master secret and its ID. In this scenario, the service provider who creates and maintains the key transmits the information to the user in a secure manner, for example, by fax or in a letter.

Shared secret-based authentication needs at least one symmetric key algorithm (such as RC5) and one hash algorithm (such as SHA-1).

Public Key-based Authentication

Public key-based authentication uses both asymmetric and symmetric cryptography. It supports server and (optionally) client authentication based on certificates. In addition, the WTLS specification allows the use of asymmetric cryptography without authentication. This means that the server sends its public key to the client, which uses it to encrypt the pre-master secret. Table 4-1 shows the different types of certificates that can be used in public key-based authentication. Note that the private key is never stored in the certificate thus allowing certificates to be freely distributed.

Table 4-1. WTLS Certificate Usage in Public Key-based Authentication.

Certificate	Stored In	Usage	How It Is Created
CA (Certificate Authority)	Client, Server, Certificate Authority (CA)	Client uses CA certificate to verify the server's certificate.	CA creates its own certificates.
Server	Server, CA	Server sends its certificate to the client.	Server Administrator creates a certificate signing request (CSR_ for signing. CA returns the certificate. Server can also sign its own certificate.
Client (Optional)	Client and optionally in server	Client sends its certificate to the server.	Will be defined in future WTLS specs.

The Future of WAP

According to the WAP Forum, "WAP 2.0 brings the wireless world closer to the Internet with a suite of specifications that utilizes technologies that will enhance the wireless user experience." The WAP Forum released the public draft of WAP 2.0 in August 2001 but implementations are still scarce. We will briefly look into the new features and enhancements.

While earlier specifications of WAP focused on utilizing the limited bandwidth and limited capabilities of the devices, this current specification focuses on providing rich features for the new 3G devices with more bandwidth and also bringing a closer convergence of Internet Technologies and Wireless Technologies.

Protocol Layers for Networks Supporting IP

WAP 2.0 provides support for protocols such as TCP and HTTP. By adding these Internet protocols and standards and providing interoperable optimizations suitable to the wireless telecommunications environment, the WAP specifications provide an environment that permits wireless devices to utilize existing Internet technologies. The support for the Internet protocols has been motivated by the emergence of high-speed wireless networks (such as 2.5G and 3G) that provide IP support directly to the wireless devices.

- ***Wireless Profiled HTTP (WP-HTTP):*** WP-HTTP specification is a profile of HTTP for the wireless environment and is fully interoperable with HTTP/1.1. The basic model of interaction between the WAP device and WAP Proxy/WAP Server is the HTTP request/response transaction. WP-HTTP supports message body compression of responses and the establishment of secure tunnels.

- ***Transport Layer Security (TLS):*** A wireless profile of the TLS protocol that will permit interoperability for secure transactions. This profile for TLS includes cipher suites, certificate formats, signing algorithms, and the use of session resume. The profile also defines the method for TLS tunneling to support end-to-end security at the transport level.

- ***Wireless Profiled TCP (WP-TCP):*** WP-TCP provides connection-oriented services. It is optimized for wireless environments and is fully interoperable with standard TCP implementations in the Internet. Research in optimizing TCP has resulted in a number of mechanisms to improve performance over wireless networks. This includes work by the IETF PILC group that has recommended the use of some of these mechanisms for TCP implementations in long thin networks. These have been documented in IETF's Standards Track RFCs and have been accepted by the Internet community as useful and technically stable.

Figure 4-8 shows an example usage of these protocols from a WAP Device to a Web Server. In this example, a TLS tunnel is enabled through an intervening WAP Proxy permitting secure, end-to-end HTTP transactions. The TCP* is a reference to end points that operate using the features of the wireless profiled TCP.

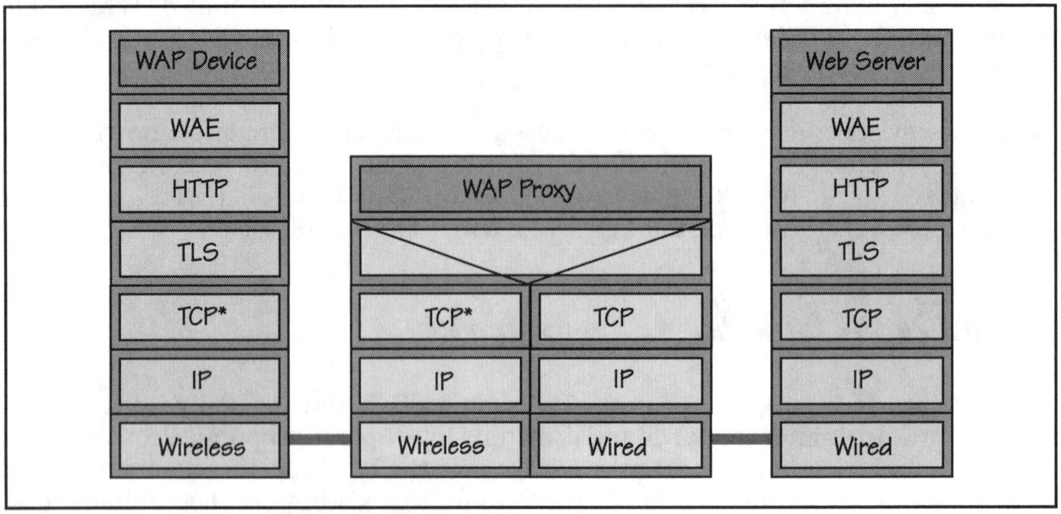

Figure 4-8. WAP HTTP Proxy with Profiled TCP and HTTP.

Since WAP 2.0 provides support for both 1.x and 2.x stacks, it is expected that these stacks would operate independently. That is, there would not be mixing and matching of protocols in accomplishing an end-to-end transaction. Figure 4-9 shows the separation of these stacks and shows the expectation that a common application environment could operate over either stack. For devices that will be supporting both old and new network types, this switching of stacks may occur as the device moves in and out of coverage of the different network areas.

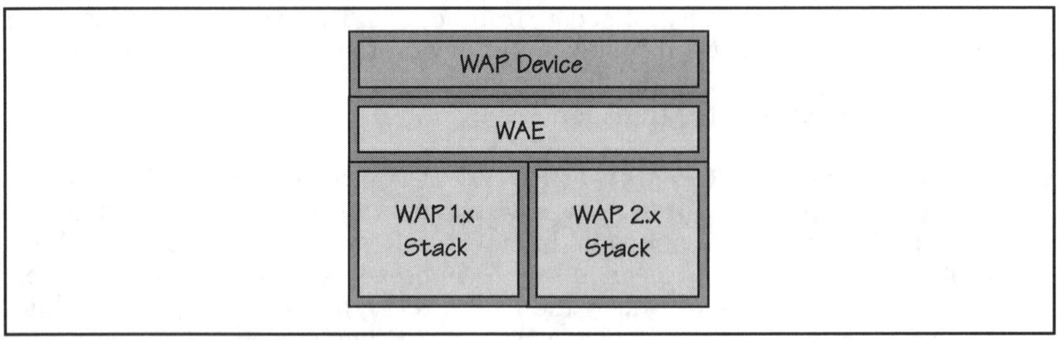

Figure 4-9. Optional Dual WAP Stack Support.

Enhancements to the Wireless Application Environment (WAE)

WAE in the earlier versions provided support for WML and WMLScript. But with the increased capabilities of the network bandwidth and devices, it is natural to support rich content types. In order to support mark-up for various content types, the WAP Forum chose to use XHTML, which is becoming the de-facto content presentation language for the Web. The WAE in WAP 2.0 continues to support an application-centric approach by embracing XHTML-Basic and several other associated technologies and includes the following elements:

- The basic markup language for the WAE in WAP 2.0, namely WML2, is based on the Basic profile of XHTML as defined by the W3C. This core was designed to be extensible and WAE takes advantage of this capability by defining additional markup features for enhanced functionality and backwards-compatibility to the original WML 1.x language. By using the XHTML modularization approach, the WML2 language is very extensible, permitting additional language elements to be added as needed. Additionally, documents written in the core XHTML Basic language will be completely operable on the WML2 browser. We will discuss more about XHTML and its profiles in Chapter 6.

- The WAE in WAP 2.0 also enhances the presentation of content by supporting stylesheets. Based on the Mobile profile of CSS (Cascading Style Sheets) from the W3C, WAP support covers both inline and external stylesheets, which are commonly supported by most Internet browsers.

- Backwards compatibility support for WML1 applications is provided in the WAE for WAP 2.0 through either native support for both languages or by a defined transformation operation. The WAP 2.0 release includes a transformation, using eXtensible Stylesheet Language Transformation (XSLT), which will permit documents defined in WML 1.x language to be converted to WML2 code, which would then be processed by the WML2 browser.

These changes will let devices accept both content written for other Internet clients in XHTML-Basic and applications and content written specifically to take advantage of the WML extension features. Enabling backwards compatibility protects investments in WAP 1.x content.

Introducing WAP Push

The WAP Programming Model, closely aligned with the Web Programming Model, uses the Pull Model, (where the client requests content from the server). However, WAP also extends the Web architecture by adding telephony support with WTA and enabling a Push Model, where a server can proactively send content to the client (see Figure 4-10). WAP Push service allows content to be sent or "pushed" to devices by server-based applications via a Push Proxy Server. This functionality has been enhanced for the WAP 2.0 release. Push functionality is especially relevant to real-time applications that send notifications to their users, such as messaging, stock price, and traffic update alerts. Without push functionality, these types of applications would require the devices to poll application servers for new information or status. In wireless environments, such polling activities would constitute inefficient and wasteful use of the resources of wireless networks. WAP's Push functionality provides control over the lifetime of pushed messages, store and forward capabilities at the Push Proxy Server, and control over bearer choice for delivery.

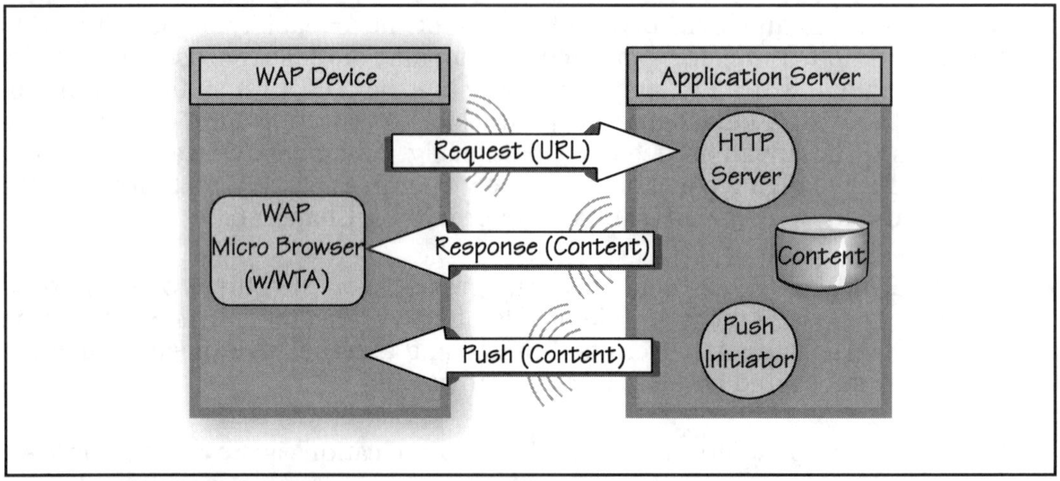

Figure 4-10.The WAP Push Model.

Proxy Requirement

In previous versions of WAP, a WAP proxy (often referred to as a WAP gateway) was required to handle the protocol inter-working between the client and the origin server. The WAP proxy communicated with the client using the WAP protocols that are based largely on Internet communication protocols, and it communicated with the origin server using the standard Internet protocols. WAP 2.0 does not require a WAP proxy, since the communication between the client and the origin server can be conducted using HTTP/1.1. However, deploying a WAP proxy can optimize the communications process and may offer mobile service enhancements, such as location-, privacy-, and presence-based services. In addition, a WAP proxy is necessary to offer Push functionality.

New Features and Services

In addition to the enhancements made for WAP Push, these are the new features and services provided in WAP 2.0.

- ***User Agent Profile (UAProf):*** This service provides a mechanism for describing the capabilities of clients and the preferences of users to an application server. It is based on the Composite Capabilities / Preference Profiles (CC/PP) work of the W3C. UAProf supports the client-server transaction model by sending client and user information to servers with the request. This information permits the servers to adapt their content accordingly in preparation for the response. This service model would also permit intervening proxies to provide value-added services by providing these adaptation services directly. In later sections we will discuss the importance of this feature in personalization of content. Recognizing the importance of user privacy, personal information submitted in these requests may be controlled by the user.

- ***Wireless Telephony Application (WTA):*** This service provides tools that provide for a range of advanced telephony applications to be operated from within the application environment that traditionally supports data functionality. These call handling services, such as making calls, answering them, placing them on hold, and redirecting them, can be seamlessly combined with other data services. This truly permits WAP WTA-enabled cell phones to be viewed as fully-integrated Internet and voice service platforms.

- ***External Functionality Interface (EFI):*** This service specifies the interface between WAE and components or entities with embedded applications that execute outside of the defined WAE capabilities. This is analogous to providing a plug-in module, which extends or enhances the capabilities of browsers or other applications. The EFI framework provides for future growth and extendibility of supported WAP devices. This framework can be used to define the specific interfaces needed for accessing external devices (such as smart cards, GPS devices, health care devices, and digital cameras).

- ***Persistent Storage Interface:*** This capability specifies a standard set of storage services that are coupled with a well-defined interface for organizing, accessing, storing, and retrieving data on the wireless device or other connected memory devices such as flash memory.

- ***Data Synchronization:*** In an approach that helps to ensure a common solution framework, the WAP Forum sought out a solution for data synchronization. As a result, WAP 2.0 acknowledges the work of the SyncML initiative by adopting the SyncML language as its choice for the data synchronization solution. The SyncML messages are supported over both the WSP and HTTP/1.1 protocols. We will discuss more about SyncML in Chapter 6 about content formats.

- ***Multimedia Messaging Service (MMS):*** This service provides the framework to implement a feature-rich messaging solution. MMS provides features and functionality that permit delivery of varied types of content. Depending on the service model, MMS permits a quick delivery paradigm (like SMS) or a store-and-forward approach (like email) or could permit both modes to operate. This flexibility allows users and operators to tailor the resulting user experience. By using other WAP services, like Push and UAProf, MMS provides an efficient messaging solution that is capable of providing message notifications that initiate adaptation services to structure the delivered content into a form that is efficiently used by the receiving device.

- ***Provisioning:*** This service provides a standard approach to providing WAP clients with information needed to operate on the wireless networks. This permits the network operator to manage the devices on its network using a common set of tools.

- ***Pictogram:*** This service permits the use of tiny images, such as 'smilies,' in a consistent fashion. Such images can be used to quickly convey concepts that permit efficient communication in a small amount of space, creating communication that transcends traditional language boundaries.

With all these enhancements and rich features, there are more opportunities for developers to design and develop applications that take advantage of these features.

WTLS Issues

WTLS, being based on an established and stable standard, is unlikely to change significantly in the near future. Most of the changes in the next few releases will be oriented towards clarifying some issues in the specification and general "housekeeping." The 1.2 specifications already did this, and added some advice about guarding against certain types of attacks. All of this information is really only of relevance to people who are developing their own WTLS implementation, who should also be aware that much of the information becomes out of date very quickly.

WAP and End-to-End Security

WAP defines the transfer of information of data between two parties, a WAP client and an origin server, passing between a third party known as a WAP gateway. However, the WAP protocol only covers the portion of the journey between the WAP client and the WAP gateway. It relies on existing Internet protocols for the rest of the journey. This works quite well in most respects; however, there is one area where there are some problems and that is in the area of security.

The WAP Forum has made it clear that they are aware of the issues around the security gap at the WAP gateway.

Details

What Is a WAP Gap?

Typically, in a WAP model, to establish secure connections, WTLS is used. WTLS establishes a secure channel between the device and the gateway and a TLS connection is established between the gateway and a Web-based service—such as a Servlet. The data communicated from the device to the gateway via WTLS has to be decoded and re-encrypted for HTTPS at the gateway. So, for that small time interval the data is in the clear at the gateway. This is known as the WAP Gap. Figure 4-11 illustrates the WAP Gap problem.

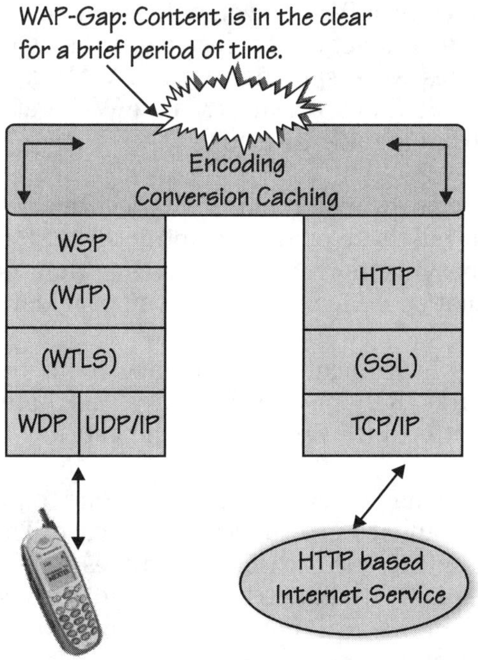

Figure 4-11. The WAP Gap on a WAP Gateway

WAP gateways try to limit the WAP gap in the following ways:

- Data is unencrypted only in the WAP gateway's memory and is not written to any permanent storage medium (such as a hard drive).

- Data is only unencrypted for the shortest possible time (typically microseconds).
- Physical access to the WAP gateway host machine is limited to authorized personnel.

Although these policies do provide a partial solution to the problem there are still serious problems they cannot overcome, such as:

- Most WAP gateways have access logs, which record user access. Many Internet services place sensitive information (such as account numbers, and passwords) in the query string of a URL. Thus, sensitive information is saved in the access logs.
- If a service is provided on a Web server, it does not necessarily know which WAP gateway was used to connect to the service. If this is the case, it cannot know what security policies were specified by the WAP gateway provider. Do they secure access to the WAP gateway host? Is the WAP gateway provider's network properly secured from outside threats?
- If the wireless service provider limits WAP gateway access to certain WAP gateway providers, will these providers notify the WSP in the event of a security breach? Will they provide information as to what their network security policies are (information that could leave the gateway provider vulnerable)?
- Many organizations have legal requirements about having unencrypted data outside their networks. Relying on third party WAP gateways fails to satisfy this requirement, no matter how well-secured the gateway is.

The WAP Forum has also made it clear that they intend to plug the gap by providing an end-to-end security standard in a subsequent release. There have been hints that they would attempt to address this through changes to WTLS. However, the issue does not arise because of any weakness in WTLS and is caused solely by the position that the gateway fulfils in the WAP communications chain. In order to address the issue, either the gateway has to be eliminated or some other solution has to be implemented, probably at a higher level in the protocol stack. The WAP Forum has also indicated that the WMLScript Crypto library may be extended in the future to include cryptographic functions. At this point in time there is only one proposed function that supports signing data but the implementation is not yet available.

WMLScript Crypto Library

There are certain aspects of security that are difficult to solve at the data transport level. One such is the criterion of non-repudiation. Non-repudiation is where no party to a transaction can deny that transaction took place. However, proving someone visited a page is not as good a proof of transaction as actually proving the transaction took place. As it happens, however, this problem can be solved at an application level by digitally signing the transaction. A digital signature is the electronic equivalent of signing a receipt.

The WAP Crypto Specification addresses digital signing at an application level. This is the WMLScript Crypto Library. This specification describes a set of security functions, which can be called using WML script. The first of these functions is signText. This function can be used by the developer to request that a user use his or her private key to digitally sign a text string. This can be useful any time it is necessary to keep persistent proof of a transaction. This function is called by the client and executed by the server to verify that the key sent by the client matches the private key.

Wireless Identity Module (WIM)

The WAP 1.2 specification has introduced the Wireless Identity Module (WIM). The WIM provides a means to offload the storage of keys and of cryptographic functions onto a smart card, or a SIM. The specification covers only the low-level capabilities of a WIM in the current specification, and doesn't present an API for making use of a WIM when present, although an API and framework may be provided in a future release. The introduction of the WIM could help to address the issues around authenticating the device as opposed to the user. This would be important because most mobile devices just walk away!

COMPARISON OF WAP VS. I-MODE

There is a natural tendency to compare WAP and i-mode. However, WAP is a standard protocol that is implemented by various parties, whereas i-mode is a proprietary service provided by NTT DoCoMo. These two cannot be compared as competing products. Table 4-2 compares the general characteristics of i-mode and WAP.

Table 4-2. WAP vs. I-mode.

Characteristics	WAP	I-Mode
Implementation	WAP is a comprehensive and scalable protocol designed for use with mobile phones using Short Message Services (SMS), General Packet Radio Services (GPRS), CDMA, and GSM as bearer protocols.	I-mode is a protocol designed by NTT DoCoMo of Japan. Code Division Multiple Access (CDMA) is used for transmission. I-mode phones are "always on" or always connected to the Internet.

Table 4-2. WAP vs. I-mode. (Continued)

Characteristics	WAP	I-Mode
Gateways	WAP needs a WAP Gateway to convert Internet data from HTML to WML so that a mobile device can access these services.	With i-mode, there is no need for a gateway because i-mode uses an overlay packet network so that the mobile device can access the Internet data directly.
Content Language	WML	cHTML
Layers vs Hyperlinks	WAP supports navigation between cards.	I-mode supports navigation through hyperlinks.
Images	WAP devices display monochrome images.	I-mode devices display multi-color images.
Availability	Available in Europe and the U.S.	Available only in Japan.

VOICEXML

VoiceXML is an XML-based, dialog-oriented markup language designed and promoted by the VoiceXML Forum. It is analogous to HTML in the sense that VoiceXML renders speech as the content format to the end user. AT&T, IBM, Lucent, and Motorola founded the VoiceXML Forum with the objective to standardize and simplify voice access to Internet services. VoiceXML technology leverages the most natural form of communication—speech—and the most pervasive and familiar communications network—the global telephone network. No special device is needed to interact with an Internet service via voice; all that is required is just a regular or a mobile phone. Enormous efforts and innovative technology form the basis for providing a seamless voice interface to Internet services.

VoiceXML does not impose special resource requirements for using new client devices; the client device is a phone—be it wired or wireless. Also VoiceXML is expected to be a significant tool because it can be used in situations where your eyes and hands are already involved in doing something else, for example, while driving.

Before plunging into the details of VoiceXML, let us briefly go over some of the advantages of voice-enabled applications. Voice-enabled applications that use VoiceXML, Java, XML, and other related technologies tend to be more flexible when compared with other *Interactive Voice Response (IVR)* systems. Most IVR systems tend to be proprietary, expensive, and inflexible. In contrast, VoiceXML-enabled IVR systems are based on open architectures and Internet technologies such as Java and XML.

So, how is speech rendered to the end user, you might wonder? We are familiar with Web browsers on our PCs and workstations and with a micro-browser on our handheld devices. A Web browser interprets HTML pages and displays graphical and textual content while a micro-browser displays WML pages. Similarly, there is an application called a *voice browser* that "renders" VoiceXML applications to the end user. The browser includes an interpreter that interprets VoiceXML pages. A user's voice is converted to commands using an Automatic Speech Recognition (ASR) engine and VoiceXML content is "spoken" by the browser using a Text-To-Speech (TTS) synthesis engine. The browser also interfaces with the telephony components. The entire gamut of components mentioned above resides in what is known as the VoiceXML Server (see Figure 4-12).

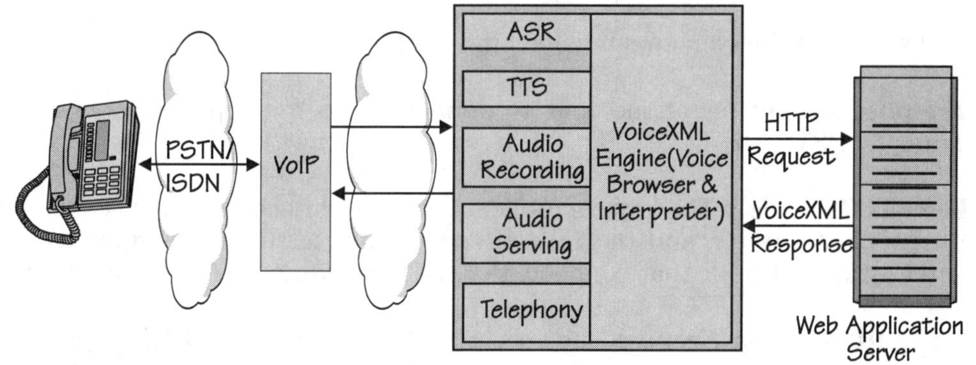

Figure 4-12. The VoiceXML Architecture.

Briefing

Browser, Browser Where Are You?

Even though we compare voice browsers with Web browsers and handheld device micro-browsers, the similarity takes a deviation. Web browsers and micro-browsers reside on the end-user device, but this is not the case with a voice browser. It is located on a VoiceXML server. ❑

An Architecture for Voice-Driven Applications

Let's delve into understanding the role that the voice and data networks and other peripheral elements play in delivering voice-driven applications. Voice-driven applications that are based on VoiceXML utilize popular existing standards such as HTTP and XML, and leverage networks such as the ubiquitous *Public Switched Telephone Network (PSTN)* and the Internet.

Referring again to Figure 4-12, let's examine the typical components involved when providing a voice interface to your application. It shows a high level generic architecture that captures the essence of any VoiceXML-based speech interaction scenario to access enterprise applications.

The following are the elements involved in typical voice-Web architecture:

- ***Telephone:*** Any telephone that is connected to a telephony network (for example, PSTN).
- ***VoiceXML Server:*** The key component in this architecture. It houses the voice browser, interpreter, and the related speech and telephony components. The gateway is the bridge that connects the PSTN network to the Internet network.
- ***Web/Application Server:*** The server that hosts the VoiceXML applications. These applications are accessible by providing a URL that points to the exact location where they reside just like for any HTML page. (Example: *http://www.yourwebserver.com/yourapplication.vxml*)

The networks involved in this architecture are described below:

- ***Telephony Network:*** PSTN, PBX systems, ISDN, VoIP (Voice over IP)—including VoIP + PSTN—all serve as telephony networks. The primary services of a telephony network include DTMF (Dual Tone Multiple Frequency) detection, call initiation, call control, and call termination. The network carries the call to the VoiceXML Server and returns the response for the gateway to the caller.
- ***Internet:*** A global communications network consisting of thousands of networks typically interconnected to each other. In the above scenario, it carries the VoiceXML server's request for VoiceXML applications and related audio and grammar files to the Web/Application server and returns them to the server.

A speech interface can be implemented in one of two ways:

- The ***Gateway model***, where a service provider hosts your VoiceXML applications. Similar to *Application Service Providers (ASPs)* for Internet applications, voice gateway providers provide a complete service for hosting your voice driven applications. This approach leverages the expertise, technology, and infrastructure of voice technology providers such as BeVocal, VoiceGenie, Tellme, Voxeo, and HeyAnita. In this model, you would develop your voice applications and then contract with the gateway providers to deploy the applications. The gateway providers are responsible for providing PSTN access to your applications (including the telephone number) and maintaining the VoiceXML and Web infrastructure required by your applications.

- The ***In-house model,*** where you provide the PSTN and networking connections plus the VoiceXML and Web servers yourself. This is an involved process requiring expert knowledge in telephony and Web technologies. In this model, you have the flexibility to choose and decide the software and the hardware tools based on the load expected.

Let's go through a simple scenario to grasp a better understanding of the architecture. We'll go through the sequence of steps involved in interacting with a VoiceXML application residing on some Web/Application server. Let's suppose a caller was interested in the current weather conditions in London. Figure 4-13 illustrates the steps involved in a typical scenario. In this figure, the VoiceXML server is a single unit but the various modules are broken up to give you a better understanding as to which modules actually perform the operation. The dotted line connecting the different modules of the VoiceXML server and the Web server should be interpreted as an interaction between the VoiceXML server and the Web server. In this figure, the sequences/interactions are marked with numbers and the section following the diagram explains these numbered interactions:

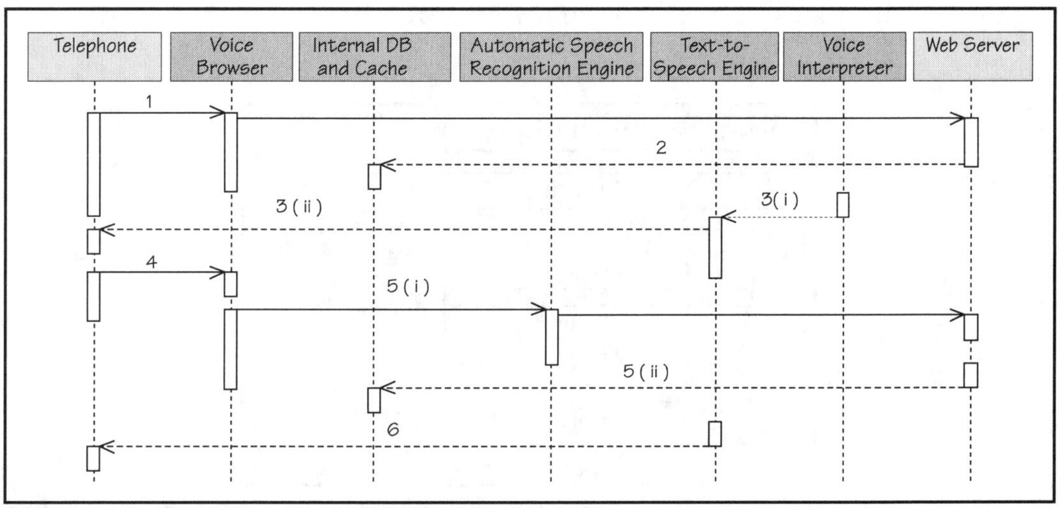

Figure 4-13. Scenario Diagram for Voice Browsing an Application.

1. The caller calls a number. This number is routed via the PSTN to the VoiceXML server.
2. Based on the dialed number, the VoiceXML server (that hosts the voice browser) searches for a URL mapped to the dialed number from its internal database. It then downloads the VoiceXML application and its related resources such as *audio files* and *grammar files* and possibly caches them.
3. The interpreter residing on the VoiceXML server runs through the VoiceXML document and, based on the document tags, interacts with the user. For example, with a weather example, it will welcome the caller and prompt for the city for which the weather information is required. For this purpose it will use a *Text-to-Speech (TTS)* engine.
4. The caller then responds by saying "London."
5. The VoiceXML server then interprets the response using the *Automatic Speech Recognition (ASR)* engine and submits the request to the Web server for data fetching.
6. The resulting data is then embedded into a VoiceXML document, which is interpreted by the VoiceXML server and returned as speech to the caller.

The VoiceXML Server

For any VoiceXML developer, the element that concerns them the most in the architecture explained above is the VoiceXML server, the heart of any VoiceXML-enabled network. It is a bridge between the PSTN and the Internet and is made up of hardware and software resources. We will briefly talk about the components that make up the VoiceXML server and the responsibilities of each. Figure 4-14 illustrates the resources of a VoiceXML server.

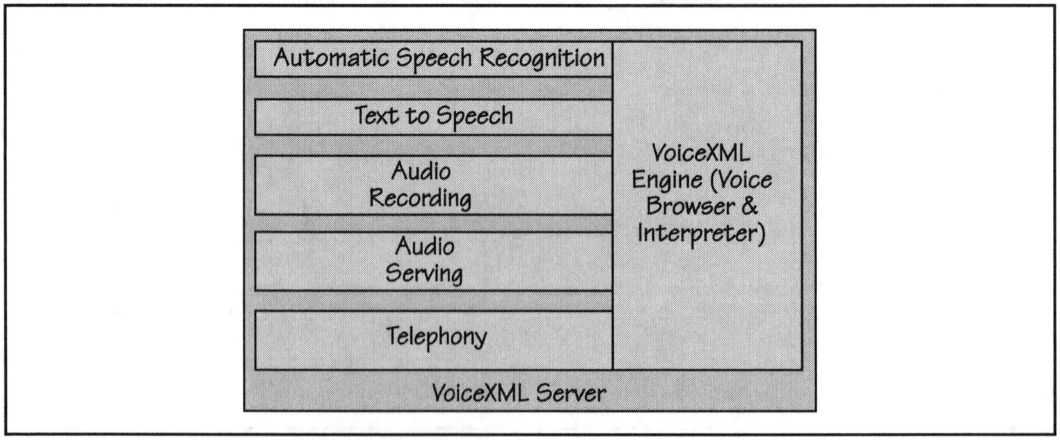

Figure 4-14. The Components of a VoiceXML Server.

Text to Speech (TTS)

TTS technology is the ability of a system to intelligently analyze input text, determine the desired pronunciations, and generate the actual sounds that accurately produce these pronunciations. To accomplish text processing and speech production tasks, text-to-speech systems have two main components: the *text module* and the *speech module*. The text module parses the input text into the appropriate linguistic units (such as *phonemes*) using algorithms and rules, while the speech module uses the linguistic structure produced by the text module to create the actual speech.

Popular TTS vendors include Nuance Communications, AT&T, IBM, Lernout & Hauspie Speech Products U.S.A. Inc., Fonix Corporation, and SpeechWorks International.

Automatic Speech Recognition (ASR)

ASR technology is the ability of a system to interpret human speech and return words, phrases, or digits that represent the input speech. The return is based on the usage of acoustic, lexical, and language models. There are several classifications of speech recognition systems:

- ***Discrete speech:*** These systems accept user input in a discrete fashion. For example, systems that accept a spoken telephone number punctuated by pauses at the end of each digit.
- ***Continuous speech:*** These systems are employed to interpret continuous speech input without any pauses. They score over discrete speech systems in terms of performance, accuracy, and scalability.
- ***Speaker-dependent:*** For speaker-dependent systems, the system is "trained" to accept a particular speaker's voice. To customize the system, the speaker is made to input speech with a varying degree of phonetic representations.
- ***Speaker-independent:*** Unlike speaker-dependent systems, these systems are trained to accept speech input from a large number of speakers. Generally, these are more complex in nature because responses from the speakers could be varied and large in number, thus making it more difficult for the system to accurately interpret the speech input.

In VoiceXML, speech recognition is driven by rules known as grammars. A valid set of utterances is grouped under a grammar element for a given dialog. When the user speaks in response to a prompt from the system, the spoken input is matched with a set of utterances. When matched, the input is stored in a predefined variable.

Popular ASR vendors include Dragon Systems, IBM, Lernout & Hauspie Speech Products USA, Philips Consumer Electronics, Nuance Communications, and SpeechWorks International.

Audio Recording and Audio Serving

These resources are responsible for playing recorded audio files of some standard format. Different platforms support different formats. The *.wav* format and its variations are the most commonly supported. The audio files must be available before the VoiceXML application that calls them is deployed.

Telephony

This component provides the interface between the VoiceXML server and the telephony network. It provides services such as echo cancellation, *Dual Tone Multiple Frequency (DTMF)* tone detection, and audio prompt and playback.

Echo cancellation is the ability to blanket the peripheral noise and provide richness to the input audio signal. DTMF tone detection is the ability to comprehend the touch-tones pressed on the telephones. Each touch-tone corresponds to a unique frequency that is a product of the column tone frequency and the row tone frequency of the telephone. Audio prompts are prerecorded prompts played to the user. Prompts can be recorded from the user as well.

Typically, the interface between the gateway and the telephony network is done using either proprietary APIs or standard APIs such as Microsoft's Telephony API (TAPI) or Sun Microsystems' Java Telephony API (JTAPI).

Voice Browser

A key component in the VoiceXML server, the voice browser supports *bookmarks*, *stop*, *back*, and other well-established Web-browsing navigation concepts like cookies and caching. A full-duplex voice browser supports a feature called "barge-in" that lets users issue prompts to direct the browser during a session, even while being prompted. This delivers significantly more flexibility to voice interfaces than older half-duplex prompt and response systems where users must wait for prompts to complete before issuing input.

In a typical scenario, when a call is made, a new session of the voice browser is started. Based on the dialed number, the voice browser fetches a VoiceXML document and other related files from a Web server using the HTTP protocol and probably caches the files depending on whether the feature is enabled. Next, the interpreter scans through the document making use of the necessary resources, such as TTS, ASR, and audio, as well as interpreting tags that necessitate actions.

Having understood the components of the VoiceXML server, let's look at some of the core responsibilities of the voice browser. They include:

- ***Interfacing with the telephony network:*** The VoiceXML server accepts calls and other telephony-related information and passes the relevant information to VoiceXML applications. It also initiates outward calls thereby bridging to an existing user if the <*transfer*> element is invoked. More on the transfer element is explained Chapter 11.

- ***Session initiation:*** Initiates new sessions of the voice browser.

- ***URL mapping:*** Based on the dialed number, the VoiceXML server does a lookup internally to match the number's corresponding URL.

- ***Accessing resources:*** Houses and invokes the resources such as the ASR, TTS, and Audio, that are primarily called by the voice browser. These resources may be located on the local host or on a remote server on a network.

- ***Caching*** VoiceXML documents, grammar, and audio files.

- ***Retrieving VoiceXML documents*** and other related files over the Internet.

- ***Interpreting VoiceXML documents*** and invoking the various resources such as TTS and ASR.

VoiceXML Security

Details

Like HTML, VoiceXML supports accompanying standards, such as HTTP, Secure Socket Layer (SSL), and cookies to deliver secure solutions for mission-critical applications. Since VoiceXML is built upon existing Internet standards, companies can reuse their existing Web-based framework for security, authentication, and personalization when extending enterprise applications to the phone.

While the above deals with Internet security, what about establishing a secure telephone line to the VoiceXML Server? Well, people are comfortable sharing confidential information over the phone with automated voice applications as well as call agents. As a matter of fact, U.S. consumers did $430 billion in commerce over the phone in 1999, more than 10 times that of global Web purchases. ❑

VoiceXML and Voice-over-IP (VoIP)

So, what's all this fuss about Voice-over-IP (VoIP) technology and what has it got to do with speech-enabling VoiceXML enterprise applications? What we have discussed above is enabling a speech interface for a pure voice network such as the PSTN in a circuit switched environment. The VoiceXML server resides at the edge of the circuit-switched network and before the packet-switched Internet. With the advent of VoIP, there has been an ever-growing demand for voice services over the IP network and existing networks. In the case of a VoIP network, packetized voice signals are transported in a packet-based environment. The advantages of enabling your VoiceXML applications over the VoIP networks are many:

- The IP network is ubiquitous.
- It is a proven fact that VoIP calls are 30–40 percent less expensive than regular PSTN calls.
- Bandwidth efficiency.
- Integrates seamlessly with existing networks.
- Computer telephony integration.
    - With VoIP networks, the VoiceXML server resides in the IP network just like the Web/Application server. Figure 4-15 shows the position of the VoiceXML server in a VoIP network that was set up at San Jose State University. The VoiceXML server is identified by an IP address.

If the VoIP network is to be integrated to a standard PSTN network, a VoIP gateway is necessary to perform the necessary translations of the voice signals to travel over the IP network. In order to enable voice to travel over an IP network, certain communication protocols are used. The common ones are the International Telecommunication Union's (ITU) teleconferencing protocol standard, *H.323*, and Internet Engineering Task Force's (IETF) *Session Initiation Protocol (SIP)* standard (RFC 2543). A brief description of the two is provided below:

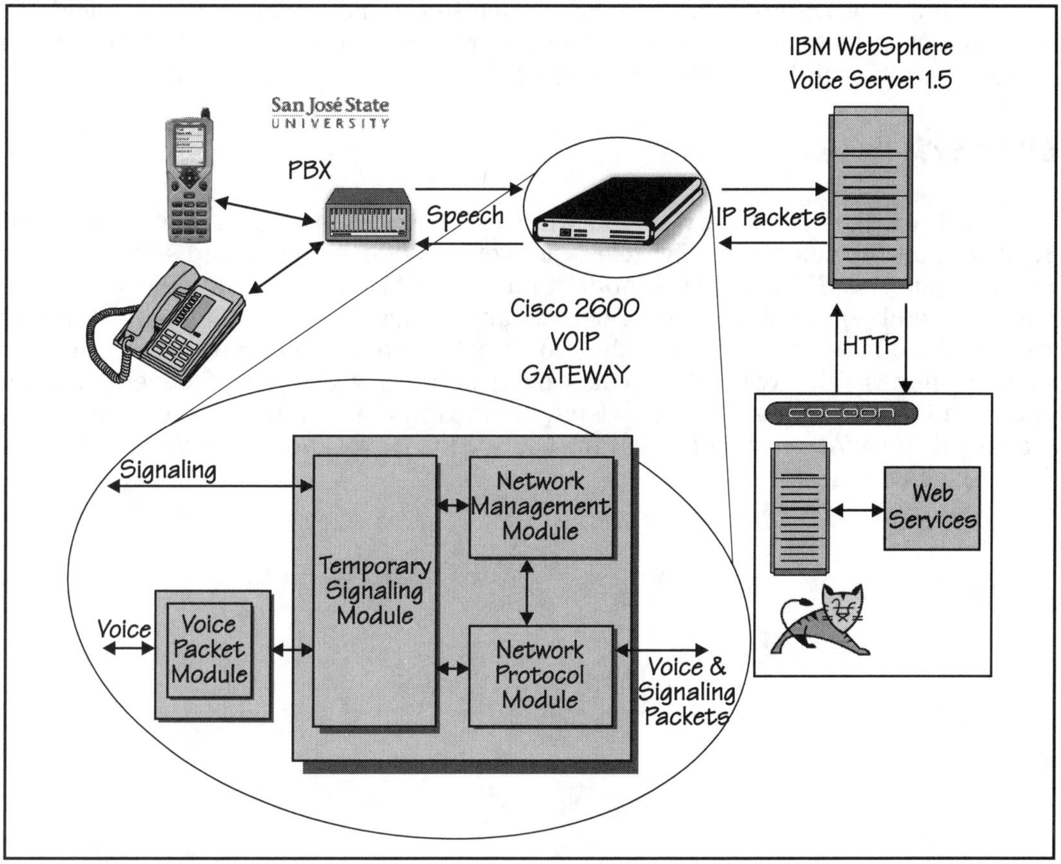

Figure 4-15. The VoIP Software Architecture.

- H.323 defines how audio and video information is formatted and packaged for transmission over the network.
  - Multimedia conferencing protocol-voice, video, and data conferencing
  - Complex, heavyweight protocol combining other protocols
- Session Initiation Protocol
  - Signaling protocol for Internet conferencing, telephony, presence, events notification, and instant messaging
  - Lightweight text-based protocol similar to HTTP

Some gateway providers such as TellMe, VoiceGenie, and BeVocal enable VoiceXML applications to be accessible over VoIP networks. Speech software vendors such as Nuance and SpeechWorks are also making their products VoIP compliant.

SUMMARY

The need for data-based services on wireless networks has led to development of services that reside on top of wireless networks. We have discussed some of these services such as SMS, USSD, i-mode, and WAP. SMS has been a very successful service used by millions of subscribers as an alternative to two-way alphanumeric paging. I-mode set a model for the world as an end-to-end Internet service for mobile phones. The WAP Forum has been doing a great job of developing the specifications for bringing Internet communications to mobile devices. We have discussed the WAP protocol stack in detail. Finally, we looked at VoiceXML as a service technology.

Chapter 5

Wireless Devices: Middleware

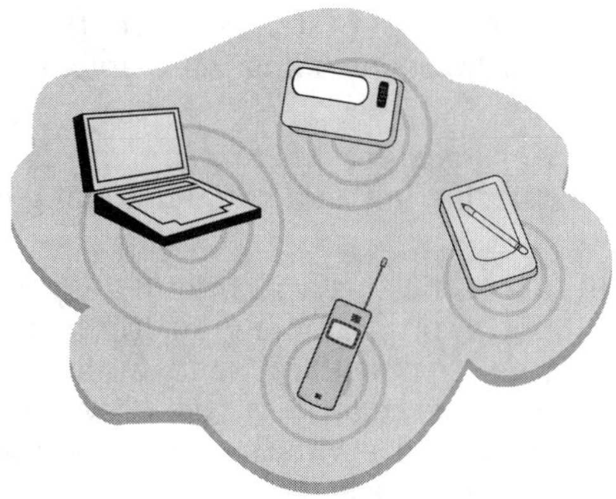

Middleware provides the software platform for the enterprise applications we develop as well as for our wireless device applications. We've described the physical platform for wireless devices in previous chapters. Now it's time to describe our software platform by describing the middleware for handheld devices. This middleware includes operating systems, communication protocols, security, and databases.

We'll find that communication protocols, security, and databases are bundled together with basic operating system function for our handheld devices. What's more, these technologies have particular importance to our use of a handheld device as a client in an enterprise application. Hence, we will also discuss some of the issues that exist for these technologies.

In this chapter we will focus on the middleware that runs on the handheld device. Of course, there is a server-side middleware counterpart to the middleware that runs on our handheld device. Further, some of this server-side middleware is specifically for supporting handheld device applications or is required to provide an end-to-end solution (for security). We will defer this server-side middleware discussion until Part 3.

OPERATING SYSTEMS FOR HANDHELD DEVICES

Operating systems for handheld devices provide several services:

- ***A software abstraction of the device's hardware features.*** This includes memory, persistent storage, the user interface, connectivity, timers, and other hardware including interrupts, sound, and expansion options.
- ***Services for managing applications.*** These include services for starting, terminating, and communicating between applications.

In this section we will briefly describe the operating systems for the devices that support a programming environment that we described in Chapter 2. Just as the devices represent a range of hardware capability, the operating systems also represent a range of functionality that matches their hardware platforms. As before, the range starts with the Blackberry two-way pager, moves to cell phones, and then PDAs. We will also take a look at operating systems for handheld PCs. Are we going to turn you into a programmer for all of these different devices? No, but we will try to give a feel for the diversity of these operating systems. We will also briefly mention the programming environments for these devices.

Blackberry OS

The operating system for the blackberry device provides a large API set and a lot of functionality for such a small device. The OS includes:

- A cooperative scheduler to simulate a multitasking operating system
- Tasks and threads to run applications
- OS-to-application and application-to-application messaging
- File and database support
- Memory management for dynamic and flash memory
- A user interface that supports the task wheel, keyboard, and LCD display
- An application "ribbon" for application selection and launch by the user
- APIs for supporting the wireless radio and serial port interfaces

In addition to the basic operating system APIs, the Blackberry OS also supports APIs for managing an address book database and for an autotext completion feature to ease the user's burden when entering text.

Table 5-1 describes the OS API features of the Blackberry device.

Table 5-1. The Blackberry RIM OS and Extended OS API.

API Category	Description
Basic OS	The Blackberry OS API provides basic file system access, keypad configuration, primitives for manipulating the LCD screen, memory allocation/deallocation, serial port manipulation, string formatting, and time management. In addition, several system APIs are included for managing tasks, intertask messages, threads, alarms, and battery status.
Messaging	The messaging API provides functions to create a message, add recipients, compose the message text, add attachments, send the message, register applications to be notified of received messages and have access to received messages.
User Interface	The Blackberry user interface API is a set of classes that provides a common look and feel for all applications developed for the RIM Wireless Handheld. Features include menu and screen handling, user fields, and dialog boxes. This functionality complements and enhances the base functionality of basic OS API. There are five groups of function: UI engine, Fields (such as Choice, Edit, Label, List, RichText, Separator, and Table), FieldManager (for dialogs and screens), Menu, and Status.
Ribbon	The ribbon provides an application launcher for the Blackberry device. Each application has an icon. The thumbwheel is used to select an icon and launch the application. When the application is launched, a message is sent to the application by the OS.
Database	The Database API provides data storage solutions for developers creating applications for the BlackBerry Device. The database library manages memory and the flash file system for the handheld. The diagram below shows the relationship of the various database classes. 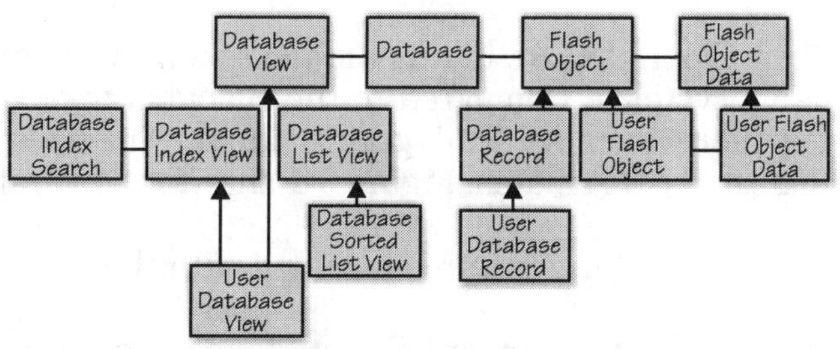
Radio	The radio communications APIs are for applications that need packet-level access to the radio network. These APIs complement the messaging APIs which can send email or fax messages.

Table 5-1. The Blackberry RIM OS and Extended OS API. (Continued)

API Category	Description
Address Book	The Address Book database in the Blackberry device maintains a list of contacts. The Address Book APIs allow an application to add, delete, and edit contact information, send messages to contacts, and send and receive contact information. APIs are also included to manage group contact lists if supported by the service provider.
Autotext	The AutoText API allows you to add auto text features to your Blackberry applications.

Developing for the BlackBerry

The native Blackberry development environment uses the C++ language. RIM recommends Microsoft Visual C++ 5.0 (or later) for development.

Cell Phone OSs

Cell phones, in general, have been sold as voice and simple Internet access devices. With the exception of the Java-enabled phones, they are not intended to provide a programming platform. So, the internal software is proprietary and not advanced as a development platform by cell phone manufacturers.

Palm OS

The Palm Operating System is the next step in the evolutionary path of handheld operating systems. Compared to the Blackberry device, a Palm PDA has a larger screen, more memory, and graffiti input. The Palm hardware requires a more sophisticated OS to match. Figure 5-1 shows the overall architecture of the Palm OS.

Here are some of the features included with the Palm OS:

- Launch, Notification, and Event Management for starting, stopping, and transferring control between applications
- Memory, File, and Database management
- User interface support for GUI elements, graffiti, and touch screen input
- Support for Add-on hardware and storage
- Support for a variety of communication mechanisms including serial port, infrared, and network communication
- Support for a number of data exchange formats

In addition to the base operating system support for on-the-device applications, there is an architecture and programming APIs for running desktop applications that exchange data with applications on the Palm device. These applications run on a desktop PC and are called *conduits*.

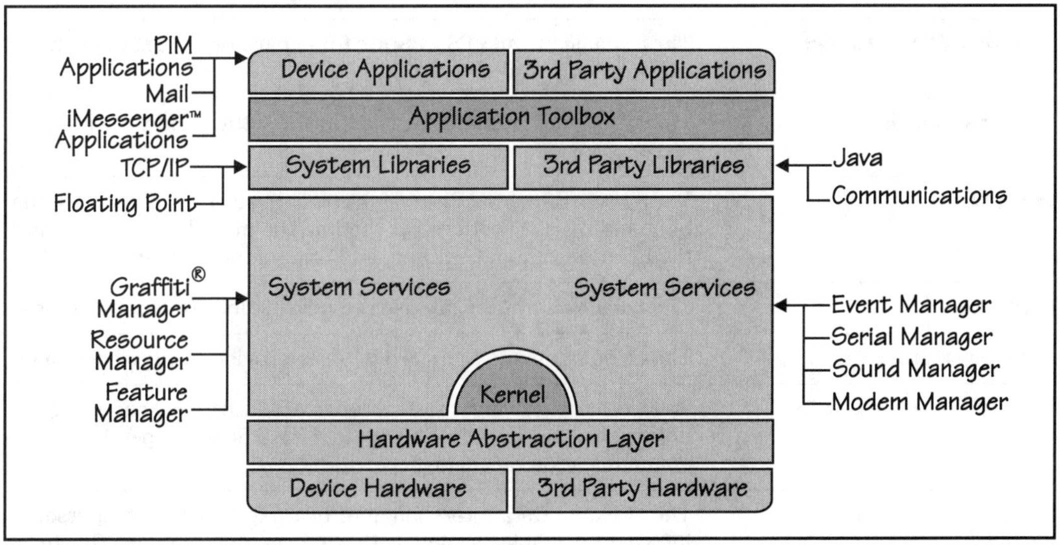

Figure 5-1. The Palm OS Architecture.

Table 5-2 provides a brief description of the APIs provided by the Palm OS.

Table 5-2. The Palm OS API Categories.

API Category	Description
Launch and Notification	Applications can be launched when the system or an application generates a launch or notification code. The Launch and Notification APIs allow an application to switch to another application or to register for or initiate a notification.
System Event Manager	The System Manager APIs are for getting top-level events (for managing the application event loop) and low-level events (for managing pen stroke to key management, pen queue management, and auto-off control).
User Interface	The User Interface APIs are for creating and managing UI elements. APIs are included for managing windows, controls, groups of controls, fonts, bitmaps, and graffiti.
Memory	The Memory APIs are for memory and heap functions such as allocating, freeing, resizing, moving and comparing.

Table 5-2. The Palm OS API Categories. (Continued)

API Category	Description
Files and Database	The File and Database APIs are for saving and accessing data. Data is stored in records, which form a database.
Expansion and Virtual File System	The Expansion and VFS APIs are for expansion hardware including external storage devices.
Attention and Alarms	The Attention and Alarm APIs allow for setting and presenting alarms.
System Support	The System Support APIs are for managing features, sound, system boot and reset, hardware interaction, the microkernel, retrieving the ROM serial number, time, and floating-point operations.
Localization	The Localization APIs are used to provide localized applications.
Debugging and Standard IO	The Debugging and Standard IO APIs provide debugging support.
Exchange Manager and Libraries	The Exchange Manager and Libraries APIs are for exchanging typed objects such as records, databases, MIME data independent of the communication mechanism
Personal Data Interchange	The Personal Data Interchange APIs are for exchanging personal information vObjects such as those formatted as vCard (business card) and vCal (calendar) objects.
Beaming	The Beaming APIs are for exchanging data using the IR port.
Serial Communications	The Serial Communication APIs are for exchanging data using a serial port.
Network Communications	The Network Communications APIs are for providing basic network services using TCP and UDP services via sockets and for Internet services such as HTTP.
Telephony	The Telephony APIs are for handling telephone calls.

Developing for the Palm OS

There are a wide variety of languages (C, C++, Java, and even Smalltalk and Forth) plus a number of third-party development environments supported for the Palm platform. A variety of tools exist for application development as well as tools for conduit, Web Clipping, and Database application development.

Windows CE

Windows CE (WinCE) is Microsoft's operating system for small footprint and mobile, 32-bit, intelligent, connected devices.[1] Windows CE works on 12 different processor architectures and on over 180 CPUs and is used in a broad range of devices including industrial automation, retail and point of sale devices, Internet access devices, "ruggedized" and handheld portable PCs, high speed data acquisition devices, and several consumer applications like game consoles and set-top boxes. Windows CE is a modular operating system that can be configured for a specific environment to minimize the memory footprint using Microsoft's Windows CE Platform Builder.

The Win CE platform builder can create any number of images, based on features the hardware developer wants to include in the image. This generally means the HW developer (the PDA developer) has to trade off between features and memory consumption.

Table 5-3 provides a brief description of the API categories supported by Windows CE.

Table 5-3. The Windows CE 2000 Operating System Function.

OS Category	Description
Kernel Services	The Kernel Services provide core operating system function such as process, thread, memory, and some file management. Real-time processing function is provided with thread priority levels, interrupt support, and timing and scheduling. The kernel also provides relatively sophisticated memory and heap management functions.
User Interface Services	The User Interface Services provide Windows-like controls as well as the GDI (Graphical Device Interface) for graphics primitives. Also included is a customizable shell, an on-screen keyboard, support for touch-sensitive screens, and handwriting recognition.
Communications and Networking	The Communication and Networking APIs provide support for serial and infrared ports, the Telephony API (TAPI), and for TCP/IP, sockets, and related technologies such as DNS, WINS, and DHCP.
Internet Services	The Internet Services APIs provide support for HTTP and HTTPS and browsing features such as a low footprint version of Internet Explorer, cookie management, and image support (GIF, JPEG, and XBM). An HTTP server can also be included.

[1] The "CE" in Windows CE is not an acronym. Microsoft thought "CE" was cool when they named the product.

Table 5-3. The Windows CE 2000 Operating System Function. (Continued)

OS Category	Description
Multimedia	The Multimedia API includes the high performance DirectX and Windows Media APIs for high performance audio, video, and streaming media.
File, Database, and Registry	The File, Database, and Registry APIs are used for data management in ROM, RAM, flash, or rotating media and for storing data about applications, drivers, and user preferences.
Advanced Runtime Services	These APIs provide support for Microsoft's COM (Component Object Model) and DCOM (Distributed Component Object Model), MSMQ (Microsoft Message Queue Service), as well as other technologies for database access, graphical applications, help, and Visual Basic program execution.
Security and Cryptography	The Security and Cryptography APIs provide function for encryption/decryption, Digital Signatures, Smart Cards, as well as other security functions.
Globalization/Localization	The Globalization/Localization provides APIs for delivering applications to international markets.
Applications	Provides APIs for Pocket Inbox (for managing IMAP4 or POP3 mail) and Pocket Word (for creating, viewing, and editing documents).
Hardware Support and Device Drivers	Provides software support for a variety of CPUs and supporting chips, expansion buses, standard platform hardware, and peripheral hardware.
Connection Services	The Connection Services APIs provide function for exchanging information with a desktop PC.

Developing for Windows CE

Microsoft provides the eMbedded Visual Tools package for developing applications for Windows CE using the Visual C++ and Visual Basic languages. Separate SDKs are included for the Handheld PC and Pocket PC platforms.

Symbian OS

Symbian is a company formed by Ericsson, Motorola, Nokia, Panasonic, and Psion to develop products and evangelize standards for wireless information devices. One of the products is the Symbian OS. The Symbian OS is currently used for the Nokia 9290 Communicator discussed in Chapter 2.

The Symbian OS has its roots in the EPOC operating system originally developed for the Psion handheld computing devices and packaged with the Eikon user interface software. Over time it has evolved, first to support the Ericsson R380 with the Ericsson Component Kit (ECK), then for a variety of devices including handheld PCs, PDAs, and Smart Phones (see Figure 5-2). The most recent version contains a core operating system function called GT for *Generic Technology.* Each of the hardware platforms has it's own UI design, dubbed *Crystal*, *Quartz*, and *Smartphone* UI.

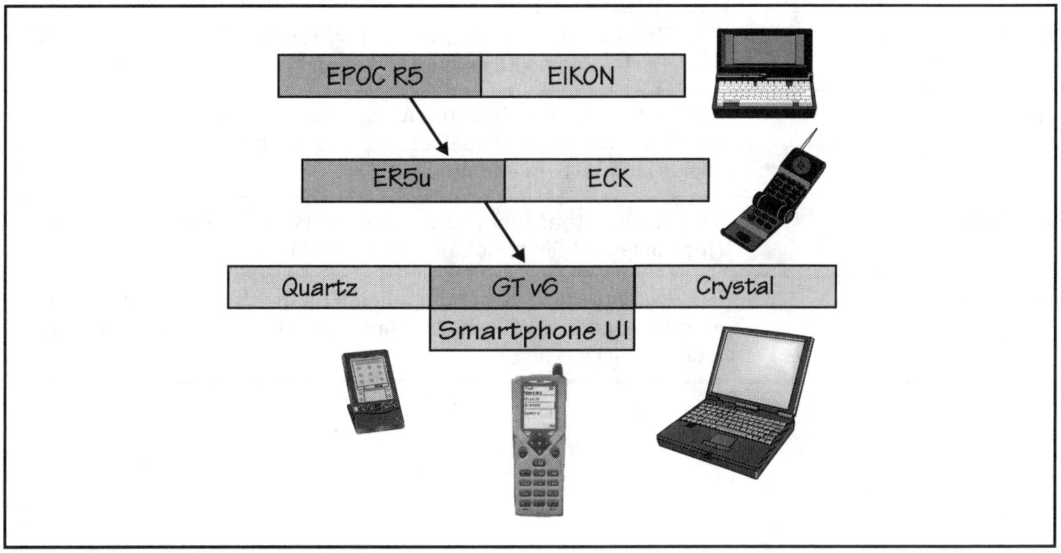

Figure 5-2. The Evolution of the Symbian OS.

Table 5-4 describes some of the key features of the Symbian OS.

Table 5-4. The Symbian Operating System Function.

OS Category	Description
Base	The base provides the runtime system, the tools required to build it, and lowest-level security. This includes a Hardware Abstraction Layer (HAL) for porting to different hardware, support for processes, threads, memory management, and program execution.
Application Framework	The Application framework components provide powerful reusable libraries for data, graphics, and text. This includes support for relational databases, clipboard, bitmap graphics, printing, a windowing system, rich text, internationalization, and Uikon (a GUI framework).
Media Server	The Media Server provides audio recording and playback plus functionality for manipulating images.

Table 5-4. The Symbian Operating System Function. (Continued)

OS Category	Description
Communication	The communication libraries provide support for a number of protocols and APIs including TCP/IP, Sockets, RS232, GSM, GPRS, Bluetooth, Infrared, and WAP.
Messaging	The Messaging Framework supports sending and receiving SMS, email (IMAP4 and POP3), and Fax messages.
Browsing	The Web Engine provides support for browsing Web sites using HTTP and HTTPS and for displaying WML and executing WMLScript.
Application Engines	Application Engines includes an agenda and contacts model (for PIM function), a sheet and spell engine (for word processing and spreadsheets), and a help engine.
Java Runtime	The Java Runtime includes support for Personal Java and JavaPhone (for managing the telephony connection).
Connectivity	Connectivity consists of components on the Symbian device, and components on a PC, which cooperate to provide connectivity and datasync services.

Developing for the Symbian OS

Symbian supports applications development in C++, Java, and OPL (a basic-like language developed for the EPOC OS).

The Linux Variants

Although scaled-down versions of Linux don't have a strong presence in the market at the time of this writing, they are beginning to appear. These include open source efforts by *handhelds.org* (an organization funded primarily by Compaq to develop open source software for handheld and wearable devices), PocketLinux by Transvirtual Technologies (available for the Compaq iPaq, the Casio Cassiopeia E-105, E-110, and E-115, the NTT DoCoMo Sigmarion, the IBM Workpad Z-50, and the Vadem Clio), and the Sharp Zaurus.

Soapbox

OSs, A Parting Thought

So where does all of this discussion about operating systems leave us? Perhaps a little overwhelmed and apprehensive. There are too many platforms, a wide variety of APIs, a number of different development languages, and a large number of tools for development. Choosing one platform implies choosing a market segment. To deliver one application to multiple market segments means learning a new operating system (never a small feat), perhaps a new language, a new style of programming for a particular device, and a new set of development tools.

Does this sound familiar? Thinking back to the early days of Java, it should. This is the problem that Java was intended to tackle. Can Java solve this problem for handheld devices? Certainly it is a large challenge due to the hardware constraints that handheld devices impose.

Some questions unique to handheld devices also remain. Will I be able to do everything in my Java application that I can do from a native application? No. But that will improve over time. The trade-off is that applications will be able to run on many devices with one development effort. Won't Java applications strain performance on a device that is already performance challenged? Possibly. But this will improve as handheld devices become more powerful. The trade-off is the ability to run on multiple devices with one development effort.

So, as with many things in life, there are trade-offs. We will begin to explore the Java possibilities in more detail in Chapter 7. ❑

HANDHELD DEVICE MIDDLEWARE CHALLENGES

We're already quite aware that our handheld devices are not the same as their desktop counterparts. This is true for both their hardware platform and middleware platform. Some areas in the handheld device middleware platform deserve special mention in light of deficiencies in communication, security, and database implementations causing limitations in the applications we would like to deliver. We will discuss these deficiencies in the following sections.

Communication Protocols

On our desktop platforms we are used to having a variety of communication protocols available. Most likely, the primary protocols you use as an application developer are TCP/IP (with its associated sockets programming APIs) or HTTP requests. Both of these protocols have APIs for client and server side programs.

Although our handheld devices will primarily be clients, there are many situations where we would like to embed server programs on our device. These include applications such as information broadcast, device provisioning, remote device control, device information query, peer-to-peer networking, multi-user games, and instant messaging.

In addition to the usual issues of limited bandwidth, memory, and processing power plus network unreliability (all particularly detrimental to server programs), a couple of software and communication infrastructure issues also inhibit our handheld devices from supporting these applications. One problem is some platforms do not include server-side communication APIs. The other problem in the proliferation of handheld devices is stressing the current IP addressing limit.

With the expectation that the devices will not host server applications, some device manufacturers chose to include only client-side communication APIs. This is particularly true for the smaller devices like cell phones. As you move up the "hardware capability curve" to PDAs, this becomes less true.

The network addressing limit imposed by the current 4 byte IP address has also pushed service providers toward solutions that are not friendly to server programs. One common solution is to support devices behind a router that supports Network Address Translation (NAT). IPv4 addressing defines a range of private IP addresses. These addresses are not used on the public Internet. Instead, they provide a range of subnet addresses that are hidden behind one public IP address. Figure 5-3 illustrates this addressing configuration.

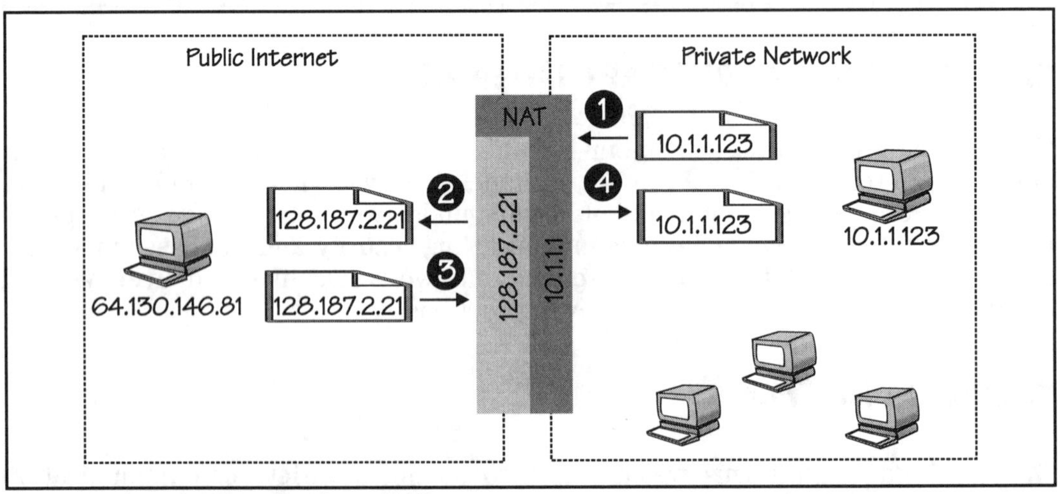

Figure 5-3. Network Address Translation (NAT).

With the NAT solution, a router tracks any traffic originating from a client in the private subnet so reply traffic can be returned to the originating client. This has one big problem. Clients in the public Internet cannot send messages to a device in the private subnet even if it is running a server program.

Is there a way out of these problems besides waiting for all of our devices to support server programming APIs and for IPv6 (which will greatly expand Internet addressing). Yes, but you might not consider it to be the most elegant of solutions. Whenever there is an addressing deficiency in our handheld devices, there is little choice but to introduce a server gateway of some type. In this case, an HTTP proxy gateway is suggested (see Figure 5-4). In this architecture, if a client wants to act like a server (without using server APIs) and receive a request from a client, it has to poll for requests from the proxy server that have been received from the client. Not the most elegant of solutions but it solves both the client-side only problem and the NAT problem.

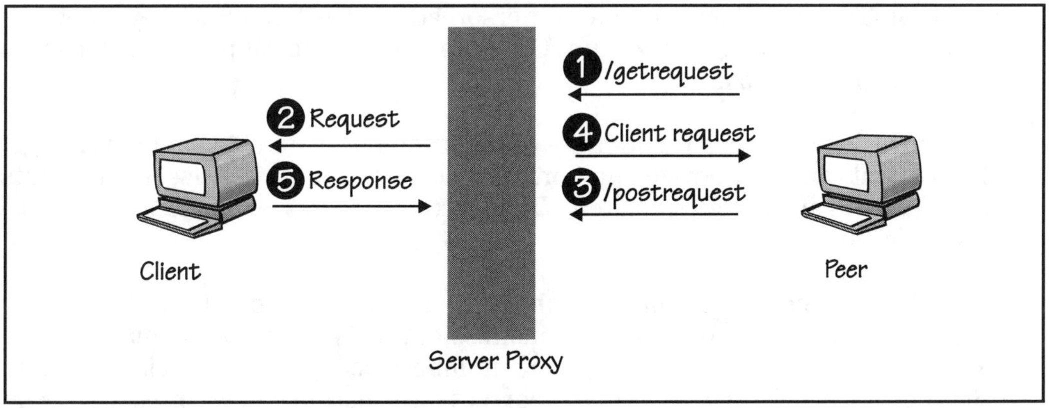

Figure 5-4. An HTTP Proxy Gateway Architecture.

Security

We depend on security to protect our commercial interests (both corporate and personal) when it comes to electronic transmissions. We must have confidence that our transmissions will not reveal our credit card and bank account numbers and that our other personal information will remain confidential. It is assumed that there are devious persons intent on doing harm and we must be protected from their efforts.

This is true for our wired connections and is no less true for the wireless connections that emanate from our handheld devices. Without confidence in our electronic transmissions, e-commerce and *m-commerce* (mobile-commerce) cannot thrive. In this section we examine some of the security issues that are specific to handheld devices. This discussion complements the WAP security discussion in Chapter 4.

First, let's remind ourselves of some of the terms and basics of secure electronic transmissions.

- ***Encryption*** applies mathematical algorithms to data to obscure the information prior to transmission. The recipient then applies a complementary algorithm to decode the obscured data. These mathematical algorithms are driven by *keys*.

- ***Keys*** are defined by the type of algorithm used, their length (typically between 40 and 128 bits in length), and whether or not the same key is used by the source and destination. In a *Secret Key* scheme, the same key is maintained by the source and destination to encrypt and decrypt data. This has an inherent problem of transmitting and maintaining the secrecy of the keys.

  An alternative is a *Public Key* scheme. In a public key scheme, keys are maintained in key pairs. One key is private (and is never transmitted or revealed). The other key is public. A person advertises his public key so others can encrypt data to be sent to him. Only the corresponding private key can be used to decrypt the data.

- ***Authentication*** verifies the identity of a participant in an electronic exchange. This prevents an unauthorized person from posing as someone else. Both clients and servers can be authenticated. *Digital Certificates* are typically used to verify identity.

- ***Digital Certificates*** contain the public key and information about the identity of a participant in an electronic transmission. This information may include name, address, email address, phone numbers, corporate associations, and other pertinent information. Since incorrect information can be presented from individuals with less than admiral intentions, certificates are verified and signed by a Certificate Authority (CA) like Verisign. A certificate is signed using the CA's private key. Anyone can then verify a certificate and its contents by using the CA's public key.

- ***Non-Tampering*** provides a mechanism to assure that data was not modified during transmission.

Wireless Security: The Challenges

Like everything else we have seen with handheld devices, security for handheld devices is also skewed by the lack of processing power, limited memory (both volatile and persistent), and low bandwidth of the devices. The causes of security issues related to the algorithms used, how to transmit and store digital certificates, and how to provide end-to-end solutions are also important.

Security algorithms use mathematical equations to encode data. The algorithms chosen can be broken but it is assumed that it would take a tremendous amount of computing power over a long period of time (many months or years) to decode the transmission without knowing the keys used. It is assumed that the longer the keys (i.e., the more bits), the more time it will take to break the transmission. However, the more bits, the longer it also takes for users of the algorithm to encrypt and decrypt data. This is not a problem for desktop machines but a big problem for handheld devices that don't have the computing power to process the larger keys.

There is hope though. New algorithms (like the *Elliptic Curve* algorithm) and their implementations are appearing on the market that are very difficult to break but are more computationally-friendly for handheld devices. Although these new cryptography algorithms are appealing, they also present a new problem.

Most Web sites are protected by *Secure Sockets Layer (SSL)*, which is now called *Transport Layer Security (TLS)*. This protocol, however, uses a different algorithm (the RSA algorithm) that does not interoperate with mechanisms based on Elliptic Curve algorithms. A potential solution to this problem is to introduce a gateway that converts the data transmission from one security mechanism to the other. The downside of this solution is a gap in the end-to-end security path similar to the WAP gap we discussed in Chapter 4. Some device manufacturers are choosing to avoid this problem and provide an SSL solution on the device. For instance, Motorola provided an upgrade to their i85s firmware to include SSL support after the initial release of the phone.

An additional challenge for handheld devices is managing certificates. Certificates can be unfriendly to handheld devices because of their size. All of the descriptive information stored in a certificate may not be needed at the handheld device. Also, certificates may be chained; a series of certificates may need to be followed and processed for verification and to determine if a certificate has been revoked or not. All of this leads to inefficient use of processing power and bandwidth on the handheld device.

One possible solution to the problem (you guessed it) is to introduce a gateway server to preprocess certificates and then only transmit a token with relevant information (like the public key) to the handheld device.

Micro Databases

There are a number of reasons to have a persistent storage mechanism that is easy to use and write database-centric applications for our handheld devices. With limited bandwidth and unreliable network connections (or even disconnected operation), it is useful to be able to capture data on the device for later offload or to synchronize data on the handheld device with a corporate database. You might imagine how applications for sales people, auditors, field technicians, doctors, real estate agents, and insurance adjusters would benefit from this type of function.

However, the persistent data mechanisms provided by the various handheld device operating systems all provide a different API set. This is already true on our desktop and server operating systems. That's why we depend on relational databases to provide a *Lingua Franca* for database access across the many platforms. Relational databases also provide a higher level of data access abstraction, making our lives as programmers a little easier. Wouldn't it be nice if we could reap such benefits on our handheld devices? The good news is you can. Many of the database vendors have released scaled-down versions of their desktop and server products for use on some of the more capable handheld devices. These products include IBM DB2 Everyplace, Oracle 9i Lite, Sybase SQLAnywhere, and Microsoft SQL Server CE.

Figure 5-5 shows the architecture and development environment of a typical relational database for a handheld device. The key parts of the architecture are:

- ***Database engine*** on the handheld device. A subset of the SQL function appropriate to a handheld device is included. This might include Join, Group By, Order By, multiple column Primary and Foreign Keys, Expression functions, Aggregate functions, and Constraints. More resource intensive capabilities like subqueries, creating views, triggers, stored procedures, and user-defined functions may not be included.
- ***Database programming API.*** Typically the programming API includes ODBC for C and C++ programs and JDBC for Java programs.
- ***A Synchronization engine*** that has components on both the handheld device and on a synchronization server. SyncML might be used as the synchronization exchange format.

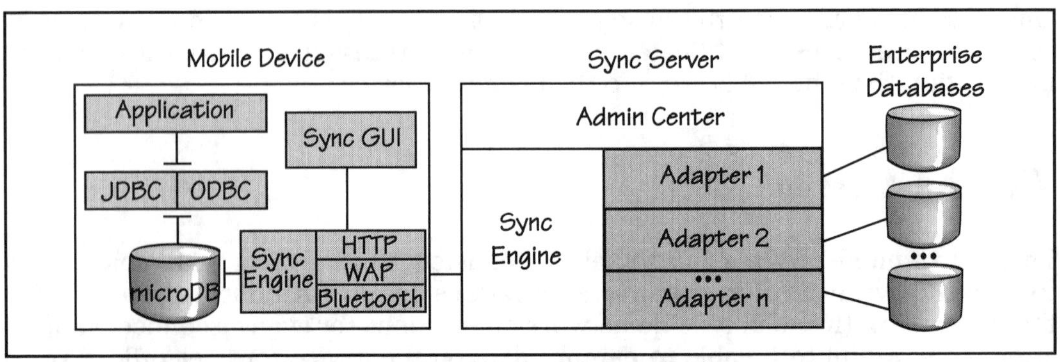

Figure 5-5. The Architecture of a Typical Handheld Relational Database System.

- ***Support for various transport APIs*** between the handheld device and the Sync Server. These include HTTP, TCP/IP, WAP, and Bluetooth.

- ***Adapters for various backend data sources.*** This gives the synchronization engine access to relational, Web server, and email data.

Micro Databases: The Challenges

Micro databases have two basic challenges. One, they require a high-end handheld device. This means they will not support the lower-end devices like the two-way pagers and cell phones introduced in Chapter 2. On the other hand, many of the applications that require a micro database also require the larger display and processing power of PDAs.

The second challenge is integration with J2ME (Java 2 Micro Edition). We will introduce J2ME in Chapter 7. The current Java development approach followed by micro database vendors is to include a complete Java runtime and development environment.

SUMMARY

We have examined four key segments of the programming platform for handheld devices: operating systems, communications, security, and micro databases. Fragmentation in the native programming platforms as well as the challenges in these areas have also been enumerated. In Chapter 7 we will examine Java 2 Micro Edition (J2ME) and explore J2ME as a possible solution to platform fragmentation.

Chapter 6

Wireless Devices: Content Delivery

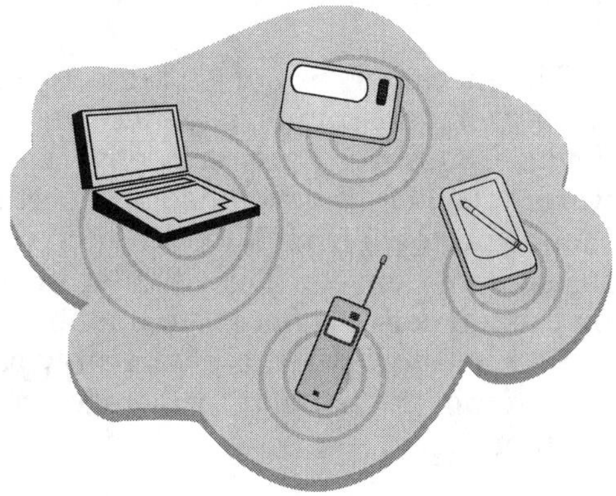

Mobile devices, such as cellular phones and PDAs, are capable of accessing the Internet. In the previous chapters we discussed the various technologies and protocols that enable devices to access the Internet. Typically a micro-browser (or equivalent software) residing on a device communicates with a Web server to fetch information, commonly known as *content*. The way content is transmitted and displayed varies from application to application. Some technologies, like WAP, enable you to directly browse the content while connected. Other technologies, like Palm Web Clipping, bundle the content and store it on your PDA in batch mode to enable browsing offline.

To achieve all these capabilities, different meta-languages are used to deliver the content to the devices. Some content languages are used for presentation purposes, while others are used for synchronization purposes. We will discuss these meta-languages (also known as markup languages) in this chapter.

CONTENT FORMATS FOR DISPLAY—MARKUP LANGUAGES

Content refers to any information obtained from Web sources. This can include news, sports, weather, and stock quotes. In the primitive stages of the Internet, content just consisted of text. With the availability of sophisticated browsers, the content now is made up of many formats—including binary formats such as images, audio, and video. These are generally known as *Multipurpose Internet Mail Extensions (MIME)*. In a typical Internet scenario, HTML is the content format used to deliver the content from the server application to the client browser. HTML not only carries the content, but also carries all the formatting information needed for visual presentation on clients.

HTML, however, is not suitable to support the wireless paradigm because of a variety of constraints like network bandwidth and the resource capabilities of devices. So, several content delivery formats have evolved along with protocols that were specifically developed for certain wireless technologies. For example, the i-mode protocol from NTT DoCoMo uses cHTML while the WAP protocol uses WML for content delivery. This chapter introduces you to the variety of formats that can be used to deliver content to a mobile device. We will be using *content format* and *markup language* interchangeably as they have the same meaning. The following are a few of the popular formats for content delivery, which we will introduce in this chapter.

- I-mode's ***Compact HTML (cHTML)*** is a subset of HTML and used by NTT DoCoMo's i-mode. cHTML does not support the resource intensive features of HTML.
- ***Handheld Device Markup Language (HDML)*** is one of the first markup languages defined for handheld devices.
- ***Wireless Markup Language (WML)*** is a markup language for WAP devices. This format is defined by the WAP Forum.
- ***eXtensible HTML (XHTML)*** is a form of HTML that strictly follows XML rules for well-formed and valid documents.
- ***VoiceXML*** is a technology from the VoiceXML Forum. The VoiceXML markup language is capable of delivering enterprise applications through a voice interface using voice recognition and text-to-speech technologies.
- ***Web Clipping***, from Palm, is for platforms that use the Palm OS. It works by optimizing HTML pages for the small screens of Palm devices. It also stores them on the device and thus reduces network traffic.

The Role of XML

eXtensible Markup Language (XML) is a meta-language used in documents by declaring and specifying new markup tags. XML is designed to only describe data and focuses on defining what the data is. XML tags are not predefined; the owner of the XML document develops them. This is how XML is extensible. Any XML is self-describing and uses a *Document Type Definition (DTD)* or schema to formally describe the document structure. When compared with HTML, we can say that HTML is about displaying information, whereas, XML is about describing information.

XML as a Meta Language

In the real world, computer systems and databases contain data in incompatible formats. One of the most time consuming challenges for developers has been to exchange data between such systems over the Internet. Converting the data to XML can greatly reduce this complexity and create data that can be read by different types of applications.

Structure vs. Format

XML supplies structure to documents. Structure by itself imposes constraints. Even though this might seem like a limitation that might prevent flexibility, in reality it encourages interoperability and lays a common framework for tools and other utilities to evolve. There are two key principles behind a structured model:

- Each element has a relationship with other elements.
- The element that describes the structure does not describe its visual appearance.

The structure of a document is determined by the elements it contains, whereas the format represents the display characteristics of the document. For example, consider a document representing a street address. In this case, the structure could contain a street number, street name, city name, state name, country, and zip code. The structure could then be formatted for display with the street number and street name on the first line of display, the remaining items of the structure on the next line, and then followed by a line.

Consider another example of a document representing some text. The structure could specify that an element is a paragraph. The format could determine how to display the paragraph using a 12-point Times font.

XML documents should be designed to be flexible when representing structure. So learning to separate structure from format is critical to using XML efficiently.

Syntax of XML

The following is an example of an XML document.

```
<?xml version="1.0"?>
<note>
<title>XML document</title>
<body>Hello World!</body>
</note>
```

The first line in the document is the XML declaration and it should always be included. It defines the XML version of the document. In this case the document conforms to the 1.0 specification of XML. The next line (the first element) is the root element of the document. In the example, the next two lines define two child elements of the root, *title* and *body.*

Well-Formed and Valid XML Documents

A *well-formed* XML document is a document that conforms to the XML syntax rules. The following are a few of these rules:

- All XML tags should have corresponding closing tags—for example, in the above sample, the <note> tag should have a corresponding </note> tag.
- XML tags are case sensitive, so the tag <Note> is different from the tag <note>.
- Tags must be properly nested. The following is an examples of incorrectly nested elements:

```
<b><i>Bold and italic</b></i>
```

These tags should be nested as follows:

```
<b><i>Bold and italic</i></b>
```

The following shows how XML tags should be nested within the root tag:

```
<root>
  <child>
    <subchild>
```

```
    </subchild>
  </child>
</root>
```

A *valid* XML document is a well-formed XML document that conforms to the rules defined in a *Document Type Definition (DTD)* or *Schema.*

Components of an XML Document

An XML document needs another document to explain its structure and organization. This information is provided via the DTD or schema file. The DTD or schema defines the order of occurrence of the elements and attributes in an XML file. Using this metadata, you could verify if an XML file is valid using any validating parser.

In addition, an XML document can be transformed to any other document format such as another XML, WML, VoiceXML, HTML, or PDF using XLink, XPath, and XSL. We will cover XSL in Part 3.

XML Attributes

XML attributes are normally used to describe XML elements, or to provide additional information about elements. Attributes follow the element name and have the following generic format:

```
<element attribute="attribute-value">
```

Attributes are always contained within the start tag of an element and the attribute value must always be quoted. In the following example, *date* is an attribute of the element *note.*

```
<?xml version="1.0"?>
<note date="8/1/2001">
<title>XML document</title>
<body>Hello World!</body>
</note>
```

XML Applications

Most of the vertical markets (application domains) have developed an XML vocabulary (XML DTD or XML Schema) for various applications such as data exchange, content exchange, content rendering, and many more. Vertical wireless applications (and some wireless services) also use XML vocabulary extensively. Wireless Markup Language (WML) is an XML vocabulary intended for presenting content to WAP-enabled devices. Similarly, VoiceXML is an XML vocabulary for enabling voice technologies. We will look into these specific vocabularies in detail in the coming sections.

Handheld Device Markup Language (HDML)

Handheld Device Markup Language (HDML) was invented in 1997 by Unwired Planet (which later became *phone.com* and then Openwave). It also provided the first micro-browser called UP.Browser. This was the first attempt to provide Internet access on mobile phones. HDML did not receive much support due to the lack of infrastructure and the lack of industry maturity. Later, Openwave founded WAP Forum and led the development of WML. WML is a fully XML-compliant descendent of HDML. But it took some time for the industry to adopt WAP and therefore there are still quite a few HDML-based handsets available in the market. One important thing to note is WML is not backward compatible with HDML. So, if you intend to support HDML phones you need to develop HDML applications. You can find a complete HDML developer guide at *http://developer.phone.com* and a list of phones that support HDML at *http://upmkt.openwave.com/dev_phones/phones.cfm*.

We will not discuss HDML in detail as we consider it obsolete. Instead, we will focus on WML applications in the coming sections.

Compact HTML

Compact HTML or *cHTML* is the markup language used by NTT DoCoMo's proprietary wireless protocol i-mode to render content on mobile devices. Compact HTML is a subset of HTML and so the two are very similar. Since cHTML is a subset of HTML, learning it is easy and the cHTML pages can be viewed from PC-based browsers. Because it is designed for resource-constrained devices with low bandwidth connections, cHTML does not support resource intensive features like tables and frames. When a cHTML page is accessed, the request passes through a cHTML gateway before it can be rendered on i-mode browsers. The gateway translates a request to the server from the i-mode protocol into HTTP and then translates the response from the server to the wireless device into cHTML.

Compact HTML: A Subset of HTML

The following are some of the features not supported in Compact HTML:

- JPEG image
- Table
- Image map
- Multiple character fonts and styles
- Background color and image
- Frame
- Stylesheet

However, Compact HTML supports GIF images. It is also possible to navigate a page using only the four buttons (up, down, right, and left) without mandating the need for a pointing device like a mouse.

Compact HTML assumes that well-designed pages will fit well in the screen space and so scrolling is not necessary. Compact HTML can also display pages using the "deck of cards" approach of WML.

Compact HTML browsers usually implement the direct selection of anchors (HREFs) using the number buttons as with micro-browsers for WAP. For example, if a list of anchor links were displayed and you choose to select the second link, then you can do so by clicking the button "2." HTML 4.0 includes similar support using the "accesskey" attribute.

A Compact HTML document would be just like an HTML page without tables, images, and frames. However, any Compact HTML program will use a line similar to the one below to specify the DTD for Compact HTML.

```
<!DOCTYPE HTML PUBLIC "-//W3C//DTD Compact HTML 1.0 Draft//EN"
```

You can also view the Compact HTML note from W3C at *http://www.w3.org /TR/1998/NOTE-compactHTML-19980209*.

Wireless Markup Language (WML)

Currently, *Wireless Markup Language (WML)* is the most popular markup language for mobile phones. WML was designed by the WAP Forum especially for micro-browsers on mobile handheld devices such as a mobile phone. The WAP Forum (*www.wapforum.org*) is an industry-wide consortium founded by Nokia, Openwave, Motorola, and Ericsson. The Wireless Application Protocol (WAP) uses WML as the markup for rendering content on micro-browsers. WML is an XML application, and its DTD used by WML can be found at *http://www.wapforum.org/DTD/wml_1.1.xml.*

The WAP Forum introduced WML instead of using HTML to optimize screen rendering. Also, WML introduces the concept of decks and cards for better network performance. WML is also designed to support variables, which are used to share information between decks and cards.

Since WML is an XML application, all tags are case sensitive. The root element for a WML document is <wml> and all document content is inside the <wml>...</wml> tags. All elements and attributes must be in lower case.

Decks and Cards

Since WML is designed for resource-constrained mobile devices, it introduces the concept of cards and decks for sending information to mobile devices. The information that has to be displayed on the client is contained in a card, and a set of cards that would need close interaction are bundled together as a deck. The concept of a deck was introduced, so that frequent network access can be avoided when high user interaction with minimal processing is involved, thus reducing network traffic. A deck is a unit of WML content that is transferred to a mobile device and is usually contained within one file. The deck is everything that is placed between the <wml> and </wml> tags. Deck size is limited by the size of a deck that a device can accept (typically just over 1 Kbyte).

Each card element has a title and an ID that is used for referencing the card. Each card in a WML document is placed inside <card>...</card> tags.

Listing 6-1 is an example of a WML document.

Listing 6-1. Hello World Using WML.

```
<?xml version="1.0"?>
   <!DOCTYPE wml PUBLIC "-//WAPFORUM//DTD WML 1.1//EN"
       "http://www.wapforum.org/DTD/wml_1.1.xml">
<wml>
  <card id="card1" title="My WML Page">
    <p>
      Hello World - Welcome to my WML page!
    </p>
  </card>
</wml>
```

The paragraph element, <p>, specifies a new paragraph. Each card must have at least one <p> element.

Unlike HTML, WML supports variables in a WML page. This gives greater flexibility in creating cards dynamically, changing display content, and navigation based on user input. WML also supports events, tables, images, check boxes, and forms. We will present a detailed explanation of these elements in Part 2.

WMLScript

WML supports a scripting language called *WMLScript*. WMLScript is also a part of the WAP specification. WMLScript belongs to the ECMAScript family and it is used in WML in the capacity that JavaScript is used in HTML. WMLScript can perform tasks like validating inputs and displaying dialogs without going to the server. The difference between JavaScript and WMLScript is that unlike JavaScript, WMLScript is not embedded in the WML page. WML pages only contain references to WML-Scripts as URLs. WMLScripts need to be compiled into bytecode on a server before they can run in a WAP browser. WMLScript byte code is invoked when a URL in a WML page references the script. We will also explore the details of WMLScript, including programming examples, in Part 2.

Extensible HyperText Markup Language (XHTML)

eXtensible HyperText Mark Language (XHTML) is a stricter and cleaner version of HTML. This means that XHTML is a well-formed and valid XML document. While HTML has all the power and capabilities to provide markup for most of the display and presentation elements, it is not XML-compliant. For instance, it allows single tags like <BR> without a closing tag. XHTML on the other hand is almost identical to HTML 4.01 and is XML compliant. XHTML 1.0 became a W3C official recommendation in January 2000.

XHTML was accepted by the WAP Forum as the standard for WAP because of its modularity and ability to scale to any type of browser. While WML did a good job of presenting decks and cards, extending WML to support rich content types and more document modules required a lot of rework. Instead the WAP Forum took another approach. XHTML 1.0 is taken as a standard implementation and WML 2.0 is implemented as a profile of XHTML. Future WML versions will be backward-compatible with this base profile and more document modules can be added as required.

XHTML 1.0

XHTML 1.0 was the first to bring in major changes to HTML since HTML 4.0 was released in 1997. It is the first in the series of XHTML recommendations and brings the usefulness of XML to the Web. XHTML shows W3C's intent in setting standards that provide richer Web pages and at the same time support a range of browser platforms including cell phones, appliances like televisions, cars, PDAs, communicators, and desktops.

XHTML is modular and this makes it easy to combine with markup tags for supporting vector graphics, multimedia, math, electronic commerce, and other special needs. This modular approach is a step in the right direction to support a variety of browsers, which could vary greatly in their capabilities. A browser in a cell phone can't offer the same experience as a browser available on a multimedia desktop machine.

XHTML is a reformulation of three versions of HTML document types as applications of XML 1.0. XHTML documents can be viewed using HTML 4.0 conforming browsers if simple guidelines are followed when designing the XHTML document. XHTML documents can use applications (e.g., scripts and pallets) that rely upon the XML Document Object Model. The old HTML documents can be converted to XHTML using W3C's Open Source HTML Tidy utility. This tool also cleans up markup errors, removes clutter, and beautifies the markup making it easier to read and maintain.

Variants of XHTML 1.0

XHTML 1.0 is available in three variants. The Document Type Definition (DTD) used in the XHTML document specifies the particular variant of XHTML that is being used (as each variant uses its own DTD).

Listing 6-2 is a simple example of an XHTML document:

Listing 6-2. Simple XHTML Document.

```
<?xml version="1.0" encoding="UTF-8"?>
<!DOCTYPE html
     PUBLIC "-//W3C//DTD XHTML 1.0 Strict//EN"
    "DTD/xhtml1-strict.dtd">
<html xmlns="http://www.w3.org/1999/xhtml" xml:lang="en" lang="en">
  <head>
    <title>Campus Portal</title>
  </head>
```

```
  <body>
    <p>Welcome to our <a href="http://sjsu.edu/portal/">Campus</a>.</p>
  </body>
</html>
```

In Listing 6-2, the DTD states that the HTML used for this document conforms to XHTML 1.0 Strict. Thus, any validating tool would know which variant of XHTML is used. The following lists the three XHTML 1.0 variants:

- ***XHTML 1.0 Strict:*** This is used when a strict XML document is needed. This variant captures structural information better than layout details. This can be used with W3C's *Cascading Style Sheet language (CSS)* to get the desired font, color, and layout effects.
- ***XHTML 1.0 Transitional:*** This variant is more lenient than the strict variant. These documents could be used when they need to be viewable by old HTML browsers that do not understand CSS, as well as by new browsers that understand new features of XHTML including CSS.
- ***XHTML 1.0 Frameset:*** This variant uses HTML Frames to partition pages into two or more frames.

XHTML Basic

XHTML Basic is the second recommendation in the series of XHTML specifications. The goal of XHTML Basic is to serve as a common language supported by various kinds of user agents (such as browsers).

The XHTML Basic document type includes the minimal set of modules required to be an XHTML Host Language document type. It also includes support for images, forms, basic tables, and objects. It is designed for Web clients that do not support the full set of XHTML features; for example, Web clients such as mobile phones, PDAs, pagers, and set top boxes. At the same time the document type is rich enough for content authoring.

XHTML Basic is designed to be a common base that may be extended. For example, an event module that is more generic than the traditional HTML 4 event system could be added or extended by additional modules from XHTML Modularization such as the Scripting Module.

The document type definition is implemented using XHTML modules as defined in W3C's document, "Modularization of XHTML". XHTML Events specification defines the XHTML Event Module, an XHTML module that provides XML languages with the ability to represent in syntax the semantics of the Document Object Model (DOM) Level 2 event interfaces.

What Is XHTML Modularization?

This is the third recommendation in the series of XHTML specifications. The modularization of XHTML defines sets of XHTML elements that can be combined and extended by document authors, document type architects, other XML standards specifications, and application and product designers to help content developers deliver content on a variety of platforms. Or, in other words, XHTML modularization is the decomposition of XHTML 1.0 into a collection of abstract modules that provide specific types of functionality. These abstract modules may be implemented using the XML Document Type Definition language, and using XML Schemas. There are rules for defining the abstract modules and for implementing them using XML DTDs.

Figure 6-1 shows how XHTML can be extended using modules. These modules may be combined to create an XHTML subset and extension document types that are also members of the XHTML family of document types.

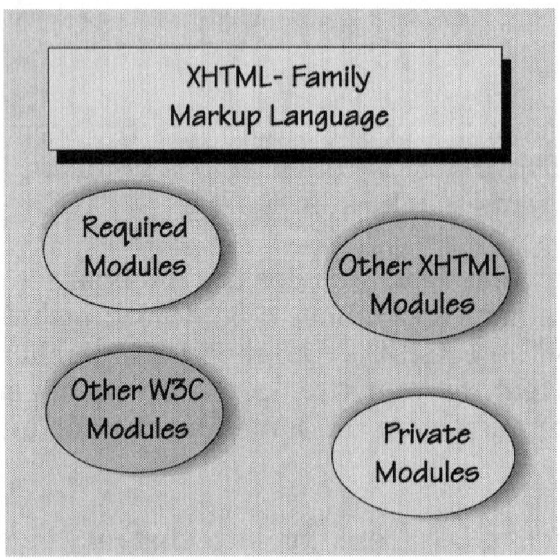

Figure 6-1. Extending XHTML Using Modules.

A document is usually an instance of one DTD, which is specified at the beginning of the document. However, XHTML modularization allows for multiple DTDs to be integrated into a new DTD (like integrating Scalable Vector Graphics into XHTML) and this new DTD can be used to create XHTML documents.

Why Modularize XHTML?

XHTML modules specify the elements that are supported by a device using standard building blocks and standard methods that specify which building blocks are used. These modules help the content generators decide the structure of the XML for their content description to serve particular devices. This relieves the content generators from generating tailored content for each and every device, which would be costly. By following standards, software processes can tailor content to a device, or the device can automatically load the software required to process a module.

Modularization also supports the extension of XHTML's layout and presentation capabilities, without breaking the XHTML standard. This provides a stable and useful framework for both content developers and publishers to generate content at a fast pace.

Although the WAP Forum is embracing XHTML 1.0 for content, WML is still the prevalent content format being used on WAP-based handheld devices. So you need not worry about WML applications becoming obsolete soon.

VoiceXML

VoiceXML is a language that is also based on XML. It is a dialog-oriented markup language that is analogous to HTML in its manner of delivery. Just as HTML is used for delivering graphical and textual content on a Web browser, VoiceXML is used to deliver voice content, in the form of dialogs, to voice browsers.

History of VoiceXML

Many companies have been working to integrate telephony and Internet technologies. As a result, these companies came up with their own proprietary voice markup languages. IBM's original voice markup was SpeechML. HP Research Labs came up with TalkML. AT&T and Lucent both had their own languages named PML (Phone Markup Language). PML from AT&T was very much different from Lucent. VoxML was another markup language developed by Motorola.

In 1999, an industry consortium led by AT&T, IBM, Lucent, and Motorola formed the VoiceXML Forum. The objective of this forum was to develop and promote a standard that enabled development of Internet-based speech applications. In March 2000, the final 1.0 specification was released after industry review and comment.

In May 2000, the Forum submitted the VoiceXML 1.0 specification to the W3C, where it was accepted by the Voice Browser working group as the basis for developing a standard language for interactive voice response applications. See Figure 6-2 for a timeline of the evolution of VoiceXML.

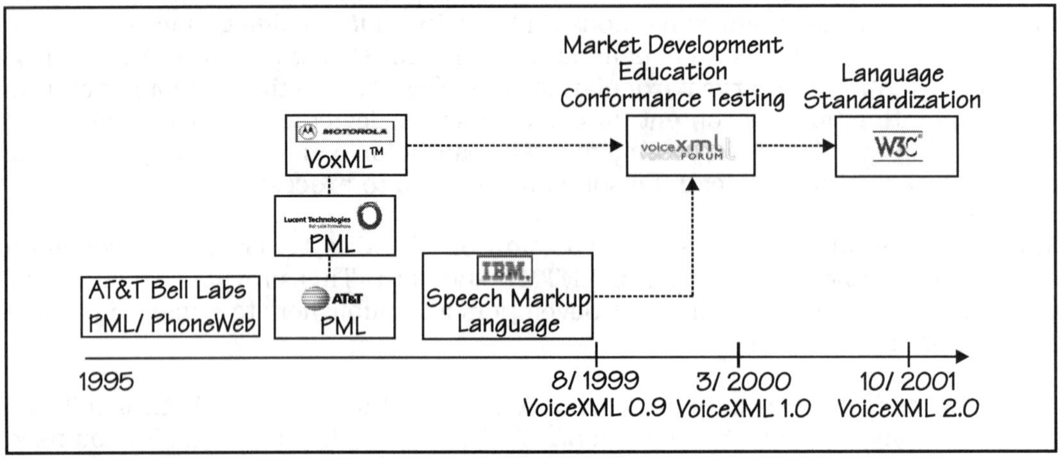

Figure 6-2. The Evolution of VoiceXML.

Elements of VoiceXML

Lets take a sneak preview of the VoiceXML markup language. The root tag for any VoiceXML document is <vxml>. VoiceXML has two principal blocks for conducting dialogs with users—one is *form*, using the <form> element, and the other is *menu*, using the <menu> element. The form and the menu elements in a VoiceXML document are linked to each other by specifying URIs.

One aspect in the development of speech applications is the ability to interpret human speech correctly. In order for the speech recognition engine to decipher speech, a vocabulary is defined. In VoiceXML applications, this vocabulary is defined by the *grammar* (<grammar>) element. There are different grammar formats that define the grammar. We will discuss these formats in Part 2. The grammar for user input in a *menu* dialog is generated dynamically (the grammar element is not defined). In a *form* dialog, the grammar is predefined.

Lets get a feel for the use of VoiceXML by looking at Listing 6-3. This example asks the user for an input and then submits it to a JavaServer Page application.

Listing 6-3. A Simple VoiceXML Example.

```
<?xml version="1.0"?>
<vxml version="1.0">
<menu>
```

```
  <prompt>
    Welcome to the library System. Please choose
  </prompt>
  <choice next="#places">
    Locations
  </choice>
  <choice next="#renew">
    Renewal
  </choice>
</menu>
<form id ="places">
  <field name="location">
  <prompt>What location are you looking for?
  </prompt>
    <grammar src="mainMenu.gram" type="application/x-jsgf"/>
  </field>
  <filled>
    <submit next="http://your.library.example/processMainMenu.jsp"/>
  </filled>
 </form>
</vxml>
```

This example has a menu and a form with the menu linked to the form based on the user input. A *field* (<field>) element is an input field. The user must provide a value for the field (in this case "location") before proceeding to the next element in the form. The *prompt* element prompts an input from the user (and stores the result in the "location" variable) that is validated by the grammar file "mainMenu.gram" defined by the *grammar* element. The "location" variable is then submitted by the submit element to the JSP located on some application server.

A sample interaction in this case would be something like this:

Computer: Welcome to library system. Please choose Locations Renewal

User: Locations

Computer: What location are you looking for?

User: Santa Clara

Computer: (submits request to http://your.library.example/processMainMenu.jsp)

We have just scratched the surface of the VoiceXML content language while walking through our simple example. We will dig deeper into the language in Part 2 where we will discuss developing and implementing a complete VoiceXML application.

CONTENT SYNCHRONIZATION TECHNIQUES

Content Synchronization allows information to be delivered from a server or desktop PC to a handheld device. This is typically done on an as-needed basis and when a connection for synchronization is available. In the next sections we will examine two content synchronization formats: Web Clipping and SyncML.

Web Clipping

Web Clipping, also known as *Palm Query Application (PQA)* is a content delivering mechanism that is a proprietary User Datagram Protocol (UDP)-based content delivery mechanism that Palm developed for their PDAs. Any Web site can be configured for browsing from a PDA using the Palm Query Application (PQA) which uses a subset of HTML 3.2 with proprietary extensions for displaying data on Palm PDAs. It can display buttons, check boxes, menus, and graphics.

In a typical scenario, the user will run a Web-Clipping application from a PDA. When the user accesses data from an HTTP content server, the proxy at Palm.net will convert the HTML pages into Web Clipping and send it to the device. These clippings will be stored to reduce the network traffic when visiting that same page again.

Web Clipping also provides security. It uses elliptic curve cryptography to authenticate between a handheld device and proxy server. On the server side, 128-bit SSL is used between proxy servers and content servers.

Developing a Web Clipping Application

There are two main components to writing successful Web Clipping applications: the Web Clipping application that you install on the device, and the backend server that returns results pages to the device.

- ***Web Clipping application:*** This is an application that you build using HTML, which is installed by end users onto their handheld organizers. A Web Clipping application is like a mini-Web site that is stored locally, so access to it is instant and free. The index page of the Web Clipping application usually contains either a form or a list of links, which are the gateways to the live data provided by your server. Static information, such as graphics, help, and "about" pages should also be placed within a Web Clipping application.
- ***Result pages (clippings):*** These pages are returned by your backend server after they receive a request from your Web Clipping application. They are

written in HTML and will often link to images and HTML documents that are stored in the Web Clipping application so that this static information doesn't need to be transmitted over the air.

Designing a successful Web Clipping application requires the careful balancing of these two components.

Writing a Web Clipping application is as simple as writing your application in a subset of HTML and then compiling it using the Palm Query Application tool. This will generate a *.pqa* file. The end users can then download this file to their Palm handheld.

Web Clipping applications are designed to be small and to give users the information they need with minimum effort and maximum speed. Likewise, results coming from your server to a handheld device should be designed to be small and to the point. Remember, this is not Web browsing, but Web Clipping. Web Clipping applications should be written to provide users with only the information they need without any extraneous items. Try to limit the size of your result pages to about 400 bytes.

HTML Features Not Available in Web Clipping

There are some HTML features that are not supported in PQAs or clippings:

- Named typefaces
- Stylesheets
- Image maps
- Frames
- Nested tables
- Scripts and applets
- Cookies

You can, however, use simple tables, grayscale color, limited font markup, lists, and images.

New Features Available in Web Clipping

The following are the four new HTML *meta* tags that were introduced for PQA applications.

- **palmcomputingplatform:** This tag indicates that the application is PalmVII-compliant. If this tag is not present, then all referenced graphics in the HTML page will be removed from the compiled version. Also, the proxy server will remove any graphics from the Web Clipping and truncate the file at 1 KB. Example:

```
<meta name="palmcomputingplatform" content="true">
```

- **historylisttext:** This tag can be used to specify which information should be kept in the history list cache of the Palm. Example:

```
<meta name="historylisttext" content="yourstring&date&time">
```

- **palmlauncherrevision:** This tag is used to determine when an upgrade is necessary. This tag is almost always included in the code. Example:

```
<meta name="palmlauncherrevision" content="X.X">
```

- **localicon:** This tag is used during compilation to instruct the compiler to include a specified icon in the compiled file. Example:

```
<meta name="localicon" content="icon.gif">
```

In addition to the above meta tags, a new tag, the <smallscreenignore> tag, has been added to filter code that need not be shown on PDAs.

Session Handling

In the HTML world, cookies are used to track sessions. However, cookies are not available on the Palm. The good news is there is a workaround available on the Palm in order to track sessions. When a proxy server authenticates each device, a unique ID is assigned to it. This device ID can be used to track sessions. This ID can be accessed by the string `%deviceid`. The following is an example that shows how this string can be accessed:

```
<form>
...
<input type="hidden" name="id" value="%deviceid">
...
</form>
```

The value for the authentication header is -1, 0, or 1. These values determine whether or not the device has been validated by the Web Clipping proxy server. If the value is 1, you can trust the device ID that follows the authentication header. If the value is 0, the device could not be validated; therefore, the device ID should be treated with caution. Finally, if the value for the authentication header is -1, validation has failed and you should not trust the value of the device ID that follows.

If you use the `%deviceid` string, you should code your server script to analyze the authentication header and ensure that it reads 1 before further processing.

The value for the device ID is a string. You can store this string in a database and use it as a key to holding information on a particular user and device pair. The format of the `%deviceid` string is undocumented and should be treated as an opaque value since it is likely to change in future devices. (Currently, it is composed of digits and periods.)

Your First PQA

Now that you know about PQAs (or Web Clipping Applications) and their capabilities, let's write a PQA and run it.

You don't need a Palm device to write a PQA. However, you need four pieces of software to build a PQA and test it. It would have been nice to have one development environment to do all of this. (Maybe you will develop one and integrate all of this!) You need a WCA (or PQA) builder that converts HTML applications into PQAs. You also need Palm Emulator software, skins, and ROM images. You can download all these from Palm developer Website.

In this example, you will create a Web Clipping application that searches the Web for movie times. This example uses basic HTML, the Query Application Builder, and the Palm OS Emulator.

First, create a new file in a text editor such as Notepad and type (or copy and paste) as shown in Listing 6-4. You can also find the file at the Web site for this book at *http://wirelessjava.engr.sjsu.edu.*

Listing 6-4. Movie Finder HTML.

```
<html>
  <head>
    <title>Movie Theatre Finder</title>
    <meta name="palmcomputingplatform" content="true">
  </head>
  <body>
    <center><img src="find3.gif" ></center>
    <h2 align=center>Movie Theater Finder</h2>
    <p align=center>Search for theatres near you:</p>
    <center>
      <form method="post"
        action="http://www.myserver.com/cgi-bin/search.pl">
       <select name="movie">
```

```
        <option>American Pie 2
        <option>Rat Race
        <option>Gladiator
      <option>Hannibal
      <option>American sweethearts
      <option selected>Rush Hour 2
      </select>
      <br><br><br>
      <input type="submit" value="Search" name="go">
    </form>
    <br>
    <a href="about.html">About Us</a>
    <a href="help.html">Help</a>
    </center>
  </body>
</html>
```

Note: The server address has been made up for the purposes of this demo. Submitting this form won't do anything useful.

1. Save the file as `index.html`.
2. Create two additional files, called `about.html` and `help.html` that contain a basic HTML document.
3. Copy the `find.gif` graphic from the book Web site. Start the Query Application Builder.
4. Choose Open Index from the File menu and select index.html and click Open. The file name is displayed in the Query Application Builder file window.

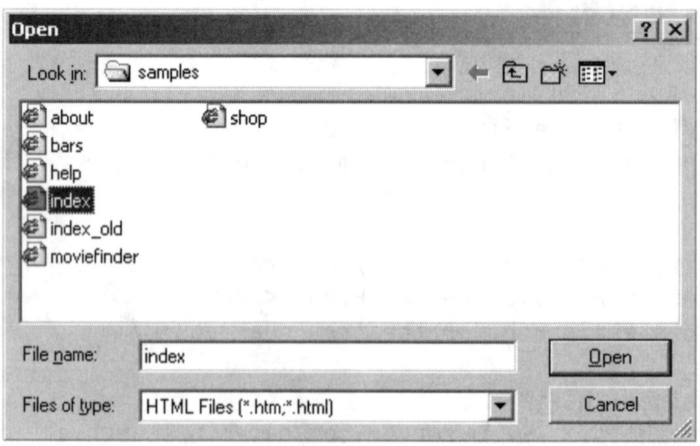

5. Choose Build PQA from the File menu.

6. Type moviefinder.pqa in the File name field and click Build.

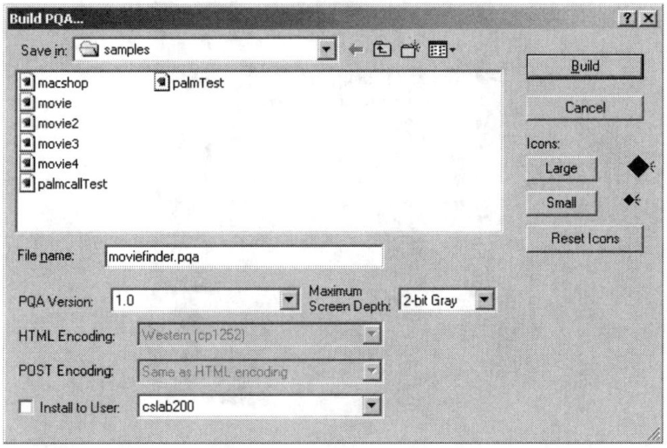

7. Install the Web Clipping application to your Palm VII handheld (or other device that supports Web Clipping applications) or onto an emulated device using the Palm OS Emulator. If you are using the emulator, you can install the PQA by just dragging the *moviefinder.pqa* file and dropping it on the emulator.

8. Tap the Applications button on your handheld. Tap the moviefinder icon to launch the application.

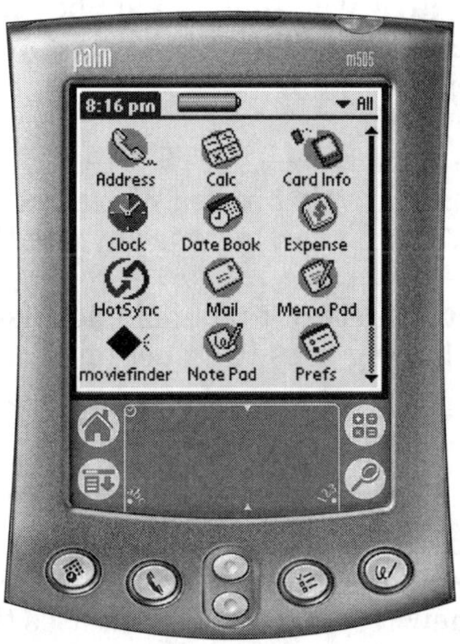

9. Your Web Clipping application is up and running!

The Difference Between WAP and Web Clipping

You might be wondering, why do we need Web Clipping when we have WAP that enables us to browse Web sites?

Palm PDAs generally are not connected. So the purpose of Web Clipping is to provide offline access of Web pages in disconnected mode. For example, one can convert driving directions into a Web Clipping application and store it on the PDA for offline browsing. This is not possible in WAP.

Other differences are:

- Using Web Clipping, you can only browse Web sites that were configured using PQA, but when using WAP, you can browse any site that delivers WML content.
- WAP works on an array of network technologies like CDMA, TDMA, Mobitex, and GPRS while Web Clipping uses Palm.net only.

SyncML

As the name suggests, *SyncML* is a protocol for data synchronization based on XML. SyncML uses XML to represent the information and HTTP as the transport protocol. SyncML does not add any new HTTP methods, but uses the existing GET and POST methods of HTTP 1.1. The specifications for SyncML were introduced in December 2000 by the SyncML consortium, founded by Ericsson, IBM, Lotus, Motorola,

Nokia, Palm, Matsushita (Panasonic), Psion, and Starfish. The SyncML consortium provides a toolkit through an open source initiative (see *syncml.org*).

It is very possible that you might carry more than one mobile device, regularly use at least two computers—one at work and one at home—to accomplish your daily tasks, and may even have a PDA. In this situation, you have a definite need to synchronize data with the other devices after making some change on one device—for example, your schedule or address book. SyncML can be used in this scenario to synchronize data on all your devices. It is also useful in applications that execute offline and later synchronize their data with other devices or systems.

SyncML Documents

SyncML can sit on top of any of several transport layers such as TCP/IP and Bluetooth. SyncML was designed to work with PCs as well as with mobile devices. As with other popular XML-based content exchange formats, a SyncML message is also is made up of a message header and a message body. The header contains the metadata and the body contains the actual payload.

In the example in Listing 6-5, the server that generated the SyncML document is identified by the `<source>` element. This source is a SyncML server permanently connected to the network and identified by a URI. The client device with which this server synchronizes its content is identified by the `<target>` element. Since the target device is not permanently connected to the network, its location identifier is proprietary. However, the `<target>` element may be set if desired to a URI like *<target>http://www.sjsu.edu/portal/syncML</target>*. The header also specifies the DTD version, the SyncML specification version, the session ID, and the message ID. In this example, this SyncML message from the server informs the client to add the data to its local database.

Listing 6-5. An Example of a SyncML Document.

```
<SyncML>
  <SyncHdr>
    <VerDTD>1.0</VerDTD>
    <VerProto>SyncML/1.0</VerProto>
    <SessionID>1</SessionID>
    <MsgID>2</MsgID>
    <Target><LocURI>IMEI:493005100592800</LocURI></Target>
    <Source><LocURI>http://www.syncml.org/sync-server</LocURI></Source>
  </SyncHdr>
  <SyncBody>
    <Add>
      <CmdID>12345</CmdID>
```

```
      <Cred>
        <Meta>
          <Type xmlns="syncml:metinf">syncml:auth-md5</Type>
          <Format xmlns="syncml:metinf">b64</Format>
        </Meta>
        <Data>OGNkNDI1ZTZjNjgwMTNiYWZkOWEyN2JjMjN1ZDM4YzENCg==</Data>
      </Cred>
      <Meta>
        <Format xmlns="syncml:metinf">chr</Format>
        <Type xmlns="syncml:metinf">text/x-vcard</Type>
      </Meta>
      <Item>
        <Source>
        <LocURI>./2</LocURI>
        </Source>
        <Data>BEGIN:VCARD
          VERSION:2.0
          FN:SJSU CA
          N:CA; SJSU
          TEL;WORK;VOICE:+1-408-555-1212
          TEL;WORK;FAX:+1-408-555-1212
          EMAIL;INTERNET:email@sjsu.edu
          END:VCARD
        </Data>
      </Item>
    </Add>
  </SyncBody>
</SyncML>
```

Table 6-1 lists some of the tags a SyncML document might contain:

Table 6-1. A Summary of SyncML Tags.

Tag	Description
Add	To add data.
Alert	To notify the target.
Atomic	To inform the originator that a set of commands should be performed as if all are just one single command. If all individual commands succeed, the entire Atomic section succeeds; if one command fails then the entire section fails.
Copy	To copy some data to the target.

Table 6-1. A Summary of SyncML Tags. (Continued)

Tag	Description
Delete	To delete some data on the target agent.
Exec	To invoke an executable on the target agent.
Get	To get some data from the target agent.
Map	To allow the originator to ask the recipient to update the identifier mapping between two data collections.
Put	To put some data on the target agent's storage.
Replace	To replace some data.
Search	To perform a query on the target agent.
Sequence	To indicate that the commands are to be executed in sequence.
Sync	To allow the originator to declare that the included commands should be treated as part of the synchronization of two data collections.

SUMMARY

With growing Internet access via various new devices, different content markup languages were designed to deliver information to these devices. We have looked at significant ones such as cHTML, HDML, WML, VoiceXML, and XHTML. We will develop some sample applications in Part 2 using WML and VoiceXML.

We also looked at different synchronization approaches for Palm devices. Finally, we looked at SyncML, an emerging standard for content exchange and synchronization. In the next chapter, we will be moving to programming platforms, which can process content, rather than just rendering content on a device.

Chapter 7

Wireless Devices and Java

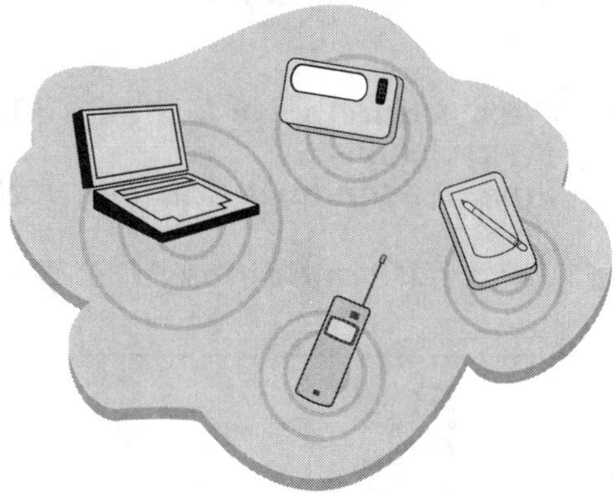

The original motive behind the invention of Java was to provide a network operating system for embedded devices. Surprisingly, Java became a very popular language for enterprise computing and the focus shifted. But recently, Java was rediscovered for use in embedded consumer devices. So now Java is available as a programming language for wireless phones, PDAs, Bluetooth-enabled devices, Smart Cards, and other embedded devices. We will explain how Java is made available on all these devices and how you can use Java to create applications for use by these devices. We will accomplish this by introducing you to J2ME, Bluetooth, and Java Card technologies. While J2ME and Bluetooth address the networking capabilities of devices, Java Card addresses the security, authentication, and authorization aspects. Even though Java Card does not have a wireless device characterization, it is used in conjunction with wireless devices.

EVOLUTION OF J2ME

The core feature of Java is its portability, well known as "Write Once Run Anywhere" (WORA). We have been taking advantage of this portability to develop applications for desktops and servers. However, the APIs designed for general programming are not appropriate for consumer devices or for pervasive computing in general.

The characteristics of pervasive devices are drastically different. To address the requirements for different environments and application domains, Java APIs along with the underlying virtual machines are delivered in three editions (see Figure 7-1). All three editions completely support the WORA principle and provide complete portability. These editions are further classified into a group of profiles and configurations, which we will discuss in the following sections. The three editions are:

- ***Java 2 Enterprise Edition (J2EE)*** for developing scalable enterprise business solutions
- ***Java 2 Standard Edition (J2SE)*** for Java application/applet development on PCs
- ***Java 2 Micro Edition (J2ME)*** for consumer and embedded devices

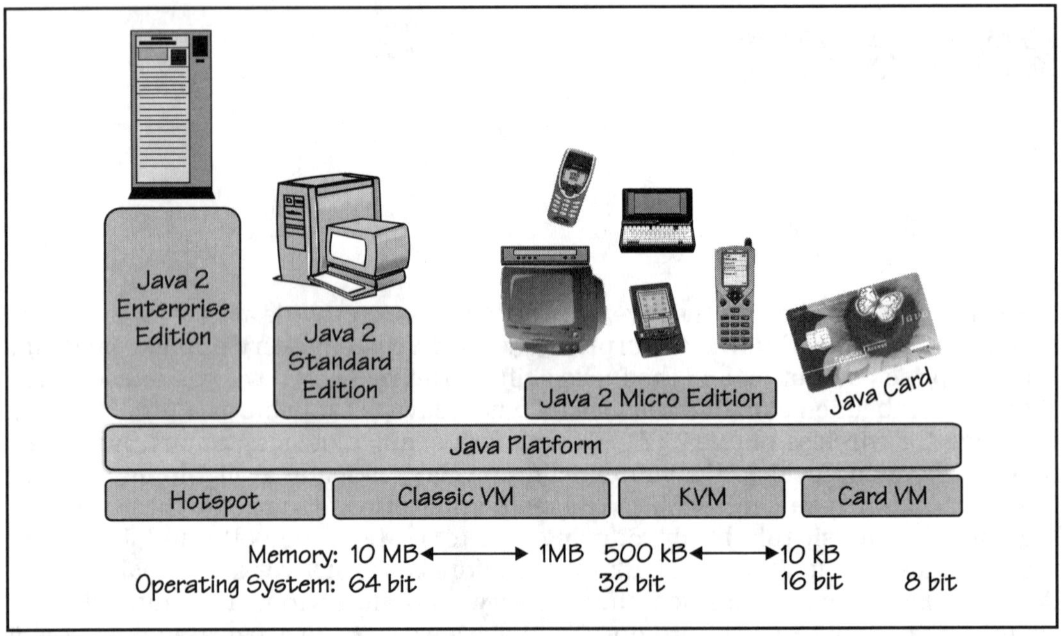

Figure 7-1. The Java Editions from Large to Small.

What Does Java Bring to Consumer Devices?

In the previous chapters we saw a progression of technologies from providing network-based message delivery (such as SMS), to bringing Internet access to mobile devices. All these technologies could be divided into three categories:

- ***Internet connectivity and content delivery technologies:*** WAP, SMS, and Palm Query Applications fall into this category. These technologies provide messaging and/or connectivity to the Internet. The most we can do is to browse static (or dynamically-generated) content that is preformatted in a particular format (such as WML or Palm Query Application).

- ***Native applications:*** These are mostly utility applications provided by the device manufacturer such as address books and games. These applications are developed in the language native to the device. They come preinstalled with the device and cannot be upgraded by the user (except by installing a firmware upgrade).
- ***Native OEM applications:*** These are the applications developed by third party vendors or independent developers. If you recall, we discussed these types of applications earlier. These applications could include micro-databases provided by various vendors for PDAs. Some of the most popular applications of this category are shareware applications developed for PDAs. While these applications achieve their purpose, they have certain limitations. The major problem with these applications is portability. For instance, a game that runs on a Palm device will not run on a mobile phone or an IPAQ device. This is due to the fact that these applications are typically written in a device-specific language.

While these technologies perform wonderful things given the limitations of the devices, Java takes these consumer devices to the next step by providing a homogenous application environment for a particular class of devices. Applications developed are then portable across devices running various proprietary operating systems.

An important thing to notice is that Java does not compete with these technologies. Java complements these technologies very well by utilizing the services they provide. For instance, Java complements WAP very well. Java itself does not provide Internet connectivity. It could, however, use the Internet connectivity provided by WAP and provide richer applications. Once you finish reading this chapter you will understand the capabilities of the Java platform on such tiny, resource-constrained devices.

J2ME specifically addresses the vast consumer space, which covers the range of extremely tiny commodities such as smart cards or a cellular phone, all the way up to the set-top box, which is an appliance almost as powerful as a personal computer. Like the other editions, J2ME maintains the qualities that Java technology has become famous for:

- Built-in consistency across products in terms of running anywhere, any time, over any device.
- Code portability.
- Leveraging of the same Java programming language.
- Safe network delivery.

- Applications written with J2ME are upwardly compatible to work with J2SE and J2EE.

J2ME: The Early Years

There were many efforts from various companies, including Sun Microsystems, to provide a Java Platform for consumer and embedded devices. PersonalJava and EmbeddedJava are the prominent platforms that have gained popularity and acceptance. Kilobyte Virtual Machine (KVM), which forms the basis for one of the current J2ME implementations, was derived from a research project called Spotless at Sun Microsystems Laboratories. The aim of the Spotless project was to implement a Java system for the Palm Connected Organizer. All of these projects were experimental, but developers adapted them to their specific needs. These projects were not really meant to address a specific class of devices and they were not developed through the community process.

J2ME evolved from the Java Community Process and reuses the effort from PersonalJava and KVM.

FYI Briefing

Java Community Process

Java Community Process is the way the Java Platform evolves.

— JCP

Starting with the early stages of Java, the development and adoption process of the platform has been a very open and a cooperative effort from various industry leaders. This process used to be informal in the early days—now all the Java Platform and API development effort evolves through a formalized process called the Java Community Process. It's an open organization of international Java developers and licensees whose charter is to develop and revise Java technology specifications, reference implementations, and technology compatibility kits.

The JCP holds the responsibility for the development of Java technology. As an open, inclusive organization with both active member and non-member public input, it primarily guides the development and approval of Java technical specifications. Anyone can join the JCP and have a part in its process, and you don't even have to join to contribute as a public participant.

The work of the Java Community under the JCP's procedures helps to ensure Java technology's standard of stability and cross-platform compatibility, enabling it to operate on hundreds of millions of devices, from desktop computers to consumer electronics to industrial robots.

Just as important, the JCP continually grows the platform's specification portfolio to meet the emerging technology needs of developers and organizations that depend on Java technology.

For a Java technology to become a standard, it has to go through the Java Community Process. Figure 7-2 illustrates the general steps involved in standardization.

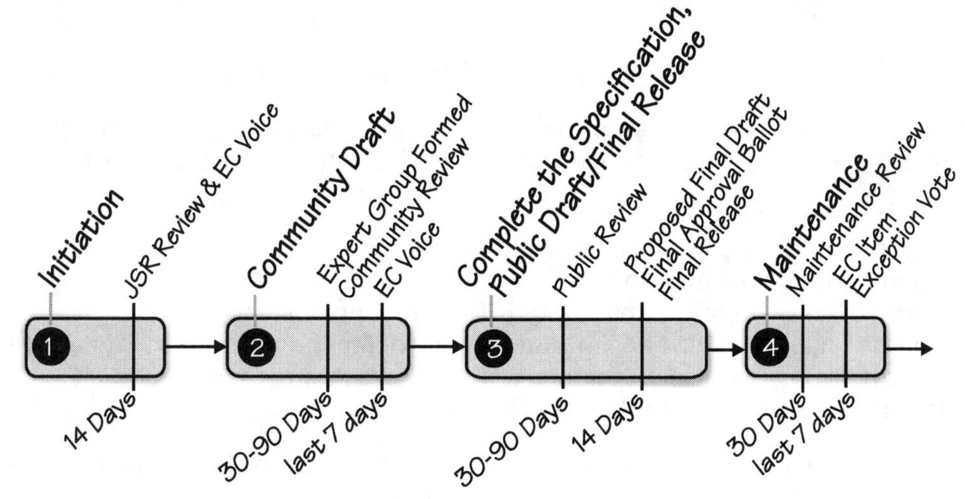

Figure 7-2. The Java Community Process JSR Life Cycle.

1. A JSR (Java Specification Request) is submitted to the JCP by one or more members to propose the development of a new specification or significant revision to an existing specification. The JSR is reviewed by the JCP Executive Committee and a vote is taken to approve or reject the JSR.

2. Upon acceptance, an Expert Group (EG) is formed to review the JSR. Ideally, the Expert Group should be large enough and diverse enough to ensure wide adoption of the resulting specification. Normally the EG is formed from companies and individuals involved in the area of interest. Experts can:

   - Actively shape the content and direction of the new and revised Java specifications.

- Review comments from members and the public and use them to improve the quality of a specification (Community Review, Public Review).
- Serve as the Specification Lead of the Expert Group and be responsible for choosing the other members of the Expert Group. The Expert Group is responsible for completing the specification, its reference implementation, and associated compatibility test suites.
- Assume responsibility for maintaining a specification after it is written (Maintenance Lead).

3. Specifications are reviewed and finalized for the public draft. A public draft is then made available to the public along with the Reference Implementation for feedback.
4. A draft is reviewed and voted for the final release of the specification.

There are two Executive Committees: one for Standard/Enterprise and another for Micro Editions. The executive committees have the following duties:

- Select JSRs for development.
- Approve draft specifications for public review.
- Approve final specifications.
- Review Technology Compatibility Kit (TCK) appeals.
- Approve maintenance revisions and maybe defer some features to a new JSR.
- Approve transfer of maintenance duties between members.
- Provide guidance to the Process Management Office.

Sun has a permanent seat on each Executive Committee. Other members are elected by the JCP members.

This approach is very similar to many open source software development efforts with some differences in the licensing and legal terms.

The good part about the JCP is that it does not accept paper-model specifications. This means that every specification needs to have a Reference Implementation (RI) and a conformance test suite that ensures that the specification is implementable and all the commercial implementations confirm to the specifications. This also means that a reference implementation would be available for developers to start designing and developing applications using the RI even before any commercial implementation is available in the market.

There are around 916 JSRs (including maintenance reviews) in various areas as of this writing. They can be accessed at *www.jcp.org*.

We will be referring to these JSRs with their JSR number when we discuss the various J2ME technologies. ❑

PersonalJava

The PersonalJava application environment was the first Micro Edition technology and has gained tremendous momentum among devices requiring full Web fidelity. The *PersonalJava Application Environment (PJAE)* addresses the software needs of networked applications running on personal consumer devices such as set-top boxes and smart phones rather than desktop computers. Currently PersonalJava is deprecated and has been replaced by CDC, Foundation Profile, and Personal Profile (see Figure 7-3). We will look into further details of these specifications and profiles in coming sections.

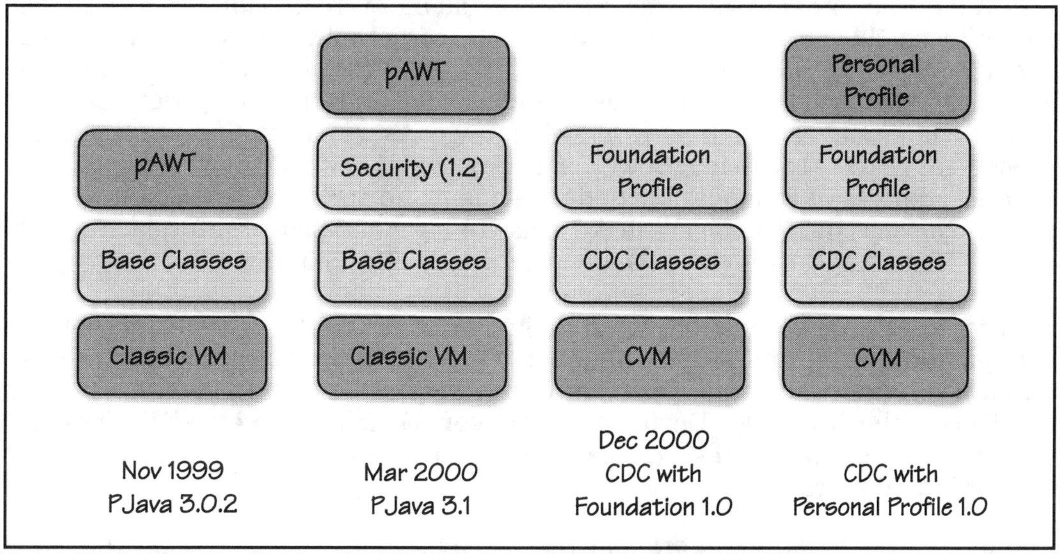

Figure 7-3. The Transition from PersonalJava Technology to J2ME Profiles and Configurations.

EmbeddedJava

EmbeddedJava technology continues into the new *configuration* and *profile* framework with little change. EmbeddedJava technology provides licensing terms and tools to vendors who want to incorporate the many benefits of Java technology in their devices, but do not require a platform-based solution. In the new framework, EmbeddedJava technology allows such vendors to utilize Java libraries outside of a J2ME profile, but only for closed, "black box" solutions—where an API is not exposed.

INTRODUCING THE J2ME PLATFORM

The J2ME platform is aimed at the consumer and embedded devices market. It specifically addresses the rapidly growing consumer space that covers commodities such as cellular telephones, pagers, PDAs, set-top boxes, and the like. All these devices are not alike, that is, they have varying functionality and capabilities. To address this issue, the J2ME platform is sliced and diced vertically and horizontally.

Configurations and Profiles

A PDA does not provide the same hardware and software capabilities as a mobile phone (even though there are devices which merge both devices) and a mobile phone does not provide the same hardware and software capabilities as a set-top box. Even though all these devices are consumer and embedded devices, their functionalities and capabilities span a wide range. As we saw in Chapter 2, there is a wide range of processing power, memory, input, and display capability for handheld devices. It is impossible to define one platform or one set of APIs that fits all these devices. It also doesn't make sense to have one platform for each type of device. There have already been many proprietary efforts in that direction.

In order to provide a standard way of defining platforms, J2ME has been split into configurations and profiles. Horizontally, a VM (Virtual Machine) provides the basis for Java program execution. Different configurations and device-specific profiles build upon the base VM. Devices are categorized into *connected* devices and *connected, limited* devices. VM capabilities are designed based on the device memory capabilities.

Configurations define the APIs and the capabilities of the VM. Even though it appears that a configuration sits on top of a VM, configurations are not separable from the VM. A configuration is defined for a class of devices with similar memory capabilities. A configuration and a VM alone do not define a platform because configurations do not define device-specific characteristics such as UI, storage, and I/O. These characteristics are defined by profiles.

Figure 7-4 shows the vertical slicing based on the functionality of the devices. In Figure 7-4, connected devices such as PCs are shown on the left. As we proceed to the right, the level of connectivity is reduced until, at the far right, we show the connected, limited devices. Each type of device has a particular profile, which is shown in the figure. Device specific characterization is based on the UI, I/O, and storage capabilities and is defined by a set of profiles. These profiles sit on top of the configurations. Profiles extend and implement functionality defined by the configurations.

Together, the VM, Configurations, and Profiles define a complete platform. There can be one or more profiles used in conjunction to form the platform.

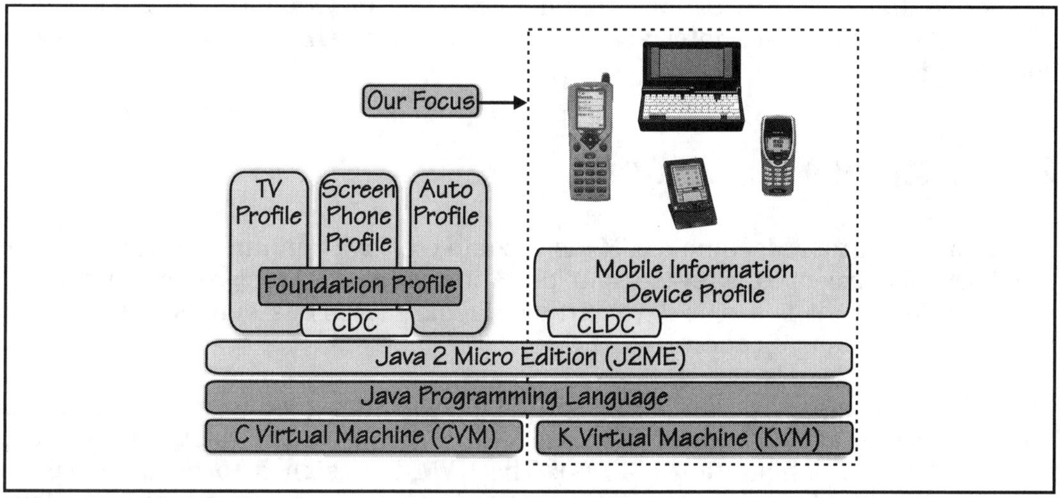

Figure 7-4. The CLDC Stack.

In addition to Profiles and Configurations there are also *Utility* packages. Utility packages are used to add additional functionality such as data structure support. This is in contrast to profiles, which provide implementation for the functionality of the device. Utility packages can be used to optionally augment the function of profiles.

Configurations

A configuration specifies VM behavior and Java programming language support for devices with a given set of capabilities. Applications confirming to a particular configuration are guaranteed to be portable across all devices that implement that configuration.

Currently there are two J2ME configurations:

- Connected Limited Device Configuration (CLDC) (JSR 30)
- Connected Device Configuration (CDC) (JSR 36)

The CLDC runs on top of the KVM (Kilobyte Virtual Machine), which is a Java virtual machine designed specifically to run on small devices. The CLDC includes several system, IO, and utility classes from J2SE and defines its own specific classes as well. On the other hand, the CDC is just a stripped-down version of J2SE (it inherits a lot from J2SE) and therefore requires a full and complete Java virtual machine. This means the KVM is not suitable for the CDC.

Neither configuration (CLDC or CDC) defines APIs for Graphical User Interfaces (GUIs). They leave user interface issues to profiles which sit on top of each configuration.

Kilobyte Virtual Machine (KVM)

The Kilobyte Virtual Machine (KVM) is a complete Java runtime environment for small devices. Except for a few specific deviations necessary for proper functioning on small devices, it is a true Java Virtual Machine (JVM) as defined by the Java Virtual Machine Specification.

The KVM is a compact, portable Java virtual machine designed from the bottom up for small, resource-constrained devices. KVM forms the basis for the Connected Limited Device Configuration (CLDC). The KVM is designed to be the smallest possible complete Java virtual machine that maintains all central aspects of the Java programming language, but runs in a resource-constrained device with only a few hundred kilobytes total memory. The J2ME specification states that the KVM is designed to be:

- Small with a static memory footprint (40 to 80 KB)
- Clean and highly portable
- Modular and customizable
- As complete and fast as possible

Connected Limited Device Configuration (CLDC)

At the implementation level, a J2ME configuration such as the CLDC defines a set of horizontal APIs for a family of products that have similar requirements for memory budget and processing power. A configuration specifies the:

- Java programming language features supported
- Java virtual machine features supported
- Java libraries and APIs supported

The high-level architecture of a typical CLDC environment is shown in Figure 7-4. At the heart of the CLDC is the KVM Java Virtual Machine, which runs on top of a host operating system.

As shown in Figure 7-4, the CLDC is intended to be used with cellular phones, pagers, and organizers, and the CDC is intended to be used with set-top boxes, Internet TVs, and car entertainment and navigation systems.

Devices supporting CLDC have the following minimum memory requirements (note that this does not include MIDP requirements—MIDP is covered in upcoming sections):

- 128 Kb of non-volatile (ROM) memory for KVM and CLDC libraries
- 32 Kb of volatile memory (RAM) for Java runtime and object memory

CLDC specifically addresses the following areas:

- Java Language and Virtual Machine features
- Core Java libraries (*java.lang.*, java.util.**)
- Input/Output
- Networking
- Security
- Internationalization

The CLDC does not specify the following areas, which are handled by profiles implemented on top of CLDC:

- Application life cycle management (application installation, launching, deletion)
- User Interface functionality
- Event Handling
- High-level application model (User interactions with application)

Networking APIs provided by J2SE (*java.net* package) are not suitable for consumer devices. In lieu of this, CLDC defines a Generic Connection Framework (defined in the javax.microedition.io package) as a common abstraction for networking APIs. We will be discussing these in detail in the programming sections of Part 2.

CVM Virtual Machine

CVM is a new Virtual Machine implementation for embedded and consumer devices that form the basis for the J2ME Connected Device Configuration (CDC) and Foundation profiles. CDC is basically targeted for devices with 2 MB or more total available memory. The current VMs for desktops do not scale well to the embedded and consumer environments. The static and dynamic memory footprint of current VMs is too large. The other factor is that current VMs are not embeddable in a chip, i.e., they are not installable in a ROM memory chip.

CVM is a complete Java2 VM with a small footprint of around 256K with JNI, reflection, and verification. Table 7-1 lists the CVM memory requirements.[1]

---

1 (Source: Sun at JavaOne, 2002).

Table 7-1. CVM Memory Requirements.

Feature	Static ROM + RAM Memory Requirements
CVM	258 KB
Natives Only	124 KB
Porting Layer Only	28 KB
Total (CVM + Natives + Porting Layer)	410 KB

Connected Device Configuration (CDC)

The Connected Device Configuration defines a minimum set of APIs for Java VM (CVM) support on consumer devices with higher memory and better network connection support.

CDC is a version of J2SE trimmed-down to meet the needs of consumer devices.

These are the Core Packages of CDC that are a subset of J2SE V1.3:

- *java.lang* for Object, Thread, etc.
- *java.util* for Utilities
- *java.net* for UDP programming and File URLs
- *java.io* for File packages
- *java.text* for Internationalization packages
- *java.security* for Access Control Packages

CDC also used the Generic Connection framework defined by CLDC to provide networking support. The Generic Connection framework is required in CDC to provide support for UDP datagrams and file support. Additional functionality is provided by the profiles on top of CDC. Table 7-2 shows the memory requirements for CDC along with CVM and also memory requirements of the Foundation Profiles based on CDC. The Foundation Profile and CDC will be discussed more in the Profiles section later in this chapter.

Table 7-2. Memory Requirements for CDC with Profiles.

Feature	Static ROM + RAM Memory Requirements
CDC Java Libraries ROMized	1011 KB
Foundation Java Libraries ROMized	1564 KB
CDC Total (CVM + Natives + Porting + Libraries)	1421 KB

Table 7-2. Memory Requirements for CDC with Profiles. (Continued)

Feature	Static ROM + RAM Memory Requirements
Java Libraries + Natives for Foundation Profile	553 KB
Java Foundation Total (CVM + Natives + Porting + Libraries)	1974 KB

Currently, CDC implementations are available only on Linux and VXWorks operating systems, but will be available on other platforms in the near future.

Details

How Is CLDC Different From CDC?

CDC is a superset of CLDC as shown in Figure 7-5. That means all the APIs present in CLDC are present in CDC. All the profiles that are on CLDC can be implemented on CDC. Applications written for CLDC are portable to CDC.

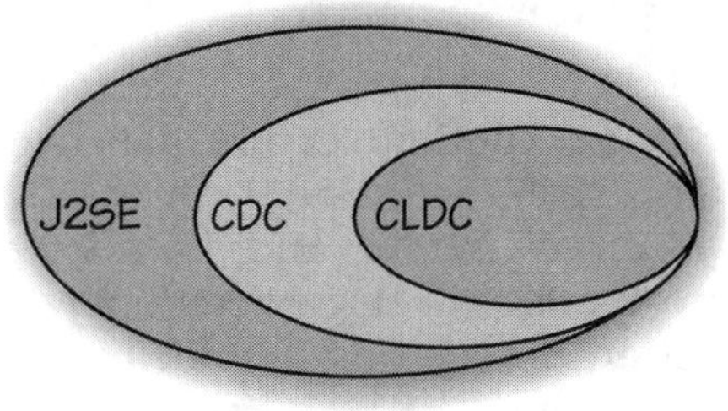

Figure 7-5. The Relationship Between J2ME Configurations.

CLDC in Detail

Configurations and Profiles reuse the APIs already defined in J2SE to maintain compatibility. CLDC is also defined by taking J2SE APIs as base APIs, eliminating the APIs that are not necessary or that could not be implemented. Some APIs were added to CLDC because J2SE APIs could not be effectively used to achieve the level of abstraction required for consumer devices. These APIs could be easily mapped to the existing APIs in J2SE, thus keeping the compatibility intact. We will describe all the APIs in CLDC that have been adopted and eliminated from J2SE as well as the new ones that have been added in the next sections.

Features Eliminated From J2SE

When creating CLDC, a number of features were eliminated from the base J2SE VM from which the CLDC was derived. These eliminations were made because the features they removed were either too expensive to implement or their presence would create security problems. Therefore, in a Java virtual machine (KVM) supporting CLDC, the following limitations exist:

- ***No floating-point support:*** Floating-point operations require hardware support to perform the computations. As the target devices might not have powerful CPUs to perform the computations, CLDC does not support floating-point numbers. Therefore, no CLDC-based application can use floating-point numbers or types such as float or double.

- ***No finalization:*** Garbage collection on the target devices is implementation-specific. Therefore, CLDC APIs do not include the *Object.finalize* method to perform final cleanup operations on object internal state before the object is garbage collected.

- ***Limited error handling:*** Runtime errors are handled in an implementation-specific manner. The CLDC defines only three error classes: *java.lang.Error*, *java.lang.OutOfMemoryError*, and *java.lang.VirtualMachineError*. Non-runtime errors are handled in a device-dependent manner that involves terminating the application or resetting the device.

- ***No Java Native Interface (JNI):*** A Java virtual machine supporting the CLDC does not implement the Java Native Interface (JNI), primarily for security reasons. Also, implementing JNI is considered expensive, given the memory constraints of CLDC target devices.

- ***No user-defined class loaders:*** A Java virtual machine supporting the CLDC must have a built-in class loader that cannot be overridden or replaced by the user. This is in place mainly to address security concerns.

- ***No support for reflection:*** Providing object introspection is expensive and consumes lots of memory. So, reflection is not supported. Because reflection is not supported, there is also no support for RMI or object serialization.

- ***No thread groups or daemon threads:*** While a Java virtual machine supporting the CLDC implements multithreading, it cannot support thread groups or daemon threads. If you want to perform thread operations for groups of threads, use collection objects to store the thread objects at the application level.

- ***No weak references:*** No application built on a Java virtual machine supporting the CLDC can require weak references.

Features Adopted From J2SE

The CLDC also adopts some features from J2SE with some modifications.

- ***Class Verification:*** In the J2SE Java virtual machine, the class verifier is responsible for rejecting invalid class files. This ensures low-level virtual machine security. A Java virtual machine supporting the CLDC must be able to reject invalid class files as well. This way the classes are ensured not to have references to memory outside the heap. However, the class verification process is expensive and time-consuming, and therefore, not ideal for small, resource-constrained devices.

  The KVM designers decided to move most of the verification work off the device and onto the desktop where the class files are compiled, or onto a server machine where applications are downloaded. Off-device class verification is referred to as *preverification*. The device is simply responsible for running a few checks on the preverified class file to ensure that it was verified and is still valid.

- ***Class Formats:*** A CLDC implementation must be able to read standard Java class files with the preverification changes mentioned above. Also, it must support compressed Java ARchive (JAR) files. Network bandwidth is very important in low-bandwidth wireless networks, and the compressed JAR format provides 30 to 50 percent greater compression over regular class files without any loss of information. Later releases of the CLDC might utilize some other class file formats, such as Java Card *CAP* files.

- ***APIs:*** The J2SE APIs require several megabytes of memory, and therefore, are not suitable for small devices with limited resources. In designing the APIs for the CLDC, the aim is to provide a minimum set of libraries useful for application development and profile definition for a variety of small devices.

  The CLDC library APIs can be divided into the following two categories:

  - ***Classes that are a subset of the J2SE APIs:*** These classes are located in the *java.lang*, *java.io*, *and java.util* packages and are derived from JDK1.3 APIs. A detailed list of the inherited classes is presented below in Table 7-3 and Table 7-4.

- ◆ ***Classes specific to the CLDC:*** These classes are located in the *javax.microedition* package and its sub packages, and are discussed in Chapters 9 and 10 in detail.

- ***Inherited Classes:*** The CLDC inherits a number of system, I/O, and utility classes from the J2SE platform.

Note: Each class that has the same name and package name as a J2SE class must be identical to, or a subset of, the corresponding J2SE class. The semantics of the classes and methods cannot be changed, and the classes cannot add any public or protected methods or fields that are not available in the corresponding J2SE class libraries. Table 7-3 lists the inherited, non-exceptional classes and interfaces from the J2SE platform:

Table 7-3. Inherited, Non-Exception Classes.

Package	Classes
java.lang	Boolean, Byte, Character, Class, Integer, Long, Math, Object, Winnable, Runtime, Short, String, StringBuffer, System, Thread, Throwable
java.io	ByteArrayInputStream, ByteArrayOutputStream, DataInput, DataOutput, DataInput-Stream, DataOutputStream, InputStream, OutputStream, InputStreamReader, Output-StreamWriter, PrintStream, Reader, Writer
java.util	Calendar, Date, Enumeration, Hashtable, Random, Stack, TimeZone, Vector

The exception and error classes listed in Table 7-4 have also been inherited from the J2SE APIs:

Table 7-4. Inherited Exception and Error Classes.

Package	Classes
java.lang	ArithmeticException, ArrayIndexOutOfBoundException, ArrayStoreException, ClassCast-Exception, ClassNotFoundException, Error, Exception, IllegalAccessException, IllegalAr-gumentException, IllegalMonitorStateException, IllegalThreadStateException, IndexOutOfBoundException, InstantiationException, InterruptedException, OutOfMemo-ryError, NegativeArraySizeException, NumberFormatException, NullPointerException, RuntimeException, SecurityException, StringIndexOutOfBoundException, VirtualMa-chineError
java.io	EOFException, IOException, InterruptedException, UnsupportedEncodingException, UTFDataFormatException
java.util	EmptyStackException, NoSuchElementException

- ***Property Support:*** In the CLDC, there is no implementation for the *java.util.Properties* class. However, accessing properties is supported by calling the *System.getProperty(String key)* method. The default set of properties shown in Table 7-5 is available.

Table 7-5. Standard Properties.

Key	Explanation	Value
microedition.platform	The host platform or device.	Default: null
microedition.encoding	The default character encoding.	Default: ISO8859_1
microedition.configurations	Current J2ME configuration and version.	Default: CLDC-1.0
microedition.profiles	Name of supported profiles.	Default: null

- ***Internationalization:*** CLDC supports translation to/from a byte sequence in a limited way. It specifies *InputStreamReader* and *OutputStreamWriter* classes identical to *Readers* and *Writers* in J2SE. *UnSupportedEncodingException* will be thrown if an encoding is not supported by implementation. Localization features are not specified by the CLDC. So, there is no built-in support for conversions of currencies, dates, and times.

- ***Class file lookup order:*** The Java VM specification and Java language specification does not specify the order in which classfiles are searched when new classes are loaded into the VM. J2SE uses the classpath environment variable to define the lookup order. CLDC does not support the classpath concept, leaving this for the implementation to handle in a platform-dependent way, with two restrictions:

  - The application programmer cannot override the system classes (CLDC and Profiles). System classes are implemented for a specific device, and changing them would lead to unpredictable behavior. Also, changing the system classes would be a compromise to the security of the device.

  - The application programmer cannot override the lookup order. Overriding the lookup order might also cause unpredictable behavior.

Features and Classes Added in CLDC

The CLDC adds features and classes to meet the needs of its target class of devices:

- ***MIDP-Specific Classes:*** In addition to the MIDP-specific classes in the *javax.microedition.rms*, *midlet*, and *lcdui* packages, the following classes, interfaces, and exception classes are available:

  - *IllegalStateException* class in the *java.lang* package
  - *Timer* and *TimerTask* classes in the *java.util* package
  - *HttpConnection* interface for HTTP protocol access over the network in the *javax.microedition.io* package

- ***Timers:*** These classes are used by MIDP applications to set and receive timer notifications. The following two classes are provided to delay or schedule activities for one-time execution or repeated execution at regular intervals:

  - *java.util.Timer*
  - *java.util.TimerTask*

- ***Generic Connection Framework:*** The target devices of CLDC such as cell phones do not provide all the network and I/O support provided by *java.io.** and *java.net.**. As it is impossible to provide all the functionality of *java.io.** and *java.net.** defined by J2SE, CLDC defines a generic connection framework to support I/O and networking in a generalized, extensible fashion. The idea is to provide a set of abstractions at the API level, rather than provide different device-specific abstractions. To expand more on this idea, it provides a generic API to make a connection. Here are some examples that show how a generic API definition is used by a specific protocol (note that the protocol implementations are provided by the profiles and not by the configuration).

  - Generic form:

    ```
    Connector.open("<protocol>:<address>;<parameters>");
    ```

  - HTTP:

    ```
    Connector.open(http://www.foo.com);
    ```

  - Socket:

    ```
    Connector.open("Socket://129.144.111.222.:9000");
    ```

  - Comm port:

    ```
    Connecter.open("comm.:0;baudrate=9600);
    ```

  - Datagram:

```
Connector.open("datagram:// 129.144.111.222);
```

- File:

```
Connector.open(file://foo.dat);
```

The interfaces, classes, and exception classes in Table 7-6 are introduced by the CLDC and are part of the *javax.microedition.io* package. These can be mapped to classes in J2SE. We will discuss these classes in more detail in the network programming for the Mobile Information Device Profile (MIDP).

Table 7-6. CLDC Specific Classes.

Package	Classes
javax.microedition.io	Connection, ConnectionNotFoundException, Connector, ContentConnector, Datagram, DatagramConnection, InputConnection, OutputConnection, StreamConnection, StreamConnectionNotifier

Profiles

A profile specifies an application platform for devices with a given set of uses.

A profile is a specification that details the Java technology APIs, built on top of and utilizing the underlying Configuration, and necessary to provide a complete runtime environment for a specific kind of device.

The J2ME platform makes it possible to define different Java platforms for various vertical markets by introducing profiles. At the implementation level, a profile is a set of vertical APIs that reside on top of a configuration to provide domain-specific capabilities. This concept is illustrated in Figure 7-4 shown previously.

Profiles provide for a helpful division of labor between platform implementers and application developers. Device manufacturers need to focus only on platform implementation, while application developers focus on application development. Also, application developers need not worry about specific device implementations. They can use emulators to develop applications, before the real devices are available.

A profile is complete, that is, the application written for a specific profile can run without the need for additional classes.

As of this writing, the Mobile Information Device Profile (MIDP) (JSR 37) is the only implemented profile. Other profiles are in the works. Table 7-7 provides a summary of profiles that are available or are in working draft status.

Table 7-7. Profiles based on CLDC.

Profile	Main Features
MIDP (JSR 37)	Defines display, persistence, networking and security. Reference Implementation and TCK (Test Conformance Kit) available.
PDA Profile (JSR 75)	MIDP + Pixel display, pointing devices, AWT-like UI.
MIDP 2.0 (JSR 118)	Defines Security, SSL, Sockets and datagrams, OTA provisioning, XML parser, sound API, push and messaging to MIDlets.
Game Profile (JSR 82)	Defines 3D modeling, rendering, animation, 2D rendering, buffering, networking, streaming media, sound, game, controllers, hardware access.
Foundation Profile (JSR 46)	Underlies TV, Screen phone, car profiles. Can include up to J2SE API.
Personal Profile (JSR 62)	Will replace PersonalJava eventually.
RMI Profile (JSR 66)	RMI capabilities for J2ME. Interoperates with J2SE RMI

We will describe MIDP in detail and mention other profiles briefly.

Mobile Information Device Profile (MIDP)

The Mobile Information Device Profile (MIDP) is a set of Java APIs, which together with the Connected Limited Device Configuration (CLDC), provides a complete J2ME application runtime environment, targeted at mobile information devices, such as mobile phones and entry-level PDAs. The MIDP specification addresses issues such as user interface, persistent storage, networking, and application life cycle.

MID Characteristics: A computing device implementing MIDP is known as a *Mobile Information Device (MID)*. Table 7-8 lists the minimum requirements that characterize a MID. These requirements are in addition to the requirements of the underlying configuration, in this case the CLDC.

Table 7-8. Mobile Information Device (MID) Requirements.

Characteristic	Minimum requirement
Display	Screen-size: 96x54 Display depth: 1-bit Pixel shape (aspect ratio): approximately 1:1
Input	One or more of the following user-input mechanisms: "one-handed keyboard," "two-handed keyboard," or touch screen.
Memory	128 kilobytes of non-volatile memory for the MIDP components. 8 kilobytes of non-volatile memory for application-created persistent data. 32 kilobytes of volatile memory for the Java runtime (e.g., the Java heap).
Networking	Two-way, wireless, possibly intermittent with limited bandwidth.

Examples of MIDs include, but are not restricted to, cellular phones, two-way pagers, and wireless-enabled Personal Digital Assistants (PDAs). Palm devices with wireless access can also be characterized as MIDs, but MIDP does not take advantage of all the features that Palm provides such as beaming, and other OS-related features. Another profile, a PDA profile, is in the works to take advantage of all the features that a PDA can provide. Until this profile is ready, one also can use MIDP for PDAs.

Architecture: MIDP is based on CLDC and provides device-specific functionality such as user interfaces, network protocols, I/O etc. The architecture of a device implementing MIDP is shown in Figure 7-6. At the lowest level, sitting just above the MID hardware, is the system software including the Operating System and libraries used by the device. On top of the system software layer sits the CLDC, which includes KVM and the supporting Java language libraries.

- The lowest level depicts the hardware. The MID hardware layer contains the processor and memory (ROM, RAM, and EEPROM).

- Native System Software contains the operating system and other native libraries used by the device to enable the interaction between input and output devices.

- On top of the Native System Software is the CLDC, which includes the KVM and the supporting Java language libraries.

- The MIDP layer contains the implementation of the profile for that particular device. This layer provides all the API implementations that we will be discussing in detail.

- MIDP Applications are known as MIDlets. MIDlets are portable across Mobile Information Devices (MIDs).

- OEM-specific classes are device-specific classes to address device-specific application requirements (such as storage for the address book).

- OEM-specific applications are Java classes and applications provided by the device manufacturer and are not portable across MIDs. Examples of OEM applications are address books, games, and calculators.

- Native applications are non-Java applications written using native libraries and are also not portable across MIDs. Examples of native applications could be Short Messaging Service (SMS) and call log applications.

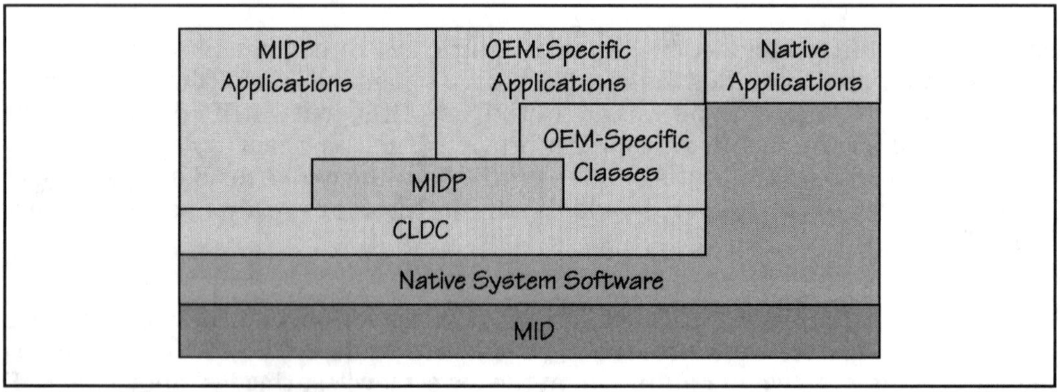

Figure 7-6. MIDP Architecture View.

MIDP APIs: MIDP defines the following set of packages in compliance with CLDC and adds classes needed to provide the functionality necessary for a MID:

- Application Life cycle APIs are provided in *javax.microedition.midlet* package.
- User Interface APIs are provided in *javax.microedition.lcdui* package.
- Persistence APIs are provided in *javax.microedition.rms* package.
- Networking APIs are provided in *javax.microedition.io.*
- Language and Util classes are inherited from *java.lang* and *java.util.*

We will briefly discuss these APIs here and in detail later in the MIDP programming chapters of Part 2.

Classes Modified in MIDP: MIDP modifies and extends CLDC. The following are the modified or extended system functions provided by MIDP:

- ***System Properties:*** MIDP defines the following additional properties that an application can access using *java.lang.System.getProperty()*.

Table 7-9. The MIDP System Properties.

System Property	Description
microedition.locale	The current locale of the device (default null).
microedition.profiles	Must contain at least MIDP-1.0

- ***Application Resource Files:*** In J2SE, application resource files are accessed by calling *getResourceAsStream(String name)* in *java.lang.Class*. In MIDP, *getResourceAsStream* is used to get resource files from the MIDlet suite's jar file.

- ***System.exit and Runtime.exit:*** Midlets cannot call *java.lang.System.exit* or *java.lang.Runtime.exit* for security reasons. If you try to call these APIs, a java.lang.SecurityException will be thrown. Instead *MIDlet.notifyDestroyed* should be called to indicate the completion of a MIDlet.

User Interface: As mentioned before, MIDP is targeted at devices such as cell phones, pagers, and PDAs. These devices differ from desktop systems in many ways, especially the way the user interacts with them. There are many factors that influence the implementation of the MIDP UI APIs. Some of the factors are:

- The form factor and UI concepts such as smaller display sizes, varying input devices such as keypads, qwerty keyboards, and pointing devices.

- Usability factors such as accommodating novice users and single-handed operation of the devices.

- Native device mappings of the MIDP UIs.

MIDP defines its own set of UI classes. AWT is not used as a basis for the following reasons:

- AWT requires a lot of garbage collection of dynamically created objects such as events. The CPU and memory constraints of a MID cannot handle this behavior.

- AWT provides window management and layout capabilities that are not appropriate for small displays.

- AWT assumes certain user interaction models such as using pointing devices, which might not be appropriate to generalize for all MIDs.

The MIDP UI is logically composed of two APIs: high-level and low-level. Applications that need to be portable across MIDs should use the high-level APIs. High-level APIs employ high-level abstractions and provide little control over the UI look and feel (such as shape, color, and font). The underlying implementation will perform necessary adaptation to the device's hardware and native UI. On the other hand, if you need precise placement and control of graphic elements, you need to use low-level APIs. Game development is one area which might use the low-level APIs. This is analogous to using AWT's layout mechanisms to provide platform independent look and feel. All the MIDP UI classes are provided in the *javax.microedition.lcdui* package. We will discuss more about UI elements and these APIs in the programming chapters.

Networking: MIDP extends the Generic Connection Framework provided by CLDC. It provides specific connection functionality by supporting a subset of HTTP that can be implemented using both IP protocols such as TCP/IP and non-IP protocols such as WAP and i-mode. MIDP provides support to access HTTP 1.1 servers and services. MIDP devices make use of a gateway for network support and other needed features such as security policies, which are not part of MIDP itself. MIDP networking capabilities are discussed more in later programming chapters. The entire network related classes are packaged in the *javax.microedition.io* package.

Persistent Storage: MIDP provides a rudimentary persistence mechanism, called *Record Management System (RMS)* to store and retrieve data. RMS is modeled after a simple record-oriented database. MIDlets can use Record Stores to store and retrieve records. Record Stores are maintained and name-scoped by the MIDlet suites. All MIDlets in a MIDlet suite (a group of related MIDlets) can share records in Record Stores belonging to that MIDlet suite. Record sharing between MIDlets in different MIDlet suites is not permitted. The platform ensures the integrity of records including during battery changes and reboots. More details are provided in the database programming sections. All the RMS-related classes are packaged in *javax.microedition.rms*.

Application packaging and deployment: MIDP defines an application model to allow the limited resources of the device to be shared by multiple MIDP applications. This model defines how to package the MIDlets to form a MIDlet suite that shares resources within the context of a single Java VM. The elements of a MIDlet suite are:

- ***Runtime execution environment:*** This is the entire environment comprised of KVM, CLDC, MIDP, and application management. All the MIDlets in the MIDlet Suite share this environment.

- ***MIDlet suite packaging:*** The JAR format is used to package Java classes for MIDlets and common classes that are shared, and for resource files used by classes (including image files), and manifest files describing the contents of the JAR.

- ***Application descriptor:*** Application management software reads a special file to identify, retrieve, install, and invoke the MIDlets. The attributes of this application descriptor file are explained later in the programming sections.

- ***Application Life cycle:*** Each MIDlet extends the *javax.microedition.MIDlet* class, thus allowing orderly starting, stopping, and cleanup of the MIDlet. This way, the life cycle of the MIDlet is controlled elegantly.

We will discuss each of these when we discuss the programming examples. Application deployment and management is discussed in deployment and dynamic provisioning in Chapter 10.

Foundation Profile

Another profile that came out of the Java Community Process (JSR46) is the Foundation Profile. The J2ME Foundation Profile is a set of APIs meant for applications running on small devices that have a reliable network connection. This is different from what MIDP provides. The Foundation Profile is intended for devices with higher capabilities than the ones that MIDP serves. The devices that the Foundation Profile targets are characterized as follows:

- 1024k minimum ROM available (application memory requirements are additional).
- 512k minimum RAM available (application memory requirements are additional).
- Connectivity to some type of network.
- No graphical user interface, unless the GUI functionality is provided by an additional profile.

MIDP specifies a graphical user interface, but the Foundation Profile does not. However, the Foundation Profile can be used in conjunction with another profile that does provide a GUI.

The Foundation Profile is based on CDC, the Connected Device Configuration (CDC). The Foundation Profile's target devices are consumer electronics and embedded devices such as smart communicators, pagers, personal digital assistants (PDAs), and interactive, digital television set-top boxes. Typically, these devices run a 32-bit microprocessor/controller and have more than 2.0 MB of total memory for the storage of the virtual machine and libraries.

Currently CDC and the Foundation Profile are available as source downloads for Linux and VxWorks operating systems. The source for CDC and the Foundation Profile has to be compiled for a specific platform. We will briefly go through the classes provided by the Foundation Profile. We will not go into programming examples for the Foundation Profile.

Foundation Profile Class Library: In general, the Foundation Profile contains the complete basic Java packages from J2SE technology, except that the GUI dependencies on *java.awt* are removed.

The basic Java packages from CDC are modified in the Foundation Profile by:

- *java.lang* rounds out the full *java.lang.** J2SE package support for the Java language (Compiler, UnknownError).
- *java.util* adds full zip support and other J2SE utilities (such as Timer).
- *java.net* adds TCP/IP Socket and HTTP connections.
- *java.io* rounds out full *java.io.** J2SE package support for Java language input/output (Readers and Writers).
- *java.text* rounds out full *java.text.** J2SE package support for I18n (Annotation, Collator, Iterator).
- *java.security* adds Code Signing and Certificates.

How Is the Foundation Profile Different From J2SE? The Foundation Profile set of libraries is based on the J2SE v1.3 software. The following classes in the Foundation Profile differ from the J2SE classes with the same name as described below:

- **java.lang.ClassLoader:** The following line was removed:

  ```
  import java.io.StringWriter;
  ```

  There is no change in behavior, because Class Loader does not use the imported class.

- **java.lang.SecurityManager:** The *java.awt* dependencies were removed, as the Foundation Profile does not include the *java.awt* package. Hard-coded references to *java.awt* were changed to use reflection to determine if the *java.awt* package is present. (It could be present in another profile.)

  If *java.awt* is present, there is no difference in Security Manager behavior compared to J2SE.

  If *java.awt* is not present:

  - *checkTopLevelWindow(Object window)* returns false.
  - *checkSystemClipboardAccess()* throws SecurityException("AWT not available").
  - *checkAwtEventQueueAccess()* throws SecurityException("AWT not available").

- **java.text.resources.LocaleElements:** The *java.awt* dependency has been removed, as the Foundation Profile does not include the *java.awt* package. Specifically, the following line is commented out:

```
import java.awt.ComponentOrientation;
```

  The following line is also commented out:

```
{ "Orientation", ComponentOrientation.LEFT_TO_RIGHT },
```

  If *java.awt* is present through an additional profile, LocaleElements will not have access to the ComponentOrientation resource.

- **java.text.resources.LocaleElements_en_US:** The *java.awt* dependency has been removed, as the Foundation Profile does not include the *java.awt* package. Specifically, the following line is commented out:

```
import java.awt.ComponentOrientation;
```

  There is no change in behavior, as **LocaleElements_en_US** does not use the imported class.

- **sun.net.www.ParseUtil:** The following line was removed:

```
import java.util.BitSet;
```

  There is no change in behavior, because **ParseUtil** does not use the imported class.

RMI Profile

The computing and memory capabilities of personal devices are increasing everyday. So, in the near future we might be able to have a complete micro Java RMI protocol stack (J2ME) that interoperates with the J2SE RMI protocol stack. The RMI profile is targeted in that direction. The J2SE RMI implementation has a very large footprint. A lightweight RMI profile is therefore being defined to fit within the limitations of consumer devices.

The J2ME RMI Profile provides Java platform to Java platform remote method invocation for Java devices and interoperates with J2SE RMI.

The J2ME RMI Profile is targeted for devices characterized with the following capabilities:

- 2.5 MB minimum ROM available
- 1 MB minimum RAM available
- TCP/IP connectivity to the network

Given the above requirements, the only configuration that could support these is CDC. The RMI profile is also based on the Foundation Profile discussed earlier.

Being able to perform Remote Method Invocations from consumer devices opens doors to more promising technologies. For instance, devices capable of running JINI technology would be able to perform RMI invocations on servers running J2SE and J2EE.

Personal Profile

We discussed PersonalJava in the earlier sections of this chapter. PersonalJava was the first J2ME technology that was adopted by many companies. The Personal Profile is intended to provide the next generation of the PersonalJava environment, and also streamline PersonalJava to the J2ME configurations and profiles. It has the explicit requirement of providing compatibility with applications developed for versions 1.1.x and 1.2.x of the PersonalJava Application Environment Specification to provide a smooth transition for those utilizing existing versions of PersonalJava technology.

The J2ME Personal Profile provides the J2ME environment for those devices with a need for a high degree of Internet connectivity and Web fidelity.

The J2ME Personal Profile provides a profile of the Java 2 Platform Micro Edition for devices characterized as follows:

- 2.5 MB minimum ROM available
- 1 MB minimum RAM available
- Robust connectivity to some type of network
- Graphical user interface, providing a high degree of Web fidelity and the capability of running Java Web applets

Given these requirements it is natural that this profile is based on J2ME CDC and the J2ME Foundation profile.

MIDP 2.0

Every technology goes through a metamorphosis from specifications to implementation to customer adaptation. This last phase again leads to revising the entire specification to incorporate the needs of customers, other technology developments, as well as the corrections required from the lessons learned from the earlier deployments.

MIDP also is going through this revision after the first few deployments of the technology. MIDP 2.0, formerly known as Mobile Information Device Profile—Next Generation (MIDP-NG), primarily focuses on the following goals:

- Backward compatibility with MIDP 1.0
- Continued focus on small, high-volume wireless phones
- Maintain tight footprint objectives to limit growth in the core APIs
- Information learned from MIDP 1.0 deployments to fine tune MIDP 1.0 APIs
- Focus on core functions needed by all devices and applications
- Focus on enabling mCommerce service-based applications

The MIDP1.0 specification provided just the bare bones to develop MIDLets. The following are the areas that MIDP 2.0 specification intends to define:

- ***Domain Security Model:*** MIDP 1.0 did not provide any security model, except for the basic VM sandbox security. It is essential to validate MIDlets and ensure that they are from a reliable source before they are used. MIDP 2.0 intends to define a domain security model, including signing of applications and verification of certificates.

- ***Support for HTTPS:*** MIDP 1.0 only supports HTTP. MIDP 2.0 addresses this shortcoming by adding support for HTTPS. This allows secure HTTP transactions.

- ***Network connectivity via sockets and datagrams:*** MIDP 1.0 provided only an HTTP implementation for the Generic Connection Framework. This is a serious limitation as a MID could therefore only communicate with Web servers. Support for sockets and datagrams would enable MIDs to perform more low-level and sophisticated network operations.

- ***Formal inclusion of OTA Provisioning:*** MIDP 1.0 did not define the way a MIDlet could be provisioned to a device. This was left to vendor implementation. But Over The Air (OTA) Provisioning was the recommended way of provisioning. MIDP 2.0 attempts to include this as a standard way for MIDlet provisioning, thus eliminating manual download of the MIDlets.

- ***Push architecture:*** MIDP1.0 did not define any push architecture. MIDP 2.0 attempts to define a push architecture for external events and messages routed to appropriate MIDlets.

- ***User Interface:*** Games have become popular applications on MIDs. To address this, MIDP 2.0 attempts to define extensions to the low-level lcdui packages to allow greater game functionality and layout control for larger screen sizes.

- ***XML Parser:*** XML is the de-facto standard for exchanging data to and from any source. As MIDP 1.0 did not define any standard XML parsing APIs, numerous parsers such as NanoXML, kXML, and TinyXML were developed to work with MIDP. (These parsers are discussed in detail in Part 3). MIDP-NG attempts to define a small, efficient XML parser to enable platform-independent data exchange.

- ***Base sound API:*** MIDP 1.0 did not define any sound APIs. As most of the MIDs are cell phones, it makes sense to provide a sound interface that could enable voice output. This way, cell phone users need not keep staring at the screen, as the output could be voice. MIDP 2.0 attempts to provide a basic sound API that could lead to more sophisticated MIDlets.

With all these sophisticated features, the MIDP 2.0 specifications seems to pave the way for killer applications that take advantage of 3G networks and sophisticated MIDs.

JAVA CARD

Smart Cards have been in the market for many years. We all carry one or more smart cards with us in the form of credit cards or access cards. These cards are not multi-purpose cards. We are using one card for each type of application. So when will we be able to use one card that replaces the bulk of cards from our wallets? That day may be approaching, thanks to Java Card. You must be wondering: What does Java Card have to do with mobile and wireless programming? Well, it is "mobile" since you can carry it with you, has a wireless characterization to it, and can be used as a Subscriber Identity Module (SIM) in a GSM mobile terminal. So, let us look at the basics of Smart Cards and Java Card. Once you are familiar with the basics, we will get to more programming details in Chapter 12.

Smart Card

Smart Cards, as the name suggests, are "smart" as they have a tiny computer embedded in the card. A Smart Card is a credit card-sized plastic card with embedded integrated circuitry. It has data storage, processing, and transmission capabilities when used in conjunction with Card Acceptance Device (CAD), sometimes called a Card Terminal or simply a card reader. ISO 7816 defines the standards for the smart card industry and this document defines the physical appearance and properties of a smart card as well.

Advantages of Smart Card

You might be wondering, why do we need these Smart Cards when we already have the cheaper magnetic cards. Here are some advantages of Smart Cards:

- ***Secure:*** Smart Cards are more difficult to tamper with than magnetic-striped cards. The data in the memory could be encrypted using cryptographic functions.
- ***Portable:*** Smart Cards could be carried in the same way as a credit card. There are special kinds of Smart Cards that could be inserted in a phone or could be worn as a ring. Also smart cards could be embedded in portable electronic devices such as phones, Personal Digital Assistants (PDAs), and PCs.
- ***Ease of use:*** A transaction could be performed simply by inserting a card into the reader.
- ***Programmable:*** Smart Cards are programmable. They can perform multiple functions in a wide range of industries and can be reused.
- ***More capabilities:*** Smart Cards currently can store a hundred times more information than a magnetic stripe card. Smart Cards are constantly evolving with more features (32-bit microcontrollers and RISC set cards are already available).

Types of Smart Cards

There are basically two types of Smart Cards: memory cards and microprocessor cards. They can also be categorized into contact cards and contactless cards based on the card access mechanism. Contact cards communicate with the contact points on the card reader when the card is inserted in the card reader. As this requires insertion of the card in the reader, a transaction might take more time. Contactless cards on the other hand, communicate through an antenna wound into the card. Contactless cards can be used in applications in which card insertion/removal may be impractical or in which speed is important. The drawbacks with these cards are the limited distance range of the communication and the security aspects. It is possible for a transaction to happen or be intercepted without the cardholder's knowledge.

Some manufacturers are making cards that function in both contact and contactless modes.

Memory cards are just used for storage purposes. They are not really "Smart" as there is no microprocessor on them, but still called smart cards as they can have non-programmable logic on them. These are low cost, non-reusable cards and can be used for simple applications using preprogrammed logic such as prepaid cards.

On the other hand, microprocessor cards contain a processor. This processor controls data handling and memory access. Some cards have support for cryptography by providing a coprocessor. These are the advantages of microprocessor cards:

- ***Increased security features:*** such as encryption/decryption, signing/verification and authentication.
- ***Multifunctional capabilities:*** multiple applications can be stored on the card based on the memory capabilities.

Smart Card System Components

Smart Cards cannot function independently. They have to work in conjunction with a host system to perform some action. We will discuss the typical components in detail:

- The ***Smart Card*** first contains the microprocessor, memory, and an Operating System. The Operating System provides system utilities such as memory management, file management, and I/O operations with the host and other security features. Then user applications can be loaded on the Smart Card. User applications could be written in assembly language (native to the microprocessor on the card) or high-level languages that the microprocessor could interpret.
- The ***Card Reader*** is used to communicate with the host system.
- The ***Host System*** contains typically two parts:
    - ***System software*** takes care of card detection and communication between the user application and the card. The system software may also provide infrastructure deployment support such as card issue management, key management, and security.
    - ***Application Software*** communicates with the card and performs transactions and other necessary actions appropriate to the application. User applications are typically written in high-level languages such as C++ and Java as these typically run on a computer with sufficient resources.

As you can see, not all of the above features are provided by one company. There are various parties involved in creating a complete smart card deployment including card manufacturers, reader vendors, host system developers, application developers, and card management infrastructure developers. Naturally there should be an understanding between all these parties, and therefore there are quite a few standards bodies and consortium of companies that have come together to define standards to govern the smart card market. Let us briefly look at the significant ones and what they have proposed.

Smart Card Standards

ISO 7816: *International Organization for Standardization (ISO)* published standards for physical characteristics, contacts, and interchange of data. These are described in 10 parts of this standard called ISO 7816 "Identification cards—Integrated circuit cards with contacts." Please visit *www.iso.org* for more details. Unfortunately, these documents are not available free of cost. Shown below is a breakdown of the various parts of the ISO 7816 standard. We will refer to these parts in later sections of our discussion:

- Part 1: Physical characteristics.
- Part 2: Dimensions and location of the contacts.
- Part 3: Electronic signals and transmission protocols.
- Part 4: Inter-industry commands for interchange.
- Part 5: Numbering system and registration procedure for application identifiers.
- Part 6: Inter-industry data elements.
- Part 7: Inter-industry commands for Structured Card Query Language (SCQL).
- Part 8: Security related inter-industry commands.
- Part 9: Additional inter-industry commands and security attributes.
- Part 10: Electronic signals and answer to reset for synchronous cards.

PC/SC: PC/SC is a consortium of companies including Microsoft, Schlumberger, and other major players in the industry. PC/SC has been working on specifications for Smart Card communication with Personal Computers. As of now only Windows implementations are available. There have been unofficial efforts to implement these on Linux. For more information about PC/SC visit *www.pcscworkgroup.com*.

OpenCard Framework: While PC/SC is Windows specific (as this is the only implementation available), there is another industry consortium called OpenCard Consortium with companies such as IBM, Sun, Schlumberger, and other major players, working on a Java-based implementation. OpenCard Framework thus is a platform independent set of APIs for interacting with card readers and applications in the card. PC/SC provides a resource manager that applications use to interact with the card. The

OpenCard Framework reuses this resource manager on the Windows environment. Please visit *www.opencard.org* for more details.

Java Card: While all the above-mentioned standards are for host applications, the Java Card specification deals with a VM on the card itself. This way we can run programs written in Java on the card, instead of writing the program in card specific assembly language. Java Card Forum is an industry consortium founded by Schlumberger. Here are some advantages of Java Card:

- ***A standard Object-Oriented Programming Language:*** Anyone who knows the Java programming language can write a Smart Card program commonly known as applets or *cardlets* (We will use *cardlets* instead of applets in order to avoid confusion with J2SE applets that run in a Web browser). As Java is an object-oriented programming language, programmers have advantages such as code reuse, superior structure, and design patterns.

- ***Security:*** The integrity and security of the Java environment is widely recognized. The following features provide program and data integrity and protection from malicious programs:

    - Java compilers provide extensive, stringent error checking when the program is compiled. For example, all the references to methods and variables are checked to make sure that the objects are of the same type. Access to uninitialized variables also is restricted.

    - Basic types and operations are well defined. All primitive types have a specific size and all operations are performed in a designated order.

    - Access modifiers control all access to methods and instance variables in a Java class file. A method can be declared to be public (no limitations), protected (accessible by methods in subclasses or in the same package), or private (no access by other classes). The default access modifier is package, which allows the method to be accessed by all classes in the package.

    - Malicious programmers cannot access the memory directly as there are no pointers accessible to programmers.

    - Local variables are unavailable outside the method as they are on the Java stack rather than the heap. This prevents malicious programs from playing with the heap.

    - The only capability that might be missing from the card VM is a garbage collector. So programmers have to be careful when allocating persistent objects on the heap (EEPROM).

- ***Multiple Cardlets on a card:*** Security features of the Java language make it possible for multiple cardlets to co-exist safely on the card. The number of cardlets is limited by the amount of EEPROM available on the card. These cardlets could be related or unrelated. For example, cardlets provided for performing bank transactions and cardlets used for storing the user's coffee preferences can be stored on the same card.

- ***Platform Independence:*** Cardlets are portable across different chip architectures and are easy to develop and maintain.

- ***Dynamic Updates:*** Cardlets can be developed and deployed incrementally, adding features as required. They can be added or replaced at any point in time.

Now you know what Java brings to the world of Smart Cards. We will go into more details of the language features of Java on these cards at the end of this section.

Smart Card Architecture and Hardware

We will now briefly look at the components of a Smart Card.[2] Here we will use a Schlumberger Cyberflex™ card as a specific example (see Figure 7-7). This card is also a Java Card.

Figure 7-7. The Schlumberger Cyberflex Java Card.

---

2 For more details, read **Java Card Technology for Smart Cards** by Zhiqun Chen (Addison-Wesley, 2001).

A Smart Card's microprocessor has two main components: the CPU and memory. The CPU receives its power and an external clock signal from the contact points that communicate with the card reader. The Cyberflex access card provides an additional cryptographic coprocessor to support public key encryption and signature validation.

There are three types of memory: ROM, EEPROM, and RAM.

- ***Read Only Memory (ROM)*** is persistent immutable memory that is written once when the chip is fabricated by a process called *masking*. After the memory is masked it cannot be written again. So this memory is used to store fixed programs such as operating system programs and other permanent data. No power is required to hold data in the ROM. This is the least expensive of all the memory types as it takes less space.

- **Electrical Erasable Programmable Read-only Memory *(EEPROM)*** is persistent mutable memory, which can preserve data content when power to the memory is turned off. It is also rewritable. This type of memory can be erased and reprogrammed repeatedly by applying electrical voltage. EEPROM retains the data for 10 years after power is removed from the microprocessor. Although EEPROM does not have an unlimited lifetime, its lifetime is so long that it is unlikely to ever become exhausted in typical Smart Card use. EEPROM is 4 times more expensive than ROM as it takes four times as much space as a ROM cell. Cyberflex access cards provide 16K EEPROM out of which 13.5K is available for the application and key storage.

- **Random Access Memory** (RAM) is non-persistent mutable memory that can be used as a temporary working space for storing and modifying data. Data is lost from this memory when power is removed.

The Communication Interface: Smart Cards communicate with a computer or a terminal when inserted into a reader. Schlumberger provides *Reflex72*, which is a smart card reader as well as writer. A reader can be connected to the serial, parallel, or USB port of a computer. Reflex72 is available with USB or serial connectors, so it draws power from the connected system (see Figure 7-8).

Smart Card applications on the card never initiate the communication. Instead, the host application initiates the communication by sending a command to the card application which processes the command and returns a response to the host. The communication between card and reader is half-duplex.

Card-to-Reader Command and Response Formats: The Smart Card and reader exchange data through input commands and output response messages. Message formats comply with ISO 7816-4 standards, which enable the cards and readers of various

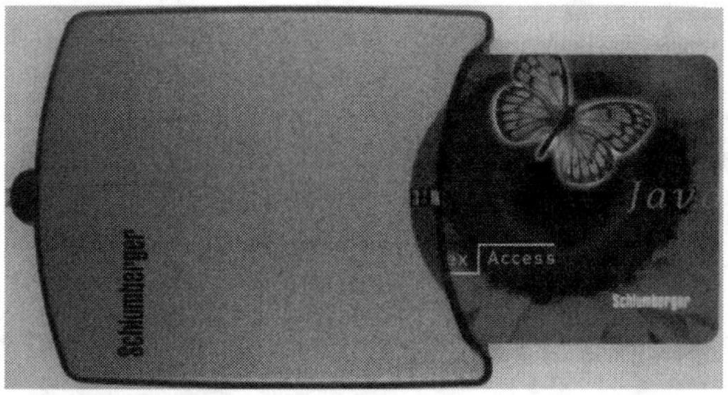

Figure 7-8. The Reflex 72 Reader with Smart Card Positioned for Insertion.

manufacturers to be interoperable (as long as they comply with this standard). Messages belong to any of these three types:

- ***Input:*** The reader sends a command and data to the card.
- ***Output:*** The reader retrieves data from the card.
- ***No data:*** The reader sends only a command to the card.

The command data is structured in a format called *Application Protocol Data Unit (APDU)*. Table 7-10 shows the components of the command-response structures.

Table 7-10. Command-Response Message Components.

Message Component	Description
CLA	Command Class.
INS	Command instruction identifier.
P1, P2	Input parameters related to the command.
P3	Length parameter that specifies the number of data bytes to transfer to or from the card as part of the command.
Ack	Acknowledgment from the card that it received the command and is ready for data transmission.
Input Data	Data the reader sends to the card.
Output Data	Data the card returns to the reader.
SW1	Card's status response related to the command (1byte).
SW2	An optional response from the card to provide additional status information.

Table 7-11 shows the input command structure:

Table 7-11. Input Command-Response Message.

Message Component								
Reader	CLA	INS	P1	P2	P3		Data	
Card						ACK		SW1, SW2
P3: Length	1 B	1 B	1 B	1 B	1 B	1 B	lgth B	2 B

P3 specifies *lgth*, or the length of card input data. If the P3 value is 00h, no data is sent to the card with the command. The reader can retrieve data from the card by using the same structure for an output command-response message as shown in Table 7-12:

Table 7-12. Input Command-Response Message.

Message Component								
Reader	CLA	INS	P1	P2	P3			
Card						ACK	Data	SW1, SW2
P3: Length	1 B	1 B	1 B	1 B	1 B	1 B	*lgth* B	2 B

Here P3 specifies *lgth* as the length of output data. The status words (SW1, SW2) indicate whether the card executed the command correctly and sent back the desired data. We will look at these messages and Java Card programming in more detail in Part 2.

APDU is application level protocol, whereas the APDUs are transmitted by the transport protocol defined in ISO 7816-3. The data structures exchanged by a host and a card are called *Transmission Protocol Data Units (TPDUs)*. There are two protocols, T=0 and T=1. The T=0 protocol is byte-oriented, which means that the smallest unit processed and transmitted by the protocol is a single byte. The T=1 protocol, on the other hand, is block-oriented. Cyberflex Access uses the T=0 protocol.

Card OS and File System: Every Smart Card vendor has its own proprietary operating system that is embedded on the card. Schlumberger has an OS called SCOS, which is ISO 7816-4 compliant. SCOS provides support for a hierarchical file system structure as defined by ISO 7816-4. This card's operating system files are divided into three types:

- ***The Master File (MF):*** The Master File (MF) is a special dedicated file that is the root of the card's file system. The MF contains dedicated files (other directories) and elementary files (data files). The MF's reserved file identifier is 3f00. The card selects the MF by default when the card's microprocessor is powered up. The structure of a MF resembles the structure of any DF, that is, a header but no data body.

- ***Dedicated Files (DFs):*** A dedicated file (DF) is a subdirectory in the file hierarchy. A DF can contain elementary files and other DFs. The DF can also hold the key files required for authentication such as RSA keys. When you create a DF, base its size on the number and size of the files it will hold. If sufficient space is not allocated for all the files needed, then the DF has to be deleted and recreated again. The size of an existing DF cannot be altered.

- ***Elementary Files (EFs):*** Elementary files can contain program data such as names, dates, and serial numbers. Cards use four types of EFs: transparent, fixed-length linear, variable-length linear, and cyclic (see Figure 7-9). The following defines the file types:

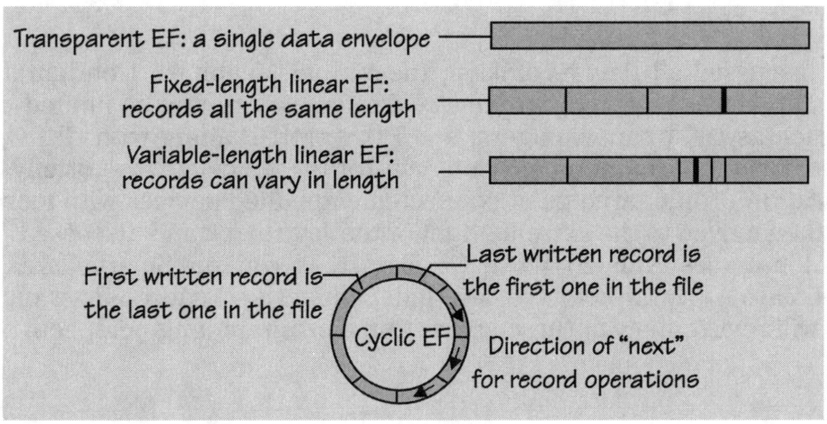

Figure 7-9. The Four Types of Elementary Files.

- ***Transparent Elementary Files*** contain a sequence of bytes. They are called transparent because they have no interior structure. These files are useful for storing objects like keys or images.

- ***Linear Fixed Elementary Files*** contain subdivisions, or records, which can be fixed- or variable-length. Each record has an identification number assigned sequentially as they are created. This record number can be used to read and write to a specific record in a linear file. Record files have a record pointer that points to the currently selected record.

  In a fixed-length linear elementary file, all the records contain equal numbers of bytes. Each file can contain a maximum of 255 records and each record can hold a maximum of 255 bytes. The precise number of bytes in a record can be specified when the file is created. Records are created sequentially.

- ***Linear Variable Elementary Files*** are the same as the fixed-length ones except that the number of bytes from one record can vary from the other record. This will conserve some space but seek times are comparatively high.

- ***Cyclic Elementary File*** is a ring of fixed length records with a maximum of 255 records per file. Each new write operation modifies the next physical record in the ring. If the ring is full, the new record overwrites the oldest record in the file. The record pointer is then set to the most recently written record, which becomes the new first record in the file. These files are useful to store transaction history.

SUMMARY

We have discussed all flavors of Java, the Virtual Machines, Configurations, and Profiles. The Mobile Information Device Profile is targeted at limited connected devices such as cell phones, pagers, and PDAs. MIDP, along with CLDC, provides a complete Java programming environment for these devices. We briefly discussed the Foundation Profile targeted at connected embedded devices with more capabilities. We then moved to the extreme right of the Java stack and discussed Java Card. Java Card provides APIs to write Cardlets that run on Smart Cards with the embedded Java VM. Cardlets are card-independent and can co-exist on the same card. We will revisit many of these areas from a programming perspective in Part 2.

Part 2
Core Wireless Device Programming

An Introduction to Part 2

In Part 2 we start our exploration of developing applications for handheld devices. We'll explore delivering these applications by writing content using markup languages—specifically *Wireless Mark-up Language (WML)* and *VoiceXML*. This content is executed on the device (for WML) or the content is interpreted by a server-side "browser," converted to sound, and delivered to the device (for VoiceXML).

We'll also look at writing Java 2 Micro Edition (J2ME) applications for two segments of J2ME: MIDP and Java Card. As you recall from Part 1, J2ME defines a Mobile Information Device Profile (MIDP) for devices such as two-way pagers, cell phones, and PDAs. MIDP defines a number of APIs for presenting GUI elements to a user, for making network calls, and for persistently storing data on the device.

To illustrate the differences between the different programming environments, we will write the same application in each of the technologies: WML, J2ME, and VoiceXML. For our programming examples in the remainder of the book, we'll follow a theme of developing applications for a university environment. For Part 2, we'll develop a university registration application that runs on a handheld device. Due to the limited nature of Java Card, we'll have to deviate from the university registration example. So, we'll present a simple application that stores a Personal Identification Number (PIN) on a Java Card and gives you the utilities to retrieve the PIN.

Here's a brief description of the chapters in Part 2:

- ***Chapter 8*** describes WML and the development environment that facilitates the development of WML content for a handheld device. We also give an overview of WMLScript which allows you to do simple programming tasks to augment your WML content.
- ***Chapter 9*** is an introduction to MIDP GUI programming. The MIDP UI APIs are explored within two categories: high-level APIs for form-based UI programming and lower-level APIs for more graphic-intensive applications.
- ***Chapter 10*** provides a more advanced look at the APIs provided by MIDP. This includes APIs for networking calls and for storing simple data items using the Record Management Store (RMS).
- ***Chapter 11*** describes the programming environment and tags for developing a VoiceXML application.
- ***Chapter 12*** completes Part 2 with a description of the programming environment for Java Card.

You'll find Part 2 to be a bit eclectic as we move from one technology to another—but never boring.

Chapter 8

Developing WML Applications

In this chapter we will delve deeper into the concepts and semantics of the *Wireless Markup Language (WML)* and *WMLScript*, which run in the Wireless Application Environment of the WAP protocol stack. We will introduce tools and SDKs that enable WML and WMLScript application development and provide a few programming tips.

WML is a markup language from the WAP Forum for rendering content on mobile devices like phones and PDAs. WML is similar to HTML, but unlike HTML, WML documents are valid XML documents. WML has also been optimized to work well on networks with poor latency and on resource-constrained devices described in earlier chapters.

THE WML USER INTERFACE

WML is targeted at devices that have a minimal user interface. Figure 8-1 shows how a WML application makes use of a handheld device's small display screen with a title, body, and softkey labels. When WML panels are too large to display on the screen, a scroll bar is also presented. WML applications are tailored to minimal user input-capability as shown in Figure 8-1. A typical Internet-enabled phone will have two softkeys, keys for navigating up, down, left, and right, a menu selection key,

Figure 8-1. The WML User Interface.

keys for initiating and terminating calls, and a numeric keypad that can also be used to enter alphabetic characters and symbols.

WML DOCUMENTS

WML documents are simple in structure and have only a few elements. Figure 8-2 shows a typical WML document and its structure. The document consists of the following:

- The ***DTD*** defines version- (and sometimes vendor-) specific elements for a WML document.
- The ***Card Deck*** includes the elements between the <wml> and </wml> tags–including the <head>, <template>, and <card> tags.
- The ***Head*** element defines metadata and access information that applies to an entire deck.
- The ***Template*** element defines default handling of events that apply to an entire deck. The event handling can be overridden in an individual card.
- The ***Cards*** define individual dialogs for display and define user interactions with the card.

```
<?xml version="1.0"?>
  <!DOCTYPE wml PUBLIC "-//WAPFORUM//DTD WML 1.3//EN"
    "http://www.wapforum.org/DTD/wml13.dtd">

<!-- Beginning of Card Deck -->
<wml>
  <head>
    <meta http-equiv="Cache-Control"
          content="max-age=3600"/>
  </head>

  <template>
    <do type="option" label="Back">
      <prev/>
    </do>
  </template>

  <!-- Card 1 -->
  <card id="welcome" title="Welcome"
        ontimer="#select" newcontext="true">
    <timer value="50"/>

    <p align="center">
      <strong>Welcome</strong><br/>
      <img src="sjsu_100.wbmp" alt="" align="middle"/>
    </p>

    <do type="accept" name="select"
        label="Next Card">
      <go href="#select"/>
    </do>
  </card>

    ....

  <!-- Card n -->
  <card id= .... >
    ...
  </card>
</wml>
```

XML document + DTD definition

Head element

Template element

First Card in Deck

Display elements in Card

Task association in Card

Additional Cards

Figure 8-2. The WML Document Structure.

Details

The WML DTD

So, which DTD should you use for your WML deck? A WML DTD represents a specific version of WML. Most WML browsers are OEM-licensed versions of an Openwave (formerly *phone.com*) Mobile Browser. A phone implements a specific version of the Openwave Mobile Browser and consequently implements a specific version of WML. Carriers in their respective geographic regions sell most phones. So, how do you know which phone manufacturer brands and models are being used in a geographic region and what version of browser they support? Take a look at the Openwave Mobile Browser Phone Reference at *http://upmkt.openwave.com/dev_phones/phones.cfm*. This reference allows you to look up the mobile browser implemented for phones by manufacturer and model for geographic regions.

Here are some of the DTDs for the various versions of WML:

- WML 1.1, Mobile Browser 4.x

```
<!DOCTYPE wml PUBLIC "-//WAPFORUM//DTD WML 1.1//EN"
     "http://www.wapforum.org/DTD/wml_1.1.xml">
```

- WML 1.3, Mobile Browser 5.x

```
<!DOCTYPE wml PUBLIC "-//WAPFORUM//DTD WML 1.3//EN"
     "http://www.wapforum.org/DTD/wml13.dtd">
```

- WML 1.3 with Openwave extensions, Mobile Browser 5.x

```
<!DOCTYPE wml PUBLIC "-//PHONE.COM//DTD WML 1.3//EN"
"http://www.openwave.com/dtd/wml13.dtd">
<!DOCTYPE wml PUBLIC "-//PHONE.COM//DTD WML 1.3//EN"
    "http://www.phone.com/dtd/wml13.dtd">
```

❑

DEVELOPING A WML APPLICATION

A WML application consists of three types of files:

- ***WML documents:*** These are card decks described using the XML document structure.

- ***WMLScript files:*** The WMLScript programming language augments the WML decks with a simple, device-centric programming language.

- ***Bitmap images:*** WML documents can include black-and-white images in the *.wbmp* format or color images using the *.png* format.

Figure 8-3 illustrates the WML application development process. Here are the steps:

1. ***Write the WML decks and WMLScript files.*** This is a simple text-editing task and can be done inside an *emulator*. An emulator provides an Integrated Development Environment (IDE) plus a simulation of handheld device execution. We will describe emulators in more detail in the next section.
2. ***Create bitmap images.*** Existing bitmap images that are not suitable for use with WML can be converted to *.wbmp* and *.png* formats using conversion utilities.
3. ***Run and test the application using the emulator.*** This includes starting and interacting with an image of the handheld device that is being tested. Like

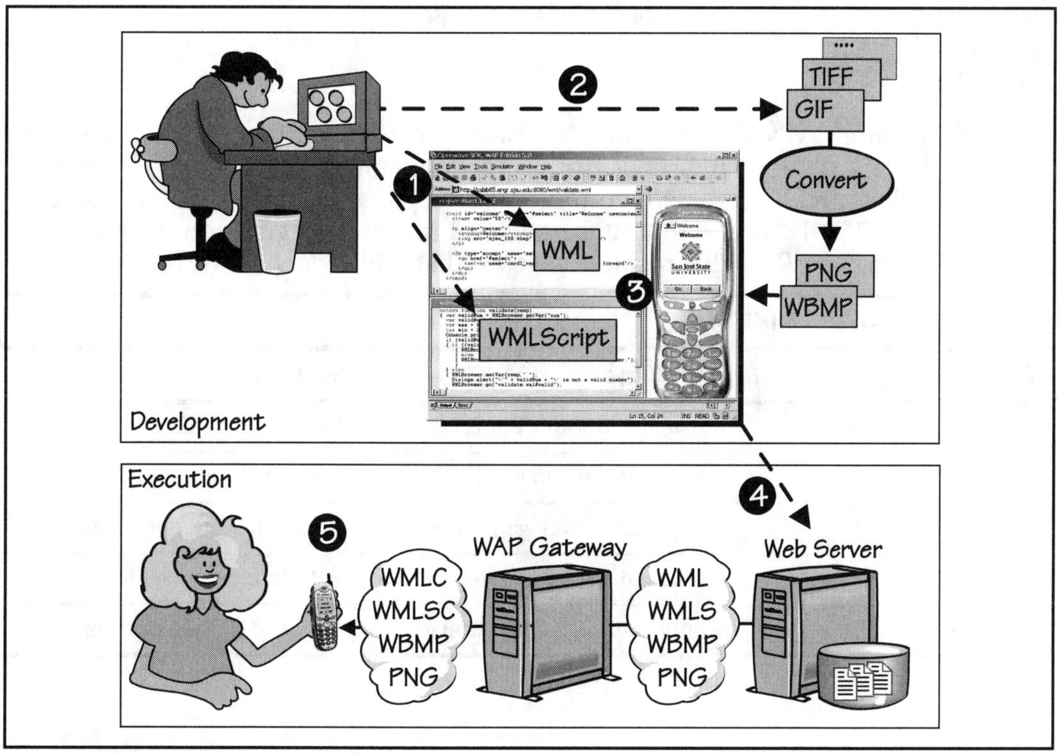

Figure 8-3. A WML Application from Developer to End User.

the test of any other GUI application, this can be a tedious process. The good news is that WML applications are usually very small.

4. ***Deploy to a production/test environment.*** This simply means making the files that are part of the WML application available via a Web server. Note that the WML files have their own MIME types so you may need to update the MIME type definitions on your Web server (see the next Details Box).

5. ***Run the application on a real handheld device.*** As the application is executed, the application files are pulled from the Web server to the WAP gateway, and the *.wml* and *.wmls* files are converted to a binary equivalent (shown as WMLC and WMLSC in Figure 8-3). These files are then delivered to the handheld device for execution and display. Keep in mind that your handheld device will cache your decks, scripts, and images so don't be surprised if your application doesn't seem to change as you modify the source code.

WML Mime Types

Details

The WAP gateway and the Mobile Browser on the handheld device require special MIME type designations to correctly identify WML documents and WMLScript source files, their compiled versions, and bitmap files. The file extension-to-MIME type mapping is specified in the configuration for the Web server where the WML files reside. Table 8-1 lists the WML MIME types.

Table 8-1. The WAP/WML MIME Types.

Extension	MIME Type	Description
wbmp	image/vnd.wap.wbmp	WBMP image file
wml	text/vnd.wap.wml	WML document
wmlc	application/vnd.wap.wmlc	Compiled WML document
wmls	text/vnd.wap.wmlscript	WMLScript source file
wmlscriptc	application/vnd.wap.wmlscriptc	Compiled WMLScript file

WML Development Emulators

Handheld devices do not provide an environment suitable for content and script program development. There is no way to introduce debuggers, memory usage tracing, network tracing, and other development aids. PC-based emulators like the one shown in Figure 8-4 fill this gap. The typical emulator provides:

- ***An image of the handheld device:*** A specific device can be selected in the emulator.
- ***Simulated interaction*** with the device by using the mouse to "press" keys.
- ***An IDE (Integrated Development Environment):*** The IDE identifies content and script errors, monitors messages sent from the device, cookie and variable watch windows, and utilities related to writing content and scripts.

After the WML decks and WMLScript programs are written, the emulator can be used to verify the execution of the application. Of course, the real application validation comes with running the application on a real device.

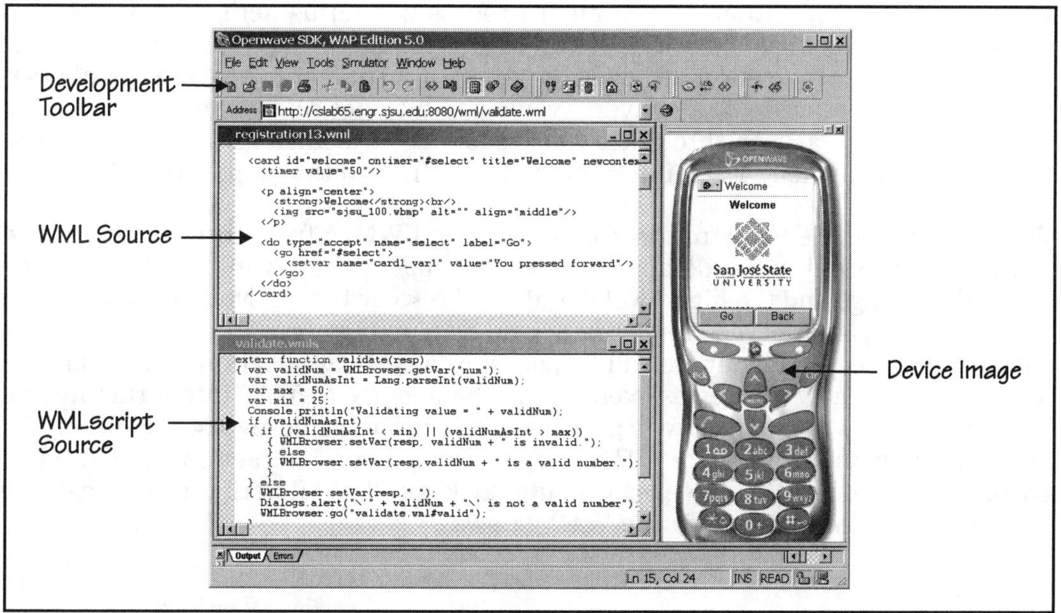

Figure 8-4. The Openwave SDK, WAP Edition 5.0 Emulator.

WML/WAP SDKs

Briefing

The Openwave SDK is not the only SDK available for developing your WML applications. Mobile phone manufacturers—such as Nokia, Ericsson, and Siemens—as well as third party tool vendors and open source efforts also provide development environments for your WML applications. These SDKs provide simulations of the execution of WML on handheld devices—including the execution of WMLScript. There are also tools available for creating *.wbmp* images or for converting existing images in other formats (such as *.gif* or *.tif*) to *.wbmp* images.

There are several portals for discovering the latest in WML/WAP development tools—such as 2ThumbsWAP.com (*http://www.2thumbswap.com*), the PaloWireless WAP Resource center (*http://www.palowireless.com*), and MobileInfo.com (*http://www.mobileinfo.com*). Other portals are only a Google search away. ❑

The University Registration Example

With the WML application development process behind us, let's take a look at a simple application that illustrates many of the features that WML provides (see Figure 8-5). The application provides a means for a student to register for courses that are being offered at a university via an Internet-ready phone. The WML application steps a user through login, registration, semester selection, course selection, and payment. Each interaction is a WML card that is part of a deck.

We will examine the Registration application (and WML) from three perspectives. First we will examine the details of a deck including the <wml>, <head>, and <template> tags and the high level details of the <card> tag. Next we will explain the display tags that cards can contain. The display tags provide the GUI widgets shown in Figure 8-5. Finally we will explain how a WML application maintains state using variables, how it receives events from the underlying WAP infrastructure and the device hardware, and how it processes selections made by user as he or she interacts with the application. These events and user interactions are used to navigate between WML screens, execute *tasks*, call WMLScript programs, and make calls to a Web server via the WAP gateway.

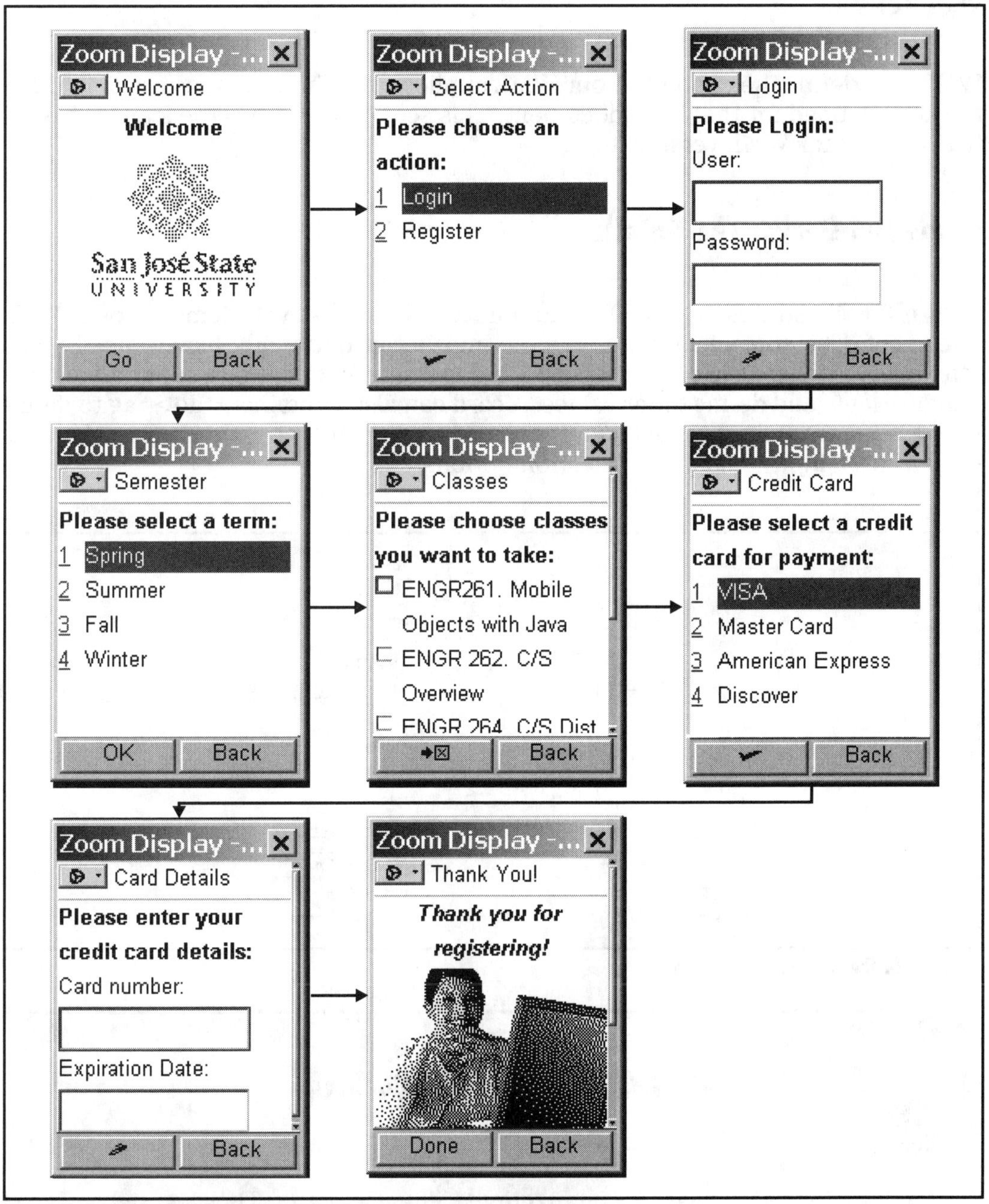

Figure 8-5. The Registration WML Example Emulator Screenshots.

WML TAGS

WML tags define the content of our WML application. The valid tags are defined by the DTD defined for the WML document. This section describes the standard WML tags defined for WML version 1.3.

Cards and Decks: The Details

Figure 8-6 illustrates the DTD definition for the high-level elements of a WML document. The root element is <wml> and it can optionally have <head> and <template> tags. At least one <card> tag is required and it can contain tags that handle events and define timer values. A card can also contain a <do> tag to define how events are bound to tasks and <p> or <pre> tags to hold display elements such as text, input fields, and selection options.

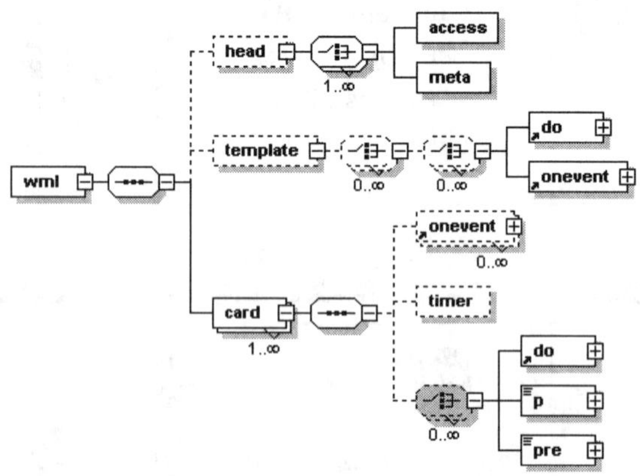

Figure 8-6. The High-Level WML Tags

FYI

WML Contexts and Cookies

Briefing

WML provides variables for managing a client's state. These variables are managed by the micro-browser and form the *context* for an executing WML application. When control is transferred to a server-side program (such as a Servlet), some (or all) of the state can be sent with the call in the form of name/value pairs.

The browser context may be initialized using the *newcontext* attribute of the card tag. This attribute takes the value *true* or *false*. When a card is traversed to using a <go> statement and the newcontext attribute is set to *true*, all context variables are removed, the history is cleared, and the device state is set to default values. When newcontext is set to *false*, the context is not modified.

Openwave also provides some extensions to the WML standard to allow the creation and management of multiple contexts. These child contexts are created and destroyed using the additional control tags <spawn>, <exit>, and <throw>.

Cookies are a common way for a client to store state between HTTP invocations. WML, however, does not store cookies on the client device. Instead, it depends on the WAP gateway to store cookies. Since, the gateway may have cookie support turned off, cookie support is not guaranteed and the application may have to resort to other techniques such as URL rewriting. ❑

WML Display Elements

WML provides tags to display visual items like text, images, tables, lists, fields, and buttons. The following sections provide a brief introduction to the display elements available in WML.

The paragraph tag <p> is used as a "container" for all WML display elements. Every WML deck must have at least one <p> element. The following shows the syntax for a paragraph tag:

```
<p align="alignment" mode="wrapmode">
  content
</p>
```

The *align* attribute can be used to position display items and can have the values *left*, *right*, or *center*. The *mode* attribute can have the values *wrap* or *nowrap* and defines whether long lines of text are wrapped to the next line or extend beyond the screen display. Text that is not wrapped can be scrolled like a "ticker-tape" by the browser.

Text

Text in a paragraph can be formatted using formatting tags that modify text size, add bolding, italics, and underlining. The use of underlining is usually discouraged as it can be confused with links.

The following shows the various formatting tags and the resulting display. Note that the <br> tag is used to force line breaks:

```
<card title="Text Formatting">
   <p align="left">Left</p>
   <p align="center">Center</p>
   <p align="right">Right</p>
   <p>
     <em>Emphasis </em>
     <strong> Strong</strong><br/>
     <big>Big </big> Normal
     <small> Small</small><br/>
     <i>Italics </i><u> Underlined</u>
   </p>
 </card>
```

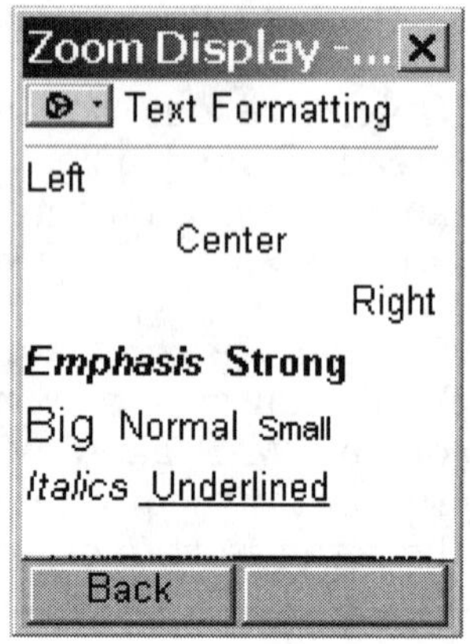

Tables

The WML table tags <table>, <tr>, and <td> can be used to provide simple row/column layout for text and images. WML tables are very simple. They do not allow you to embed tables in other tables or to specify column widths. The syntax for a typical table is shown below:

```
<table title="name" align="left|right|center"
       columns="number of columns">
  <tr>
    <td>content</td>
    ...
  </tr>
  ...
</table>
```

Note that there can be multiple rows defined using the <tr> tag. The number of columns is specified using the *columns* attribute of the <table> tag. Tabular data is specified using the <td> tag and can be text, images, or anchors.

Images

Bitmap images are displayed using the <img> tag. The syntax is shown below:

```
<img alt="text" src="url" localsrc="icon"
     align="alignment" height="n" width="n"
 vspace="n" hspace="n"/>
```

The *src* attribute specifies the image location. The *alt* attribute is used to specify text to be displayed should the device not support image rendering or if the image cannot be found. The *localsrc* attribute specifies an icon from a library of icons defined by the WAP architecture. The remaining attributes specify size and spacing values. The following shows an image tag and the resulting display:

```
<p align="center">
<strong>Welcome</strong><br/>
<img src="sjsu_100.wbmp"
  alt="SJSU" align="middle"/>
</p>
```

The localsrc icon library has around 200 icons defined that are 9 pixels high and 13-15 pixels wide. The following shows a few of the icons and the tags used to display them:

```
<card id="wmlimages"
      title="Local Images">
  <p align="center">
    <img src="" alt=":-)"
         localsrc="smileyface"/>
    <img src="" alt=""
         localsrc="flower"/>
      ...
  </p>
</card>
```

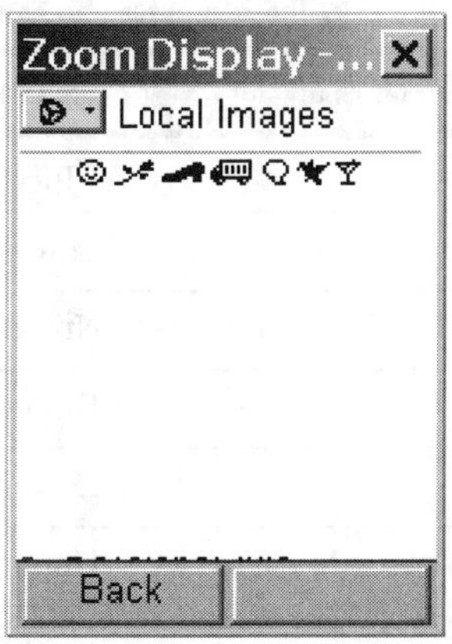

Input

The <input> tag is used to specify a field for user text entry. The following is the syntax for the <input> tag:

```
<input name="variable" title="label" type="type"
       value="value" format="specifier" emptyok="boolean"
       size="n" maxlength="n" tabindex="n"/>
```

The *name* attribute specifies the name of a variable that the browser uses to capture the user's input. The *title* attribute provides a brief label for the input field. It might provide a way for summary display and navigation to the field in situations where not all of the content in a card can be displayed on the screen. The *type* attribute indicates if the field is *text* or *password*. If the type is password, the characters the user types are converted to asterisks. The *value* attribute specifies the initial value for the field. The *format* attribute is a string that specifies the format of the data that is entered. Table 8-2 lists the values for the format specifier characters. If a number precedes the character, then the number indicates the number of characters for the type specified. For example, a value of "5A" for the *format* attribute would limit user input to five uppercase letters or symbols only. Other characters can be included in the format attribute to provide the desired display appearance. For instance, format="NNN-NN-NNNN" could be used to display a social security number.

The *size* attribute indicates the length of the display field that is rendered by the device. The *maxlength* attribute indicates the number of characters the user can enter into the field.

Table 8-2. The <input> Tag Formatting Characters for the Format Attribute.

Format Character	Comments
A	Uppercase letters or symbols
a	Lowercase letters or symbols
N	Numbers
X	Uppercase letters, symbols or numbers
x	Lowercase letters, symbols or numbers
M	Uppercase letters, symbols or numbers (changeable to lowercase)
m	Lowercase letters, symbols or numbers (changeable to uppercase)

The following shows an example of display fields:

```
<p>
  <b>Enter Information:</b> <br/>
Name: <input name="name" size="10"
       emptyok="false"/>
SSN: <input name="ssn" size="10"
       emptyok="false"
       format="NNN-NN-NNNN"/>
Email: <input name="email" size="10"
   emptyok="false"/>
Password: <input name="password"
       type ="password" size="10"
       emptyok="false"/>
Phone: <input name="phone" size="10"
       emptyok="true"
       format="(NNN) NNN-NNNN"/>
</p>
```

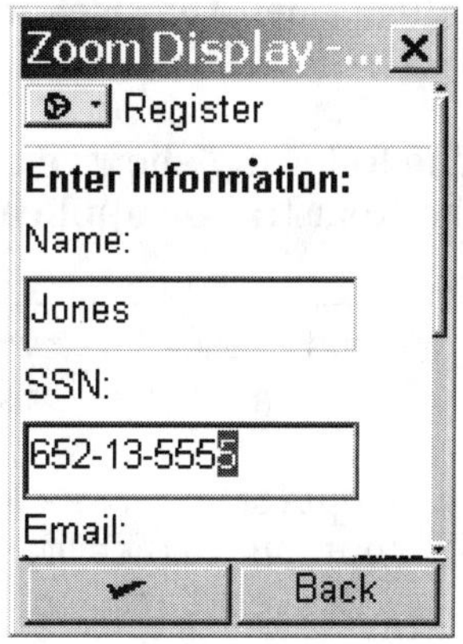

Option Lists

One or more <option> tags can be specified within a <select> tag to display a list of options. The selection list can be either single- or multiple-select. Here's the syntax of a selection list:

```
<select title="label" multiple="boolean" name="variable"
        value="default" iname="index_var" ivalue="default"
        tabindex="n">
  <option title="label" value="value" onpick="url">
    content
 </option>
   . . .
</select>
```

The *title* attribute provides a brief label for the selection list. It might provide a way for summary display and navigation to the field in situations where not all of the content in a card can be displayed on the screen. The *multiple* attribute, if true, indicates multiple items in the list can be selected. The *name* attribute specifies the variable name the browser will use to store the selections the user makes. The *value* attribute specifies the default value for the *name* variable. The *iname* and *ivalue* attributes are used to store the index value of the items selected.

The *title* attribute of the <option> tag specifies the name of the accept key label. The *value* attribute specifies the value that is stored in the *name* variable of the <select> tag. If the list is multiple-select, the values are stored as a list of semicolon-separated values. The *on-pick* attribute specifies the location that is called when the user makes a selection (or deselects an item for multiple-select lists).

The following shows examples of single- and multiple-selection lists that are part of the Registration application:

```
<select name="cc_type">
  <option value="visa">
    VISA
  </option>
  <option value="master">
    Master Card
  </option>
  <option value="american">
    American Express
  </option>
  <option value="discover">
    Discover
  </option>
</select>
```

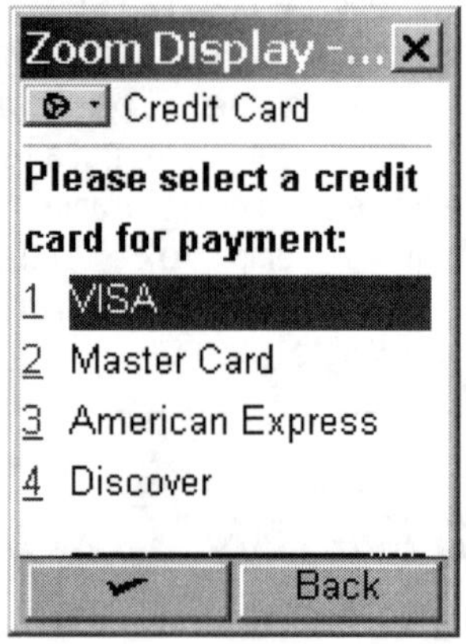


```
<select name="class" multiple="true">
  <option value="java">
    ENGR261. Mobile Objects with Java
  </option>
  <option value="overview">
    ENGR 262. C/S Overview
  </option>
  <option value="corba">
    ENGR 264. C/S Dist. Objects
  </option>
    . . .
 </select>
```

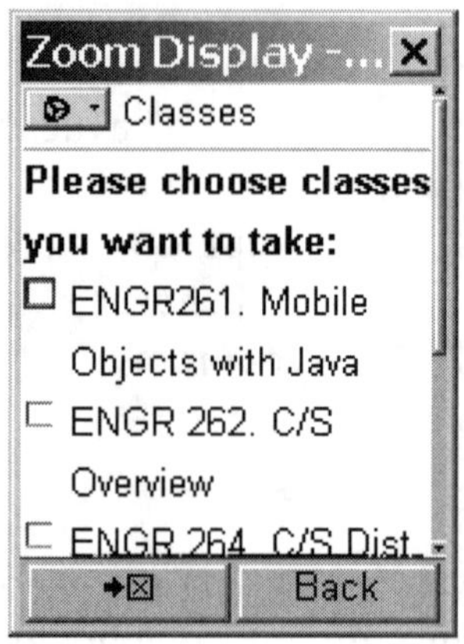

Openwave also provides some WML extensions to enhance the WML application user interface. The most obvious extensions are to the <select> tag. A new attribute called *type* has been added which can have the values *radio*, *popup*, and *spin* in addition to the value *list* (which provides the standard list behavior). Figure 8-7 shows the resulting lists when the new type attributes are used. Of course, the mobile browser that will display your WML application must support these extensions.

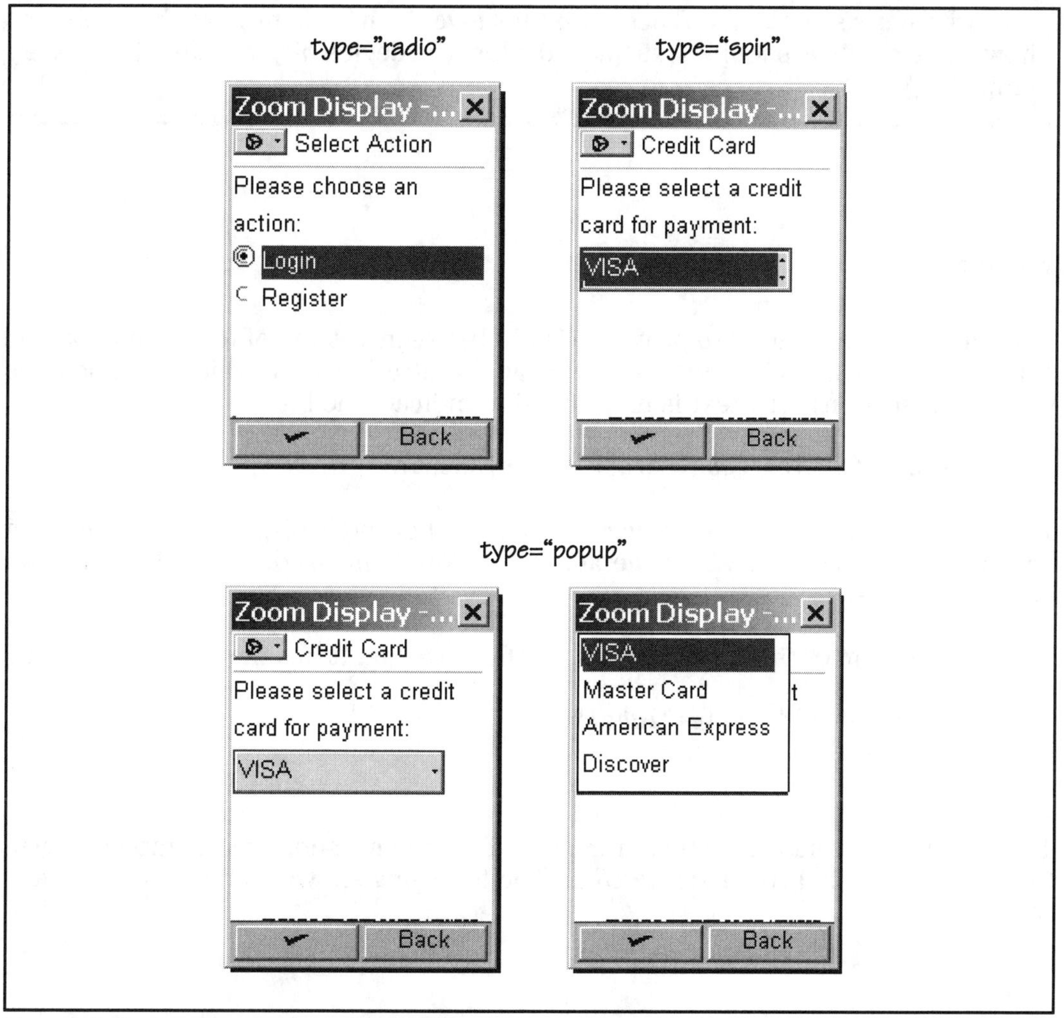

Figure 8-7. The Openwave <select> Tag GUI Extensions.

Details

Grouping Fields and Options

The <fieldset> tag is used to group <input> and <select> fields and can also contain other <fieldset> tags. The <optgroup> tag is used to group <option> tags within a <select> list. These two tags give the mobile browser hints about how fields and lists should be displayed when the device display capabilities are limited. ❑

Anchors

The <anchor> tag is used to provide a "link" between a string of formatted text and a task. Tasks can include <go>, <prev>, and <refresh> tags (which are discussed in the next section). The text is underlined to indicate the link.

```
<anchor title="label">task text</anchor>
```

The *title* attribute is used as the accept softkey label and is displayed when the user navigates to the anchor. When the accept softkey is pressed, the mobile browser executes the task.

A shorter version of the <anchor> tag is the <a> tag. It has the following syntax:

```
<a href="url" title="label">
  content
</a>
```

The <a> tag content can be text or images. The tag only supports an implied <go> task to the specified *ref* attribute *URL*. The following shows <anchor> tag examples:

```
<card id="links" title="Links:">
    <p>
      <anchor title="News">
         <go href="news.wml"/>News
      </anchor><br/>
      <anchor title="Sports">
         <go href="sports.wml"/>Sports
      </anchor><br/>
           ...
    </p>
  </card>
```

This ends our discussion of the display tags that WML supports. The next sections discuss the mechanisms supplied by the mobile browser that supports capturing user input into variables, handling user interaction and browser-generated events, and associating those events with tasks that the browser performs.

WML Variables

WML variables are named pieces of information maintained by the mobile browser. Variables capture the user's input, and in general maintain the state of the WML application. Variables can also be used to create the content that cards display and can be used to navigate to other cards based on user input. Variables can be implicitly declared and their values assigned as part of display tags (like <select> or <input>). They can also be explicitly declared using the <setvar> tag (which declares a variable for use by the mobile browser) or by using the <postfield> tag (which is used to declare variables that are passed as name/value pairs when a HTTP request is made to a Web server).

WML variables are stored as untyped strings. WML variable names must start with either an underscore or a US ASCII character and can be followed by zero or more letters, digits, or underscores.

The value of a variable can be substituted for text in a card, for URL references, and for tag attribute values. The value of the variable is referenced by preceding the variable name with a "$" sign. The variable name must be enclosed in parentheses when the end of the variable string cannot be discerned from the text that follows.

When a variable name is followed by a colon, a conversion algorithm specified by the name after the colon is applied as the variable value is dereferenced. This is useful when the value of a variable needs to be included into a URL. This is also known as URL-escaping. The possible values for the conversion are *escape*, *unescape*, and *noescape* (the letters 'e', 'u' and 'n' can be used as abbreviations).

WML Events

Events mark an occurrence in the mobile browser such as the expiration of a timer, a state transition in the WML application (such as movement from one card to another), or a user interaction with the WML application. Events are bound to tasks, which are processes that can be executed as the result of an event.

There are three tags that can be used to bind tasks to events: <anchor> (and its abbreviated form <a>), <onevent>, and <do>. These are described in more detail in the following sections.

The <onevent> Tag

The <onevent> tag is used to handle intrinsic events. Intrinsic events are triggered by the mobile browser during card-to-card navigation, timer expiration, and option selection. For example, this could be an event that was generated when the user navigates to a card because of executing a <go> task. The following is the syntax of <onevent> tag:

```
<onevent type="type">task</onevent>
```

The task can be one of <go>, <prev>, <noop>, or <refresh>.

There are four types of intrinsic events:

- ***oneventforward*** is generated when the user enters a card as a result of executing a <go> task.
- ***oneventbackward*** is generated when the user enters a card because of executing a <prev> task or other similar mechanism.
- ***ontimer*** is generated when a timer expires. Timers are defined using the <timer> tag. The timer value is specified in tenths of seconds.
- ***onpick*** is generated when the user selects or deselects an <option> item.

These event names are used as values for the *type* attribute of the <onevent> element).

The <timer> Tag

WML provides timers, which are very useful in many scenarios. Timers can be used to reduce the number of user inputs, to display splash screens, and to generate timed events. The scope of a timer is the card in which it is defined. Once the timer expires, an event is generated. This event can be caught and any desired operations can be performed—just as in any high level programming language. A timer is defined using the <timer> element. Every timer can have a name and a time interval value. The name is specified using the *name* attribute of the <timer> tag, and the time interval after which the timer expires is specified using the *value* attribute of the <timer> tag. The granularity for time value in WML is tenths of seconds. Since the time interval has a unit of measure of one tenth of a second, to create a time interval of 5 seconds you would specify 50 for the time value. The *name* attribute is optional and the *value* attribute is mandatory. The following illustrates the syntax of the <timer> element:

```
<timer name="variable" value="value"/>
```

The <do> Tag

The <do> tag is used to bind user-triggered events to tasks. User-triggered events are those that are triggered when users activate user interface widgets (for example, "accept" or "cancel" widgets). These widgets are mandated by the WML specification but the actual implementation of these widgets is left up to the device manufacturers. For example, when considering "accept" widgets, one manufacturer could choose to provide a real button on a phone while another could choose to provide an icon on a touch-sensitive screen. Thus, when the user activates any of these widgets, an event is generated that can be handled using the <do> element. By specifying the type attribute of the <do> element, the application developer can delegate control to the appropriate event handler. A type value of "unknown" can be used as a generic type element value. The following are the events defined by the WML specification and these also become the values for the *type* attribute of the <do> tag:

- **accept:** Widget to acknowledge or accept.
- **prev:** Widget to request to go to previous card that was accessed.
- **help:** Widget to request help.
- **reset:** Widget to reset the context of the device.
- **options:** Widget to check if other additional operations are available.
- **delete:** Widget to delete current item.

Events can have a scope associated with them. The scope defines if the events can occur at a card-level or at the deck-level. The <onevent> and <do> events can occur at the deck-level. The <onpick> event can occur only at the card-level. The scope can be specified by the placement of the applicable event elements. To specify events at deck-level a <template> tag has to be used to enclose the event-task binding.

WML Tasks

There are four standard WML tasks: <noop>, <refresh>, <prev> and <go>. Let's take a look at each one.

The <noop> Tag

As the name implies, this tag performs "no operation." It is used to override any deck-level tasks that are defined in the <template> tag. The following is the syntax for the <noop> tag:

```
<noop/>
```

The <prev> Tag

This is used to transfer control to the previous card that was accessed. The following is the syntax for the <prev> tag:

```
<prev>
  content
</prev>
```

The <prev> tag content can include a <setvar> tag to set variable data prior to transferring control to the new card.

The <refresh> Tag

The <refresh> tag is used to refresh the variables specified in the tag's content. The following is the syntax of the <refresh> element:

```
<refresh>
  content
</refresh>
```

If the variables that are refreshed are used to generate the current card, then the display is also refreshed.

The <go> Tag

The <go> tag is used to transfer control to a specified URL or card. If the *method* attribute of this tag is specified, it can be used to specify if the GET or POST method is to be used when submitting the request to an HTTP server. Typically a <postfield> tag is used in this case to specify the name/value pairs to be submitted. If <postfield> is used, then the values are evaluated, properly transcoded to the appropriate character set, properly URL-encoded, appended to the query part of the HTTP request, and submitted using the specified HTTP method to the HTTP server. The following is an example of the use of a <go> tag.

```
<go href="url" sendreferer="boolean" method="method"
    accept-charset="charset"
<postfield name="data" value="value"/>
  content
</go>
```

The <postfield> tag is used to specify the variables to be submitted to the server application as name value pairs. The *method* attribute of the <go> tag is used to specify which HTTP method is to be used (typically either GET or POST).

THE REGISTRATION WML LISTING

Listing 8-1 shows the complete "WML Registration" example. We hope that you find it to be an easy read after going through the previous sections.

Listing 8-1. The Registration WML Application.

```
<?xml version="1.0"?>
  <!DOCTYPE wml PUBLIC "-//WAPFORUM//DTD WML 1.3//EN"
    "http://www.wapforum.org/DTD/wml13.dtd">
<wml>
  <template>
    <do type="option" label="Back">
      <prev>
        <setvar name="card1_var1" value="You pressed back"/>
      </prev>
    </do>
  </template>
```

```
<card id="welcome" ontimer="#select" title="Welcome" newcontext="true">
    <timer value="50"/>

    <p align="center">
      <strong>Welcome</strong><br/>
      <img src="sjsu_100.wbmp" alt="" align="middle"/>
    </p>

    <do type="accept" name="select" label="Go">
      <go href="#select">
        <setvar name="card1_var1" value="You pressed forward"/>
      </go>
    </do>
  </card>


  <card id="select" title="Select Action">
    <p><b>Please choose an action:</b><br/>
      <select name="action">
        <option value="login" onpick="#login">Login</option>
    <option value="register" onpick="#register">Register</option>
      </select>
    </p>
  </card>


  <card id="login" title="Login">
    <p>
      <b>Please Login:</b><br/>
      User: <input name="user" size="10"/>
      Password: <input name="password" type="password" size="10"/>
    </p>

    <do type="accept">
      <go href="#semester"/>
    </do>
  </card>


  <card id="register" title="Register">
    <p>
      <b>Enter Information:</b> <br/>
      Name: <input name="name" size="10" emptyok="false"/>
      SSN: <input name="ssn" size="10"
           emptyok="false" format="NNN-NN-NNNN"/>
```

```
    Email:<input name="email" size="10"
          emptyok="false"/>
    Password:<input name="password" type="password"
          size="10" emptyok="false"/>
    Phone:<input name="phone" size="10" emptyok="true"
          format="NNN-NNN-NNNN"/>
  </p>

  <do type="accept">
    <go href="#semester"/>
  </do>
</card>


<card id="semester" title="Semester">
  <p mode="nowrap">
    <b>Please select a term:</b><br/>
    <select name="semester">
      <option title="OK" value="spring">Spring</option>
      <option title="OK" value="summer">Summer</option>
      <option title="OK" value="fall">Fall</option>
      <option title="OK" value="winter">Winter</option>
    </select>
  </p>

  <do type="accept">
    <go href="#classes"/>
  </do>
</card>


<card id="classes" title="Classes">
  <p mode="wrap">
    <b>Please choose classes you want to take:</b>
    <select name="class" multiple="true">
      <option value="java">ENGR261. Mobile Objects with Java</option>
      <option value="overview">ENGR 262. C/S Overview</option>
      <option value="corba">ENGR 264. C/S Dist. Objects</option>
      <option value="database">ENGR 266. C/S Databases</option>
      <option value="components">ENGR 268. C/S Components</option>
    </select>
  </p>

  <do type="accept">
    <go href="#creditCard"/>
```

```
  </do>
</card>


<card id="creditCard" title="Credit Card">
  <p>
    <b>Please select a credit card for payment:</b>
    <select name="cc_type">
      <optgroup title="card">
        <option value="visa">VISA</option>
        <option value="master">Master Card</option>
        <option value="american">American Express</option>
        <option value="discover">Discover</option>
      </optgroup>
    </select>
  </p>

  <do type="accept">
    <go href="#cardDetails"/>
  </do>
</card>


<card id="cardDetails" title="Card Details">
  <p>
    <b>Please enter your credit card details:</b><br/>
    Card number: <input name="cc_number" size="10" emptyok="false"/>
    Expiration Date: <input name="cc_date" size="10" emptyok="false"/>
  </p>

  <do type="accept">
    <go href="#done"/>
  </do>
</card>


<card id="done" ontimer="#welcome" title="Thank You!">
  <timer value="50"/>

  <p align="center">
    <em>Thank you for registering!</em>
    <br/><br/><br/>
    <img src="register.wbmp" alt="" align="middle"/>
    <br/>
  </p>
```

```
    <do type="accept" label="Done">
      <go href="#welcome"/>
    </do>
  </card>

</wml>
```

INTRODUCTION TO WMLSCRIPT

WMLScript is loosely based on ECMAScript and provides scripting capability for the WAP architecture. WMLScript is primarily used to augment the capabilities of WML. Like JavaScript, its cousin in the desktop world, WMLScript can be used to provide client-side functionality such as:

- Validating user inputs
- Generating alerts, confirmations, and dialogs locally so that a network trip all the way to the back-end Internet service is avoided
- Accessing device features like the device address book, sending messages, and accessing the SIM card
- Configuring devices and extending device software after the device has been deployed

This section is intended to provide a brief introduction to the many features WMLScript provides.

WMLScript has its own byte code and interpreter. Also, WMLScript supports locally installed libraries.

WMLScript provides the core ECMAScript features including basic types, variables, expressions, and statements. Special features that are not part of the core of ECMAScript are also available to better utilize low-bandwidth network channels. For example, WMLScript is compiled into byte code before it is sent to the client. From a programming perspective, WMLScript does not support arrays while ECMA-Script does.

Let's begin our exploration of WMLScript with a simple application called *validate*. The *validate* example has one WML card with a single input field. It is expected that a number between 25 and 50 will be entered into the field. A WMLScript function is used to validate the user's input. Figure 8-8 shows the initial screen for the Validate card and three possible scenarios that might occur during the validation of the input field.

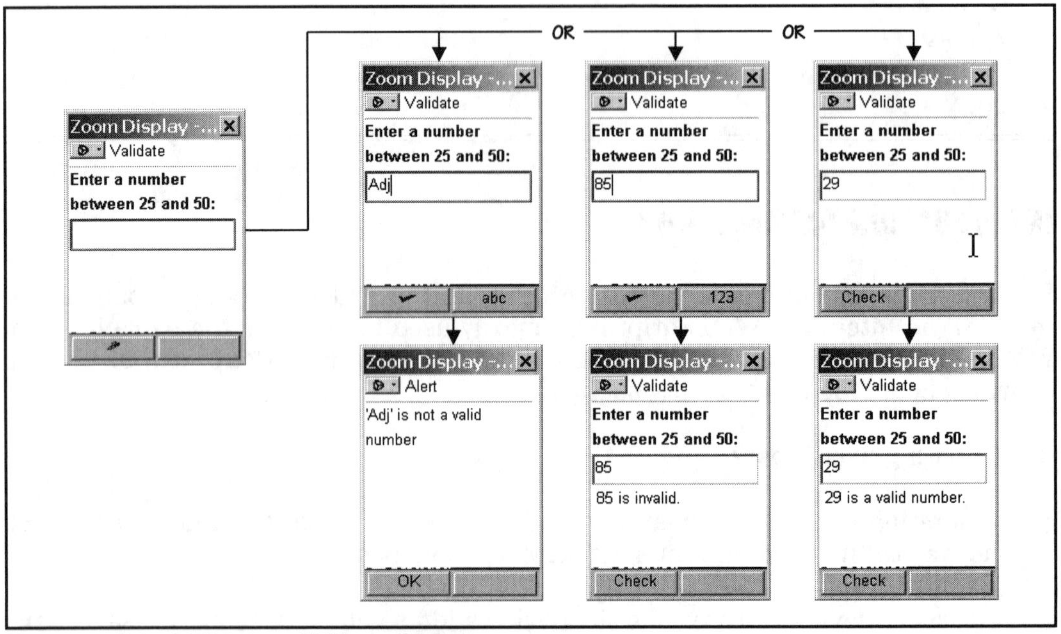

Figure 8-8. Using WMLScript to Validate a User's Input.

Listing 8-2 shows the WML document for the *validate* application. The single card in the *validate.wml* file provides one input field. When the user enters some data into the field and presses the accept softkey, the WMLScript function *validate* is invoked.

Listing 8-2. A WML Card, validate.wml, that Calls the validate.wmls Script.

```
<?xml version="1.0"?>
  <!DOCTYPE wml PUBLIC "-//WAPFORUM//DTD WML 1.3//EN"
    "http://www.wapforum.org/DTD/wml13.dtd">

<wml>
  <card id="valid" title="Validate">
    <p>
      Enter a number between 25 and 50:
      <input name="num" format="nn"/>
      $resp<br/>
      <do type="accept" label="Check">
```

```
        <go href="validate.wmls#validate('resp')"/>
      </do>
    </p>
  </card>
</wml>
```

Listing 8-3 shows the WMLScript that does the validation of the user's input. The *validate* function is defined as an external function in the file validate.wmls. This function checks if the number is a valid integer. If the entry is not a number, then an alert is displayed with an error message. The alert is presented using a WMLScript library function.

If the number is an integer, then it is checked to see if it falls in the range 25 through 50. A variable is returned with an appropriate message indicating if the number is valid or not. This variable is displayed as part of the WML card.

Listing 8-3 illustrates how WMLScript supports common features contained in most languages such as function declaration, variable declaration and initialization, console output, conditional and assignment statements, operators, and library calls.

Listing 8-3. The validate.wmls Function.

```
extern function validate(resp)
{ var validNum = WMLBrowser.getVar("num");
  var validNumAsInt = Lang.parseInt(validNum);
  var max = 50;
  var min = 25;
  Console.println("Validating value = " + validNum);
  if (validNumAsInt)
  { if ((validNumAsInt < min) || (validNumAsInt > max))
     { WMLBrowser.setVar(resp, validNum + " is invalid.");
     } else
     { WMLBrowser.setVar(resp,validNum + " is a valid number.");
     }
  } else
  { Dialogs.alert("The number " + validNum + " is not a valid number");
    WMLBrowser.go("validate.wml#valid");
  }
  WMLBrowser.refresh();
}
```

The following sections describe WMLScript in more detail.

Variables and Data Types

WMLScript supports variables that are declared inside functions or passed as function parameters. Variables must be declared before they are used. The scope of any declared variables is the function within which they are declared. Variables are declared by placing *var* before the variable name and may optionally be initialized. Any uninitialized variables are automatically initialized to the empty string ("").

WMLScript variables are untyped. However, internally four basic data types are maintained: *boolean*, *integer*, *float*, and *string*. A variable takes on a specific type based on the value assigned to it. WMLScript automatically converts between types as required. If an invalid value is assigned to a value, then the variable type is *invalid*. Table 8-3 shows the variable types and some declaration examples.

Table 8-3. The WMLScript Variable Types.

Variable Type	Example	Comments

Integer	`var postitiveNumber = 40`
	`var negativeNumber = -41`
	`var hexNumber = 0x4A`
	`var octalNumber = 032`
		Integer literals indicated by numeric values.

Floating-Point	`var fp = 2.5`	Floating point numbers contain a decimal point.
String	`var myString = "Hello"`	Strings are delimited by double quotes (" ") or single quotes (' ').
Boolean	`var myBoolean = false`	Boolean variable take on values *true* or *false*.
Invalid	`var myVar = someliteral`	The right-hand side of the expression does not contain a valid literal.

Pragmas

WMLScript pragmas provide special directives to the compiler. They are specified at the beginning of the file before any function declarations. Pragmas have the form:

```
use pragma-name pragma-attributes …
```

Table 8-4 shows some WMLScript pragma examples.

Table 8-4. The WMLScript Pragmas.

Pragma	Example	Comments
url	`use url myScript http://wirelessjava.engr.sjsu.edu/script`	Provides a symbolic reference to an external resource.
access	`use access domain http://wirelessjava.engr.sjsu.edu" path "/script"`	Restricts access specified domain and path.
meta	`use meta name "Created" "10-March-2002"`	Defines a property name and value.

Operators

WMLScript defines a variety of operators to perform assignment, arithmetic operations, logical evaluations, and comparisons. Operators, in combination with constants, literal values, functions, and variable names, are used to form expressions within WMLScript. Table 8-5 enumerates the WMLScript operators.

Table 8-5. The WMLScript Operators.

Operator Type	Operator	Description
Assignment	=	assign
	+=	add (numbers)/concatenate (strings) and assign
	-=	subtract and assign
	*=	multiply and assign
	/=	divide and assign
	div=	divide (integer division) and assign
	%=	remainder or modulus and assign
	<<=	bitwise left shift and assign
	>>=	bitwise right shift with sign and assign
	>>>=	bitwise right shift zero fill and assign
	&=	bitwise AND and assign
	|=	bitwise OR and assign
	^=	bitwise XOR and assign
Arithmetic	+	add (numbers)/concatenate strings
(Basic binary)	-	subtract
	*	multiply
	/	divide
	div	integer division

Table 8-5. The WMLScript Operators. (Continued)

<table>
<tr><th>Operator Type</th><th>Operator</th><th>Description</th></tr>
<tr><td>Arithmetic
(Complex binary)</td><td>%
<<
>>
>>>
&
|
^</td><td>remainder or modulus
bitwise left shift
bitwise right shift
bitwise right shift zero fill
bitwise AND
bitwise OR
bitwise XOR</td></tr>
<tr><td>Arithmetic
(Basic unary)</td><td>+
-
--
++
~</td><td>plus
minus
pre-or-post decrement
pre-or-post increment
bitwise NOT</td></tr>
<tr><td>Logical</td><td>&&
||
!</td><td>logical AND
logical OR
logical NOT (unary)</td></tr>
<tr><td>Comparison</td><td><
<=
==
>=
>
!=</td><td>less than
less than or equal
equal
greater than or equal
greater than
inequality</td></tr>
<tr><td>Comma</td><td>,</td><td>Combines multiple evaluations into one expression.</td></tr>
<tr><td>Conditional</td><td>x ? a : b</td><td>Assigns a value based on Boolean result of initial value.</td></tr>
<tr><td>Typeof</td><td>typeof</td><td>Returns an integer value representing the type of the expression.</td></tr>
<tr><td>Invalid</td><td>invalid</td><td>Returns false if the type of an expression is invalid.</td></tr>
</table>

Functions

WMLScript functions declare a name, parameters, and a block statement that is executed when the function is called. Function declarations cannot be nested and all function names must be unique within one file. Function parameters are passed by value to the function and behave like local variables to the function that have been initialized prior to the function call. A function always returns a value. If no return value is specified, the empty string ("") is returned. A function can be declared *extern* to make it available outside of the file. Table 8-6 lists the function types.

Table 8-6. The WMLScript Function Types.

Function Type	Example	Comments
Local	`function localFunction(parm)` `{ . . . }`	Local functions declare the function name and parameters.

External	`use url ext "http://host.com" path "/script"`
	
	`function callExternal(parm1)`
	`{ return ext#extFunction("test")`
	`}`
		The *use* pragma defines the external file. The call to the external function is preceded by the name declared by the use pragma and the hash (#) symbol.

Library	`Dialogs.alert("Invalid!");`	Library function calls are preceded by the name of the library and a period (.).

Statements

Statements define expressions, variable declaration, and execution control for WMLScript programs. Table 8-7 enumerates the WMLScript statements.

Table 8-7. WMLScript Statements.

Statement	Example	Comments
Empty	`while (!test(bit))` `;`	Used when statement is needed but no operation.

Expression	`var hello = "Hello World";`
	`alert("Invalid");`
	`area = pi * r * r;`
		Expressions assign values, calculate values, call functions, and so on.

Block	`{ var j = 2;`
	`j =  j + 4;`
	`Console.print("j = " + j);`
	`}`
		A group of statements contained inside curly braces ({}). Used to form a compound statement.

Variable	`var result = 0;`	Declares and initializes variables.

If	`if (happy)`
	`{ smile = true`
	`} else`
	`  {smile = false;`
	`  }`
		Conditional branching execution depending on Boolean test value.

Table 8-7. WMLScript Statements. (Continued)

Statement	Example	Comments

While	`while (i < 3)`
	`{ total++;`
	`}`
		Conditional looping execution depending on Boolean test value.

For	`for (var i=0; i<10; i++)`
	`{ total++;`
	`}`
		Conditional looping execution depending on Boolean test value.

Break	`if (i == 3) break;`	Terminates a for or while loop.
Continue	`if (i == 3) continue;`	Terminates execution of a for or while loop and resumes execution with the next iteration.
Return	`return "Invalid Number";`	Returns a value for a function call.

Libraries

Libraries are named collections of functions. Library functions are called using the following syntax:

```
Lib-name.func-name(parm1,…);
```

Table 8-8 lists the WMLScript libraries and their functions.

Table 8-8. WMLScript Libraries and Their Functions.

Library	Functions	Comments
Lang	abort, abs, characterSet, exit, float, isFloat, isInt, max, maxInt, min, minInt, parseFloat, parseInt, random, seed	Provides additional core language function.
Float	ceil, floor, int, maxFloat, minFloat, pow, round, sqrt	Provides typical floating-point functions.

Table 8-8. WMLScript Libraries and Their Functions. (Continued)

Library	Functions	Comments
String	`charAt, compare, elementAt, elements, find, format, insertAt, isEmpty, length, removeAt, replace, replaceAt, squeeze, subString, toString, trim`	Provides typical string functions.
URL	`escapeString, getBase, getFragment, getHost, getParameters, getPath, getPort, getQuery, getReferer, getScheme, isValid, loadString, resolve, unescapeString`	Provides functions for handling absolute and relative URLs.
WMLBrowser	`getCurrentCard, getVar, go, newContext, prev, refresh, setVar`	Provides functions for accessing the WML context.
Dialogs	`alert, confirm, prompt`	Provides a set of user-interface functions.
Console	`print, println`	Provide functions for writing debug information to the console.

PROGRAMMING TIPS

In this section we will highlight some of the issues to keep in mind while developing a WAP/WML application. Though conventional software engineering methodologies still apply, the approach must be shifted to focus on device constraints.

When designing WAP/WML applications, keep the focus on the end user's needs and the user experience. Since most of these applications will be run from mobile phones, the user experience and UI design determine the overall usefulness of the application and whether the application provides the right amount of the right kind of information. Users find it confusing to navigate in deep menus, so try not to use more than three levels of menus. Place the most frequently used items at the top of a card. Also, try to limit the number of key clicks required to get the end result. Try to build your application with multiple navigation paths so a first time user and an experienced user can use it with ease.

Try to provide lists and not input fields, to get user inputs. If applicable, provide a way to remember and reuse input from their previous user visits. If you have tried entering your street address using the twelve tiny keys normally available on a phone, you'll appreciate this advice.

Maintain consistency throughout if you are providing a suite of applications, like a portal. Last, but not least, you should always test your WAP/WML application with real devices before deploying because there will be features that are implemented differently on different devices. Also, only by testing with real devices, can you find errors or issues (if any) caused by network and device limitations—for example, the size of a deck that can be downloaded by a device, and so forth.

SUMMARY

In this chapter, we have explained the various elements of WML and WMLScript and how they can be used together to build an application. Even though WML is a markup language like HTML, it has certain features of a programming language that make it very powerful: providing support for variables, tasks, events, and timers. It is better suited for use on mobile devices with resource constraints because it has many features designed to enhance performance, such as the concept of decks and cards. Also, WAP has been carefully designed to offer better performance than traditional markup languages like HTML. The ease and simplicity of WMLScript further enhances WML programming. All these reasons explain why WML is a popular markup language for mobile devices.

Chapter 9

Introduction to MIDP Programming

We're now ready to develop MIDP J2ME programs based on the configurations and profiles presented in Part 1. The *Mobile Information Device Profile (MIDP)* that is implemented on top of the *Connection Limited Device Configuration (CLDC)* provides a complete J2ME programming platform for consumer devices such as cell phones, two-way pagers, and personal digital assistants. In this chapter we will look at the development process for MIDP applications and then jump into developing a MIDP-based GUI. We'll look at the MIDP GUI from two aspects: high-level, for extremely portable MIDP business applications, and low-level for less portable but highly graphical applications such as games.

We will defer the more advanced aspects of MIDP such as network programming and persistence until the next chapter.

THE J2ME MIDP USER INTERFACE

Like WML, MIDP is targeted at devices that have a minimal user interface. Figure 9-1 shows how a MIDP application makes use of a handheld device's small display screen. Like WML, MIDP screens are frequently presented with a title, body, and softkey labels. Unlike WML, MIDP applications have a great deal of flexibility in the user interface. This includes the ability to place MIDP widgets and draw simple vector graphics. When MIDP panels are too large to display on the screen, a scroll bar is also presented.

MIDP applications are also tailored to limited user input capability such as shown in Figure 9-1. A typical Java-enabled phone will have two softkeys, navigating keys (such as up, down, left, and right), a menu selection key, keys for initiating and terminating calls, and a numeric keypad that can also be used to enter alphabetic characters and symbols. Here's a summary of the keys:

- The ***softkeys*** are used to invoke application commands. They are identified by labels that appear near them on the screen. In Figure 9-1, the left soft button is mapped to the *Exit* command and the right button to the *Next* command. The softkey labels are shown in the display just above the keys.

- The ***control keys*** are the keys with the "on-hook" and "off-hook" telephone handset icons on them. Control keys are found on most hand-held phones and are sometimes labeled *Talk* and *End*. Control keys are not used for any call control from within the MIDlet; however, the End key terminates the currently running MIDP application.

- The ***navigation keys*** allow you to move around and select items in the display area.

- The ***alphanumeric keys*** let you enter numbers, letters, and symbols into the display input fields. When a text entry screen is showing, a typical phone lets you use the button on the alphanumeric keypad with the pound sign and the up arrow to cycle through several different input modes to allow entry of lowercase letters, uppercase letters, numbers, or symbols. The top of the display area indicates which mode is active.

    - ***Letters Active:*** With uppercase or lowercase letters active, you cycle through the three letters on each button by repeatedly clicking the button until the letter you want on that button appears.

    - ***Symbols Active:*** With symbols active, the symbols appear in the display area and you use the navigation keys to locate the one you want. Click the center navigation key with the dot on it to select a symbol.

    - ***Erasing:*** The key with the left arrow lets you backspace over numbers, letters, or symbols.

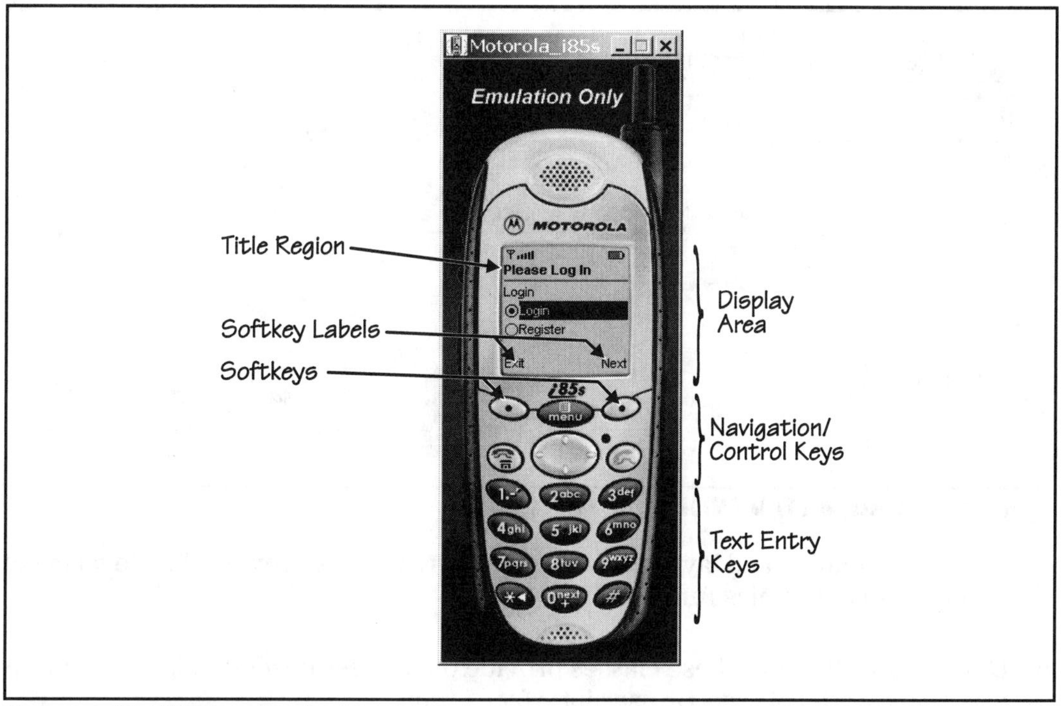

Figure 9-1. The MIDP User Interface.

MIDP APPLICATIONS

Unlike WML, which is a type of content and is represented by an XML document, MIDP applications are Java programs. So, instead of a text file structured by the rules of an associated DTD, we have Java classes and their associated hierarchies. Figure 9-2 shows some of the classes that a typical GUI MIDP application might use. The classes shown are a subset of the classes required to form a user interface for a MIDP application. We will discuss the more advanced classes for networking and persisting program data in the next chapter.

The classes shown in Figure 9-2 fall into three categories:

- ***MIDlet class***. This class provides the base class for all MIDP applications. Like an Applet class, the MIDlet class provides simple life cycle semantics such as start, pause, and destroy for the MIDP application as well as some callback methods for your MIDlet class.

- ***GUI widgets and low-level graphic classes***. These classes provide the raw user interface elements. The GUI widgets include text fields, string items, alerts, select choices and so on. Base classes such a **Screen** and **Displayable** provide

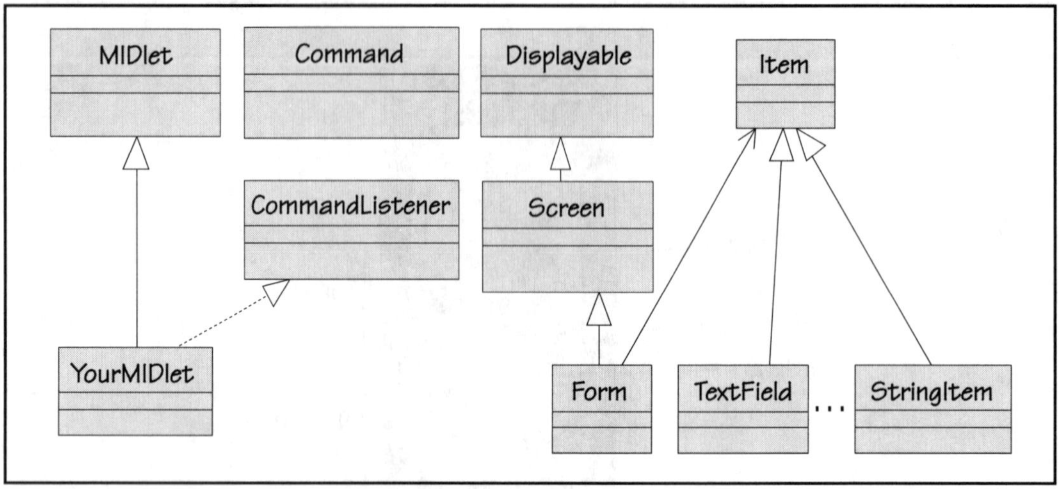

Figure 9-2. The Classes of a Typical MIDP Application.

basic containment and layout function. A **Canvas** class provides low-level event and primitive graphics functions.

- ***Command classes***. These classes provide the classes needed to handle actions associated with softkeys or user interface items.

DEVELOPING A MIDP APPLICATION

A MIDP application consists of three types of files:

- ***Java files:*** These files include the *.java* program source files, the *.class* compiled source files, and the preverified *.class* files.

- ***Bitmap images:*** MIDP applications can use PNG bitmaps to enhance the usability of the application.

- ***Deployment files:*** As with other Java applications, the Java ARchive (JAR) format is used to package and deploy the MIDP Java class files. In addition, a *Java Application Descriptor (JAD)* file is used to describe the deployment characteristics of a MIDP application.

Figure 9-3 illustrates the development, deployment, and execution of a MIDP application. Here are the steps:

1. ***Write the MIDP application.*** This is a programming task done using your favorite text editor and is similar to the creation of any other Java program. A

JAD file also needs to be created. Typically this is done using a wizard that your IDE or SDK provides.

2. ***Compile the application***. The Java files are compiled and prepared for testing in an emulator. This step is typically done in an IDE environment or a SDK with just the press of a button.
3. ***Test using one or more device emulators.*** All development environments for MIDP applications include device emulators that simulate a real device. As always, beware of the differences between an emulator and the real device.
4. ***Deploy to an application vending machine.*** Most MIDP devices are supported by a wireless carrier for network connectivity. These carriers also provide services for a customer to select and download MIDP applications (as well as other content) to their devices. These are called *vending machines* and are part of the *Over the Air (OTA)* provisioning strategy for MIDP applications. As the name implies, OTA provides a way for MIDP applications to be downloaded to the device over the wireless network.
5. ***Deploy the application to the device.*** Once a user selects an application for download, the MIDP program is downloaded and installed into the device. The JAD file is used to verify that the program is suitable for downloading to the device.
6. ***Run the MIDP application.*** After the application is deployed to the device, the program can be invoked and run by the user. The program does not have to be downloaded again for subsequent invocations by the user.

MIDP Development and Deployment: The Details

MIDP Java program development is not unlike other Java development. Java files are created, they are compiled, and then they are packaged. There are a few steps that are different though. Figure 9-4 illustrates the tasks necessary to take a MIDP Java file from creation to deployment. Here are the steps:

1. ***Create*** the Java (*.java*) file using an editor.
2. ***Compile*** the *.java* file using the javac command. This is the same javac command that the Java 2 Standard Edition (J2SE) JDK provides. The javac command produces a *.class* file.
3. ***Preverify*** the *.class* file. The class file produced by javac is not usable on a J2ME device. The preverification step simplifies the class-loading step on a J2ME device. A new *.class* file with preverification information is produced.
4. ***Package*** the verified *.class* file into a JAR (*.jar*) file.
5. ***Deploy*** the JAR file and the JAD file to the device. A JAD file that describes the MIDP program deployment is also required.

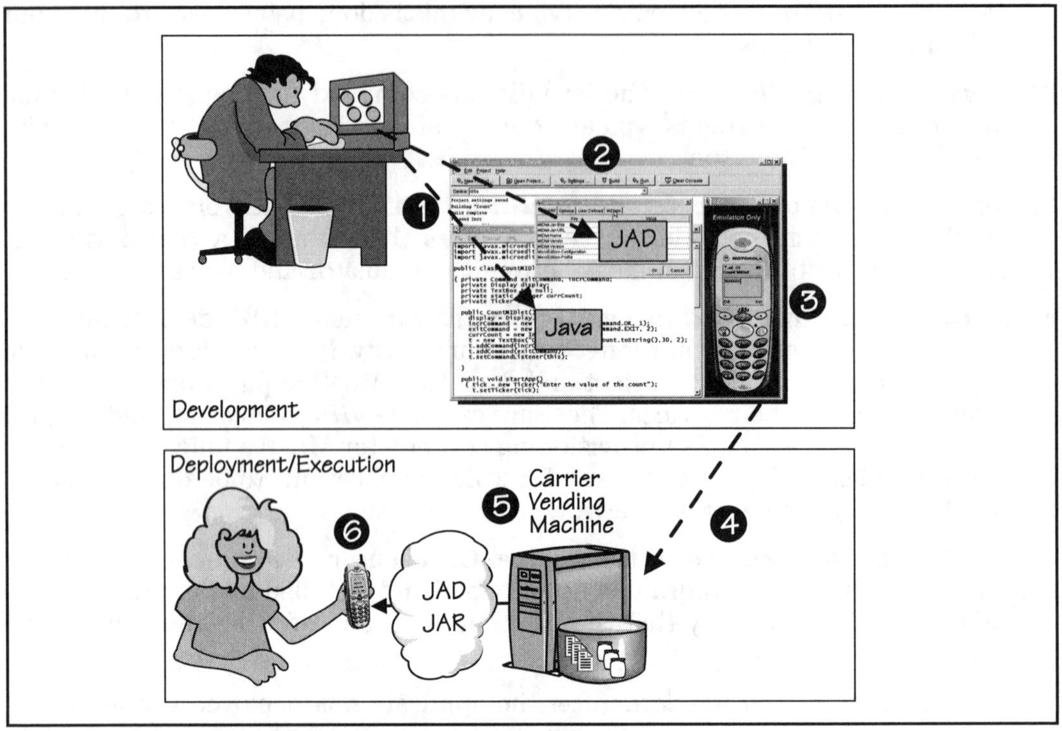

Figure 9-3. A J2ME MIDP Application from Developer to End User.

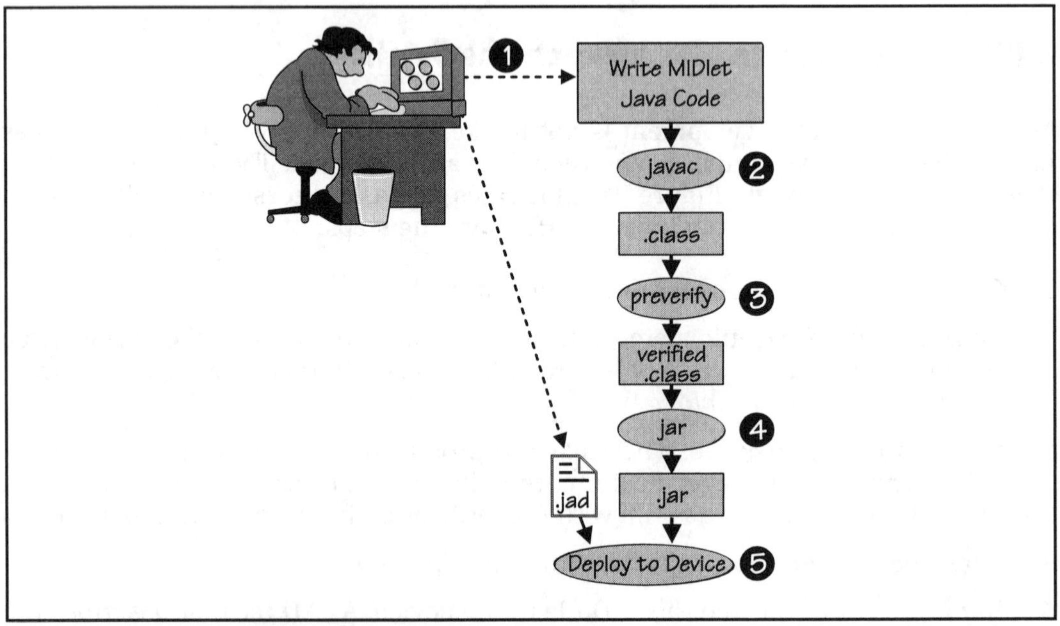

Figure 9-4. The Details of MIDP Application Development.

We're ready to take our first MIDlet through the development process. We will use a "University Registration" example developed using the MIDP APIs that is similar to the WML registration example. For the moment we will ignore the details of the program and focus on the development steps using Sun's J2ME Wireless Toolkit (J2MEWTK). We'll examine the code for the Registration example later in the chapter.

FYI

MIDP: The Tools of the Trade

Briefing

There are many devices in the market that provide MIDP capabilities from a variety of mobile handset vendors. Many of the handset vendors as well as Java tool vendors provide Software Development Kits (SDKs) and Integrated Development Environments (IDEs) that include device emulators to help you with your MIDP application development. Here is a sample of the development environments that are available:

- ***J2ME Wireless Toolkit from Sun Microsystems (java.sun.com/j2me):*** This toolkit is the reference implementation provided by Sun, available as a free download for the Windows, Linux, and Solaris platforms. This works as a standalone toolkit or in conjunction with Sun's Forte IDE. Currently the toolkit provides an implementation for the CLDC 1.0.3 and MIDP 1.0.3 specifications. This toolkit provides application development features such as compiling, preverification, packaging, and debugging. Source level debugging is supported when used in conjunction with Forte.
- ***JBuilder MobileSet from Borland Software Corp. (www.borland.com):*** This toolkit provides implementation for MIDP 1.0.3 that integrates with JBuilder 6.0 or later.
- ***WHITEboard SDK from Zucotto Wireless Inc. (www.zucotto.com):*** Apart from providing a J2ME toolkit, this toolkit extends MIDP access to the Bluetooth stack.
- ***WebSphere Studio Device Developer from IBM Corp. (http://www.embedded.oti.com/wdd/):*** Previously known as VisualAge Micro Edition, this toolkit provides a MIDP compatible development environment for handheld devices. It is capable of generating JAR and JAD files that can be deployed to a MIDP-compatible MID. The toolkit is based on IBM's J9 VM, instead of Sun's KVM.

It is very easy to develop MIDP compliant applications using these toolkits. But, be sure to treat a device emulator for what is: an emulator. It won't represent the true operational characteristics of the device it is emulating.

The only way to observe your MIDP program as your customers will see it is to run your program on all devices to which it will be deployed. Many of the handset and PDA vendors such as Nokia, Siemens, Motorola, and Palm have developer programs with discounted and loaner devices for testing. They also provide application deployment support. If you are serious about MIDP programming, then you should sign up for the developer programs for the devices to which you will be deploying. ❑

Create Your MIDlet

Your first task is to write a MIDlet program. Our program is called **RegMIDlet**. We will use the J2MEWTK to first create a project that will hold the Java source code as well as other files related to the RegMIDlet application.

The Toolkit is started by invoking the KToolbar application that comes with the J2ME Wireless Toolkit 1.0.3. From the menu bar click File->New Project. Figure 9-5 shows the New Project dialog that is presented. MIDlets that are related or share resources are normally packaged as a suite. A MIDlet suite forms the name space for all of the MIDlets contained in the suite. We will use Registration as the MIDlet Suite name and the Project Name. Our MIDlet suite will only contain one MIDlet named RegMIDlet. Enter the class name in the MIDlet Class Name field.

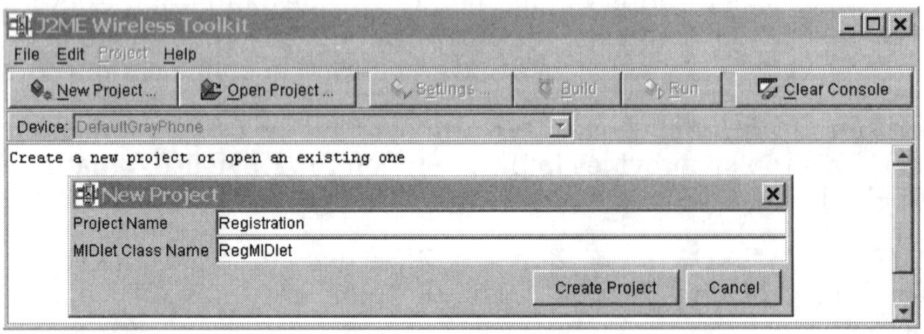

Figure 9-5. Creating a Project using the J2ME Wireless Toolkit.

When the "Create Project" button of the New Project dialog is clicked, the tabbed dialog, "Settings for project Registration" appears (see Figure 9-6). You can change any values necessary for your MIDlet suite using this dialog. The "Required" tab pane displays values necessary for the MIDlet Suite. Change the MIDlet name to RegMIDlet. The "Optional" pane shows values that can be added to the MIDlet as needed. The "User Defined" pane shows properties that can be accessed from within the application as properties, or name-value pairs. Finally, the MIDlets pane allows other MIDlets to be added to, removed from, or edited in the suite.

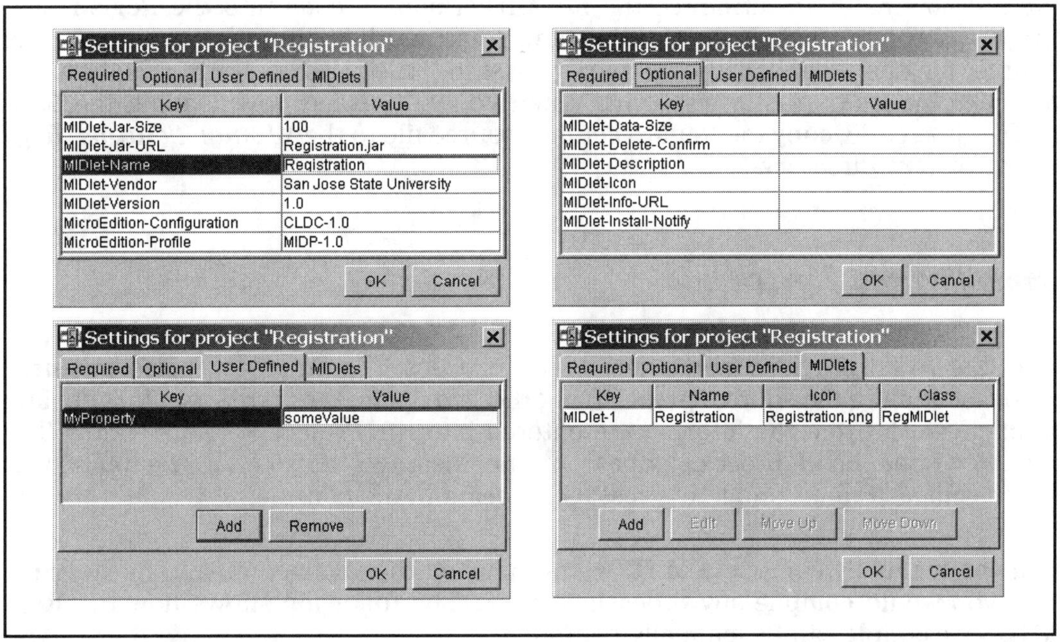

Figure 9-6. The MIDlet Suite Settings.

When you have completed entering the desired values in the dialog, click OK. All projects are created in the *apps* directory of the J2MEWTK installation directory with the name of the project you specified (in this case, *Registration)*. Informational messages about project creation (including where project files should be placed) appear in the message console when the project is created (see Figure 9-7).

Figure 9-7 illustrates the directory structure created for the Registration project. The Java source files are placed in the *src* directory. Application resources (such as images) are placed in the *res* directory. The JAR and JAD files are placed in the *bin* directory. Preverified class files are placed in the *classes* directory.

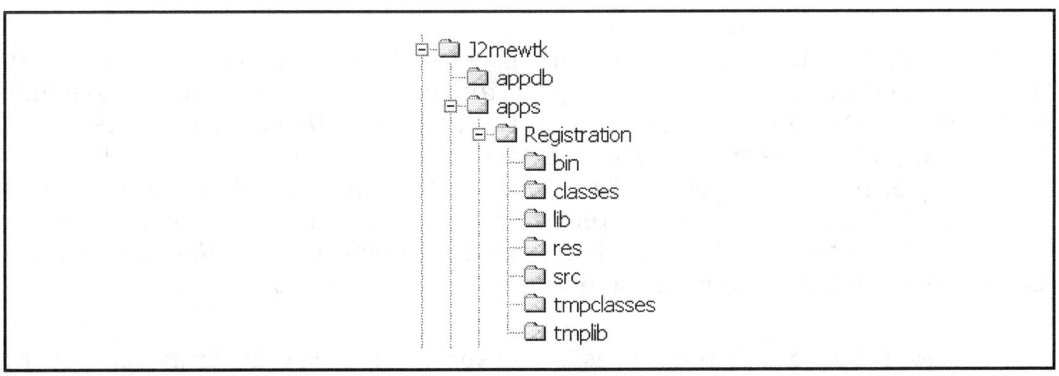

Figure 9-7. The J2MEWTK Project Directories.

We are now ready to introduce the RegMIDlet source code into the Registration project. You can create a source file using your favorite editor. You can obtain the source for the RegMIDlet example by visiting the Web site for this book at *http://wirelessjava.engr.sjsu.edu*. When you've finished downloading the RegMIDlet.java code, copy the RegMIDlet.java source file to the *\J2mewtk\apps\Registration\src* directory.

Compiling and Preverifying

The J2MEWTK provides a one-step "build" process. The build step performs three steps: compiling the source, preverifying the *.class* file, then writing a JAD file and a manifest file from the information entered into the Project Settings dialog. The results of the build process appear in the message console. If the build was successful, a "Build Complete" message appears in the console.

You can manually compile a MIDP class by using the javac command in the same way you would compile any other Java class. The following shows how the **RegMIDlet** example can be manually compiled:

```
javac -d ..\classes –bootclasspath c:\J2mewtk\lib\midpapi.zip
RegMIDlet.java
```

The build step also performs preverification. Preverification processes the *.class* file so the Java virtual machine that is part of CLDC can reject invalid or malicious classes without having to provide a complex class loader. Many of the steps required to assure that a class file's integrity is maintained is done during preverification and not while loading on the device. The device is simply responsible for running a few simple checks on the preverified class file to ensure that it is valid.

After compiling your program, you must process it with the preverify command. This command preprocesses the program for use by the CLDC virtual machine (KVM) without changing the name of a class. The *preverify* command takes a class or a directory of classes and preprocesses them. You can specify an output directory for the preverified classes; otherwise, a default directory named *output* is created and the classes are written to this directory. Skipping this step will produce an error if you try to run your application. The following shows how the *RegMIDlet.class* file can be manually preverified using the preverify command:

```
preverify –d ..\tmpclasses -classpath c:\J2mewtk\lib\midpapi.zip;
RegMIDlet.class
```

Packaging

The packaging of your MIDP application is done by invoking Project->Package from the menu bar. A JAR file with all of the preverified class files, resource files, and the manifest file for your MIDlet suite is created.

To manually package the RegMIDlet example, put the RegMIDlet class file into a JAR file as follows:

```
jar cvf Registration.jar RegMIDlet.class
```

The next step in packaging is creating a manifest file (or application descriptor), which provides information about the contents of the JAR file. Manifest attributes that start with *MIDlet-* are passed to the MIDlet when requested. The application descriptors are used by the MIDP *Application Management System (AMS)* to manage the MIDlet. The AMS provides a runtime environment for an executing MIDlet. We will describe the AMS in more detail later in this chapter. The application descriptor is also used by the MIDlet to configure specific attributes. The file extension of the application descriptor must be *.jad*.

There is a predefined set of attributes that can be used in every application descriptor. Table 9-1 describes all the application descriptors available for use in the manifest file.

Table 9-1. MIDP Application Descriptors.

Attribute Name	Description
MIDlet-name	Name of the MIDlet suite that identifies the MIDlets to the user.
MIDlet-version	The version number of the MIDlet suite. The format is major.minor.micro as described in the JDK product versioning specification. It can be used by the application management software for install and upgrade issues, as well as communication with the user.
MIDlet-Vendor	The company that provides the MIDlet suite.
MIDlet-Icon	The name of the PNG file within the jar file used to represent the MIDlet suite. Application management software can use this icon to identify the suite.
MIDlet-Description	The description of the MIDlet suite.
MIDlet-Info-URL	A URL for information further describing the MIDlet suite.
MIDlet-<n>	The name, icon, and class for the nth MIDlet in the jar file separated by a comma. The lowest value of <n> must be 1 and consecutive ordinals must be used. This is explained in detail later.

Table 9-1. MIDP Application Descriptors. (Continued)

Attribute Name	Description
MIDlet-Jar-URL	The URL from which the jar file can be loaded.
MIDlet-Jar-Size	The number of bytes in the jar file.
MIDlet-Data-Size	The minimum number of bytes of persistent data required by the MIDlet. The device may provide additional storage according to its own policy. The default is zero.
MicroEdition-Profile	The J2ME profile required, using the same format and value as the System property microedition.profiles (for example MIDP-1.0).
MicroEdition-Configuration	The J2ME configuration required, using the same format and value as the System property microedition.configuration (for example CLDC-1.0).

Listing 9-1 shows the descriptor file for the RegMIDlet.

Listing 9-1. The Application Descriptors for RegMIDlet.

```
MIDlet-1: Registration, Registration.png, RegMIDlet
MIDlet-Jar-Size: 10251
MIDlet-Jar-URL: Registration.jar
MIDlet-Name: Registration
MIDlet-Vendor: San Jose State University
MIDlet-Version: 1.0
MyProperty: someValue
```

The MIDlet-1 attribute has three parameters: name, icon, and class. The name appears in the device display so the MIDlet can be selected, the icon (if one is supplied) appears next to the name, and the class is the executable called to launch the MIDlet. The attribute name MIDlet-1 indicates this is the first, and in this case only, MIDlet. Other MIDlets, if any, referenced by this descriptor would be added as MIDlet-2, MIDlet-3, and so on.

Running

You are now ready to see your program in action. Select an emulation device from the Device menu and then click Run. Note that the system and tracing output is displayed in the message console as you interact with your program. You can also test with other devices by selecting them from the J2MEWTK device drop-down list.

Debugging

As with other programs, sometimes you can't understand why a program doesn't work as expected. The J2MEWTK gives you some insight into the behavior of your program by allowing the trace of method calls, exceptions, class loading, and garbage collection. You can enable the debug output from the toolkit's preferences menu, but be aware that this might produce a lot of output.

The J2MEWTK also provides source level debugging that can be used in conjunction with a debugger, such as one that might be included with an IDE (see Figure 9-8). This is implemented using the Java Debug Wire Protocol (JDWP). If you are familiar with any other tool that implements JDWP, you can just point it to the J2MEWTK debug port (the default is 5000). You can change this by hitting the preferences menu in the toolkit. The debugger allows you to step into the code, set break points and watches.

Of course, when it comes to debugging, there's always the old standby: *System.out.println*. Don't forget to remove these calls before shipping the application, as there is no way of viewing the output generated from these calls on real devices.

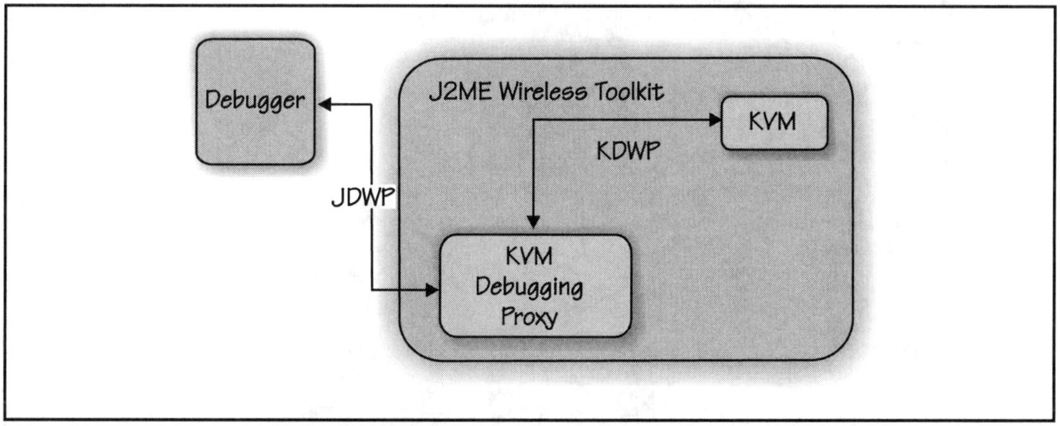

Figure 9-8. The J2MEWTK Debugging Architecture.

Deployment

Once you are satisfied with your MIDlet, you can deploy it to a Web server or to an Application Vending Machine Server by uploading the JAR and JAD files to the server. Once deployed to a Web server, your MIDlet is downloadable; however, you may need to add the following MIME type to your server's configuration:

```
text/vnd.sun.j2me.app-descriptor jad
```

Consumers who want to use the MIDlet can download the MIDlet to their handheld device if supported by their network provider. Some networks support Over The Air (OTA) provisioning of MIDlets, that is, the MIDlet could be downloaded by the device over the wireless network and installed on the device. In other cases, the MIDlet has to be installed by transferring the MIDlet to the device using a serial or USB cable.

The University Registration Example

We're now ready to revisit our Registration example (see Figure 9-9). The Registration example provides similar function to the WML registration example but

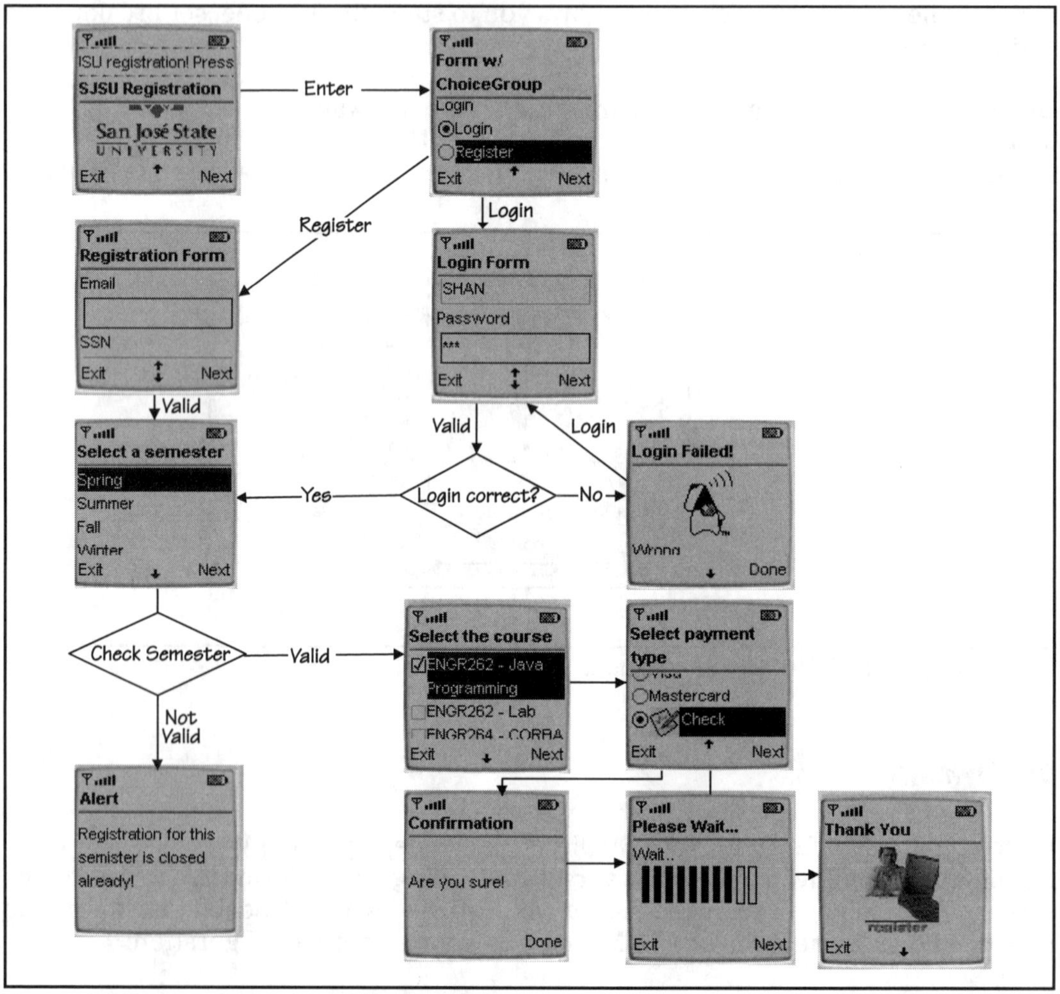

Figure 9-9. The Course Registration MIDlet Application.

provides a few new features that the MIDP classes allow us to provide (such as a progress gauge). The program also takes advantage of Java programming to provide more complex screen-to-screen navigation. With that quick introduction to the RegMIDlet application, let's take a look at the details of MIDP programming.

THE MIDP CLASSES

The bulk of the MIDP classes are grouped into four packages:

- The ***javax.microediton.midp*** package contains the **MIDlet** and **MIDlet StateChangeException** classes. The **MIDlet** class defines the interaction of a MIDP application with the runtime environment, called the Application Management System (AMS), and also defines the life cycle of the MIDlet.
- The ***javax.microedition.lcdui*** package contains classes that are used to provide the user interface for a MIDP application.
- The ***javax.microedition.io*** package provides networking support for MIDP applications.
- The ***javax.microedition.rms*** package provides classes that allow a MIDP application to persistently store and retrieve data.

The *io* and *rms* packages will be discussed in the next chapter.

The MIDlet Class

Every MIDP application must extend the MIDlet class. This class allows the Application Management Software (AMS) to control the life cycle of all of the MIDlets that are executing. The AMS provides the runtime execution environment for the all MIDlets running on a device. A MIDlet goes through a clearly defined life cycle when it is executing. Several methods are defined by the **MIDlet** class that allow the AMS to control the MIDlet and for the MIDlet to communicate with the AMS.

Figure 9-10 shows the MIDlet class. A MIDlet overrides the methods *destroyApp*, *pauseApp*, and *startApp* so it will be notified of state changes by the AMS. The *notifyDestroyed*, *notifyPaused*, and *resumeRequest* are methods that a MIDlet can use to communicate desired state changes to the AMS. The *getAppProperty* is used to query the AMS about properties contained in the JAD file.

MIDlet Life Cycle

Figure 9-11 illustrates the MIDlet life cycle state diagram. The MIDlet enters the *paused* state when first created by the AMS. The MIDlet can move back and forth between the *paused* state and the *active* state and is notified of the changes by the

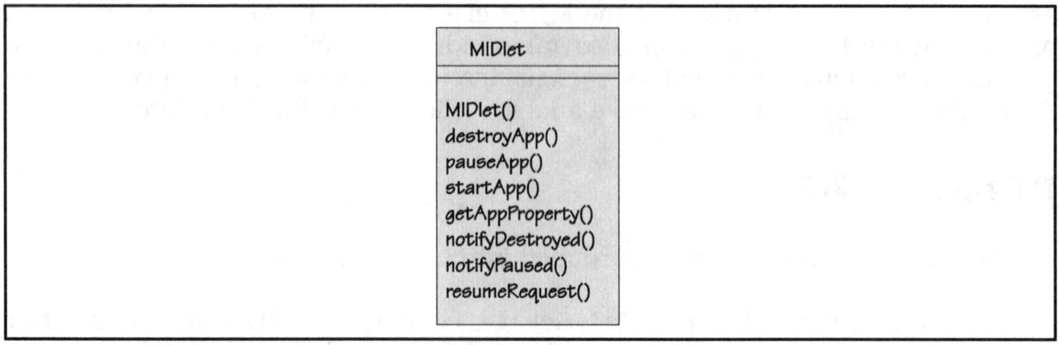

Figure 9-10. The MIDlet Class.

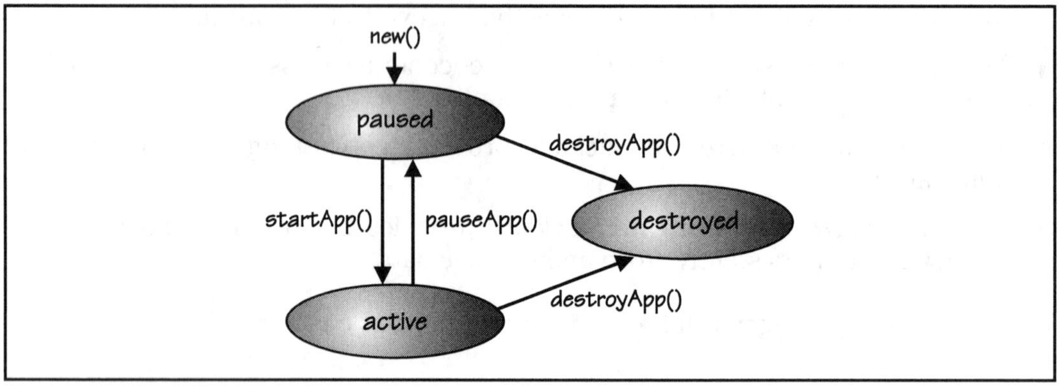

Figure 9-11. The MIDlet Life Cycle State Diagram.

startApp and *pauseApp* methods. If the MIDlet is destroyed, it is notified by the AMS via the *destroyApp* method and enters the *destroyed* state.

Table 9-2 describes the details of the MIDlet life cycle.

Table 9-2. The MIDlet Life Cycle Details.

State	Description
Paused	The MIDlet is initialized and is quiescent. It should not be holding or using any shared resources. This state is entered: ■ After the MIDlet has been created using *new*. The public no argument constructor for the MIDlet is called and returns without throwing an exception. The application typically does little or no initialization in this step. If an exception occurs, the application immediately enters the *Destroyed* state and is discarded. ■ From the *Active* state after the *pauseApp* method returns successfully. ■ From the *Active* state when the *notifyPaused* method returns successfully to the MIDlet. ■ From the *Active* state if *startApp* throws a **MIDletStateChangeException**.

Table 9-2. The MIDlet Life Cycle Details. (Continued)

State	Description
Active	The MIDlet is functioning normally. This state is entered just prior to calling the *startApp* method.
Destroyed	The MIDlet has released all of its resources and terminated. This state is entered: ■ When the *destroyApp* method returns except in the case when the unconditional argument is false and a **MIDletStateChangeException** is thrown. The *destroyApp* method shall release all resources held and perform any necessary cleanup so it may be garbage collected. ■ When the *notifyDestroyed* method returns successfully to the application. The MIDlet must have performed the equivalent of the *destroyApp* method before calling *notifyDestroyed*. Note: This state is only entered once.

Listing 9-2 shows an abbreviated view of the RegMIDlet Java source. The listing should give you an idea of the typical structure of a MIDlet application. Here are the parts:

- The ***import statements*** define the packages used by the MIDlet, in this case the lcdui and midlet packages.
- The ***class definition*** defines the name of the class, extends the **MIDlet** class, and defines any listener interfaces that need to be implemented.
- ***Private variables and constants*** are used to maintain references to any GUI constructs and command definitions that are created.
- The ***constructor*** is used to create any GUI constructs and initialize other resources that are needed by the MIDlet.
- The ***life cycle methods*** are used to notify the MIDlet of state changes.
- The ***event methods*** are methods that the MIDlet overrides to intercept events and commands related to the GUI interface. The event methods typically contain the logic for moving from screen to screen.

Listing 9-2. The Abbreviated Registration Midlet (RegMIDlet).

```
import javax.microedition.lcdui.*;
import javax.microedition.midlet.*;

public class RegMIDlet extends MIDlet implements CommandListener
```

```
{
  // variable definitions
  private Form form_welcome;

  // Command Definitions
  private static final Command nextCommand = new Command("Next",
                                        Command.SCREEN, 1);
      ...

// Constructor
public RegMIDlet()
{ // Load images
    try
    { image = Image.createImage("/sjsu.png");
      ...
    } catch (java.io.IOException e)
    { System.out.println("Error loading Image: " + e.toString());
    }

    // Instantiate Forms and Screens
    form_welcome = new Form("SJSU Registration");
      ...
}

  // get Display to be used by the MIDlet
  public void startApp() throws MIDletStateChangeException
  {
    display = Display.getDisplay(this);
    display.setCurrent(form_welcome);
  }


  // Pause the MIDlet
  public void pauseApp()
  {  }


  // Destroy the MIDlet
  public void destroyApp(boolean unconditional)
         throws MIDletStateChangeException
  { notifyDestroyed();
  }

  // Respond to commands
  public void commandAction(Command c, Displayable s)
```

```
  { if (c==exitCommand) // Always exit if user selects the exitCommand
    { try
      { this.destroyApp(true);
      } catch(MIDletStateChangeException e)
      {}
    }
    else if (s==form_welcome) // Splash Screen
      { display.setCurrent( form_select );
      }
    else if ...

  }
}
```

MIDP GUI Classes

With the development process and the high-level details of a MIDP application behind us, we're now ready to examine the details of the classes that are used to present the GUI interface for a MIDP application.

As we discuss the MIDP GUI classes, we'll use portions of the Registration MIDlet to illustrate the GUI components. Note that this MIDlet is a type of business application that collects information from a student, performs some logic based on the user input, and displays the output. These applications have to be highly portable across all devices that support MIDP. To achieve this portability, the Registration MIDlet is implemented using only high-level UI APIs provided by the *lcdui* package. These high-level APIs take advantage of the consistent look and feel defined by MIDP. In other words:

- The actual drawing to the MID's display is performed by the MIDP implementation. Applications do not define the visual appearance (such as shape, color, or font) of the components.
- Navigation, scrolling, and other primitive interactions are encapsulated by the MIDP implementation, and the application is not aware of these interactions.
- Applications cannot access concrete input devices like specific individual keys.
- When using the high-level APIs, it is assumed that the underlying implementation of MIDP will do the necessary adaptation to the device's hardware and native UI style.

LCDUI Classes: An Overview

Figure 9-12 illustrates the class and interface hierarchy for the LCDUI package. The classes in the diagram can be divided into the following categories:

- ***Base display classes:*** This includes the **Display** and **Displayable** classes. These classes abstract the device implementation including the device display and input keys and provide an abstract representation that will be displayed on the device screen.

- ***Command classes and interfaces:*** These include the **Command** class and the **CommandListener** and **ItemStateListener** interfaces. The **Command** class represents the semantics of an action such as the press of a softkey button. The **CommandListener** interface is implemented by a MIDlet so a **Command** action can be delivered to the MIDlet when an action occurs. The **ItemState-Changed** interface can also be implemented by a MIDlet so that it can be notified when the state of a widget that is part of a form changes. This might include choice field selections, entries into text and date fields, or interaction with a gauge.

- ***Screen classes:*** These classes include **TextBox**, **Alert**, and **List**. These classes encapsulate a complex user interface component such as a text field, alert, or list. The structure of these screens is predefined, and the application cannot add other components to these screens, except for a **Ticker** component.

- ***Form and item classes:*** The **Form** class is also a derivative of the **Screen** class but unlike the other screen classes, it can act as a container for widgets

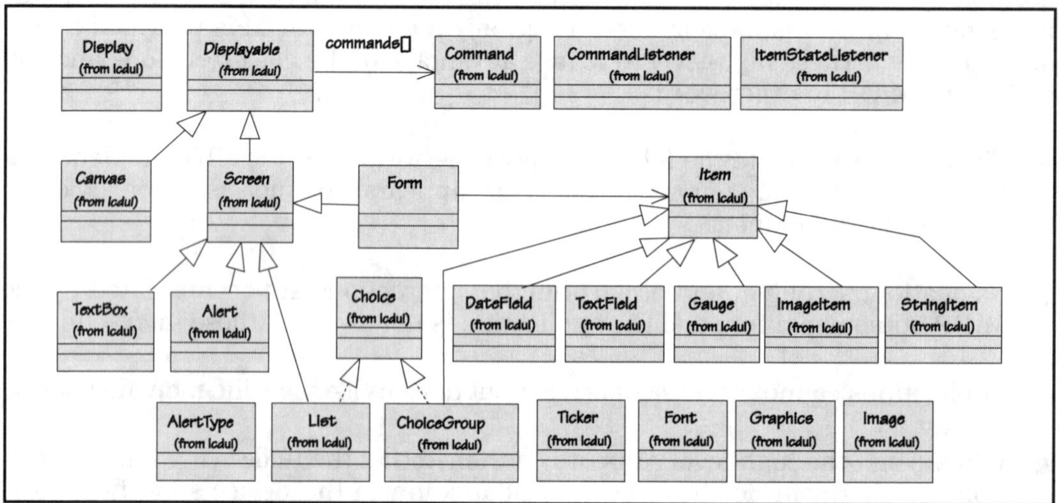

Figure 9-12. The MIDP UI Classes and Interfaces.

represented by classes that are derivatives of the **Item** class. These classes include **ChoiceGroup**, **TextField**, **DateField**, **ImageItem**, **StringItem**, and **Gauge**. A MIDP application uses these **Item** derivatives to populate a form with text, images, and simple sets of related UI components to represent a simple business UI construct.

- ***Low-level classes:*** The low-level classes include the **Canvas**, **Graphics**, **Font**, and **Image** classes. These classes represent the low-level display primitives and events and may not be completely portable between devices.

The following sections describe the details of the individual classes and provide programming details for the LCDUI classes.

The Display and Displayable Classes

The **Display** class represents the display and input manager of the device. Exactly one **Display** class is instantiated per MIDlet. The *getDisplay* method returns the **Display** object. This object is valid from the beginning of the *startApp* method to the end of the *destroyApp* method.

A screen or canvas (which we will discuss later in this chapter) is made visible by calling the *setCurrent* method of the **Display** class. The *getCurrent* method returns the current **Displayable** object. The *isColor* and *numColors* methods return information about the device's display (see Figure 9-13).

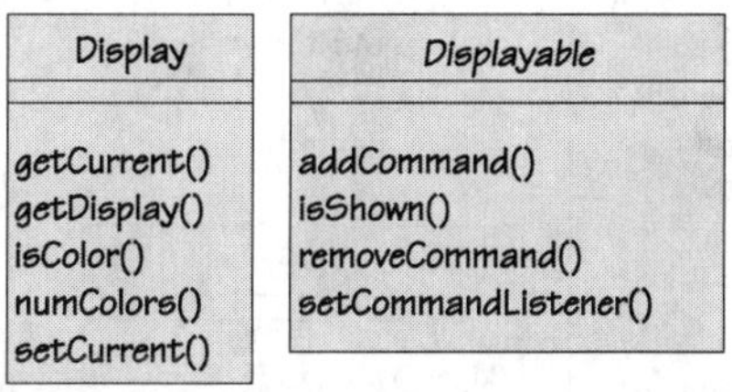

Figure 9-13. The MIDP LCDUI Display Classes.

The following is an example *startApp* method that gets the current **Display** object then sets the current screen to be the welcome form:

```
// get Display to be used by the MIDlet
public void startApp() throws MIDletStateChangeException
{
  display = Display.getDisplay(this);
  display.setCurrent(form_welcome);
}
```

The Screen Class

Given the resource constraints of the devices, a MID can only display one screen at a time. So, all the UI components of MIDP are screen oriented and all the components are displayed as part of a screen. The **Screen** class takes care of all events that occur as the user navigates within the screen. Only the high-level events are passed on to the application. For example, consider a screen with two text fields and a submit button. The navigation from one text field to the other and the highlighting or selection of text is handled by the screen. Simple events, such as when the user navigates from one text field to another, are not handed to the application. However, any action associated with the submit button is passed to the application. We will discuss event handling in an upcoming section. The **Screen** class hides implementation-specific functionality such as component layout, scrolling, and focus traversal.

Figure 9-14 shows the classes and the methods for the screen-based user objects. The **Screen** class provides methods for getting and setting a title for the screen and for associating a ticker with the screen.

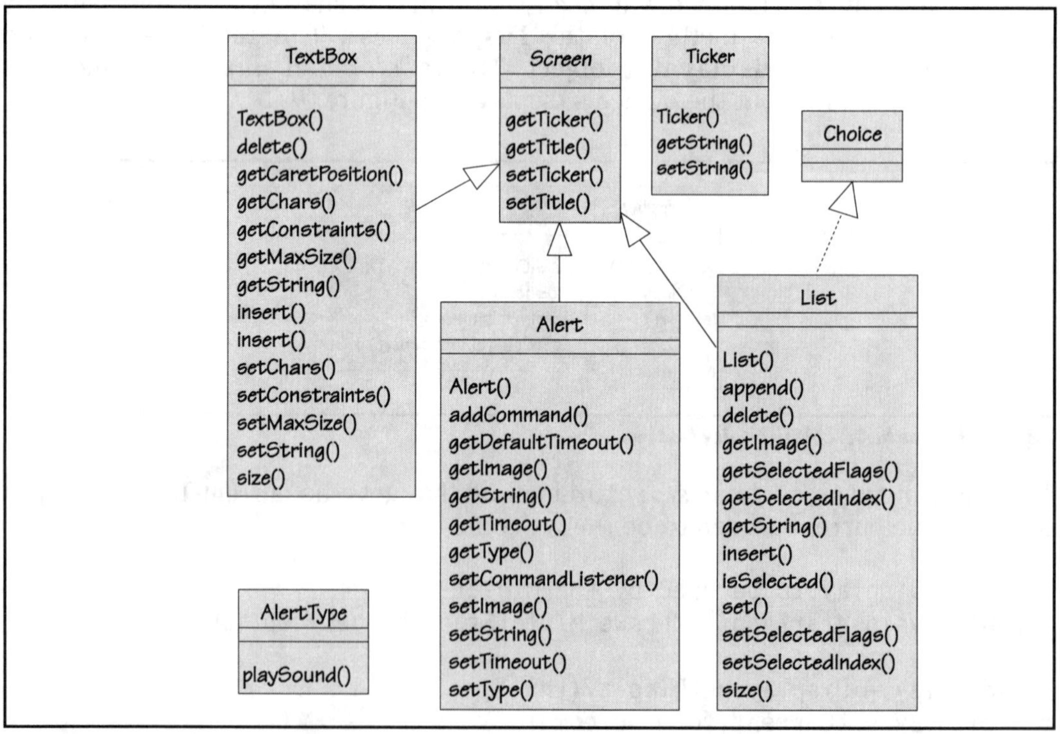

Figure 9-14. The Screen-based UI Classes.

The Ticker Classes

The **Ticker** class provides a piece of text that continuously moves across a screen. The class provides methods for setting and getting the string text for the ticker. The ticker can be associated with one or more screens and will be shown in the same location for all screens on the device display. The following shows an example of a ticker that is associated with a form that is the Welcome page for the Registration example:

```
// Welcome: Form with ticker and image
form_welcome = new Form("SJSU Registration");
form_welcome.setTicker(
new Ticker("Welcome to SJSU registration!" +
Press next " +"to login or register"));
```

The Alert Class

An alert presents error and other information about the execution of a MIDP application to the user. The alert can contain text and an image. The alert is presented for a specified time to the user before proceeding to the next screen. Methods are provided to set the text string, to set the image for the alert, to set the time-out value, and for setting how commands are handled. When the alert is constructed, an AlertType can be specified. The AlertType can be ALARM, CONFIRMATION, ERROR, INFO, or WARNING. The AlertType class also defines a *playSound* method that can cause the device to produce a device-specific sound.

The following shows an example of an error alert that is produced if a login fails for the Registration example:

```
alert_error = new Alert("Login Failed!",
                  "Wrong Username/passwd", null,
                   AlertType.ERROR);
```

The List Class

The **List** class presents a screen with a list of choices. The user selects a choice with the device's select key. The choices can be specified to be:

- IMPLICIT: Selection causes immediate notification to the MIDlet via the **CommandListener** interface.
- EXCLUSIVE: The state of the selected item is changed but the MIDlet is not notified.
- MULTIPLE: The state of the focused item is changed. The MIDlet is not notified.

In the latter two cases, the application can query the state of the selected items when the user presses a softkey that causes an action.

The **List** class provides methods to add items to the list, to get and set the state of the list items, to remove items from the list, and to query the length of the list. The following shows the creation of an implicit list that allows the selection of a semester for the Registration example:

```
list_semester = new List("Select a semester",
                         Choice.IMPLICIT );
list_semester.append("Spring", null);
list_semester.append("Summer", null);
list_semester.append("Fall", null);
list_semester.append("Winter", null);
```

The Textbox Class

A MIDlet **TextBox** class defines three areas: title, contents, and command. When you create a **TextBox** component, you specify text for the title and the initial text contents. The **TextBox** maintains a text string of a specified maximum size. Methods are provided by the **TextBox** class to set and get the text string, move and locate the text entry cursor, determine the cursor location, and to insert text into the current string. The type of string that the **TextBox** holds (such as numeric, URL, and password), its length, and its initial value can be specified during construction.

The following is an example of a TextBox that hold a numeric text field that is 30 characters long and is initialized to a string that represents the integer 0:

```
currCount = new Integer(0);
t = new TextBox("Count MIDlet",
     currCount.toString(), 30, TextField.NUMERIC);
incrCommand = new Command("Incr", Command.OK, 1);
exitCommand = new Command("Exit", Command.EXIT, 2);
t.addCommand(incrCommand);
t.addCommand(exitCommand);
```

The Form Class

The **Form** class is a screen that can hold an arbitrary mixture of elements that are derivatives of the **Item** class. The **Form** class provides the layout and event handling that allows the navigation between the **Form** elements without the intervention of the MIDP application.

Figure 9-15 shows the **Form** classes and the **Item** classes that can be placed into a **Form**. The **Form** class provides methods to edit the contents of a **Form** using *insert*, *append*, *set*, and *delete* methods. The *setItemStateListener* method can be used to indicate the class that will handle state changes of the items within the form. The *size* method returns the number of items in the form.

The Choice and ChoiceGroup Classes

The **ChoiceGroup** class is used to add a list of selectable items to a form. The **ChoiceGroup** class implements a number of methods defined in the **Choice** interface. The list items may contain text and an optional image. The **ChoiceGroup** class has list types EXCLUSIVE and MULTIPLE similar to the **List** screen type. The list type IMPLICIT is invalid for **ChoiceGroup** objects.

The **ChoiceGroup** class (like the **List** class) provides methods to add items to the list, to get and set the state of the list items, to remove items from the list, and to query the length of the list. The following two examples show the creation of an EXCLUSIVE and a MULTIPLE list for the Registration MIDlet:

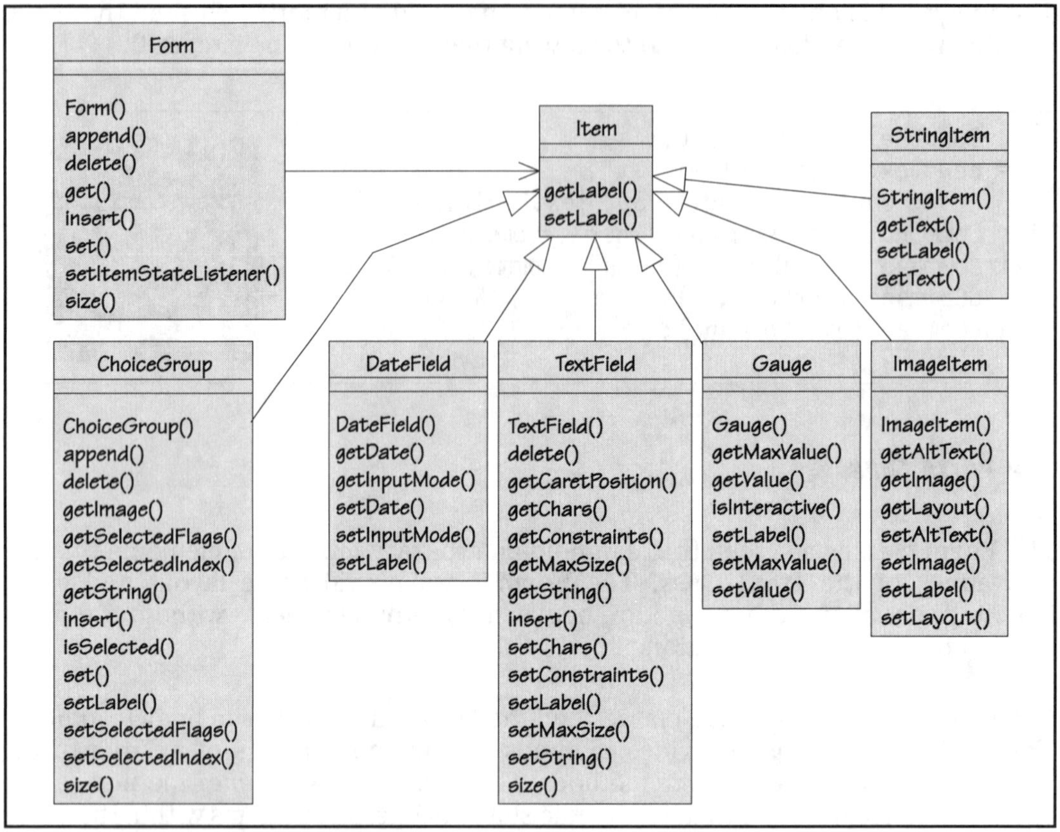

Figure 9-15. The Form and Related LCDUI Classes.

```
// Select: Form with exclusive choice group
form_select = new Form("Please Log In");
cg1 = new ChoiceGroup(Login", Choice.EXCLUSIVE );
cg1.append("Login", null );
cg1.append("Register", null);
form_select.append(cg1);
```

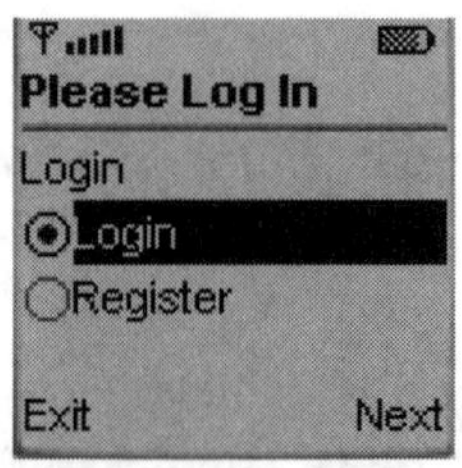


```
// Courses: Explicit multiple list box
list_courses = new List("Select a course",
                        Choice.MULTIPLE);
list_courses.append("ENGR262 - Java Programming",
                    null);
list_courses.append('ENGR262 - Lab', null);
list_courses.append("ENGR264 - CORBA and Java",
                    null);
```

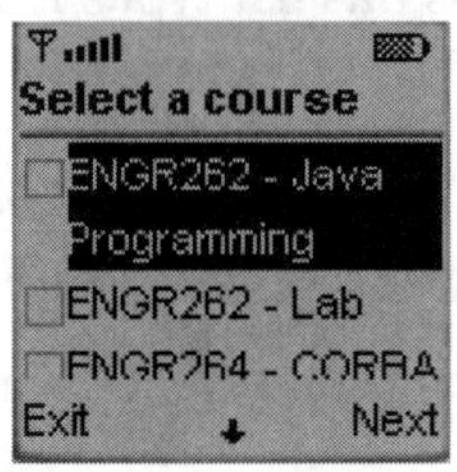

The StringItem Class

The **StringItem** class allows you to place a text string within a form. A **StringItem** holds a string and a label that cannot be edited by the user. However, the application program can change both the label and the string using methods provided by the **StringItem** class. The example below shows a form with a **TextField** and a **StringItem**:

```
form = new Form("CountMIDlet");
countInput = new TextField("Enter Count Value",
    currCount.toString(), 50, TextField.NUMERIC);
outText = new StringItem("Avg Time=","0");
```

The TextField Class

A **TextField** defines an editable text item that can be placed in a form. A **TextField** can have a label, a maximum size for the text string that it can store, and the type of input constraints that it supports (such as NUMERIC, PHONENUMBER, PASSWORD, and URL). The following is an example of a form that has two text fields. One field is 50 characters long with a label of UserName and accepts any type of text. The other field is also 50 characters long with a label of Password and the field type is password. Both fields are initialized to the empty string initially.

```
// Login: Form with Text Fields
form_login = new Form("Login");
userName = new TextField("UserName", "", 50,
                        TextField.ANY);
form_login.append(userName);
passwd = new TextField("Password", "", 50,
                      TextField.PASSWORD);
form_login.append( passwd );
```

The DateField Class

The **DateField** defines a field that can be put into a form for editing dates. The field can represent date, time, or date and time. The class provides methods to get and set the date and time. The following shows the definition of a **DateField** and the interaction a user might experience on a MIDP device. When the user selects the field, the device will provide a device-dependent method for date and time entry.

```
form_register =
  new Form("Registration Form");
        . . .
form_register.append(
new TextField("URL", "",
       50,TextField.URL) );
form_register.append(
new DateField("Birth Date",
       DateField.DATE));
```

The ImageItem Class

The **ImageItem** class allows you to place an image on a form. The image can have a label and can be placed on the form according to layout constants defined by the ImageItem class. Images can be placed horizontally (left, right, or center) and can have a new line start before or after the image. Alternate text can also be defined for the image. This text is displayed when the device is unable to display the image.

The following shows an example of an image that is used as a closing screen for the Registration example:

```
// Thanks: Form with text and image
form_thanks = new Form("Thank You");
form_thanks.append(new ImageItem( "", endImg,
        ImageItem.LAYOUT_CENTER +
        ImageItem.LAYOUT_NEWLINE_BEFORE +
        ImageItem.LAYOUT_NEWLINE_AFTER,
        "Thank You"));
```

The Gauge Class

The **Gauge** class displays a bar graph with a specific value. The **Gauge** class can be added to a form and can be used to display a progress indicator or display an application value in a graphical manner. Gauges can be interactive (the user can set and retrieve the gauge value) or non-interactive (the user cannot change the value). Gauges have a maximum value. The number of bars shown for a gauge is device dependent.

The following shows a class from the Registration MIDlet that extends the **Gauge** class. This class implements a progress indicator. The class includes a thread to manipulate the progress indicator's values.

```
class NonInteractiveGauge extends Gauge
                           implements Runnable
{
// Indicates the number of units
// to move per iteration
private int movement = 1;

public NonInteractiveGauge(String title)
{ super(title, false, 30, 0);
  new Thread(this).start();
}

// This method moves the gauge left and right.
public void run()
{ while (true)
  { int newValue = this.getValue() + movement;
    if(newValue >= 30)
    { movement = -1;
    } else
    { if(newValue <= 0)  movement = 1;
    }
    this.setValue(newValue);
    try
    { Thread.currentThread().sleep(1000);
    } catch(InterruptedException err)
    {}
  }
 }
}
```

The MIDlet High-Level Events

User interfaces are meant to interact with the user. So when a user interacts with a MIDlet loaded on a mobile information device, application events are generated. For example, in the Registration example, the user enters some information and then presses a softkey. The application is notified of this action and responds to the generated event.

An application is notified of such events through *callback* methods. Callbacks are invocation of methods that are executed by the application in response to actions taken by a user at run time. These callback methods are defined in listener interfaces that are implemented by the MIDlet.

MIDP defines two levels of user interface APIs: high-level and low-level. This is true for events also. MIDP defines high-level events (such as selecting an item from a list) and low-level events (such as pressing a key on the device). There are four kinds of user interface callbacks in MIDP:

- Abstract commands that are part of the high-level API.
- Low-level events that represent single key presses and releases.
- Calls to the *paint* method of a **Canvas** class.
- Calls to a **Runnable** object's *run* method requested by a call to the *callSerially* method of the **Display** class.

In this section we will discuss the high-level events and defer the discussion of low-level events until later in the chapter.

Handling events in the high-level API is based on a listener model. The **Screen** and **Canvas** objects can have listeners for commands. For an object to be a listener, it must implement the **CommandListener** interface.

You add commands to a displayable object with the *addCommand* method and register a listener using the *setCommandListener* method. Both of these methods are part of the **Displayable** class and they are inherited by the **Screen** and **Canvas** classes. Figure 9-16 illustrates the **Command** class and the **CommandListener** and **ItemStateListener** interfaces.

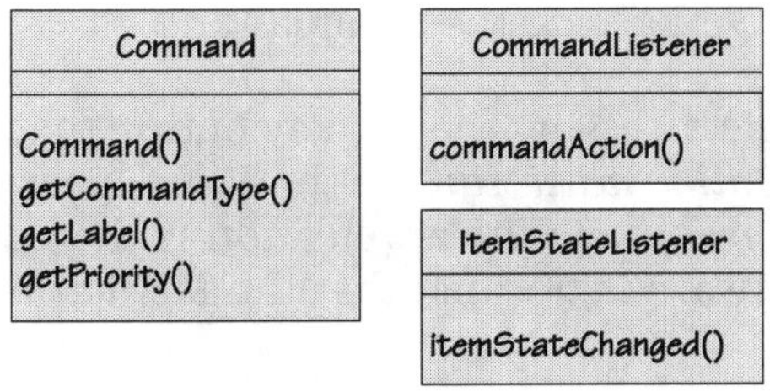

Figure 9-16. The Command and Related Listener Interfaces.

The Command Class

The **Command** class encapsulates the semantic information of an action that a user performs. The command itself contains only information about a command, but not the actual action that happens when a command is activated. The action is defined in a **CommandListener** object associated with the screen. Here are the **Command** objects defined for the RegMIDlet example:

```
private static final Command nextCommand = new Command("Next",
                                            Command.SCREEN, 1);
private static final Command exitCommand = new Command("Exit",
                                            Command.EXIT, 1 );
```

As you can see, the **Command** class constructor takes three parameters, and therefore, contains the following three pieces of information: label, type, and priority.

- The label is a string used for the visual representation of the command. The position of this label is implementation specific. Normally they are placed on top of the soft buttons we discussed in our RegMIDlet example.

- The type specifies the command's intent. The defined types are:

    - BACK, CANCEL, EXIT, HELP, ITEM, OK, SCREEN, and STOP.

- The priority value describes the importance of this command relative to other commands on the screen. A priority value of 1 indicates the most important command, and higher priority values indicate commands of lesser importance.

A higher priority command will most likely be configured by the MID to respond to one of the MID's more easily accessible keys. A lesser priority command might be mapped to an infrequently used key, or a multiple key combination. The command-to-key mapping is device dependent and handled by the device according to the type specified when you create the command.

In the RegMIDlet example, pressing the softkey button that has the *Exit* label above it calls the MIDlet's *commandAction* method and passes it a reference to the command called *exitCommand*. The application is then destroyed, and the application management software is notified that the MIDlet has entered the destroyed state (*notifyDestroyed*).

Pressing the Next button calls the MIDlet's *commandAction* method and passes it a reference to the command called *nextCommand*. If the command action is a command of type *Next*, then a screen-specific operation is performed. No notification is required and the application remains in the active state.

For some devices, pressing a control key calls the MIDlet's *destroyApp(true)* method. The environment returns to the application manager when either the EXIT softkey button or the Command button is pressed.

The CommandListener Interface

The **CommandListener** interface defines the callbacks for MIDlets that need to receive high-level events from the implementation. This interface has one method that a listener must implement, which is the *commandAction* method.

```
public void commandAction(Command c, Displayable d)
```

The *c* parameter is a command object that identifies the command (if any) that has been added to **Displayable** with the *addCommand* method. The *d* parameter is the displayable where the event occurred. The following code snippet adds the RegMIDlet commands to the welcome form:

```
form_welcome.addCommand(exitCommand);
form_welcome.addCommand(nextCommand);
form_welcome.setCommandListener(this);
```

Notice that we are setting the class that implements the **CommandListener** interface as the event handler. In this case, it is the RegMIDlet class (this) where the implementation for the *commandAction* method is provided. The following is an abbreviated view of the RegMIDlet *commandAction* method:

```
// Respond to commands
public void commandAction(Command c, Displayable d)
{ if (c==exitCommand) // Always exit if user selects the exitCommand
  { try
```

```
        { this.destroyApp(true);
        } catch(MIDletStateChangeException e)
        {}
      }
      else if (s==form_welcome) // Splash Screen
      { display.setCurrent( form_select );
      }
      else if ...
   }
```

The RegMIDlet performs actions based on the command invoked and based on the screen where command is invoked. When the *Exit* key is pressed the MIDlet is destroyed and the application exits regardless of which screen is current. If the command is a Next command, an action is performed such as traversing to the next screen or making a network call.

The ItemStateListener Interface

The **ItemStateListener** interface provides a callback mechanism for events associated with the change in state of widgets that are derivatives of the **Item** class. Recall that derivatives of the **Item** class are placed in a form. Applications implement this interface's *itemStateChanged* method to trap the following events when the user:

- Changes the set of selected values in a **ChoiceGroup**.
- Adjusts the value of an interactive **Gauge**.
- Enters or modifies the value in a **TextField**.
- Enters a new date or time in a **DateField**.

The following is an example of setting a listener for item state changes:

```
form.setItemStateListener(this);
```

MIDP Applications and Threading

Our registration example is not multithreaded. But even if it were to be multithreaded, the MIDP user interface APIs are thread-safe. Events are handled serially. This means the user interface callbacks are called as soon as the previous callback returns. For example let us assume there are two keys: one associated with selection of some value and the other to save the value to the database. When the user presses both the keys in succession, the action associated with the second key (saving to the database) happens only after the action associated with selection is completed.

MIDP also provides a mechanism for event synchronization. An application can use the *callSerially* method of the **Display** class to execute an operation serially with events. This is more useful when dealing with graphics, which we will be our next topic of discussion.

The Registration Example Listing

That was a whirlwind overview of the MIDP application programming and the MIDP user interface APIs. Listing 9-3 shows the complete code example for the RegMIDlet application. Hopefully you will be able to read it without explanation having survived the previous sections.

Listing 9-3. The Registration MIDlet Code.

```
import javax.microedition.lcdui.*;
import javax.microedition.midlet.*;

public class RegMIDlet extends MIDlet implements CommandListener
{
  private Form form_welcome;
  private Form form_select;
  private Form form_login;
  private Form form_register;
  private Form form_gauge;
  private Form form_thanks;
  private List list_semester;
  private List list_payment;
  private List list_courses;
  private List list_alerts;
  private Alert alert5;
  private Alert alert_confirmation;
  private Alert alert_error;
  private Alert alert_warning;
  private Display display;
  private ChoiceGroup cg1;
  private TextField userName;
  private TextField passwd;
  boolean payed = false;

  // Commands
  private static final Command nextCommand = new Command("Next",
                                                Command.SCREEN, 1);
  private static final Command exitCommand = new Command("Exit",
```

```
                                          Command.EXIT, 1 );

// Initialize all of the screens used by the MIDlet
public void startApp() throws MIDletStateChangeException
{ display = Display.getDisplay(this);
  Image endImg = null;
  Image errImg = null;
  Image image = null;
  Image check = null;

  // Load all of the images from the jar file
  try
  { image = Image.createImage("/sjsu.png");
    endImg = Image.createImage("/reg.png");
    check = Image.createImage("/check.png");
    errImg = Image.createImage("/err.png");
  } catch (java.io.IOException e)
  { System.out.println("Error loading Image: " + e.toString());
  }

  // Welcome: Form with text and image
  form_welcome = new Form("SJSU Registration");
  form_welcome.setTicker(newTicker("WelcometoSJSUregistration!Press
         next " + "to login or register"));
  form_welcome.append(new ImageItem("", image,
      ImageItem.LAYOUT_CENTER + ImageItem.LAYOUT_NEWLINE_BEFORE +
      ImageItem.LAYOUT_NEWLINE_AFTER, "alt text"));
  form_welcome.addCommand(exitCommand);
  form_welcome.addCommand(nextCommand);
  form_welcome.setCommandListener(this);
  display.setCurrent(form_welcome);

  // Select: Form with exclusive choice groups -- Radio Buttons
  form_select = new Form("Please Log In");
  cg1 = new ChoiceGroup( "Login", Choice.EXCLUSIVE );
  cg1.append( "Login", null );
  cg1.append( "Register", null );
  form_select.append( cg1 );
  form_select.addCommand(exitCommand);
  form_select.addCommand(nextCommand);
  form_select.setCommandListener(this);

  // Login: Form with Text Fields
  form_login = new Form("Login");
  userName = new TextField("UserName", "", 50, TextField.ANY);
```

```
form_login.append(userName);
passwd = new TextField("Password", "", 50, TextField.PASSWORD);
form_login.append( passwd );
ChoiceGroup cg2 = new ChoiceGroup( "Multiple", Choice.MULTIPLE );
cg2.append( "Save Login", null );
form_login.append(cg2);
form_login.addCommand(exitCommand);
form_login.addCommand(nextCommand);
form_login.setCommandListener(this);

// Register: Form with all types of text fields
form_register = new Form("Registration Form");
form_register.append(new TextField("Name","", 50, TextField.ANY) );
                                      form_register.append(new
TextField("Email","",50,TextField.EMAILADDR));
form_register.append(new TextField("SSN","",50,TextField.NUMERIC));
form_register.append(new TextField("Password", "",50,
                    TextField.PASSWORD) );
form_register.append(new TextField("Phone", "", 50,
                    TextField.PHONENUMBER) );
form_register.append(new TextField("URL", "", 50,TextField.URL) );
form_register.append(new DateField("Birth Date",DateField.DATE) );
form_register.addCommand(exitCommand);
form_register.addCommand(nextCommand);
form_register.setCommandListener(this);

// Thanks: Form with text and image
form_thanks = new Form("Thank You");
form_thanks.append(new ImageItem( "", endImg,
                  ImageItem.LAYOUT_CENTER +
                  ImageItem.LAYOUT_NEWLINE_BEFORE +
                  ImageItem.LAYOUT_NEWLINE_AFTER, "alt text" ) );
form_thanks.addCommand(exitCommand);
form_thanks.setCommandListener(this);

// Guage: Form with Non-interactive Gauge
form_gauge = new Form("Please Wait...");
form_gauge.append( new NonInteractiveGauge("Wait..") );
form_gauge.addCommand(exitCommand);
form_gauge.addCommand(nextCommand);
form_gauge.setCommandListener(this);

// Semsester: implicit list box
list_semester = new List( "Select a semester", Choice.IMPLICIT );
list_semester.append("Spring", null);
```

```
list_semester.append("Summer", null);
list_semester.append("Fall", null);
list_semester.append("Winter", null);
list_semester.addCommand(exitCommand);
list_semester.addCommand(nextCommand);
list_semester.setCommandListener(this);

// Courses: Explicit multiple list box
list_courses = new List( "Select the course", Choice.MULTIPLE);
list_courses.append("ENGR262 - Java Programming", null);
list_courses.append("ENGR262 - Lab", null);
list_courses.append("ENGR264 - CORBA and JAVA", null);
list_courses.append("ENGR264 - Lab", null);
list_courses.append("ENGR296J - Mobile Java Programming", null);
list_courses.append("ENGR296J - Lab", null );
list_courses.append("ENGR268 -  Enterprise Java Beans", null);
list_courses.append("ENGR268 - Lab", null);
list_courses.addCommand(exitCommand);
list_courses.addCommand(nextCommand);
list_courses.setCommandListener(this);

// Payment: Exclusive list box
list_payment = new List( "Select payment type", Choice.EXCLUSIVE);
list_payment.append( "American Express", null);
list_payment.append( "Visa", null);
list_payment.append( "Mastercard", null);
list_payment.append( "Check", check);
list_payment.addCommand(exitCommand);
list_payment.addCommand(nextCommand);
list_payment.setCommandListener(this);

// 15 second alert
alert5 = new Alert("Alert");
alert5.setString("Registration closed for this semester closed!");
alert5.setTimeout(15000);

// Different Alert types
alert_confirmation = new Alert("Confirmation",
                     "Are you sure!", null, AlertType.CONFIRMATION);
alert_confirmation.setTimeout(Alert.FOREVER);
alert_error = new Alert("Login Failed!",
              "Wrong Username/passwd", null, AlertType.ERROR);
alert_error.setImage(errImg);
alert_error.setTimeout(Alert.FOREVER);
alert_warning = new Alert("You need to register for lab also!",
```

```
                              "", null, AlertType.WARNING);
  alert_warning.setTimeout(Alert.FOREVER);
}


// Pause the MIDlet
public void pauseApp() { /* Not implemented */ }

// Destroy the MIDlet
public void destroyApp(boolean unconditional)
       throws MIDletStateChangeException
{ notifyDestroyed();
}

/**
 * Respond to commands on input screens
 */
public void commandAction(Command c, Displayable s)
{ if (c==exitCommand) // Always exit if user selects the exitCommand
  { try
    { this.destroyApp(true);
    } catch(MIDletStateChangeException e)
    {}
  }
  else if (s==form_welcome) // Splash Screen
    { display.setCurrent( form_select );
    }
  else if (s==form_select) // Selection form
    { int index= cg1.getSelectedIndex();
      if (index == 0)
      { display.setCurrent( form_login); // Login form
      }
      else
      { display.setCurrent( form_register);// Registration form
      }
    }
  else if (s== form_login) // Form with choice groups
    { if (userName.getString().equals("")) //Username cannot be null
      { display.setCurrent(alert_error);
        alert_error.getType().playSound(display);
        } else
        { display.setCurrent( list_semester ); // List Semesters
        }
    }
  else if (s==form_register )
```

```
    { display.setCurrent( list_semester ); //List Semesters
    }
  else if (s==list_semester)
    { switch (list_semester.getSelectedIndex())
      { case 0: display.setCurrent( alert5 );// Registration is closed!
                break;
        default:display.setCurrent( list_courses );
      }
    }
  else if (s==list_courses) // List the courses offered
    { if ( (list_courses.isSelected(0)) &&
          !(list_courses.isSelected(1)))
      { display.setCurrent(alert_warning);// Enforce Lab
      } else
      { display.setCurrent( list_payment);//List payment methods
      }
    }
  else if (s==list_payment) // Show the confirmation & progress bar
    { if (payed)
      { display.setCurrent( form_gauge);
      } else
      { display.setCurrent(alert_confirmation);
        payed = true;
      }
    }
  else if (s==form_gauge)
    { display.setCurrent(form_thanks);
    }
  else if (s==form_thanks) // Exit the App
    { try
      { this.destroyApp(true);
      } catch (MIDletStateChangeException e)
      {}
    }
}

// Implements a NonInteractiveGauge
class NonInteractiveGauge extends Gauge implements Runnable
{
  // Indicates the number of units to move per iteration
  private int movement = 1;

  public NonInteractiveGauge(String title)
  { super(title, false, 30, 0);
    new Thread(this).start();
```

```
    }

    // This method moves the gauge left and right.
    public void run()
    { while (true)
      { int newValue = this.getValue() + movement;
        if(newValue >= 30)
        { movement = -1;
        } else
        { if(newValue <= 0)  movement = 1;
        }
        this.setValue(newValue);
        try
        { Thread.currentThread().sleep(1000);
        }
        catch(InterruptedException err) {}
      }
    }
  }
}
```

Low-Level APIs and Event Handling

The MIDP low-level APIs provide the primitives to render two-dimensional graphics on the MID. Handheld devices vary greatly from each other when it comes to display characteristics such as screen size, pixel ratio, depth, and color. Applications written using these APIs are not guaranteed to be portable. These APIs are most appropriate for graphic-rich applications (such as a game) targeted to a specific device.

Low-level APIs also provide low-level event handling. Recall that in high-level APIs we did not have any facility to associate an action to a specific key. Low-level APIs provide facilities to directly handle key events and also pointer events.

Let us look at one example to see how low-level APIs could be used. Figure 9-17 shows several screens generated on a cell phone emulator using **LowLevelMIDlet** APIs. This MIDlet example is part of the Whiteboard Wireless Toolkit from Zucotto Wireless, Inc.

Figure 9-18 shows the output when run on a Palm PDA emulator. Note that the PDA device can handle some device-specific events such as the location of a pointer (see lower right-hand screen).

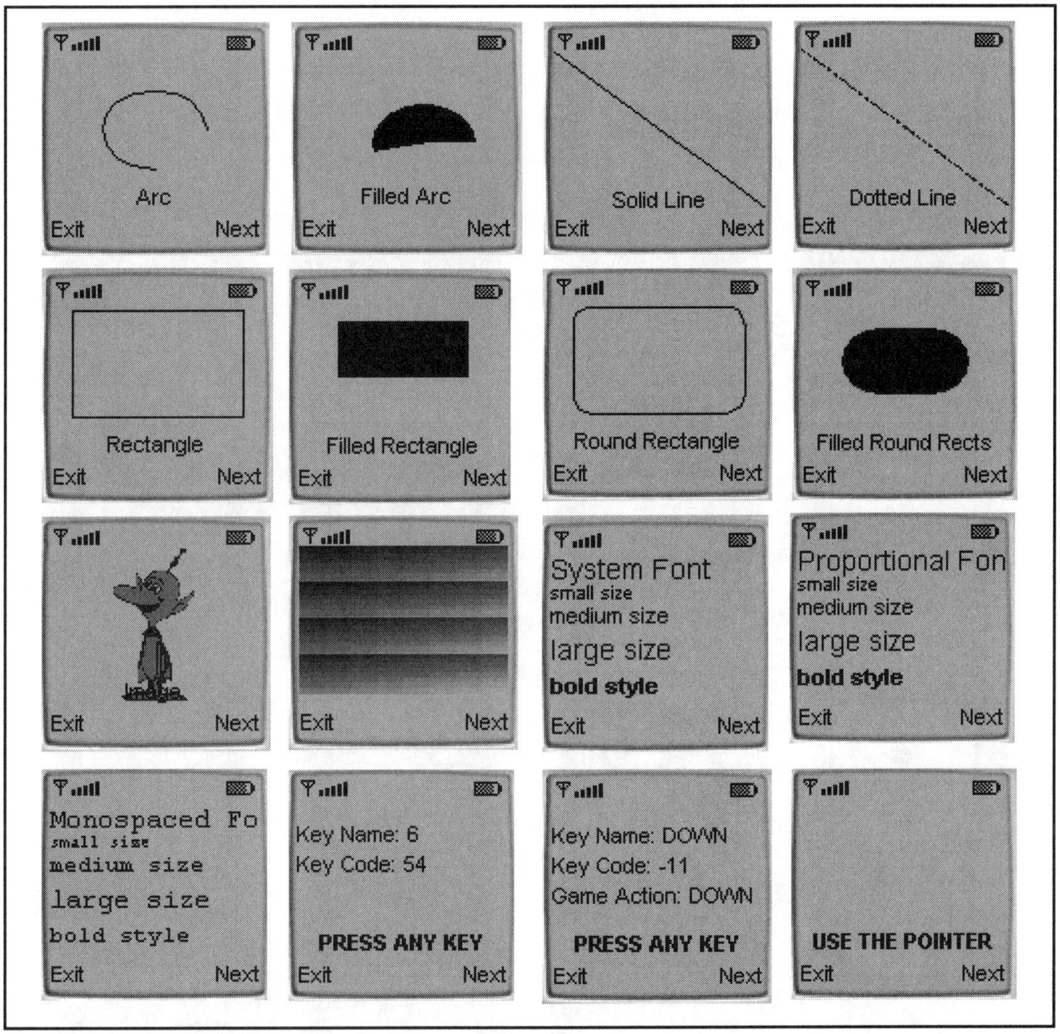

Figure 9-17. LowLevelMIDlet Output on a Java-enabled Phone Emulator.

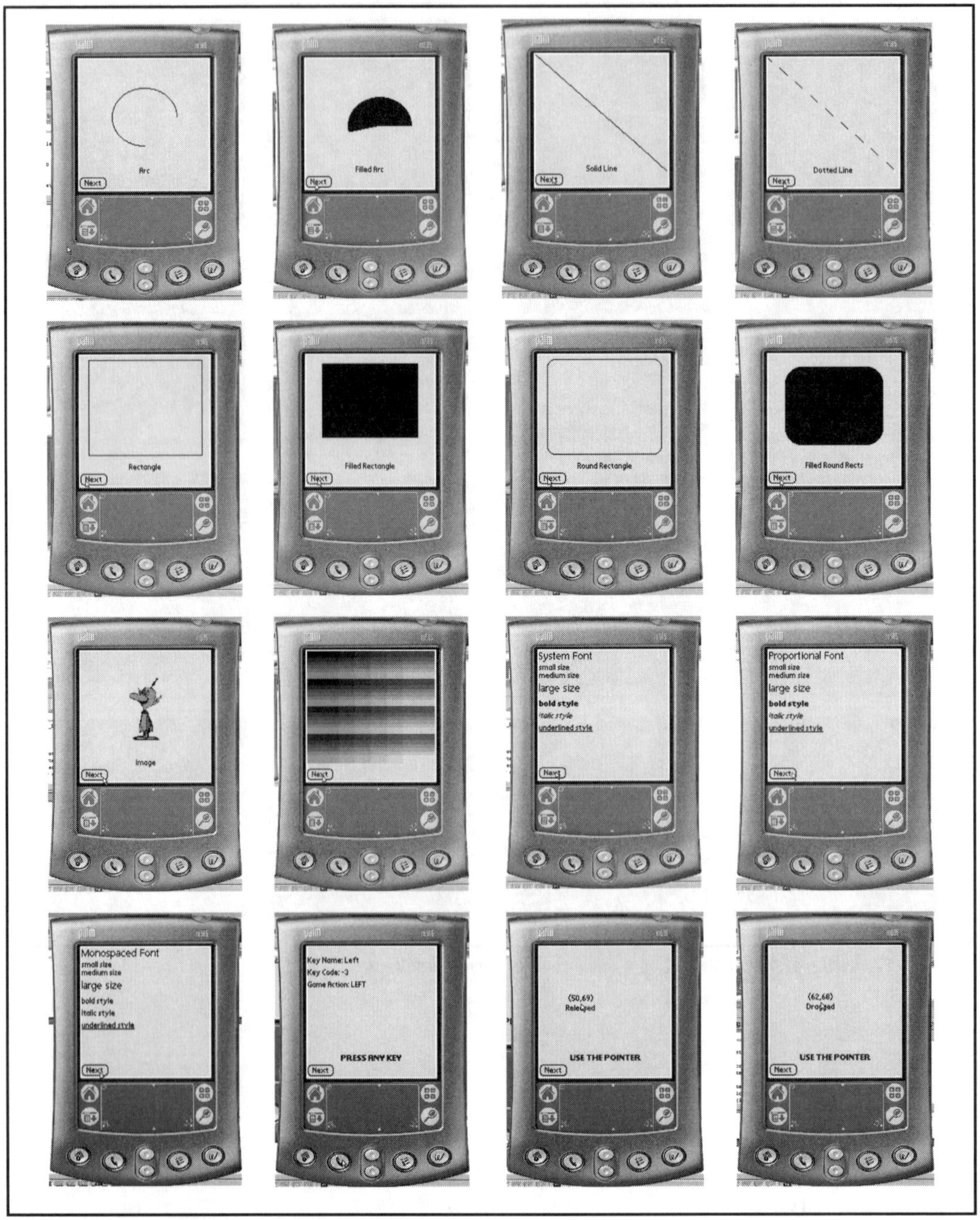

Figure 9-18. LowLevelMIDlet Output on a Palm Emulator.

As you can see from the screen shots, it is possible to render sophisticated two-dimensional graphics on a MID. This enables application developers to develop graphic intensive applications such as games and maps. Figure 9-19 gives an overview of the MIDP classes that provide the MIDP low-level APIs. The classes are described in more detail in the following sections.

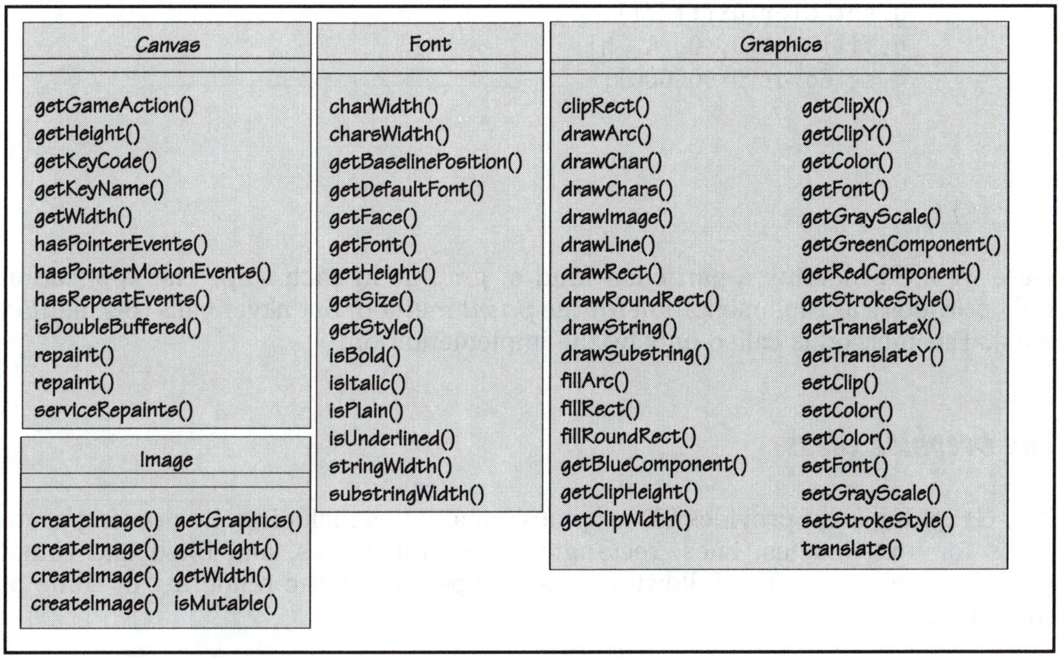

Figure 9-19. The Classes Related to Low-Level APIs and Event Handling.

The Canvas Class

The **Canvas** class is a base class for writing applications that need to handle low-level events and to issue graphics calls for drawing to the display. It defines the *paint* method that renders the **Canvas**.

```
protected abstract void paint(Graphics g)
```

As this is an abstract method, the application has to provide the implementation. In our LowLevelMIDlet, we are implementing **Canvas** with another class, called **LowLevelCanvas**. Here's an abbreviated view of the **LowLevelCanvas** class:

```
public class LowLevelCanvas extends Canvas
{
```

```
  . . .
Public void paint(Graphics g)
{ // get Screen width and height variables
    w = getWidth();
    h = getHeight();
    // Clear the entire screen
    g.setColor(0xffffff);
    g.fillRect(0, 0, w, h);
    g.setColor(0x000000);
   . . .
}
  . . .
}
```

Here we are rendering a particular kind of graphic in each step. The application only provides the implementation for the *paint* method, but never calls this method itself. This method is called only by the implementation.

The Graphics Class

The **Graphics** class provides 2D graphics capabilities including the drawing primitives for text, images, lines, rectangles, arcs, filled arcs, and filled and round cornered rectangles. The following code snippet shows the rendering of some of these objects:

```
switch (step)
{ case 0:
    g.drawArc(w/4,h/4,w/2,h/2,0,270);
    g.drawString("Arc", w/2,h-2,g.HCENTER|g.BASELINE);
    break;
  case 1:
    g.fillArc(w/3,h/3,w/2,h/2,0,193);
    g.drawString("Filled Arc", w/2,h-2,g.HCENTER|g.BASELINE);
    break;
  case 2:
    g.drawLine( 0, 0, w, h );
    g.drawString("Solid Line", w/2,h,g.HCENTER|g.BASELINE);
    break;
  case 3:
    g.setStrokeStyle( g.DOTTED );
    g.drawLine( 0, 0, w, h );
    g.drawString("Dotted Line", w/2,h,g.HCENTER|g.BASELINE);
    break;
```

```
case 4:
  g.drawRect( 10, 5, w-20, h-25 );
  g.drawString("Rectangle", w/2,h-2,g.HCENTER | g.BASELINE);
  break;
case 5:
  g.fillRect( 20, 10, w-40, h-50 );
 g.drawString("Filled Rectangle", w/2,h-2,g.HCENTER|g.BASELINE);
  break;
case 6:
  g.drawRoundRect( 10, 5, w-20, h-25, 20, 20 );
  g.drawString("Round Rectangle", w/2,h-2,g.HCENTER|g.BASELINE);
  break;
case 7:
  g.fillRoundRect( 20, 15, w-40, h-45, 40, 40 );
  g.drawString("Filled Round Rects",
               w/2,h-2,g.HCENTER|g.BASELINE);
  break;
```

The only drawing operation provided is pixel replacement. The destination pixel value is simply replaced by the current pixel value specified in the graphics object being used for rendering. No facility for combining pixel values, such as raster-ops or alpha blending is provided. This class provides many features such as supporting a 24-bit color model. But, because all the MIDs do not have the same capabilities, an application developer needs to query the device's capabilities before using these features.

There are two ways of rendering the graphics object:

- ***Direct rendering to the display.*** This is the technique used in the LowLevelMIDlet. The **Graphic** object is passed to the paint() method to render the graphics to the display.
- ***Rendering to an off-screen image buffer.*** A graphics object for rendering to an off-screen image buffer may be obtained by calling the *getGraphics()* method on the desired image. This object can be held indefinitely by the application, and requests may be issued on this graphics object at any time.

The Font Class

The **Font** class represents fonts and their metrics. An application uses the **Font** class to specify the name and metrics of a desired font. The MIDP system then matches the characteristics with the closest available font provided by the device. A font is specified by face, style, and size. Face can designate a system, proportional, or monospace font. Style can specify such font attributes as bolding and italics. The **Font** class provides methods to manipulate and query these attributes.

Event Handling

If you use the **Canvas** class to write applications to access low-level input events or to issue graphics calls for drawing to the display, you must handle low-level events. Game applications are likely to use the **Canvas** class because it provides methods to handle game actions and key events. The key events are reported with respect to key codes that are directly bound to concrete keys on the device.

The **Canvas** class, which is a subclass of **Displayable**, allows the application to register a listener for commands, but it requires applications to subclass it first. Also, unlike screens that allow the application to define listeners and register them with instances of the **Screen** class, the **Canvas** class does not allow this because several new listener interfaces need to be created with one for each kind of event.

Keys are a source of low-level events and every key for which events are reported is assigned a key code. The MIDP defines the following key codes in the **Canvas** class:

```
KEY_NUM0 - KEY_NUM9, KEY_STAR, and KEY_POUND.
```

These are the keys 0..9, *, and # found on MIDP devices like cell phones. Other keys might exist on some devices, but for portability, applications should use only the standard key codes.

If your application needs arrow key and gaming-related events, then game action are used instead of key codes. MIDP defines the following game actions:

```
GAME_A - GAME_B, UP, DOWN, RIGHT, LEFT
```

While each key code is mapped to one game action, a game action can be associated with more than one key code. The translation between the two is done with the *getKeyCode* and *getGameAction* methods.

If your application uses game actions and you want it to be portable, you should translate key events into game actions with the *getGameAction* method and test the result. For example, the game actions UP, DOWN, LEFT, and RIGHT can be mapped differently on different devices. The *getGameAction* method would return the RIGHT game action, for example, when the user presses the key that is a natural right on the device.

So how are these key and game codes delivered to an application? The following methods are available for handling low-level events.

```
protected void keyPressed(int keyCode)
```

```
protected void keyReleased(int keyCode)
protected void keyRepeated(int keyCode)
protected void pointerPressed(int x, int y)
protected void pointerDragged(int x, int y)
protected void pointerReleased(int x, int y)
protected void showNotify()
protected void hideNotify()
protected abstract void paint(Graphics g)
```

The *commandAction* method of the **CommandListener** interface can also handle low-level events.

In our LowLevelMIDlet application, the implementation for most of these methods is in the **LowLevelCanvas** class.

```
protected void keyPressed(int keyCode)
{ key = keyCode;
  repaint();
}
```

Note that this method doesn't have any action on the phone, as there is no pointer device present, whereas this action is triggered on the PDA.

```
protected void pointerPressed(int x, int y) {
    pointerEvent = "Pressed";
    pointerX = x;
    pointerY = y;
    repaint();
}
protected void pointerReleased(int x, int y) {
    pointerEvent = "Released";
    pointerX = x;
    pointerY = y;
    repaint();
}
protected void pointerDragged(int x, int y) {
    pointerEvent = "Dragged";
    pointerX = x;
    pointerY = y;
    repaint();
}
```

Note the following for key and pointer events:

- The *keyRepeated* method might not be available on all devices. Your application should check the availability of repeat actions by calling the *hasRepeatEvents* method.
- The pointer events may not be present on all devices, so before using the *pointerPressed*, *pointerDragged*, and *pointerReleased* methods, your application should check if a pointer mechanism is available by calling the *hasPointerEvents* and *hasPointerMotionEvents* methods first.

The LowLevelMIDlet Listings

You should be comfortable with MIDP low-level APIs and events after the next few sections. The following are the complete listings for the **LowLevelMIDlet** for your viewing pleasure, starting with Listing 9-4.

Listing 9-4. The LowLevelMIDlet Class.

```
import javax.microedition.lcdui.*;
import javax.microedition.midlet.*;

public class LowLevelMIDlet extends MIDlet implements CommandListener
{ private LowLevelCanvas canvas; // The low level screen for this MIDlet
  private Display display;
  private static Command exitCommand= new Command("Exit",Command.EXIT, 1);
  private static Command nextCommand =
                                   new Command("Next",Command.SCREEN,1);
  /*** Start the MIDlet   */
  public void startApp() throws MIDletStateChangeException {
    display = Display.getDisplay(this);
    canvas = new LowLevelCanvas(display); // Create the display canvas.
    canvas.addCommand( exitCommand );          // associate commands
    canvas.addCommand( nextCommand );
    canvas.setCommandListener( this ); // register  listenter
    display.setCurrent(canvas);
  }
  public void pauseApp() { /* not handled*/ }

  /*** Called by the app manager before the MIDlet is destroyed.  */
  public void destroyApp(boolean unconditional)
  { notifyDestroyed();
  }

  /*** Listener for high level command action events.  */
  public void commandAction(Command c, Displayable s) {
    if (c==exitCommand) {
```

```
        destroyApp(true);
    } else {
        if (canvas.step<15)       // increment through the screens
        { canvas.step++;
        } else
        { canvas.step = 0;
        }
        canvas.repaint();
        canvas.serviceRepaints();
        // Calculating screen which gets drawn over by colors when done
        if (canvas.step==9)
        { canvas.step++;
          canvas.repaint();
        }
    }
  }
}
```

This LowLevelMIDlet just handles commands and shows the appropriate screen. The main implementation of all the low-level APIs is done in the **LowLevelCanvas** class shown in Listing 9-5.

Listing 9-5. The LowLevelCanvas Class.

```
import javax.microedition.lcdui.*;
public class LowLevelCanvas extends Canvas {
  private int key = -1;          // last pressed key
  protected int step; // state variable for tracking current screen
  private String pointerEvent = ""; // variable for tracking pointer status
  private int pointerX = 0;
  private int pointerY = 0;
  private Display display;
  private int h=0;
  private int w=0;

  /*** Constructor, store display variable */
  public LowLevelCanvas(Display display) {
      this.display = display;
  }
  /*** Called when a key is pressed.*/
  protected void keyPressed(int keyCode)
  { key = keyCode;
    repaint();
  }
  /*** Called when a key is repeated (held down) */
```

```
protected void keyRepeated(int keyCode)
{ // Not used
}
/*** Called when a key is released.*/
protected void keyReleased(int keyCode)
{ // Not used
}
/*** Called when the pointer is pressed.           */
protected void pointerPressed(int x, int y)
{ pointerEvent = "Pressed";
  pointerX = x;
  pointerY = y;
  repaint();
}
/*** Called when the pointer is released */
protected void pointerReleased(int x, int y)
{ pointerEvent = "Released";
  pointerX = x;
  pointerY = y;
  repaint();
}
/*** Called when the pointer is dragged.*/
protected void pointerDragged(int x, int y) {
    pointerEvent = "Dragged";
    pointerX = x;
    pointerY = y;
    repaint();
}
/*** Renders the Canvas. The application must implement this method in or
der to paint any graphics.*/
public void paint(Graphics g)
{ // Screen width and height variables
  w = getWidth();
  h = getHeight();
  g.setColor(0xffffff); // Clear the entire screen
  g.fillRect(0, 0, w, h);
  g.setColor(0x000000);
  switch (step) // Increment through all of the demonstration
  screens
  { case 0:
        g.drawArc(w/4,h/4,w/2,h/2,0,270);
        g.drawString("Arc", w/2,h-2,g.HCENTER|g.BASELINE);
        break;
    case 1:
        g.fillArc(w/3,h/3,w/2,h/2,0,193);
```

```
        g.drawString("Filled Arc", w/2,h-2,g.HCENTER|g.BASELINE);
        break;
    case 2:
        g.drawLine( 0, 0, w, h );
        g.drawString("Solid Line", w/2,h,g.HCENTER|g.BASELINE);
        break;
    case 3:
        g.setStrokeStyle( g.DOTTED );
        g.drawLine( 0, 0, w, h );
        g.drawString("Dotted Line", w/2,h,g.HCENTER|g.BASELINE);
        break;
    case 4:
        g.drawRect( 10, 5, w-20, h-25 );
        g.drawString("Rectangle", w/2,h-2,g.HCENTER | g.BASELINE);
        break;
    case 5:
        g.fillRect( 20, 10, w-40, h-50 );
        g.drawString("Filled Rectangle", w/2,h-
        2,g.HCENTER|g.BASELINE);
        break;
    case 6:
        g.drawRoundRect( 10, 5, w-20, h-25, 20, 20 );
        g.drawString("Round Rectangle", w/2,h-
        2,g.HCENTER|g.BASELINE);
        break;
    case 7:
        g.fillRoundRect( 20, 15, w-40, h-45, 40, 40 );
        g.drawString("Filled Round Rects", w/2,h-
        2,g.HCENTER|g.BASELINE);
        break;
    case 8:
        Image image = null;
        try {
            image = Image.createImage("/image.png");
        } catch (java.io.IOException e) {
            System.out.println("Error loading Image:
            "+e.toString());
        }
        g.drawImage(image,w/2,h/2,g.HCENTER|g.VCENTER);
        g.drawString("Image", w/2,h-2,g.HCENTER | g.BASELINE);
        break;
    case 9:
        g.drawString("Calculating...", w/2,h/2,g.HCENTER |
        g.BASELINE);
        break;
```

```
case 10:
    int red;
    int green;
    int blue;
    // cycle through all colors
    switch (display.numColors()) {
      case 2:              // draw gray scales for 2 grays
          drawGray( g, 0, 2 );
          drawGray( g, 1, 2 );
          break;
      case 4:              // draw gray scales for 4 grays
          for (int i=0;i<4;i++) {
              drawGray( g, i, 4 );
          }
          break;
      case 16: // draw gray scales for 16 grays
          for (int i=0;i<16;i++) {
              drawGray( g, i, 16 );
          }
          break;
      case 256: // Draw color screen for 256 colors, 8 bit
          for (int i=0;i<4;i++)
          { // calculate row height of each color swath
            int swathHeight = h/8;
            swathHeight = swathHeight*2;
            blue=i*255/3;     // set basic blue colors
            for (int j=0;j<8;j++)
            { green = j*255/7;// set basic green colors
              for (int k=0;k<8;k++)
              { red = k*255/7; // set basic red colors
                g.setColor( red, green, blue );
                g.fillRect( k*w/8, i*swathHeight+j*h/32,
                w/8, w/32 );
               }
            }
          }
          break;
      case 16777216://Draw color screen for 16777216 colors, 24bit
               for (int y=0;y<=(h/3);y++)       // draw red box
          {   red = (y*255*3)/h;
              blue = 255 - (y*255*3)/h;
              for (int x=0;x<=w;x++)
              {  green = 255 - (x*255)/w;
                   drawColor( g, red, green, blue, x, y );
              }
```

```
                }
                for (int y=h/3;y<=(2*h/3);y++)
                {   red = 255 - ((y-h/3)*255*3)/h;
                    green = ((y-h/3)*255*3)/h;
                    for (int x=0;x<=w;x++)
                    {   blue = 255 - (x*255)/w;
                        drawColor( g, red, green, blue, x, y );
                    }
                }
                for (int y=2*h/3;y<=h;y++)
                {   green = 255 - ((y-2*h/3)*255*3)/h;
                    blue = ((y-2*h/3)*255*3)/h;
                    for (int x=0;x<=w;x++)
                    {   red = 255 - (x*255)/w;
                        drawColor( g, red, green, blue, x, y );
                    }
                }
                break;
            }
            g.setColor( 0 );
            break;
        case 11:
            g.setFont( Font.getFont( Font.FACE_SYSTEM, Font.STYLE_PLAIN,
            Font.SIZE_LARGE ) );
            g.drawString( "System Font",0,0,g.LEFT | g.TOP );
            g.setFont( Font.getFont( Font.FACE_SYSTEM, Font.STYLE_PLAIN,
            Font.SIZE_SMALL ) );
            g.drawString( "small size",0,15,g.LEFT | g.TOP );
            g.setFont( Font.getFont( Font.FACE_SYSTEM, Font.STYLE_PLAIN,
            Font.SIZE_MEDIUM ) );
            g.drawString( "medium size",0,25,g.LEFT | g.TOP );
            g.setFont( Font.getFont( Font.FACE_SYSTEM, Font.STYLE_PLAIN,
            Font.SIZE_LARGE ) );
            g.drawString( "large size",0,40,g.LEFT | g.TOP );
            g.setFont( Font.getFont( Font.FACE_SYSTEM, Font.STYLE_BOLD,
            Font.SIZE_MEDIUM ) );
            g.drawString( "bold style",0,60,g.LEFT | g.TOP );
            g.setFont( Font.getFont( Font.FACE_SYSTEM,
            Font.STYLE_ITALIC,
            Font.SIZE_MEDIUM ) );
            g.drawString( "italic style",0,75,g.LEFT | g.TOP );
            g.setFont( Font.getFont( Font.FACE_SYSTEM,
            Font.STYLE_UNDERLINED, Font.SIZE_MEDIUM ) );
            g.drawString( "underlined style",0,90,g.LEFT | g.TOP );
            break;
```

```
case 12:
    g.setFont( Font.getFont( Font.FACE_PROPORTIONAL,
    Font.STYLE_PLAIN, Font.SIZE_LARGE ) );
    g.drawString( "Proportional Font",0,0,g.LEFT | g.TOP );
    g.setFont( Font.getFont( Font.FACE_PROPORTIONAL,
    Font.STYLE_PLAIN, Font.SIZE_SMALL ) );
    g.drawString( "small size",0,15,g.LEFT | g.TOP );
    g.setFont( Font.getFont( Font.FACE_PROPORTIONAL,
    Font.STYLE_PLAIN, Font.SIZE_MEDIUM ) );
    g.drawString( "medium size",0,25,g.LEFT | g.TOP );
    g.setFont( Font.getFont( Font.FACE_PROPORTIONAL,
    Font.STYLE_PLAIN, Font.SIZE_LARGE ) );
    g.drawString( "large size",0,40,g.LEFT | g.TOP );
    g.setFont( Font.getFont( Font.FACE_PROPORTIONAL,
    Font.STYLE_BOLD, Font.SIZE_MEDIUM ) );
    g.drawString( "bold style",0,60,g.LEFT | g.TOP );
    g.setFont( Font.getFont( Font.FACE_PROPORTIONAL,
    Font.STYLE_ITALIC, Font.SIZE_MEDIUM ) );
    g.drawString( "italic style",0,75,g.LEFT | g.TOP );
    g.setFont( Font.getFont( Font.FACE_PROPORTIONAL,
    Font.STYLE_UNDERLINED, Font.SIZE_MEDIUM ) );
    g.drawString( "underlined style",0,90,g.LEFT | g.TOP );
    break;
case 13:
    g.setFont( Font.getFont( Font.FACE_MONOSPACE,
    Font.STYLE_PLAIN, Font.SIZE_LARGE ) );
    g.drawString( "Monospaced Font",0,0,g.LEFT | g.TOP );
    g.setFont( Font.getFont( Font.FACE_MONOSPACE,
    Font.STYLE_PLAIN, Font.SIZE_SMALL ) );
    g.drawString( "small size",0,15,g.LEFT | g.TOP );
    g.setFont( Font.getFont( Font.FACE_MONOSPACE,
    Font.STYLE_PLAIN, Font.SIZE_MEDIUM ) );
    g.drawString( "medium size",0,25,g.LEFT | g.TOP );
    g.setFont( Font.getFont( Font.FACE_MONOSPACE,
    Font.STYLE_PLAIN, Font.SIZE_LARGE ) );
    g.drawString( "large size",0,40,g.LEFT | g.TOP );
    g.setFont( Font.getFont( Font.FACE_MONOSPACE,
    Font.STYLE_PLAIN, Font.SIZE_MEDIUM ) );
    g.drawString( "bold style",0,60,g.LEFT | g.TOP );
    g.setFont( Font.getFont( Font.FACE_MONOSPACE,
    Font.STYLE_PLAIN, Font.SIZE_MEDIUM ) );
    g.drawString( "italic style",0,75,g.LEFT | g.TOP );
    g.setFont( Font.getFont( Font.FACE_MONOSPACE,
    Font.STYLE_UNDERLINED, Font.SIZE_MEDIUM ) );
    g.drawString( "underlined style",0,90,g.LEFT | g.TOP );
```

```
    break;
case 14:
    if (key!=-1)
    { g.drawString( "Key Name:
      "+getKeyName(key),0,10,g.TOP|g.LEFT );
      g.drawString( "Key Code: " + key,0,25, g.TOP|g.LEFT );
      String s = "Game Action: ";
      switch (getGameAction(key)) // draw game action key
        {   case UP:
                g.drawString( s + "UP",0,40, g.TOP|g.LEFT );
                break;
            case DOWN:
                g.drawString( s + "DOWN",0,40, g.TOP|g.LEFT );
                break;
            case LEFT:
                g.drawString( s + "LEFT",0,40, g.TOP|g.LEFT );
                break;
            case RIGHT:
                g.drawString( s + "RIGHT",0,40, g.TOP|g.LEFT );
                break;
            case FIRE:
                g.drawString( s + "FIRE",0,40, g.TOP|g.LEFT );
                break;
            case GAME_A:
                g.drawString( s + "GAME_A",0,40, g.TOP|g.LEFT );
                break;
            case GAME_B:
                g.drawString( s + "GAME_B",0,40, g.TOP|g.LEFT );
                break;
            case GAME_C:
                g.drawString( s + "GAME_C",0,40, g.TOP|g.LEFT );
                break;
            case GAME_D:
                g.drawString( s + "GAME_D",0,40, g.TOP|g.LEFT );
                break;
        }
    }
    g.setFont( Font.getFont( Font.FACE_PROPORTIONAL,
    Font.STYLE_BOLD, Font.SIZE_MEDIUM ) );
    g.drawString("PRESS ANY KEY", w/2,h-2,g.HCENTER,g.BASELINE);
    break;
case 15:
    g.drawString( "("+pointerX+","+pointerY+")",
                  pointerX, pointerY, g.BOTTOM|g.HCENTER );
    g.drawString( pointerEvent,
```

```
            pointerX,pointerY,g.TOP|g.HCENTER );
            g.setFont( Font.getFont( Font.FACE_PROPORTIONAL,
                       Font.STYLE_BOLD, Font.SIZE_MEDIUM ) );
            g.drawString("USE THE POINTER", w/2,h-2,g.HCENTER |
            g.BASELINE);
            break;
        }
    }

  /*** helper method for drawing grayscale rectangles on screen */
  private void drawGray(Graphics g, int currentGray, int totalShades)
  { g.setGrayScale( (currentGray) * (255/(totalShades-1)) );
    g.fillRect( 0,h/totalShades*currentGray, w,h/totalShades );
  }

  /*** helper method for drawing color pixels on screen. Ensures col-
  ors are in acceptable range.   */
  private void drawColor(Graphics gr, int r, int g, int b, int x, int y )
  { r = r > 255 ? 255 : r;
    r = r < 0 ? 0 : r;
    g = g > 255 ? 255 : g;
    g = g < 0 ? 0 : g;
    b = b > 255 ? 255 : b;
    b = b < 0 ? 0 : b;
    gr.setColor( r, g, b );
    gr.drawLine( x,y, x,y );
  }
}
```

SUMMARY

What a wild ride! We hope you have survived the quick introduction to MIDP development and the MIDP APIs—including high-level APIs and low-level APIs. High-level APIs are provided for writing portable applications that are appropriate for the enterprise. Low-level APIs provide primitive 2D graphic capabilities and low-level event handling for things like keys and pointers. These are necessary for games and other graphics-intensive applications.

In the next chapter we will move on to the MIDP classes that perform networking and allow persistent data storage on our handheld device. Without networking, our device would be isolated and not qualify as a participant in an enterprise application. Being able to store data allows us to optimize the use of our device as a personal participant in enterprise applications.

Chapter 10

Advanced MIDP Programming

Having survived the basics of MIDP programming in the last chapter, we're now ready to tackle some of the more advanced features of MIDP: networking and persistent storage on the MIDP device. We'll first take a look at the classes and interfaces that are part of the *javax.microedition.io* package. This package includes the interfaces and classes necessary for HTTP-based communication.

Next we'll look at the *java.microedition.rms* package. This package includes the primitive classes and interfaces for managing a simple Record Management Store (RMS). Don't expect anything fancy. The limited facilities of a MIDP device dictate that the storage and retrieval of data to the device's persistent storage (most likely some type of EEPROM) is going to be done with only basic classes.

We'll use several MIDlet examples to demonstrate the *io* and *rms* package classes.

NETWORK PROGRAMMING

Java is the preferred programming language to build networked applications. It is very well established that Java provides powerful features for building network applications for the desktop (Java applications), for the enterprise (using Servlets, JSPs, and EJBs) and the Internet (using Servlets, JSPs, and Applets). These inherent network capabilities of the Java programming language and its extensions make Java a logical choice for mobile devices. We will discuss the networking facilities provided by CLDC and MIDP in this section and provide some detailed examples.

You may recall from Chapter 7 that CLDC does not provide all of the networking support classes and functions from the *java.net* and *java.io* packages in J2SE. Instead, CLDC provides a new set of APIs packed in *javax.microedition.io* for networking and I/O. You must be wondering: Java already has so many abstractions—such as **Socket**, **ServerSocket**, **DatagramSocket**, and **URLConnection**—why not use them instead of introducing new ones? Here are some of the obvious reasons:

- J2SE networking classes are based on TCP/IP or UDP, which might not be available on all wireless devices.
- These classes are designed for servers and desktop systems. So they obviously require a lot of memory.
- A significant part of the standard I/O and networking functionality is not directly applicable to today's small devices, which often need to support specific types of connections such as infrared or Bluetooth, or which do not provide TCP/IP support.

In general, the requirements for the networking and storage libraries vary significantly from one resource-constrained device to another (refer to Chapter 3 for a review of the network protocols supported by devices). For instance:

- Devices based on a circuit-switched cellular network work better with stream-based connections (TCP/IP).
- Devices based on packet-switched data networks work better with datagram-based communication mechanisms (UDP).
- Some of the devices have traditional file systems, while many others have highly device-specific mechanisms.
- Some devices have other networking support such as IEEE 802.11, CDPD, or Bluetooth.

Due to strict memory limitations, devices supporting certain kinds of input/output, networking, and storage capabilities generally do not support multiple mechanisms. Networking classes in J2SE do not have support for the variety of wireless network protocols. So instead of reusing the APIs from *java.net* and *java.io*, CLDC defines its own set of APIs that can be easily mapped to the J2SE implementation.

As an application programmer, you want to write network-capable MIDlets that work on all the devices that support MIDP. You should not be dealing with the low-level details of these protocols and networks. In order to separate the programmer from the low-level details, CLDC provides a clean abstraction for different types of connections. This layer of abstraction is known as the *Generic Connection Framework (GCF)*. MIDP extends and provides the implementation for some of the interfaces provided by this framework. As an application programmer all you need to do is use these APIs and voila, your MIDlet will work on all MIDs! Let us look into the details of these abstractions before we start writing code.

Generic Connection Framework

The Generic Connection Framework defines a hierarchy of seven interfaces that group together classes of protocols with the same semantics as shown in Figure 10-1. At the implementation level, a minimum of one class is needed for implementing each supported protocol. Often, each protocol implementation class simply contains a number of wrapper functions that call the native functions of the underlying host operating system.

This collection of interfaces forms a hierarchy that becomes progressively more capable as the hierarchy progresses from the root **Connection** interface.

- The **Connection** interface is the most basic connection type. It can only be opened and closed.
- The **InputConnection** interface represents a device from which data can be read.
- The **OuputConnection** interface represents a device to which data can be written.
- The **StreamConnection** interface combines the input and output connections.
- The **ContentConnection** is a sub-interface of **StreamConnection**. It provides access to some of the basic metadata information provided by HTTP connections, such as the HTTP headers, content type, and content length.
- The **StreamConnectionNotifier** interface waits for a connection to be established. It returns a **StreamConnection** on which a communication link has been established.
- The **DatagramConnection** interface represents a datagram endpoint.
- The **HTTPConnection** interface represents an HTTP endpoint.

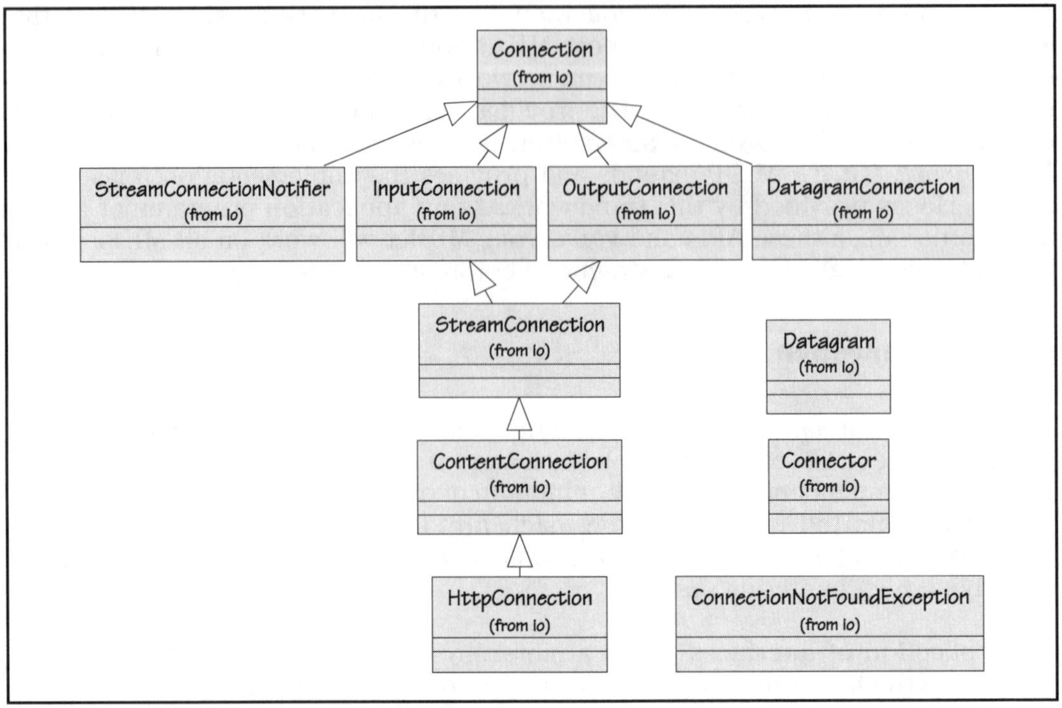

Figure 10-1. The MIDP Connection Framework Hierarchy.

This arrangement allows application programmers to choose the optimal level of cross-protocol portability for the code they are writing. These interfaces are defined in the CLDC configuration and the implementation is provided by the profile. However, profiles are not required to provide implementations for all of the interfaces. They might choose to provide implementation for some of these or even might extend these interfaces when providing the implementation. MIDP, for example provides an **HTTPConnection** interface (and the implementation for it), but does not provide implementation for the **DatagramConnection** interface. We will describe the MIDP facilities shortly.

Another important class that the GCF provides is **Connector**. None of the above interfaces provide an *open* method to establish a connection. The **Connector** class provides static methods that enable an application to establish a connection. For example, to open a connection the syntax looks like this:

```
Connector.open("<protocol>:<address>;<parameters>");
```

The parameter that is passed in is a **String**, which embodies a *Uniform Resource Indicator (URI)*. As you can see, the URI format is very generic in representing any type of connection. Let us look at some examples, as shown in Table 10-1.

Table 10-1. URI Formats for Connections.

Protocol	Syntax
HTTP	Connector.open("http://www.sjsu.edu");
Sockets	Connector.open("socket://129.144.111.222:9000");
Datagrams	Connector.open("datagram://129.144.111.333");
Serial ports	Connector.open("comm:0;baudrate=9600");
Files	Connector.open("file://map.dat");

The key advantage of this syntax is that only the parameter string determines the protocol that will be used by the connection. Most of the application code remains the same regardless of the protocol.

The implementation of the protocol is provided by the profile. When the *open* method is called, the protocol implementation is determined by the protocol identifier in the string (such as http: or file:) and loaded at runtime. This is analogous to the how device drivers are used by applications on desktop systems.

Enough of this abstract discussion. Let's look at what networking facilities MIDP provides.

The ResultsMIDlet: Networking With MIDP

Let's illustrate simple networking with another university example. Professor Stuckinthesixties has posted the grade results of a course on his Web site. He still prefers to use a prehistoric comma-separated file for this purpose. Let's write a MIDlet that fetches the results from this file.

Go ahead and run the MIDlet in Listing 10-1. If you are connected to the Internet and you don't have any problems connecting to the Web site, you will see the output shown in Figure 10-2. Or, you can change the URL to point to your own Web site or a local Web server to test the MIDlet.

The ResultsMIDlet is a simple MIDlet that fetches a page from a Web server and displays the content (for brevity there are no commands in this MIDlet, not even an *exit* command). Here is what is happening in the MIDlet:

- A **StreamConnection** is opened (this provides both input and output streams) to the results file residing on the Web server. The static *open* method of the

Figure 10-2. The ResultsMIDlet Output.

Connector is called, which returns a **StreamConnection** object (and its implementation). The protocol used is the "http" protocol supplied by the MIDP profile. It also provides a new connection type that extends what CLDC defined, which we will look at in a moment.

```
String url =
"http://mobilejava.engr.sjsu.edu/tests/fall01_results.txt";
. . .
conn = (StreamConnection)Connector.open(url);
```

- Once we have the **StreamConnection** object, an **InputStream** is opened and the file is read, character-by-character, until the end of the file.

```
InputStream is = conn.openInputStream();
while((ch = is.read()) != -1)
{ buff.append((char) ch);
       . . .
}
```

- Remember, our MIDlet may be using a wireless connection to fetch the file from the Web server. Therefore, the possibility of the connection being dropped is very high. Because of this potential problem, we want to handle the exceptions elegantly and free the resources we have allocated when an exception occurs. Thus the connection and the **InputStream** are closed in a *finally* clause.

```
. . .
finally {
      if(is != null) {
        is.close();
      }
      if(conn != null) {
        conn.close();
      }
```

Listing 10-1 shows the complete code for the MIDlet.

Listing 10-1. The ResultsMIDlet Code.

```
import java.io.*;
import javax.microedition.io.*;
import javax.microedition.lcdui.*;
import javax.microedition.midlet.*;

public class ResultsMIDlet extends MIDlet
{ private Display display;
  String url =
  http://wirelessjava.engr.sjsu.edu/tests/fall01_results.txt";
  public ResultsMIDlet () {
    display = Display.getDisplay(this);
  }

  public void startApp()
  { try
    { getResults(url);
    } catch (Exception e)
    {Alert alert_error = new Alert( "Error",
         "Could not connect.. Please try later!", null, AlertType.ERROR );
        alert_error.setTimeout( Alert.FOREVER );
        display.setCurrent(alert_error);
      //Handle Exceptions any other way you like.
        System.out.println("IOException " + e);
        e.printStackTrace();
    }
  }

  public void pauseApp() {/* not handled */ }

  public void destroyApp(boolean unconditional) { /* do nothing */}
    /*** read url via stream connection  */

  void getResults(String url) throws Exception {
    StreamConnection conn = null;
    InputStream is = null;
    StringBuffer buff = new StringBuffer();
    TextBox restext = null;
    try
    { conn = (StreamConnection)Connector.open(url);
      is = conn.openInputStream();
      int ch;
```

```
      int cnt = 0;
      while((ch = is.read()) != -1)
      { buff.append((char) ch);
        cnt++;
}

// display the contents of the file in a text box.
      System.out.println(buff.toString());
      restext = new TextBox("Your Grades!!....",
                     buff.toString(),cnt , 0);
    } finally
    {
      if(is != null)
      { is.close();
      }
      if(conn != null)
      { conn.close();
      }
    }
    display.setCurrent(restext);
  }
}
```

The HTTPConnection Interface

You noticed in the ResultsMIDlet that we used "http" as the protocol but we did not do any "http" operations. Instead we just read a file using a stream. MIDP provides another interface called **HttpConnection**, which extends the **ContentConnection** interface defined by CLDC.

```
public interface HttpConnection
extends javax.microedition.io.ContentConnection
```

MIDP provides only this interface to do network programming. It does not provide support for other connection types such as datagrams or sockets. This means all vendors that support MIDP have to provide an implementation of this interface. Some vendors might choose to provide support for other connection types such as datagrams (the Siemens toolkit and MID phones support datagrams). However, programming to these connections would make your application non-portable. In other words, programming to **HttpConnection** (and its super interfaces) makes your MIDlets portable across all devices that support MIDP.

Why HTTP?

The current wireless network infrastructure varies in a lot of factors from vendor to vendor. Some support IP and some do not. Nevertheless, HTTP is ubiquitous. It is one protocol, which can be implemented on the majority of the existing networks regardless of the bearer technology. Also, MIDP attempts to provide a common functionality that can be supported by the majority of vendors. To accomplish this goal, MIDP chose to provide a subset of HTTP 1.1 functionality. The bottom line is if your MIDlet is written to MIDP APIs, it will run on any MIDP implementation, whether it is on a GSM phone with a WAP stack, a phone with i-mode, a Palm VII wireless device, or a hand-held device with Bluetooth.

Note that MIDP 2.0 will include datagram connection support as a standard. This might take some time to hit the market, so there is no need to jump on it immediately. If you know ahead of time that the device you are targeting has these extra features you might want to take advantage of them. For the purposes of this chapter, we will limit our discussion here to HTTP connections only.

Also note that the latest version of MIDP has support for “https” or http over SSL. No changes are required to your program to take advantage of it. All you need to do is change the URL to “https” and include the keys required for encryption. We will show how to do this a little later.

Going back to our university scenario, Professor Withthetimes realized that Professor Stuckinthesixties’ grade retrieval program allows a student to see everyone else’s grade. So he put his grade results in a database and added a Servlet to his Web site to access a grade by student ID (SSN) in order to fetch the results. Listing 10-2 shows the Servlet code.

Listing 10-2. The Grade Finder Servlet.

```
public class gradeFinderServlet extends HttpServlet
{
  // Get method calls Post method
  public void doGet(HttpServletRequest request,
               HttpServletResponse response)
                  throws ServletException, IOException
  {
    doPost(request,response);
  }

  public void doPost(HttpServletRequest request,
                     HttpServletResponse response)
                     throws ServletException, IOException
  { //get the input parameters
```

```
    String courseId = request.getParameter("courseid");
    String studentId = request.getParameter("studentid");
    HttpSession session = request.getSession();
    String grade = fetchGradefromDB(courseId,studentId);
    PrintWriter out = response.getWriter();
    response.setContentType("text/plain");
    response.setContentLength(grade.length());
    out.println(grade);
  }

  String fetchGradefromDB(String courseId, String studentID)
  {
    // do the database operations - stub this operation for now
    . . .
  }
}
```

The Servlet code is self-explanatory. It takes two parameters, *courseid* and the *studentid* (which is the SSN), fetches the results from the database, and outputs the results. The code remains the same regardless of how you access it. You can access it from an HTML page with a form input or from a MIDlet. So, let us write the MIDlet to access this Servlet. This Servlet is hosted at *http://wireless-java.engr.sjsu.edu:8080/gradeFinder*.

All that is needed is a MIDlet that can call the Servlet. We will call this MIDlet **GradeFinder**. When you run the above MIDlet you will see output similar to the one in Figure 10-3.

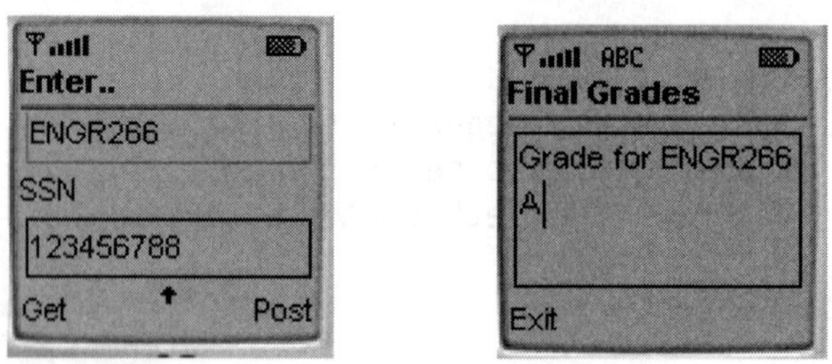

Figure 10-3. The GradeFinder MIDlet.

Listing 10-3 shows the code for the MIDlet. The explanation follows.

Listing 10-3. Grade Finder MIDlet.

```
import java.io.*;
import javax.microedition.io.*;
import javax.microedition.lcdui.*;
import javax.microedition.midlet.*;

public class GradeFinder extends MIDlet implements CommandListener
{ private Display display;
  private String baseURL = "http://wirelessjava.engr.sjsu.edu:8080";
  private static final Command postCmd =
                             new Command("Post",Command.SCREEN, 1);
  private static final Command getCmd =
                             new Command("Get", Command.SCREEN, 1);
  private static final Command exitCommand =
                             new Command("Exit", Command.EXIT, 1 );
  Form form_text;
  String url = baseURL+"/examples/servlet/gradeFinderServlet";

  public GradeFinder()
  { form_text = new Form("Enter..");
    form_text.append(new TextField("CourseId", "", 50, TextField.ANY));
    form_text.append(new TextField("SSN", "", 50, TextField.NUMERIC));
    form_text.addCommand(postCmd);
    form_text.addCommand(getCmd);
    form_text.setCommandListener(this);
  }

  public void startApp()
  { display = Display.getDisplay(this);
   display.setCurrent(form_text);
  }

  public void pauseApp() { }
  public void destroyApp(boolean unconditional)
          throws MIDletStateChangeException
  {  notifyDestroyed();
  }

  // Retrieve a grade
  void getGradeusingPost(String ssn, String courseid) throws IOException
  { HttpConnection c = null;
    InputStream is = null;
    OutputStream os = null;
    StringBuffer b = new StringBuffer();
```

```
    TextBox t = null;
    try
    { c = (HttpConnection)Connector.open(url);
      c.setRequestMethod(HttpConnection.POST);
    c.setRequestProperty("IF-Modified-Since","10 Nov 2000 17:29:12 GMT");
      c.setRequestProperty("User-Agent",
                           "Profile/MIDP-1.0 Confirguration/CLDC-1.0");
      c.setRequestProperty("Content-Language","en-US");
      os = c.openOutputStream();
      String str = "?courseid="+courseid+"&"+"ssn="+ssn;
      os.write(str.getBytes());
      os.flush();
      os.close();


      // Get the length and process the data
      int len = (int)c.getLength();
      if (len > 0)
      { byte[ ] data = new byte[len];
        int actual = is.read(data);
        String res = new String(data);
        b.append(  res);
      } else
      { int ch;
        while ((ch = is.read()) != -1)
        {  b.append((char) ch);
        }
      }
      t = new TextBox("Final Grades", b.toString(), 1024, 0);
    } finally
    { if(is!= null)
      { is.close();
      }
      if(os != null)
      { os.close();
      }
      if(c != null)
      { c.close();
      }
    }
    t.addCommand(exitCommand);
    t.setCommandListener(this);
    display.setCurrent(t);
}
```

```
void getGradeusingGet(String ssn, String courseid) throws IOException
{
  HttpConnection c = null;
  InputStream is = null;
  StringBuffer b = new StringBuffer();
  TextBox t = null;
  try
  { String str = "?courseid="+courseid+"&"+"ssn="+ssn;
    String getURL = url + str;
    c = (HttpConnection)Connector.open(getURL);
    c.setRequestMethod(HttpConnection.GET);
    c.setRequestProperty("IF-Modified-Since",
                           "10 Nov 2000 17:29:12 GMT");
    c.setRequestProperty("User-Agent",
                      "Profile/MIDP-1.0 Confirguration/CLDC-1.0");
    c.setRequestProperty("Content-Language","en-CA");
    is = c.openDataInputStream();

    // Get the length and process the data
    int len = (int)c.getLength();
    if (len > 0)
    { byte[ ] data = new byte[len];
      String res = new String(data);
      b.append(res);
    } else
    { int ch;
      while ((ch = is.read()) != -1)
      { b.append((char) ch);
      }
      }
      t = new TextBox("Final Grades", b.toString(), 1024, 0);
    } finally
    { if(is!= null)
      { is.close();
      }
      if(c != null)
      { c.close();
      }
    }
    t.addCommand(exitCommand);
    t.setCommandListener(this);
    display.setCurrent(t);
}
```

```
  public void commandAction(Command c, Displayable s)
  { if (c==exitCommand) // Always exit if user selects the exitCommand
    {  try
       {  this.destroyApp(true);
       } catch(MIDletStateChangeException e) {}
    } else if (c==postCmd)
    { try
      { String courseid = ((TextField) form_text.get(0)).getString();
        String ssn = ((TextField) form_text.get(1)).getString();
        getGradeusingPost(ssn,courseid);
      } catch (IOException e)
      { e.printStackTrace();
      }
    } else if (c==getCmd)
      { try
        { String courseid = ((TextField) form_text.get(0)).getString();
          String ssn = ((TextField) form_text.get(1)).getString();
          getGradeusingGet(ssn,courseid);
        } catch (IOException e)
        { e.printStackTrace();
        }
      }
}
```

You have just seen a MIDlet that could communicate with an enterprise application, though it is a simple one.

Before we delve into details of this MIDlet, let us look at how HTTP works. HyperText Transfer Protocol (HTTP) is a stateless request-response protocol. A client sends a request to the Web server and the Web server does the necessary operations such as invoking a Servlet before returning a response back to the client. Many network operations happen under the covers to provide the HTTP implementation. When you try to access the Web server with its URL, URL-to-IP conversion has to be done by doing a DNS lookup and the request has to be forwarded to the appropriate Web server. Some MIDP implementations might do this on the device itself or have a proxy connected to the network, which does these operations. This is implementation-specific and transparent to the application developer.

HTTP defines two methods to send data to the Web server, GET and POST. The *GET* method encodes all the input parameters in the URL that you access. This method is simple, but has some disadvantages. The URL string has a size limit, so you cannot send as much data as you might need. Also, the URL you are accessing is exposed, so it is a problem if you are sending sensitive data. The *POST* method, on the other hand, sends the data as the body of the HTTP packet. There is no limitation for the size of the data and you can encrypt the data you are sending to enhance security.

When you send the request to the server, the header that accompanies the request gives the server some information about what you are sending. Table 10-2 shows a sample header for a GET request.

Table 10-2. An HTTP Header for a Get Request.

Header Information	Description
GET /examples/servlet/gradeFinderServlet?courseid=&ssn= HTTP/1.1	Request type.
IF-Modified-Since: 10 Nov 2000 17:29:12 GMT	Modification time for content.
User-Agent: Profile/MIDP-1.0 Confirguration/CLDC-1.0	Server can recognize the type of client the request is coming from and do special processing if required.
Content-Language: en-CA	Language encoding.
Content-Length: 0	There is no body as this is a GET request.
Host: localhost:8080	Host to which the call is made.

Some of the header information has to be set by the application before doing any operations on the connection and sending the request. MIDP provides the following APIs to set these properties:

- *setRequestMethod*
- *setRequestProperty*

Going back to our GradeFinder MIDlet (see Listing 10-3), this is what we are doing:

- Get the student SSN and the course ID from the MIDlet form.
- Based on the button pressed, make a GET or POST request to the server and fetch the result.

The code Listing 10-3 is complete with all the commands and actions handled appropriately. For our discussion we are interested only in the *getGradeusingGet* and *getGradeusingPost* methods.

Performing an HTTP GET

The first step is to construct the URL string to make the get request. The entire URL would look like this:

```
http://myhost:8080/pathto/gradeFinderServlet?courseid=ENGR266&ssn
=123456788
```

Here's the code that creates the URL string:

```
String str = "?courseid="+courseid+"&"+"ssn="+ssn;
String getURL = url + str;
```

The next step is to open the HTTP connection using the static *open* method of **Connector** and set the request type to GET.

```
c = (HttpConnection)Connector.open(getURL);
c.setRequestMethod(HttpConnection.GET);
```

With the connection open and the request type set, the next step is to set the request properties. Note the User-Agent property as this plays an important role for the server side programming.

```
c.setRequestProperty("IF-Modified-Since",
      "10 Nov 2000 17:29:12 GMT");
c.setRequestProperty("User-Agent",
      "Profile/MIDP-1.0 Confirguration/CLDC-1.0");
c.setRequestProperty("Content-Language","en-CA");
```

Once the connection is obtained and request is made, the following methods can be invoked:

```
openInputStream              openOutputStream
openDataInputStream          openDataOutputStream
getLength                    getType
getDate                      getExpiration
getURL                       getProtocol
getHost                      getPort
```

In the GET example a **DataInputStream** is used to read the response. The input parameters of the request are sent as part of the URL when we opened the connection, so there is no need to open any output streams. Next, an **InputStream** is opened and the response is read. The response can be read in two ways: by getting the length of the response from the connection (*c.getLength()*) and reading that many bytes of data from the input stream. Or, read the **InputStream**, character-by-character until the end.

```
is = c.openDataInputStream();
// Get the length and process the data
int len = (int)c.getLength();
```

```
if (len > 0)
{ byte[ ] data = new byte[len];
  int result = is.read(data);
  String res = new String(data);
  b.append(res);
} else
{ int ch;
  while ((ch = is.read()) != -1) {
  b.append((char) ch);
}
```

Performing an HTTP POST

Using the *POST* method varies a little bit from the GET. First, the HTTP connection to the Servlet is opened.

```
String url = http://myhost:8080/servlet/gradeFinderServlet;
 . . .
c = (HttpConnection)Connector.open(url);
```

Next the request type and some header information is set.

```
c.setRequestMethod(HttpConnection.POST);
c.setRequestProperty("IF-Modified-Since",
                     "10 Nov 2000 17:29:12 GMT");
c.setRequestProperty("User-Agent",
                     "Profile/MIDP-1.0 Confirguration/CLDC-1.0");
c.setRequestProperty("Content-Language", "en-CA");
```

Then an **OutputStream** is opened and the HTTP body is written.

```
os = c.openOutputStream();
String str = "?courseid="+courseid+"&"+"ssn="+ssn;
os.write(str.getBytes());
os.flush();
```

Calling *flush* will make sure that everything is written from the output buffer. Next the **InputStream** is opened and the response is read using the same technique as in the *GET* method.

```
is = c.openDataInputStream();

// Get the length and process the data
int len = (int)c.getLength();
```

```
if (len > 0)
{ byte[ ] data = new byte[len];
  int result = is.read(data);
  String res = new String(data);
  b.append(res);
} else
{ int ch;
  while ((ch = is.read()) != -1)
  { b.append((char) ch);
}
```

When finished, the resources that were allocated are freed in a *finally* clause.

```
} finally
{ if(is!= null)
  { is.close();
  }
if(os != null)
  { os.close();
  }
if(c != null)
  { c.close();
  }
}
```

Isn't MIDP network programming easy? Coming back to our discussion about HTTP, we mentioned that HTTP is stateless. This means when you make requests to the server, it is not aware of the previous requests and thus cannot maintain any state between the requests. This is a major problem when we want to maintain a session between requests. There are three solutions to this problem:

- ***Cookies:*** This is a piece of information that the server sends to the client as a name-value pair. The client stores it and when subsequent requests are made to the same server, all the cookies related to that server are sent to the server in the header. The server can then read the cookies and use them to maintain session state.

- ***URL rewriting:*** The cookie feature is an optional feature that is supported by browsers (and gateways) and can be turned off. Further, cookies tend to be a security threat when sensitive information is stored as cookies on the client computer. So when cookies are not available, URL rewriting helps maintain the session state. When a request is made, the existing URL is also encoded and transmitted to the server. This way the server knows that this is a follow-up request made to the server. This mechanism works in all scenarios.

- ***SSL client id:*** This facility can be used only when using HTTP over SSL. When communicating over SSL, the client is assigned a unique ID that can be used to maintain the session. As SSL is not always available, this solution is not suggested.

Cookies in MIDP

Lets look at one example where you want to send your login and password to the server as a cookie and the server returns a computed cookie. You might then want to store the cookie the server sent in the local record store so that you can send it to the server on subsequent calls. As we haven't introduced the persistence APIs yet, for now we will just send a cookie, and then receive a cookie and display it. The Servlet example from the Tomcat Web server is used as the cookie Servlet with only one change: the content type is changed from text/html to text/plain. The Servlet code is not shown.

Figure 10-4 shows the screens for our CookieMIDlet.

Figure 10-4. The Cookie MIDlet Screens.

The name and value entered in the first screen are sent as cookies.

```
String cookiestring = name + "=" + value;
c.setRequestProperty("cookie",cookiestring);
```

The Servlet reads the cookie and then computes the cookie it wants to send back and sets that cookie in the response header. After the **InputStream** is opened to fetch the response, the incoming cookie can be retrieved from the response header using the following code snippet:

```
String cookieFromServer = c.getHeaderField("Set-cookie");
```

Listing 10-4 shows the complete code listing for the CookieMIDlet example.

Listing 10-4. The CookieMIDlet Example.

```
import java.io.*;
import javax.microedition.io.*;
import javax.microedition.lcdui.*;
import javax.microedition.midlet.*;
public class CookieMIDlet extends MIDlet implements CommandListener
{ private Display display;
  private static final Command setCmd = new Command("Set",
                                                      Command.SCREEN, 1);
  private static final Command nextCmd = new Command("Next",
                                                      Command.SCREEN, 1);
  private static final Command getCmd = new Command("Get",
           Command.SCREEN, 1);
  private static final Command exitCommand = new Command("Exit",
                  Command.EXIT, 1 );
  Form form_text;
  String cookieFromServer;
  String url = "http://localhost:8070/examples/servlet/MyCookieExample";

  public void startApp()
  { display = Display.getDisplay(this);
    form_text = new Form("Set cookie..");
    form_text.append( new TextField("name", "", 50, TextField.ANY) );
    form_text.append( new TextField("value", "", 50, TextField.ANY) );
    form_text.addCommand(setCmd);
    form_text.addCommand(exitCommand);
    form_text.setCommandListener(this);
    display.setCurrent(form_text);
  }

  public void pauseApp() { }

  public void destroyApp(boolean unconditional) throws
                                MIDletStateChangeException
  { notifyDestroyed();
  }

  // Retrieve a grade
  void setCookie(String name, String value) throws IOException
  { HttpConnection c = null;
    InputStream is = null;
    OutputStream os = null;
    StringBuffer b = new StringBuffer();
    TextBox t = null;
```

```
  try
  { c = (HttpConnection)Connector.open(url);
    c.setRequestMethod(HttpConnection.POST);
    c.setRequestProperty("IF-Modified-Since",
                                  "10 Nov 2001 12:29:12 PST");
    c.setRequestProperty("User-Agent",
                      "Profile/MIDP-1.0 Confirguration/CLDC-1.0");
    c.setRequestProperty("Content-Language","en-US");
    String cookiestring = name + "=" + value;
    c.setRequestProperty("cookie",cookiestring);
    os = c.openOutputStream();
    String str = "?cookieName="+name+"&"+"cookieValue="+value;
    os.write(str.getBytes());
    os.flush();
    is = c.openDataInputStream();
    int len = (int)c.getLength(); // Get length and process data
    if (len > 0)
    { byte[ ] data = new byte[len];
      int actual = is.read(data);
      String res = new String(data);
      b.append(res);
      System.out.println(res);
    } else
    { int ch;
       while ((ch = is.read()) != -1)
       { b.append((char) ch);
         System.out.println((char)ch);
       }
    }
    cookieFromServer = c.getHeaderField("Set-cookie");
    System.out.println("Cookie" + cookieFromServer);
    t = new TextBox("Cookies sent by me..",
                     b.toString(), 1024, 0);
  } finally
  { if(is!= null)
    { is.close();
    }
    if(os != null)
    { os.close();
    }
    if(c != null)
    { c.close();
    }
 }
 t.addCommand(nextCmd);
```

```
    t.addCommand(exitCommand);
    t.setCommandListener(this);
    display.setCurrent(t);
}

  public void showCookies()
  { TextBox t = new TextBox("Cookies sent by server ..",
                                  cookieFromServer, 1024, 0);
    t.addCommand(exitCommand);
    t.setCommandListener(this);
    display.setCurrent(t);
  }

  public void commandAction(Command c, Displayable s)
  { if (c==exitCommand)
    { try
      { this.destroyApp(true);
      } catch(MIDletStateChangeException e) {}
    } else if (c==setCmd)
    { try
      { String name = ((TextField) form_text.get(0)).getString();
        String value = ((TextField) form_text.get(1)).getString();
        setCookie(name,value);
      } catch (IOException e)
      { System.out.println("IOException " + e);
        e.printStackTrace();
      }
    } else if (c==nextCmd)
      { showCookies();
      }
  }
}
```

The Generic Connection Framework provided by CLDC has proven to be a useful framework in a number of scenarios—such as networking with **StreamConnection**, using HTTP *GET* and *POST* methods, and finally using Cookies. Let's turn our attention from network programming and move to the persistent storage of data on our MIDP device.

MIDP DATABASE PROGRAMMING

One of the important features that distinguishes MIDP from other wireless programming environments (such as WAP) is persistent storage. Why do we need persistence on the hand-held devices? Here are some reasons:

- It allows you to create applications that can store information on the client. Examples of such applications include address books and numerous games.
- Sometimes you might need the objects to exist even after the application that created the object is destroyed. In such cases you can store the state of the object in local persistent storage and retrieve the state later.
- Mobile devices are mostly personal devices. Given the resource constraints of the devices, it makes sense to have some level of personalization of applications. Local persistence helps in creating personalized applications. For instance, generic information such as username and address can be persisted and used instead of expecting them to be repeatedly reentered. We will discuss more about personalization in Part 3.

Local persistence can also help in other ways:

- It can reduce the number of network calls, thus requiring less bandwidth.
- It can reduce the number of inputs from the user.

The persistence mechanisms available in MIDP are not the same as the persistence mechanisms provided by J2SE. J2SE provides two types of persistence: access to Relational Databases using JDBC and object serialization to persist object state to the local file system. Mobile devices do not have the storage resources to have a relational database or a file system. Some vendors have lightweight databases for PDAs such as IBM's DB2 Everyplace and Sybase's IAnyWhere. But these databases are proprietary and do not fit on phones. As MIDP provides APIs for commonly available resources, it does not provide access to any of these databases. In the future, when devices become more capable, there might be support for JDBC-like APIs. But nothing is in the works yet.

So let us take a look at the persistence capabilities of MIDP, provided by a facility called Record Management System (RMS).

Record Management System (RMS)

RMS provides a record store, a simple record-oriented database. This database can hold a collection of records that remain persistent between MIDlet invocations. These records are also persistent even when the device is turned off. When the MIDlet is invoked again it can retrieve the records from the database.

Record stores are stored as binary files. The location of these files and their implementation are implementation specific. MIDlets within a single application suite (a MIDlet suite) can create multiple record stores (database files) with different names. The RMS APIs provide the following functionality:

- Allow MIDlets to manipulate (add and remove) records within a record store.
- Allow MIDlets in the same application to share records (access one another's record store directly). Different MIDlet suites cannot share record stores.

AddressMIDlet Using RMS

RMS APIs are packaged in the ***javax.micoedition.rms*** package. This package defines four interfaces, one class (**RecordStore**), and four exceptions classes (see Figure 10-5). Table 10-3 describes the function of the classes and interfaces.

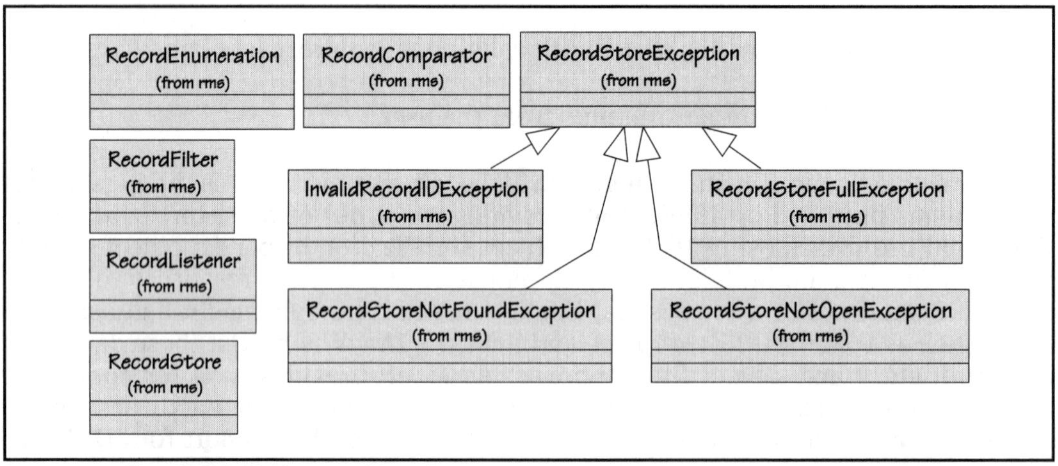

Figure 10-5. The MIDP RMS Classes and Interfaces.

Table 10-3. The RMS Packages.

Interfaces	Description
RecordComparator	Defines a comparator to compare two records.
RecordEnumeration	Represents a bi-directional record enumerator.
RecordFilter	Defines a filter to examine a record and checks if it matches criteria defined by the application.
RecordListener	Receives records that were added, changed, or deleted from a record store.
Classes	

Table 10-3. The RMS Packages. (Continued)

Interfaces	Description
RecordStore	Represents a record store.
Exceptions	
InvalidRecordIDException	Thrown to indicate the RecordID is invalid.
RecordStoreException	Thrown to indicate a general exception was thrown.
RecordStoreNotFoundException	Thrown to indicate the record store could not be found.
RecordStoreFullException	Thrown to indicate the record store file system is full.
RecordStoreNotOpenException	Thrown to indicate an operation on a closed record store.

Let's illustrate the use of RMS with a Contact database MIDlet called AddressMIDlet. The AddressMIDlet allows you to store contact information on a MIDP device. Your phone might already have a Contact application where you can store phone numbers but lets assume you cannot add any other information to that. So this AddressMIDlet has the capability to:

- ***Create*** a contact database (record store).
- ***Add*** a contact (name, email address and phone number) record.
- ***Browse*** a contact.
- ***Modify*** a contact.
- ***Delete*** a contact.

Figures 10-6 – 10-8 illustrate the functions that the AddressMIDlet provides.

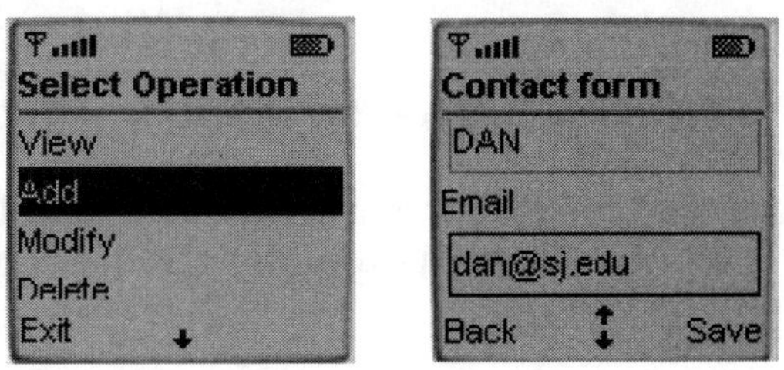

Figure 10-6. Adding a Contact.

With visuals of the Contact application firmly in mind, let's work our way through the Java code for the ContactMIDlet application.

Figure 10-7. Modifying a Contact.

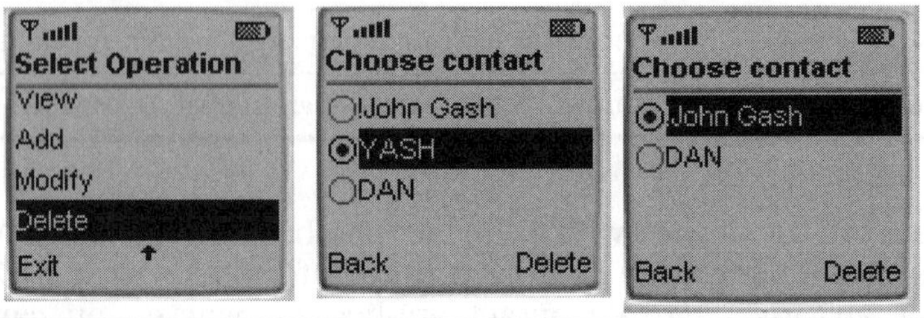

Figure 10-8. Deleting a Contact.

The Contact Record Structure

RMS is very primitive. It does not provide any representation of rows and columns like you would expect from a relational database. Instead, it stores data in records. A collection of records forms a record store. Each record is an array of bytes. So if you plan to have more than one field in a record, it is important to design your record structure properly. Here is the structure of the Contact record we have chosen to use. For our application purposes we consider Name to be the primary key.

```
Name(10)|Email(10)|Phone(10)|record-id(3)   - Total fixed length 36
```

All of the fields—Name, Email, and Phone are fixed-length with 10 characters. We are using the symbol "|" as delimiter. The total length of the record is fixed at 36 characters.

The **Contact** class shown in Listing 10-5 is used to represent a contact. This class is just a utility class to extract the fields in a record and does not use any RMS functionality.

Listing 10-5. The Contact Class Code.

```
public class Contact
{ private static String name, phone, email;
  private static int recordid;

  public static void parse(String data)
  { int index = data.indexOf('|');
    name = data.substring(1, index);
    index++;
    email = data.substring(index,(index=data.indexOf('|',index)));
    index++;
    phone = data.substring(index,(index=data.indexOf('|',index)));
    index++;
    recordid = Integer.parseInt(data.substring(index));
  }

  public static String getName(String record)
  { parse(record);
    return(name);
  }

  public static String getPhone(String record)
  { parse(record);
    return(phone);
  }

  public static String getEmail(String record)
  { parse(record);
    return(email);
  }

  public static int getRecordid(String record)
  { parse(record);
    return(recordid);
  }
}
```

The database operations are provided in one class to keep the implementation simple. The **ContactDB** class in Listing 10-6 provides the necessary database operations, which we will discuss in detail in the following sections.

Listing 10-6. The ContactDB Class.

```
import javax.microedition.rms.*;
```

```
import java.util.Enumeration;
import java.io.*;

public class ContactDB
{ RecordStore recordStore = null;
  public ContactDB() {}
  public ContactDB(String dbName )
  { // Open the record store, create it if it does not exist
    try
    { recordStore = RecordStore.openRecordStore(dbName, true);
    } catch(RecordStoreException rse)
    { rse.printStackTrace();
    }
  }

public void close()
      throws RecordStoreNotOpenException, RecordStoreException
  { if (recordStore.getNumRecords() == 0)
    { String dbName = recordStore.getName();
      recordStore.closeRecordStore();
      recordStore.deleteRecordStore(dbName);
    } else
    { recordStore.closeRecordStore();
    }
  }

  public synchronized void addContact(String record)
  { ByteArrayOutputStream baos = new ByteArrayOutputStream();
    DataOutputStream outputStream = new DataOutputStream(baos);
    try
    { outputStream.writeUTF(record);
    } catch (IOException ioe)
    { System.out.println(ioe);
      ioe.printStackTrace();
    }
    byte[ ] b = baos.toByteArray();
    try
    { recordStore.addRecord(b, 0, b.length);
      System.out.println("successfully added record");
    } catch (RecordStoreException rse)
    { System.out.println(rse);
      rse.printStackTrace();
    }
  }
```

```
public synchronized RecordEnumeration enumerate()
                  throws RecordStoreNotOpenException
  { return recordStore.enumerateRecords(null, null, false);
  }

  public synchronized void deleteContact(int recordid) throws
           RecordStoreNotOpenException, InvalidRecordIDException,
                          RecordStoreException
  { recordStore.deleteRecord(recordid);
    System.out.println("Successfully deleted record at" + recordid);
  }

  public synchronized void updateContact(int recid,String record)
                          throws RecordStoreNotOpenException,
                         InvalidRecordIDException, RecordStoreException
  { System.out.println("updateContact" + record);
    ByteArrayOutputStream baos = new ByteArrayOutputStream();
    DataOutputStream outputStream = new DataOutputStream(baos);
    try
    {  outputStream.writeUTF(record);
    } catch (IOException ioe)
    {  System.out.println(ioe);
       ioe.printStackTrace();
    }
    byte[ ] b = baos.toByteArray();
    recordStore.setRecord(recid,b,0,b.length);
  }

  public synchronized int getRid() throws
RecordStoreNotOpenException,RecordStoreException
  { return recordStore.getNextRecordID();
  }
}
```

The **AddressMIDlet** class is shown in Listing 10-7. This is the main MIDlet class that provides all the UI implementation and handles the commands. You may want to revisit Chapter 9 if you have trouble reading this code.

Listing 10-7. The AddressMIDlet Class.

```
import javax.microedition.rms.*;
import javax.microedition.lcdui.*;
import javax.microedition.midlet.*;
import javax.microedition.io.*;

public class AddressMIDlet extends MIDlet implements CommandListener
```

```
{ Display display = null;
  List choose = null;
  static final Command backCommand = new Command("Back", Command.BACK, 0);
  static final Command main_menuCommand = new Command("Main",
                                              Command.SCREEN, 1);
  static final Command saveCommand = new Command("Save", Command.OK, 2);
  static final Command deleteCommand = new Command("Delete",
                                              Command.OK, 2);
  static final Command exitCommand = new Command("Exit", Command.STOP, 3);
  String currentDisplay = null;
  List main_menu = null; // main menu
  TextField nameField, phoneField, emailField;
  Form contact_form = null; // Form to hold the contact for add and modify
  List view_list = null;
  List delete_list = null; // list to select a contact for deletion

  // Contact has name, phone and email- name is the primary key
  String name, email, phone ;
  ContactDB db = null;

  public AddressMIDlet() { }
  public void startApp() throws MIDletStateChangeException
  { display = Display.getDisplay(this);

    // Open the db, create one if it does not exist
    try
    { db = new ContactDB("mycontacts");
    } catch(Exception e) {}

    // Display the main menu Add/modify/delete/view
    main_menu = new List("Select Operation", Choice.IMPLICIT);
    main_menu.append("View", null);
    main_menu.append("Add", null);
    main_menu.append("Modify", null);
    main_menu.append("Delete",null);
    main_menu.addCommand(exitCommand);
    main_menu.setCommandListener(this);
    mainmenu();

    // Initialize the contact form
    contact_form = new Form("Contact form");
    nameField = new TextField("Name", "", 10, TextField.ANY);
    emailField = new TextField("Email", "", 10, TextField.EMAILADDR);
    phoneField = new TextField("phone", "", 10, TextField.PHONENUMBER);
    contact_form.append(nameField);
```

```
    contact_form.append(emailField);
    contact_form.append(phoneField);
    contact_form.addCommand(saveCommand);
    contact_form.addCommand(backCommand);
    contact_form.setCommandListener(this);
  }

  public void pauseApp()
  { // Free the resources
    // Close the db
    display = null;
    try
    {  db.close();
       db = null;
    } catch(Exception e) {}
  }

  public void destroyApp(boolean unconditional)
  { try
    { db.close();
    } catch(Exception e) {}
    notifyDestroyed();
  }

  void mainmenu()
  { display.setCurrent(main_menu);
    currentDisplay = "main_menu";
  }

  public void addContact()
  {  nameField.setString("");
     emailField.setString("");
     phoneField.setString("");
     display.setCurrent(contact_form);
     currentDisplay = "Add";
  }

  public void listContacts()
  { view_list = new List("Choose contact", Choice.IMPLICIT);
    view_list.addCommand(backCommand);
    view_list.setCommandListener(this);
    try
    {  RecordEnumeration re = db.enumerate();
       System.out.println("Records " + re);
       while(re.hasNextElement())
```

```
     {System.out.println("Found Records");
        String curContact = new String(re.nextRecord());
        view_list.append(Contact.getName(curContact), null);
     }
   } catch(Exception ex) {}
     display.setCurrent(view_list);
     currentDisplay = "view_list";
}

public void deleteContacts()
{ delete_list = new List("Choose contact", Choice.EXCLUSIVE);
  delete_list.addCommand(deleteCommand);
  delete_list.addCommand(backCommand);
  delete_list.setCommandListener(this);
  try
  {  RecordEnumeration re = db.enumerate();
     System.out.println("Records " + re);
     while(re.hasNextElement())
     {  System.out.println("Found Records");
        String curContact = new String(re.nextRecord());
        delete_list.append(Contact.getName(curContact), null);
     }
   } catch(Exception ex) {}
   display.setCurrent(delete_list);
   currentDisplay = "delete_list";
}

public void viewContact(String name)
{ try
  { RecordEnumeration re = db.enumerate();
    System.out.println("Records " + re);
    while(re.hasNextElement())
    { System.out.println("Found Records");
      String curContact = new String(re.nextRecord());
      if (name.equals(Contact.getName(curContact)))
      { nameField.setString(name);
        phoneField.setString(Contact.getPhone(curContact));
        emailField.setString(Contact.getEmail(curContact));
        break;
      }
    }
  } catch(Exception ex) {}
  display.setCurrent(contact_form);
  currentDisplay = "contact_form";
}
```

```
public void commandAction(Command c, Displayable d)
{ String label = c.getLabel();
  if (label.equals("Exit"))
  { destroyApp(true);
  } else if (label.equals("Save"))
  { int rid = 0;
    boolean add = false;
    String name = nameField.getString();
    String email = emailField.getString();
    String phone = phoneField.getString();
    try
    { RecordEnumeration re = db.enumerate();
      System.out.println("Records " + re);
      while(re.hasNextElement())
      { System.out.println("Found Records " );
        String curContact = new String(re.nextRecord());
        if (name.equals(Contact.getName(curContact)))
        { rid = Contact.getRecordid(curContact);
          break;
        }
      }
     } catch (Exception e) {}
     if (rid == 0)
     { add = true;
       try
       {  rid = db.getRid();
       } catch(Exception e) {}
     }
     String record = name + "|" +  email + "|" + phone + "|" + rid;
     String junk = new String(new char[35-record.length()]);
     String rec = record + "|" + junk;
     System.out.println(record);
     if (add)
     { db.addContact(record);
     }
     else
     { try
        { System.out.println("name" + name);
          db.updateContact(rid,record);
        } catch(Exception e) {}
     }
     mainmenu();
   } else if (label.equals("Back"))
   {  mainmenu();
```

```
    } else if (label.equals("Delete"))
    { try
       { int i =0;
  String name =
  delete_list.getString(delete_list.getSelectedIndex());
         RecordEnumeration re = db.enumerate();
         System.out.println("Records " + re);
         while(re.hasNextElement())
         {  System.out.println("Found Records");
            String curContact = new String(re.nextRecord());
            if (name.equals(Contact.getName(curContact)))
            {  i = Contact.getRecordid(curContact);
               break;
            }
         }
         System.out.println("Before calling delete at" + i);
         db.deleteContact(i);
       } catch (Exception e)
       {  e.printStackTrace();
       }
       mainmenu();
   }
   else
   { if (currentDisplay.equals("main_menu"))
     { switch(main_menu.getSelectedIndex())
         { case 0: listContacts();break;
           case 1: addContact();break;
           case 2: listContacts();break;
           case 3: deleteContacts();break;
         }
      } else if (currentDisplay.equals("view_list") )
     { System.out.println(
            view_list.getString(view_list.getSelectedIndex()));
        viewContact(view_list.getString(view_list.getSelectedIndex()));
        System.out.println("Currently in the view list");
      }
   }
 }
}
```

The AddressMIDlet, when invoked, creates an instance of **ContactDB**, which in turn creates a database with given name. For example:

```
RecordStore recordStore = null;
```

```
 . . .
try
{ recordStore = RecordStore.openRecordStore(dbName, true);
} catch(RecordStoreException rse)
{ rse.printStackTrace();
}
```

The RecordStore and RecordStoreEnumeration Classes

The **RecordStore** class provides the semantics for the MIDP RMS (see Figure 10-9). The class provides methods to create or open a record store using the static *openRecordStore* method. The second parameter, if set to true, creates the database if it does not already exist. The reference to any given record store is a singleton. This means if the *openRecordStore* method is called by a MIDlet when the record store has already been opened by the same MIDlet (or another MIDlet in the same MIDlet suite), the method returns a reference to the same RecordStore object.

The **RecordEnumeration** class maintains a sequence of RecordIDs for the records in a **RecordStore**. The **RecordComparator** class compares two records to see if they match or what their relative sort order is. The *compare* method is called by **RecordEnumeration** to sort and return records in an application-specified order.

The following sections describe the details of the Contact application.

Adding a New Contact

When Add is selected from the MIDlet main menu, the contact form is presented in order to fetch contact information. When Save is pressed, the **AddressMIDlet** prepares the record to be added and passes it to **ContactDB** for writing to the record store.

The following is a code snippet for a record in the **RecordStore**:

```
try
{rid = db.getRid();
} catch(Exception e) {}

String record = name + "|" +  email + "|" + phone + "|" + rid;
String junk = new String(new char[35-record.length()]);
String rec = record + "|" + junk;
```

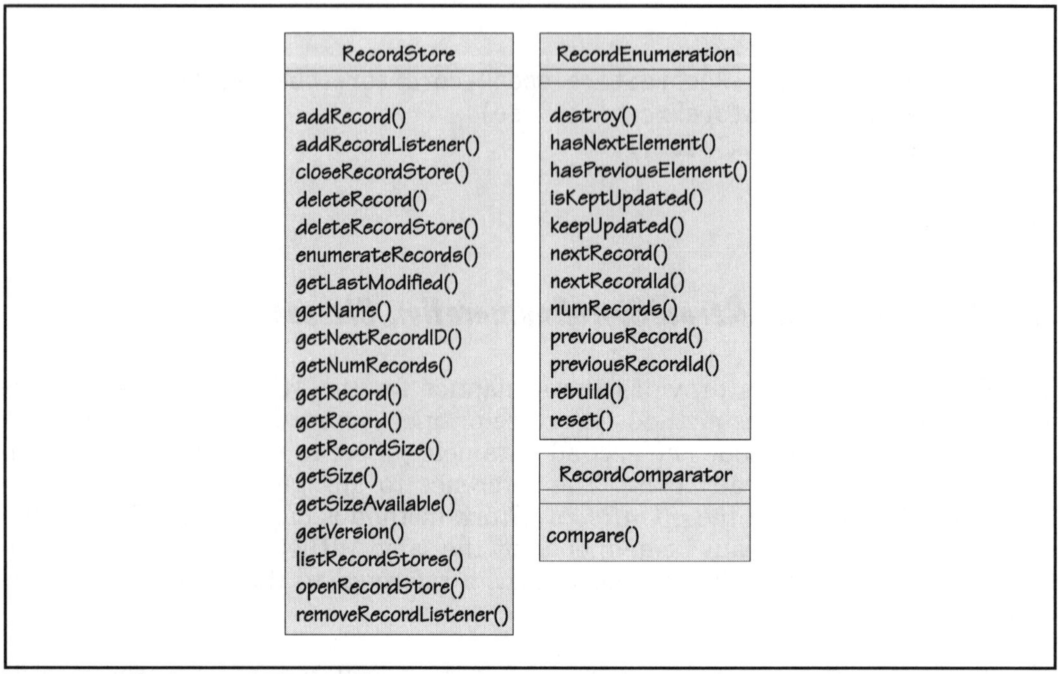

Figure 10-9. The Primary MIDP RMS Classes.

```
db.addContact(record);
```

Note that the "|" is used as delimiter between the fields. Also note that to keep the record to a fixed length of 36 characters, we are filling the record with blank characters. We are also adding a record ID as one of the fields in the record. We will explain its use a little later. The following snippet shows the **addContact** method:

```
public synchronized void addContact(String record)
{
  ByteArrayOutputStream baos = new ByteArrayOutputStream();
  DataOutputStream outputStream = new DataOutputStream(baos);
  try
  { outputStream.writeUTF(record);
  } catch (IOException ioe)
  { System.out.println(ioe);
    ioe.printStackTrace();
  }
  byte[ ] b = baos.toByteArray();
  try
  { recordStore.addRecord(b, 0, b.length);
```

```
            System.out.println("successfully added record");
        } catch (RecordStoreException rse)
        { System.out.println(rse);
          rse.printStackTrace();
        }
    }
```

As mentioned earlier, a record is an array of bytes, so we need to convert our record string to a byte array before writing it. We can use the **DataInputStream**, **DataOutputStream**, **ByteArrayInputStream**, and **ByteArrayOutputStream** classes to pack and unpack data types into and out of the byte arrays.

So in this case we have created the byte array *b* and called the *addRecord* method of **RecordStore**. This call will allocate bytes in the device's persistent storage. There is the possibility that the storage is full. So, we need to handle any exceptions that might occur. The **RecordStoreException** is a generic exception that is handled here.

The **RecordStore** class assigns a unique ID to every record that is stored. This record ID can be treated as a primary key. The record ID starts with 1 and every subsequent record is assigned a record ID incremented by 1. The record ID of the next record to be added to the record store can be obtained as follows.

```
recordStore.getNextRecordID();
```

In the Contact application, this is used to add an ID field to the record.

Viewing the Contacts

When the user requests a record to view, **AddressMIDlet** makes a call to the **ContactDB** class to fetch all the records. The following is a snippet of the **AddressMIDlet** code that fetches the records from the **RecordStore**:

```
try
{ RecordEnumeration re = db.enumerate();
  System.out.println("Records " + re);
  while(re.hasNextElement())
  { System.out.println("Found Records");
    String curContact = new String(re.nextRecord());
    view_list.append(Contact.getName(curContact), null);
  }
}
```

The following is a snippet of the **ContactDB** code that fetches all the records from the record store:

```
public synchronized RecordEnumeration enumerate()
                             throws RecordStoreNotOpenException
{ return recordStore.enumerateRecords(null, null, false);
}
```

The *enumerateRecords* method returns all the records in the record store. Once all the records have been fetched, we can loop through each record and fetch the fields from the record based on the delimiter.

Modifying a Contact

We have seen that adding a record and viewing records are very simple. Modifying a record is little tricky. Here is the API provided by RMS to modify a record.

```
public void setRecord(int recordId,
                      byte[ ] newData,
                      int offset,
                      int numBytes)
               throws RecordStoreNotOpenException,
                      InvalidRecordIDException,
                      RecordStoreException,
                      RecordStoreFullException
```

If we want to modify the entire record we need the recordId and the number of bytes of the record. We will show how to get the recordId in the delete section. But if the length of the new record is more than the original record, then the chances are that you would be overwriting the next record. There are two solutions to overcome this problem:

1. Fix the size of the record. This way we are sure that the new record and the existing record are the same size so as not to step on the next record. This is the technique used in the Contact application.
2. When the record size is variable, the solution is to delete the existing record and add the modifying record as a new one.

When modifying a particular field in the record, we need to know the offset at which the field starts. Even in this case it would be easy if the field length is fixed. Otherwise, overwriting the entire record is much easier, which is what we are doing in our application.

In the **AddressMIDlet**, we are using name as the primary key (logical primary key for the application). So when the any field changes and the record is saved, the entire record is updated (rather than just the changed field).

```
public synchronized void updateContact(int recid,String record)
                     throws RecordStoreNotOpenException,
                   InvalidRecordIDException, RecordStoreException
{ System.out.println("updateContact" + record);
  ByteArrayOutputStream baos = new ByteArrayOutputStream();
  DataOutputStream outputStream = new DataOutputStream(baos);
try
  {outputStream.writeUTF(record);
} catch (IOException ioe)
  {System.out.println(ioe);
   ioe.printStackTrace();
  }

  byte[ ] b = baos.toByteArray();
  recordStore.setRecord(recid,b,0,b.length);
}
```

The record is converted into a byte array just like when a record is added to the record store. Note that *b.length* will always be 36 in this case and ensures that we are not stepping on the next record. The offset is always "0" as we are updating the entire record.

For the Contact example, the name, email, and phone number are part of each Contact. If records are not going to be modified, then variable size records are not difficult to implement. However, if the records will be modified, then it helps to have fixed-size records. This is because the number of bytes to be changed has to be specified when modifying a record. With a variable record length, there is a possibility to overwrite the next record, if not updated carefully

Deleting a Contact

Deleting a record is much trickier. This is the API provided to delete a record:

```
public void deleteRecord(int recordId)
                  throws RecordStoreNotOpenException,
                         InvalidRecordIDException,
                         RecordStoreException
```

When deleting a record, the record must be known. Unfortunately, there is no method to get the recordId of a record. So a workaround is needed to determine the record ID of the record we want to delete.

The simplest way to do this is to store the recordId as part of the record itself. If you look back at our record format, the recordId is one field in the record. Before we store the record, we are fetching the next recordId and adding it to the record. When we want to delete a record, we can enumerate through the records and compare the record we are trying to delete, fetch the recordId from the matching record, and then delete it. This code snippet shows an implementation of record comparison by implementing the **Comparator** interface.

```
public myClass implements Comparator
{ public int compare(byte[ ] record1, byte[ ] record2) {
   ByteArrayInputStream bais1 =
            new ByteArrayInputStream(record1);
   DataInputStream dis1 =
            new DataInputStream(bais1);
   ByteArrayInputStream bais2 =
            new ByteArrayInputStream(record2);
   DataInputStream dis2 =
            new DataInputStream(bais2);

   String name1 = dis1.readUTF();
   String name2 = dis.readUTF();
   int num = name1.compareTo(name2);
if (num > 0)
{ return RecordComparator.FOLLOWS;
} else if (num < 0)
       { return RecordComparator.PRECEDES;
} else
{ return RecordComparator.EQUIVALENT;
          }
  }
}
```

The constants FOLLOWS, PRECEDES, and EQUIVALENT are defined in the Re**cordComparator** interface and have the following meanings:

- **FOLLOWS**: Its value is 1 and means the left parameter follows the right parameter in terms of search or sort order.
- **PRECEDES**: Its value is -1 and means the left parameter precedes the right parameter in terms of search or sort order.
- **EQUIVALENT**: Its value is 0 and means the two parameters are the same.

This method is not used in the Contact application. Recall that name is defined to be the primary key. So we are enumerating through the records and comparing just the name to fetch the matching recordId. Here's a code snippet:

```
int i = 0;
String name =
delete_list.getString(delete_list.getSelectedIndex());
RecordEnumeration re = db.enumerate();
System.out.println("Records " + re);
while(re.hasNextElement())
{ System.out.println("Found Records");
  String curContact = new String(re.nextRecord());
if (name.equals(Contact.getName(curContact)))
{ i = Contact.getRecordid(curContact);
    break;
  }
}
System.out.println("Before calling delete at" + i);
db.deleteContact(i);
```

Closing the RecordStore

When a MIDlet is paused or exited, it is important that the **RecordStore** is closed. The following snippet closes the record store:

```
recordStore.closeRecordStore();
```

Multithreading in RMS

You might have noticed from our **ContactDB** class that all our methods are *synchronized*. The MIDP RMS implementation ensures that all individual record store operations are atomic, synchronous, and serialized, so no corruption occurs with multiple access. However, if MIDlets use multiple threads to access a record store, it is the programmer's responsibility to synchronize this access. Otherwise the data will be corrupted. The **AddressMIDlet** application does not use threads.

MIDLET PROVISIONING

MIDlet provisioning is a process that defines the way the MIDlet is transferred from a server to the device and the way the device makes the MIDlet available to the user. The MIDP specification does not specify this process and left this as implementation-specific behavior. The initial implementations of MIDP on phones provided a manual process. For instance, Motorola provides a desktop utility called *idenUpdate* for its *iden* phones. Users can utilize this tool to download MIDlets to the device by means of cable. Similarly, other vendors provided their own way of MIDlet provisioning.

This is not a very convenient way of downloading MIDlets as this involves using cables and custom desktop software. Since we are dealing with wireless devices here, one should be able to download MIDlets "over the air" and be able to use them.

Although there is no standard that specifies this, there is a recommended specification defined in the document *Over The Air User Initiated Provisioning Recommended Practice for the Mobile Information Device Profile* available at *http://java.sun.com/products/midp/OTAProvisioning-1.0.pdf.*

This specification defines in detail the following functions that a device needs to perform to support Over the Air deployment of MIDlet suites:

- ***Discovery:*** Able to browse and locate MIDlet suites' application descriptors in the network.
- ***Downloading:*** Able to select and transfer a MIDlet suite and its associated application descriptor to the device from a server using HTTP 1.1 or a session protocol that implements the HTTP 1.1 protocol.
- ***Authentication:*** Servers that host the MIDlet suites might restrict the download of MIDlets to authorized users. So the device needs to prompt the user for user name and password in such cases and pass on the credentials to the server.
- ***Installation and Invocation:*** Allow installing the downloaded MIDlet suite on the device and be able to invoke the MIDlets.
- ***Deletion:*** Allowing the user to delete MIDlet suites stored on the device. Single MIDlets cannot be deleted since the MIDlet suite is the unit of transfer and installation.

Note: In order for the MID to detect the MIDlet suite, the following MIME type needs to be added to the Web server configuration.

```
"text/vnd.sun.j2me.app-descriptor"
```

BLUETOOTH REVISITED

Bluetooth, with its promise of a desktop without wires, is being included in many new types of wireless devices. In a short time, we might see Bluetooth hardware installed in high end desktop PCs, notebooks, PDAs, cellular phones, and everything in between. Even consumer devices like headsets and PC peripherals like keyboards and mice can benefit from Bluetooth's wireless *Personal Area Network (PAN)* technology.

The following sections will introduce you to some of the details behind the Bluetooth standard and the recently approved Java APIs for Bluetooth. We'll look in detail at the various modes of Bluetooth wireless transmission, and how Bluetooth devices locate each other in the PAN. We'll even write a simple Bluetooth application using MIDP.

Technology Overview

We'll start the chapter with a look at the technology fundamentals that comprise the Bluetooth specification. These fundamentals will make it easier to understand the API and sample program later in the chapter. We've intentionally omitted some parts of the specification that aren't relevant to this book's subject matter. We'll make you aware of these omissions as we go along—you can read up on the omitted technologies on your own if your application needs to make use of them.

Bluetooth Device Classes

Every type of Bluetooth device, whether it is a PC, a PDA, a printer, a headset, or anything else, is classified using a numbering scheme defined by the Bluetooth SIG. Each device has a particular (although not necessarily unique) code that identifies the type of device and its capabilities.

The code, called the *Class of Device* code (CoD), is a 24-bit number. Bits 23 (MSB) to 13 comprise the Service Class of the device. A device's Service Class is a high-level abstraction of what the device does, such as printing, telephony, networking, and the like.

The Service Class bits are arranged as follows:

- Bit 23 – Information device (Web server, WAP server)
- Bit 22 – Telephony (cellular phones, modem, headset for phones)
- Bit 21 – Audio (speaker, microphone)
- Bit 20 – Object transfer (inbox, folder, file storage)
- Bit 19 – Capturing (digital camera, scanner)
- Bit 18 – Rendering (printer)
- Bit 17 – Networking (LAN, WAN)
- Bit 16 – Positioning (GPS, location identification)
- Bits 14-15 – Reserved
- Bits 13 – Limited discoverable device

A device's Bluetooth implementation defines the appropriate Service Class for the device by combining these bits in the pattern that represents the type of service the device provides. Bit 13, the Limited discoverable device, represents a device that is not discoverable by other devices through broadcasting (or is discoverable in a restricted fashion).

In addition to the Service Class, each device has a *Device Class* that describes the device in more detail. The Device Class is broken up into a *Major Device Class* and a *Minor Device Class*. The Major Device Class occupies bits 12-8 of the CoD. These 5 bits provide up to 32 possible Major Device Classes (2^5), but not all 32 possibilities have yet been defined. The Major Device Classes that have been defined are:

- 00000 – Miscellaneous Device
- 00001 – Computer (desktop PC, notebook, PDA)
- 00010 – Phone (cellular, cordless, pay phone, modem)
- 00011 – LAN hardware
- 00100 – Audio/Video (speaker, microphone, video display, DVD, VCR)
- 00101 – Peripheral (mouse, keyboard, joystick)
- 00110 – Imaging (camera, printer, display)
- 11111 – Uncategorized

For each of the above Major Device Classes, there is a set of Minor Device Classes. Each Minor Device Class value only has meaning when interpreted together with the device's Major Device Class. The Minor Device Class occupies bits 7-2 of the CoD, which gives a total of 64 Minor Device Classes for each Major Device Class.

The Minor Device Classes defined for the Computer Major Device Class are:

- 000000 – Uncategorized
- 000001 – Desktop workstation
- 000010 – Server
- 000011 – Laptop
- 000100 – Handheld PC/PDA
- 000101 – Palm or Pocket sized PC/PDA
- 000110 – Wearable computer

The Minor Device Classes defined for the Phone Major Device Class are:

- 000000 – Uncategorized
- 000001 – Cellular
- 000010 – Cordless
- 000011 – Smart phone
- 000100 – Wired modem or voice gateway (PBX)
- 000101 – ISDN
- 000110 – SIM card reader

The Minor Device Classes defined for the Networking Hardware Major Device Class are:

- 000000 – Fully available (0% utilized)
- 001000 – 1%-17% utilized
- 010000 – 17%-33% utilized
- 011000 – 33%-50% utilized
- 100000 – 50%-67% utilized
- 101000 – 67%-83% utilized
- 110000 – 83%-99% utilized
- 111000 – Fully utilized (no service available)

The capacity utilizations shown above are used when a Bluetooth device wishes to connect to a LAN gateway of some kind. By locating all Networking devices within its range, it can choose the gateway with the lowest load by examining each device's Minor Device Class.

The Minor Device Classes defined for the Audio/Video Major Device Class are:

- 000000 – Uncategorized
- 000001 – Headset
- 000010 – Hands free headset
- 000011 – Reserved
- 000100 – Microphone
- 000101 – Loudspeaker
- 000110 – Headphones
- 000111 – Portable Audio
- 001000 – Car Audio
- 001001 – Set-top Box
- 001010 – HiFi Audio Device
- 001011 – VCR
- 001100 – Video Camera
- 001101 – Camcorder
- 001110 – Video Monitor
- 001111 – Video Display and Loudspeaker
- 010000 – Video Conferencing System
- 010001 – Reserved
- 010010 – Game or Toy

The Minor Device Classes defined for the Peripheral Major Device Class are:

- 010000 – Uncategorized Keyboard
- 010001 – Joystick + Keyboard
- 010010 – Gamepad + Keyboard
- 010011 – Remote Control + Keyboard

- 010100 – Sensing device + Keyboard
- 010101 – Digitizer tablet + Keyboard
- 100000 – Uncategorized pointing device
- 100001 – Joystick + Pointing device
- 100010 – Gamepad + Pointing device
- 100011 – Remote Control + Pointing device
- 100100 – Sensing device + Pointing device
- 100101 – Digitizer tablet + Pointing device
- 110000 – Uncategorized Keyboard/Pointing device
- 110001 – Joystick + Keyboard/Pointing device
- 110010 – Gamepad + Keyboard/Pointing device
- 110011 – Remote Control + Keyboard/Pointing device
- 110100 – Sensing device + Keyboard/Pointing device
- 110101 – Digitizer tablet + Keyboard/Pointing device

Finally, the Minor Device Classes defined for the Imaging Major Device Class are:

- XXX100 – Display
- XX1X00 – Camera
- X1XX00 – Scanner
- 1XXX00 – Printer

The bits listed above for the Imaging Minor Device classes can be combined (ORed together) for multifunction devices.

Discovering Other Devices

The large numbers of major/minor devices, paired with service classes, make it possible for our Bluetooth devices to locate, discover, and use the exact device they require. How does this discovery and location work? The Bluetooth stack present inside each device's Bluetooth transmission module handles broadcasting over the air and locating the devices within range. Thankfully, the Java API for Bluetooth (which we'll discuss shortly) handles the low-level RF transmission details for us. With just a few lines of code, we can ask the Bluetooth stack to locate desired devices for us.

RF Communication and Connections

Recall that the MIDP networking layer uses a connection framework to manage stream-based connections in a standard way. Our MIDlets ask for an HTTP connection using an HTTP-style URL, and the MIDP implementation handles opening the correct type of TCP/IP socket, establishing a connection, and returning the stream handle to our MIDlet.

FYI Briefing

The Devil Is In The Details

We're not going to go into great detail about the Bluetooth protocol stack. Rather, we're going to give you just enough to get started writing Bluetooth-enabled MIDlets. Although you will probably eventually want to read more about the lower-layer protocols provided by Bluetooth, an entire volume could be written describing them.

The Bluetooth SIG (*http://www.bluetooth.org*), has voluminous documentation describing the other aspects of the protocols involved, including RF transmission details and packet structure. Although these are interesting topics, you won't need them to understand the forthcoming sample application. ❑

The Bluetooth connection model works in the same way, but instead of asking for an HTTP connection, we ask for a L2CAP connection. L2CAP, the *Logical Link Control and Adaptation Protocol*, provides a connection between two Bluetooth devices, in the same way HTTP provides a connection between a Web server and Web browser. The Java API for Bluetooth specifies a *L2CAPConnection* interface that is returned to our MIDlets when a Bluetooth L2CAP connection is requested from the connections framework.

Java API for Bluetooth (JSR-82)

In general, the Bluetooth program flow proceeds as follows:

1. Your MIDlet is started and configures the Bluetooth environment.
2. Your device locates (or is located by) another device on the PAN.
3. Your device opens (or accepts) a connection to/from the other device.
4. Your device sends (or receives) data over the connection.

Steps 2-4 could be asynchronous, meaning multiple connections could be open, with data being transferred over each one at different times. The connections might have different lifetimes as well. Conceptually, the steps above comprise the same behavior as a normal TCP/IP connection, but the connections established in this case use L2CAP over a wireless connection (more precisely, over the lower-level Bluetooth protocols). The API provides listeners and notifiers that can be registered to be called when incoming connections are requested, or when new devices are discovered by the Bluetooth stack.

How do we code the steps described above? Let's take a look at the Java Classes that provide the functionality we'll need. All the classes we'll be covering reside in the *javax.bluetooth* package, except where noted. Once done, we'll write a simple Bluetooth MIDlet.

The LocalDevice Class

The **LocalDevice** class provides the means to interface with your local device's Bluetooth stack. The **LocalDevice** class provides methods for:

- Retrieving your device's service class and major/minor device class by using the *getDeviceClass* method. The *getDeviceClass* method returns a **DeviceClass** object with which we can extract the service class, and major/minor device classes. We'll cover the **DeviceClass** class in more detail in a subsequent section.

- Retrieving your device's Bluetooth address (a 12-digit number unique to each Bluetooth device) via the *getBluetoothAddress* method. The *getBluetoothAddress* method returns a **String** that represents the address of the local device.

- Examining and changing the *discoverability mode* of your device by using the *getDiscoverableMode* and *setDiscoverableMode*, respectively (see the next Details box).

- Locating the *Discovery Agent* for this device. The Discovery Agent is the part of the Bluetooth stack that is responsible for locating other Bluetooth devices in the PAN. Your device uses its own Discovery Agent to locate other Bluetooth devices with which it wishes to communicate. The local Discovery Agent is obtained via the *getDiscoveryAgent* method, which returns a **DiscoveryAgent** object. We'll look in more detail at the **DiscoveryAgent** class in a forthcoming section.

- Querying system properties (like the Bluetooth version supported, the maximum number of simultaneous connections, and whether or not the device is currently connected to another device) through the *getProperty* method.

Discoverability

Details

Your Bluetooth device can be in one of three discoverability modes:

- ***Generally Discoverable:*** Meaning other devices can discover your device by asking their Bluetooth stacks to scan for your device's address or characteristics.
- ***Limited Discoverable:*** In a limited discoverable mode, devices are only discoverable for a short time period. Typically, the user of the device causes the device to enter this mode in response to some event. This mode could be used when discovery of the device is done infrequently or when many devices of the same type are present in the PAN.
- ***Non-Discoverable***: In this mode, the local device is not discoverable by other devices.

The Bluetooth SIG specifies certain constants that represent the discoverable mode of a given device:

- GIAC – For Generally Discoverable devices. The GIAC constant is 0x9E8B33.
- LIAC – For Limited Discoverable devices. The LIAC constant is 0x9EB800.
- NOT_DISCOVERABLE – The NOT_DISCOVERABLE constant is 0x00. ❑

The L2CAPConnection Interface

As stated earlier, the MIDP specification abstracts the notion of connections by using a standard connection framework based on URLs that indicate which protocol to use and where to connect.

The Java API for Bluetooth adds a new form of URL that is used when connecting to Bluetooth devices. The URL takes the form:

```
btl2cap://device_address:UUID
```

The device_address represents the 12-digit Bluetooth address of the remote device, and UUID represents the service we want to connect to. The UUID is a unique hexadecimal identifier—each service has a different one.

So what is a Bluetooth service? A service is simply an operation that a device can perform for clients. When a Bluetooth application informs the Bluetooth stack it is ready to accept connections, it passes along the UUID of the service it is providing.

You can roughly equate UUIDs (when used to identify Bluetooth services) with TCP/IP ports. On a standard workstation using TCP/IP, a service is bound to a particular port. For example, a Web server is typically bound to TCP/IP port 80. An FTP server is typically bound to port 21. Our Bluetooth applications can provide multiple services, each bound to a different "port" or UUID. With standard TCP/IP, there are certain conventions that are adhered to, and certain service-to-port mappings that the industry accepts as the norm. Similarly, there are accepted UUID-to-service mappings defined by the Bluetooth SIG.

Details

UUID to Service Mapping

There are hundreds of accepted mappings already in place. We won't describe them here. If you would like to review the mapping, visit:

http://www.bluetooth.org/assigned-numbers/sdp.htm. ❑

When requesting a connection to a remote device, we use MIDP's **Connector** class to open a connection, passing in the URL that identifies the remote device and service being requested. The return value from this method is typecast to the **L2CAPConnection** interface.

The **L2CAPConnection** interface, as it extends the standard MIDP **Connection** interface, is what our application uses to read and write bytes over the connection to the remote device, and it provides the standard *receive* and *send* methods.

Warning

MTUs

L2CAP connections do not queue data for later retrieval. When using the **L2CAPConnection** interface's *receive* or *send* method, you have to be aware of a concept called *MTU (Maximum Transmission Unit)*. The MTU specifies the maximum number of bytes that can be exchanged in one packet. It is an integer value that has different instances for receiving data (*ReceiveMTU*) and sending data (*TransmitMTU*).

If you attempt to read or write more bytes than the MTU specifies, the overflow will be discarded, and your message will be truncated.

This restriction is cumbersome, but can be worked around. The **L2CAPConnection** interface provides two methods, *getTransmitMTU* and *getReceiveMTU*, which allow our MIDlet to determine the maximum number of bytes that can be sent (or received) at one time. If we are aware of these limits ahead of time, our MIDlet can chop the data up into segments and send the segments one at a time. Each segment would then be restricted to the size of the TransmitMTU limit. A similar scheme could be used for reading data.

The TransmitMTU and ReceiveMTU values can be specified when opening the connection by modifying the connection URL as follows:

```
btl2cap://device_address:UUID;ReceiveMTU=x;
TransmitMTU=y
```

Where x and y are the desired values for the connection's ReceiveMTU and TransmitMTU, respectively. Each connection can have its own separate MTU values. ❑

The L2CAPConnectionNotifier Interface

The **L2CAPConnection** interface (just described) is what a device would use to connect to another device's service that is accepting connections. The **L2CAPConnectionNotifier** interface handles the server side of the connection. We use the term "server" here rather loosely—the "server" we're referring to is the device that is accepting incoming connections. The "client" is the device that is requesting outgoing connections.

Just like the standard client side, we create a URL that represents the address of the local device (remember we're actually speaking in the context of the server device now, so the local device represents the device that is offering the service) and the UUID of the service being provided. We can also append ReceiveMTU and TransmitMTU values if desired.

Once we create the URL, we again call the **Connector** class' *open* method to create an unconnected socket. This is uncharted waters for most MIDP programmers, as MIDP does not provide server-side sockets when using regular TCP/IP. We again take the return value from the open method and this time typecast it to an **L2CAPConnectionNotifier** object.

The **L2CAPConnectionNotifier** object will await incoming connections on the URL specified when the connection was opened. Our application can wait for the next connection to be established by calling the **L2CAPConnectionNotifier** object's *acceptAndOpen* method. The *acceptAndOpen* method returns an **L2CAPConnection** object, with which our server application can communicate with the client application that connected with it.

The RemoteDevice Class

We detailed the **LocalDevice** class earlier. Our Bluetooth applications can also retrieve information about the device at the other end of a connection, once the connection is established. The **RemoteDevice** class provides a static method, *getRemoteDevice*, which takes as an argument the **L2CAPConnection** we established earlier. The *getRemoteDevice* method returns a **RemoteDevice** object, with which our application can query characteristics of the remote device.

In particular, the **RemoteDevice** class supports:

- ***Authentication and authorization of devices:*** The **RemoteDevice** class provides methods that can be used to authenticate the identity of a remote device, to prevent unauthorized access to services, and to prevent hijacking of connections. Authentication and authorization of devices is beyond the scope of this text, as it is implementation-dependent. It involves a bi-directional challenge between devices against a common stored PIN.

- ***Retrieval of the remote device's Bluetooth address:*** This is provided via the *getBluetoothAddress* method of the **RemoteDevice** class.

The DiscoveryAgent and DiscoveryListener Classes

Let's now look in more detail at how devices can locate each other. Locating devices can be done in several ways:

- By passively scanning for devices in the local proximity.
- By actively searching for devices that support a given UUID (service).
- By listing already discovered devices.

A Bluetooth device can be placed in inquiry mode by using the **DiscoveryAgent** object's *startInquiry* method. Calling this method will cause the device to actively scan for new Bluetooth devices in the area. When a device is found, a callback will be made to a **DiscoveryListener** object (supplied as a parameter to the *startInquiry* method). The *startInquiry* method will call the DiscoveryListener object's *deviceDiscovered* method every time a new device is located. The *deviceDiscovered* method is passed a **RemoteDevice** object and a **DeviceClass** object that can be used to learn more information about the remote device that was just located.

When the inquiry is complete, the **DiscoveryListener's** *inquiryCompleted* method is called.

Another way to discover devices is by actively searching for them. In this scenario, the **DiscoveryAgent's** searchService method is called. The application passes to the searchService method a list of UUIDs being searched for, and a DiscoveryListener that will be called when a candidate device or devices is located.

When searching for services, the *searchService* looks for devices which support all of the UUIDs supplied as search criteria. If a device is located that contains some, but not all, of the UUIDs desired, it is eliminated from consideration. When a device is located that does provide all the UUIDs desired, the **DiscoveryListener** object's *servicesDiscovered* method is called, and passed a list of the candidates.

When the search is complete (either because all the known devices have been searched or an error occurred), the **DiscoveryListener**'s *serviceSearchCompleted* method is called.

A final way to discover devices is to examine the list of devices that have already been discovered previously. For this scenario, the DiscoveryAgent class provides two methods:

- *retrieveDevices:* This method is used to obtain a list of either previously discovered devices (from previous searches or inquires), or from a well-known list of services created by the user (perhaps in a control panel or configuration file).

- *selectService:* This method is used to select any device in the PAN that supports a particular service, by passing the service's UUID. Which device is returned (if there is more than one that supports the service) is undefined.

Exceptions

Several new Bluetooth-related exceptions are defined in the API to provide robust error trapping for problems that might crop up during a Bluetooth application's lifetime:

- **BluetoothStateException:** This exception is thrown when the state of the device would not support a particular operation. For example, some devices don't allow an inquiry to be initiated if there is currently a connection established. In this situation, the Bluetooth stack would throw a **BluetoothStateException**.

- **BluetoothConnectionException:** This exception is thrown when a connection cannot be established successfully. There are various reasons this can happen, and the reason any particular connection failed is passed as part of the exception. Your application can inspect this reason and take the appropriate course of action.

A SIMPLE BLUETOOTH APPLICATION

We now have enough background to illustrate a very simple Bluetooth application. Keeping with our university theme, we'll create a homework retrieval system that could be used in a typical university course.

The application will consist of two parts:

- A server application, into which an instructor can enter a homework or reading assignment (shown in Figure 10-10).

- A client application, which students can use to retrieve the homework assignment from the instructor (shown in Figure 10-11).

We'll make the assumption that both the student and instructor are using Bluetooth-enabled MIDP cell phones. What? You don't have Bluetooth-enabled MIDP cell phones? Don't worry—we'll show you a way in which you can experiment with Bluetooth Java programming without even having hardware!

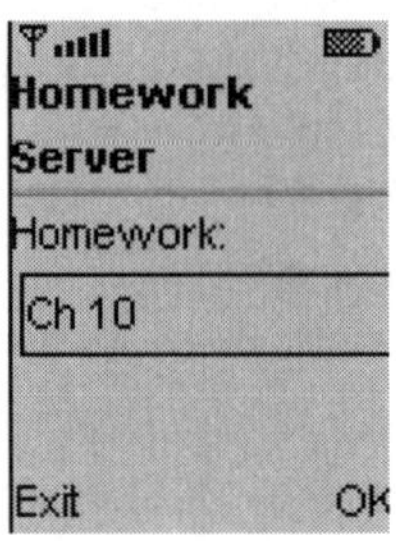

Figure 10-10. The Server Side MIDlet.

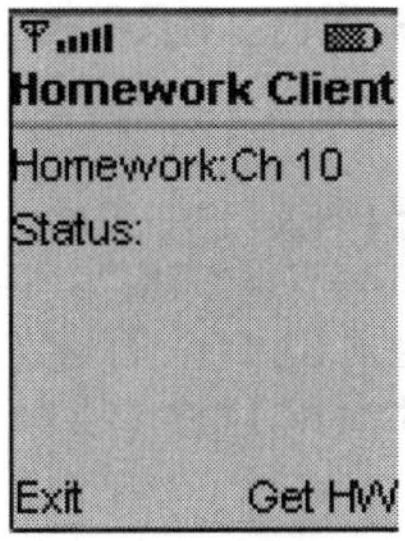

Figure 10-11. The Client MIDlet.

Listing 10-8 contains the HomeworkClient MIDlet. Since you're all MIDP UI experts by now, we'll skip the discussion about how the GUI is constructed and focus on the Bluetooth portion of the application.

We declare a fictitious UUID of 0x12345678 in our MIDlet. This is the UUID which will represent our homework retrieval service. Once the user presses the "Get HW" button, control is transferred to the *commandAction* method (since the MIDlet itself was registered as a **CommandListener** for the button).

The first thing we do is obtain a reference to the local Bluetooth device by calling *getLocalDevice*. Once we have this **LocalDevice** object, we find its **DiscoveryAgent** by calling its *getDiscoveryAgent* method.

We then ask the **DiscoveryAgent** to locate for us a device which provides the service corresponding to our UUID. We're going to take the easy way out and just assume we only have one device in our PAN providing the service. If we had multiple devices, calling *selectService* as we are doing in our client application would return a random device in the PAN that satisfies the service requirements.

The selectService method returns a URL that we can use to connect to the device that will supply our service. We pass this URL to the **Connector** object's *open* method. This returns an **L2CAPConnection** object that we use to read the bytes representing the homework assignment. We format the bytes into a **String** and finally display it on the screen for the user to see.

Listing 10-8. The HomeworkClient Bluetooth MIDlet.

```
import java.io.*;
import javax.microedition.midlet.*;
import javax.microedition.lcdui.*;
import javax.microedition.io.*;
import javax.bluetooth.*;
```

```
public class HomeworkClient extends MIDlet implements CommandListener
{   Display display;
    Form form;
    Command exit = new Command("Exit", Command.EXIT, 1);
    Command gethw = new Command("Get HW", Command.SCREEN, 1);
    StringItem assignment = new StringItem("Homework:", "");
    StringItem status = new StringItem("Status:", "");

    // ficticious UUID
    final static UUID uuid = new UUID(0x12345678);
    LocalDevice device;

    public HomeworkClient()
    {
      display = Display.getDisplay(this);
    }

    public void startApp()
    { form = new Form("Homework Client");
      form.append(assignment);
      form.append(status);
      form.setCommandListener(this);
      form.addCommand(gethw);
      form.addCommand(exit);
      display.setCurrent(form);
    }

    public void destroyApp(boolean force) { }

    public void pauseApp() { }

    public void commandAction(Command c, Displayable s)
    { String url;
      byte[] buffer=new byte[100];
      L2CAPConnection conn=null;

      if (c == exit) {
          destroyApp(false);
          notifyDestroyed();
          return;
      }

      try
          {
```

```
            device = LocalDevice.getLocalDevice();

            DiscoveryAgent agent = device.getDiscoveryAgent();
            url = agent.selectService(uuid,
                    ServiceRecord.NOAUTHENTICATE_NOENCRYPT, false);
            conn = (L2CAPConnection)Connector.open(url);

            conn.receive(buffer);
            conn.close();

            assignment.setText( new String(buffer) );
         }
      catch(BluetoothStateException BTSE)
         {  status.setText("BT Exception: "+BTSE.toString());
         }
      catch(IOException IOE)
         {  status.setText("IO Exception: "+IOE.toString());
         }
   }
}
```

Our server side MIDlet is shown in Listing 10-9. Again, we'll focus solely on the Bluetooth-related aspects of the program, and leave the GUI discussion out. In our server side implementation, we create the URL ourselves, by appending the UUID of our service (notice that it has to match the UUID in the client MIDlet, which it does) to the base URL "btl2cap://localhost". After asking the instructor to type the homework assignment into the text field, the code transfers control to the *commandAction* method.

Our server MIDlet first asks the Bluetooth API to open a server connection using the URL we constructed. This returns an **L2CAPConnectionNotifier** object that we use to wait for connections. When a connection is made from some client device, we send the bytes representing the homework assignment entered by the instructor over the connection to the client. We then close the connection and wait for the next client.

Listing 10-9. The HomeworkServer Bluetooth MIDlet.

```
import java.io.*;
import javax.microedition.midlet.*;
import javax.microedition.lcdui.*;
import javax.microedition.io.*;
import javax.bluetooth.*;
```

```
public class HomeworkServer extends MIDlet implements CommandListener
{   Display display;
    Form form;
    Command exit = new Command("Exit", Command.EXIT, 1);
    Command ok = new Command("OK", Command.SCREEN, 1);
    TextField assignfield = new TextField("Homework:",
                                 "Read Chapter 10", 160, TextField.ANY);
    String assignment=null;

    // ficticious UUID
    final static UUID uuid = new UUID(0x12345678);
    final static String SERVER_URL = "btl2cap://localhost:";
    LocalDevice device;

    public HomeworkServer()
    { display = Display.getDisplay(this);
    }

    public void startApp()
    { form = new Form("Homework Server");
      form.append(assignfield);
      form.setCommandListener(this);
      form.addCommand(ok);
      form.addCommand(exit);
      display.setCurrent(form);
    }

    public void destroyApp(boolean force) { }

    public void pauseApp() { }

    public void commandAction(Command c, Displayable s)
    { String url;
      byte[] buffer=null;
      L2CAPConnection conn=null;

      if (c == exit) {
          destroyApp(false);
          notifyDestroyed();
          return;
      }

      try
          { device = LocalDevice.getLocalDevice();
            assignment = assignfield.getString();
```

```
            url = SERVER_URL + uuid.toString();

            L2CAPConnectionNotifier n=
                (L2CAPConnectionNotifier)Connector.open(url);
            for (;;)
                {  conn = n.acceptAndOpen();
                   buffer = assignment.getBytes();
                   conn.send(buffer);
                   conn.close();
                }
          }

      catch(BluetoothStateException BTSE)
          { System.out.println("BT Exception: "+BTSE.toString());
          }
      catch(IOException IOE)
          { System.out.println("IO Exception: "+IOE.toString());
          }
    }
}
```

Bluetooth Simulators

As real Bluetooth-enabled MIDP cell phones and other devices are hard to come by, there have been several Bluetooth simulators developed that let us develop our code without having to have real hardware at hand. If you can't afford, or can't find hardware, you may want to take a look at the simulator products listed below:

- Rococo's Impronto Simulator
- Zucotto's XJB 100 Bluetooth Host Stack
- Atinav's Bluetooth Protocol Stack 2.0
- Motorola's Bluetooth Reference Implementation

There are new simulators being released all the time. For more information on these simulators, or to find out if any new ones have been announced, visit the Java APIs for Bluetooth Wireless Technology (JABWT) Web site at *http://www.jabwt.com*.

Details

Rococo's Impronto Simulator

The sample MIDlets shown above were tested using Rococo's Impronto Bluetooth Simulator (shown in Figure 10-12). The simulator provides a virtual Bluetooth environment that supports both J2SE and J2ME platforms. Using the simulator, you can test your Bluetooth applications before deploying them to real devices.

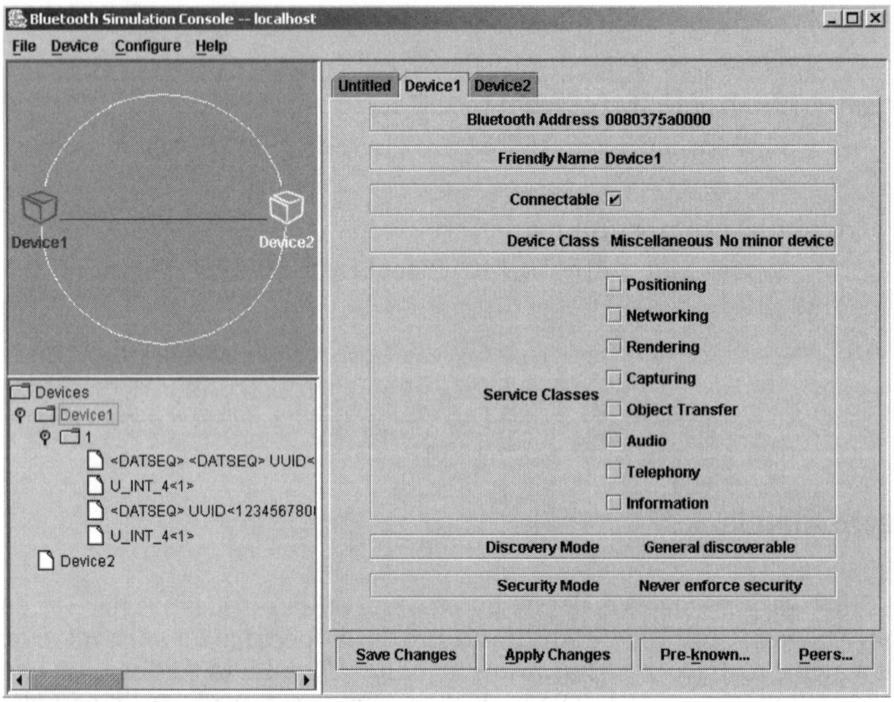

Figure 10-12. Rococo's Impronto Bluetooth Simulator.

You can use the simulator to browse the devices in your simulated Bluetooth world and inspect the services they provide. The simulator shows a pictorial view of the connections being made between devices, so you can quickly tell which device is communicating with another.

Although we chose to use the Impronto simulator, the other simulators above offer similar features. You are urged to evaluate each and make the decision as to which simulator to use based on your business needs. ❑

SUMMARY

Once more, you are a MIDP survivor. The record store APIs are very primitive in nature and some effort was required to overcome some of the API limitations. Some vendor tools, such as Borland's MobileSet, provide utility classes to make the use of the record store simpler. A layer of abstraction is added above the MIDP APIs.

In this chapter we covered MIDP networking and persistence. You might want to combine the two APIs and make your Contact application a networked application. This would allow you to synchronize your contact database with a backend address book, providing a unified address book that is available online and offline. It would also test your newly acquired MIDP networking and record store skills. As an "extra-credit exercise," you should try to use OTA provisioning to deploy your midlets.

We also revisited Bluetooth in more detail. We covered the Bluetooth API, and showed how we can develop MIDlets that use the Bluetooth stack to communicate with each other. We developed a sample Bluetooth application that put to work what we have learned so far.

Chapter 11

Developing VoiceXML Applications

In this chapter we will present the details of developing a VoiceXML application. This will add one more technology to our repertoire of wireless client development skills. A few of the popular developer tools and environments for developing VoiceXML applications are also discussed in this chapter. We will also introduce ECMAScript, a scripting language supported by the VoiceXML Forum. Some programming tips for developing effective VoiceXML applications are provided at the end of the chapter.

VOICEXML APPLICATIONS

Figure 11-1 shows the structure of a VoiceXML application. A *root document* provides an overall structure for a voice driven application. Interactions between the system and the users are performed by what are known as *dialogs*. Dialogs can be a *form* or a *menu*. One or more of these dialogs can be bundled into a VoiceXML document; many such documents form an application.

So, let's look at the programmatic structure of a simple VoiceXML document. This is shown in Figure 11-2. The programming structure shows the basic tags that make up a VoiceXML document.

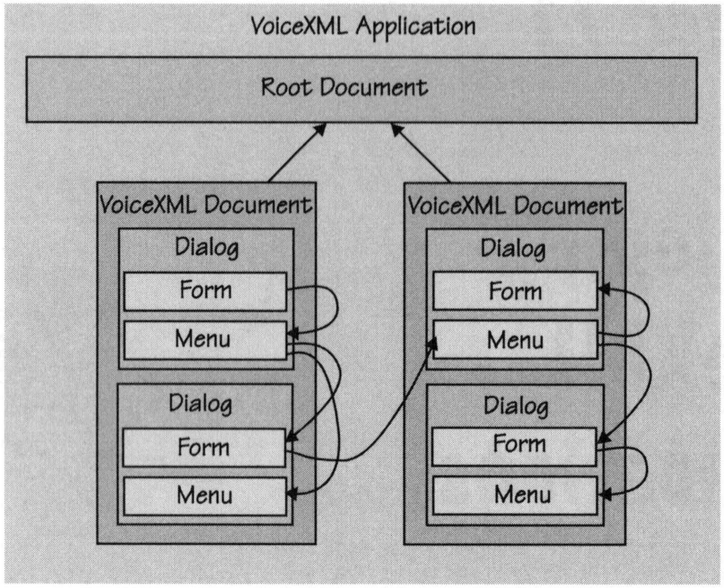

Figure 11-1. The Structure of a VoiceXML Application.

DEVELOPING A VOICEXML APPLICATION

A simple VoiceXML application consist of three types of files:

- ***VoiceXML documents:*** These consist of dialogs (form or menu) described using the XML document structure.
- ***Grammar files:*** The validation of spoken user input can be encapsulated within a separate file that can be accessed by the VoiceXML document. (Note: It is not always necessary that grammar be stored in an external file. Grammar can be also defined within a VoiceXML document).
- ***Audio Files:*** Pre-recorded audio files consisting of various formats (such as *.wav*).

Figure 11-3 illustrates the VoiceXML application development process. Here are the steps involved in the creation of a VoiceXML application:

1. ***Write the VoiceXML documents and grammar files.*** This is a simple text-editing task and can be done in any regular text editor. Another alternative is using voice toolkits that provide an IDE development environment plus a simulation of the voice application execution.

```
<?xml version="1.0"?>
<!--Your DTD goes in here (optional)-->

<vxml version="1.0"
 application="app_root.vxml">

<meta name="author" content="Dan Harkey"/>

<!Global variables and event handlers go here-->

<var name="name" expr="Joe"/>
<catch event="noinput">
Please say something
</catch>

<!--Dialogs (Form or Menu)-->
<!--Menu-->
<menuid="example" scope="dialog" dtmf="false">
<prompt> Hello World!
You have reached the main menu.
</prompt>
<choice dtmf="1" next="#newForm">
First Choice
</choice>
<choice dtmf="2" next="#anotherForm">
Second Choice
</choice>
...
</menu>

<!--Form-->
<form id-"newForm">
<block>
You reached my form. Good Bye!
</block>
</form>

<form id="anotherForm">
...
<submit next="http://wirelessjava.engr.sjsu.edu/
                processMainmenu.vxml"/>
</form>

...
</vxml>
```

XML document + DTD definition

Root tag with version and root document

Global Variables and Event Handlers

First Dialog Menu

Transitioning within dialogs within a document

Second Dialog Form

Transitioning to another document

Figure 11-2. The Structure of a VoiceXML Document.

2. ***Grouping audio files.*** Pre-recorded audio files are grouped and placed for invocation from within the VoiceXML documents.

3. ***Run and test the application using an SDK.*** The VoiceXML documents are tested using a Software Development Toolkit (SDK) before being deployed on to a real time environment. This facilitates easier debugging, testing, and faster development cycles.

Figure 11-3. A VoiceXML Application From Developer to End User.

4. ***Deploy to a production/test environment.*** This simply means making the files that are part of the VoiceXML application available via a Web server. VoiceXML files have their own MIME types, so you may need to update the MIME type definitions on your Web server. Of course, outside of the development framework, one needs to take into account the setting up of the network that connects the telephone (PSTN) network, Voice Server, and the Web server.

5. ***Run the VoiceXML application.*** Once we have the network connections, the Voice Server and the Web server ready and configured, all that we need to do is pick up the telephone, call a number, and listen to the VoiceXML application.

VoiceXML Cookie Support

Details

Cookies enable a browser to save information across multiple HTTP connections. Cookies are typically used to store user preferences and data that the user submits while browsing a Web application. As with HTML applications, you can use cookies with VoiceXML applications to track the state of your VoiceXML application by sharing cookies between a voice browser and a server-side application running on a Web application server. Cookies persist only for the duration of a voice browser session, and can only be sent back to the originating server.

When the voice browser invokes a server application, it executes the necessary operations and prepares a response. As part of preparing the response, it sets the content type (the MIME type). Usually the MIME type is *application/x-vxml.* If cookies are needed, the server application also creates them (in the form *name=value)* and adds them to the response. This response is then returned to the voice browser.

The voice browser receives the VoiceXML document from the server and extracts the cookies from the HTTP header. When the voice browser makes a subsequent call to the same server, it passes the cookie back in the HTTP header. ❑

VoiceXML MIME Types

Details

The file extension-to-MIME (Multi-Purpose Internet Mail Extensions) type mapping is specified in the configuration for the Web server where the VoiceXML files reside. The extension *.vxml* has to be mapped to the MIME-type *text/vxml.* (Other possible MIME types include *application/x-vxml*, *application/vxml*, and *text/x-vxml).* Since ECMAScript is embedded in a VoiceXML document, there is no separate MIME type setup for ECMAScript (contrary to the case of WMLScript).

For example, in case of Apache Tomcat, these mappings are set in the file $TOMCAT_HOME/conf/web.xml.

For the various grammar formats (to be discussed in the VoiceXML tags section) the appropriate MIME types are listed in Table 11-1. ❑

Table 11-1. The VoiceXML MIME Types.

Grammar Format	MIME type
Java Speech Grammar Format (JSGF)	application/x-jsgf
Grammar Specification Language (GSL)	application/x-gsl

Table 11-1. The VoiceXML MIME Types. (Continued)

Grammar Format	MIME type
Speech Grammar Recognition Format (SRGF) XML form	application/grammar+xml
Speech Grammar Recognition Format (SRGF) Augmented Backus-Naur Form (ABNF) form	application/grammar

VoiceXML Tools

For a convenient way to develop, debug, and test VoiceXML applications there are several tools. These tools enable greater productivity when compared to real time development and testing. VoiceXML tools are broadly divided within three categories:

- ***Text editors.*** Since VoiceXML documents are regular XML files, you can write these documents using regular text editors such as TextPad, WordPad, etc. These editors, in conjunction with standalone Voice SDKs (such as IBM's Voice Server SDK), provide a simple and convenient way of testing VoiceXML applications.

- ***Web-based development tools.*** These tools are available for you to develop, test, debug, and run VoiceXML applications online for no cost. All you have to do is to register with the developer network and make use of their facilities. Once you build a VoiceXML application, you can test your applications by calling their toll-free numbers and interact with your application. Table 11-2 gives a list of some of the Web-based development tools.

Table 11-2. Web-based Development Tools.

Company	Tool	URL
VoiceGenie Technologies Inc.	VoiceGenie Developer Workshop	*developer.voicegenie.com*
TellMe. Inc.	TellMe. Studio	*studio.tellme.com*
HeyAnita Inc.	FreeSpeech Developer Network	*freespeech.heyanita.com*
BeVocal Inc.	BeVocal Café	*cafe.bevocal.com*

- ***Software development toolkits (SDKs).*** Another alternative to develop, test, and debug VoiceXML applications is to use SDKs that are bundled with the necessary VoiceXML resources—such as the Automatic Speech Recognition

(ASR) and Text To Speech (TTS) systems that are included with your VoiceXML toolkit. When using these toolkits, you can run your applications in a desktop or a PC using a microphone and a headset. Most SDKs provide debugging facilities and provide an IDE for building VoiceXML applications, as shown in Figure 11-4. The popular SDKs include:

- ***IBM's WebSphere Voice Toolkit.*** The toolkit offered by IBM offers features such as a VoiceXML editor, a grammar editor, a pronunciation builder, and a basic audio recorder. When used in conjunction with the IBM WebSphere Voice Server SDK, it makes VoiceXML development easy and quick. This toolkit can be downloaded from *http://www-4.ibm.com/software/speech/enterprise/vtoolkit.html.*

- ***IBM's WebSphere Voice Server SDK.*** You can download the WebSphere Voice Server SDK from IBM's website at *http://www-4.ibm.com/software/speech/enterprise/ep_11.html.*

- ***Nuance Communications' V-Builder.*** This comprehensive VoiceXML development tool from Nuance Communications can be downloaded for free from their site *http://extranet.nuance.com/login.html* once you sign up for the Nuance Developer Network (NDN) program.

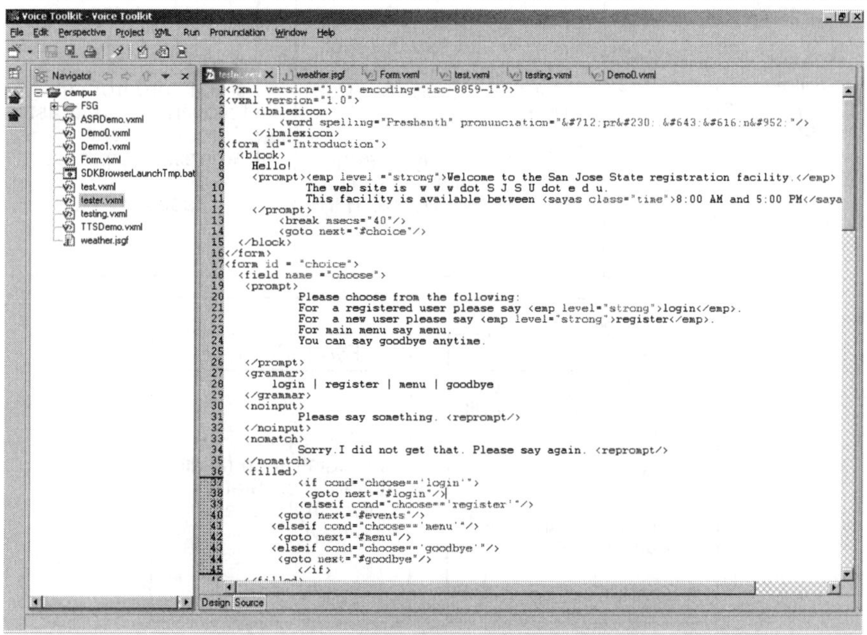

Figure 11-4. The IBM WebSphere Voice Toolkit.

The University Registration Example

So, let's press forward and revisit the familiar registration example that was introduced to you in the WML chapter. The same functionality (login, registration, semester selection, course selection, and payment) is provided using VoiceXML as the content format. Each interaction is a dialog that is a part of a VoiceXML document. The individual interactions are depicted in Figure 11-5.

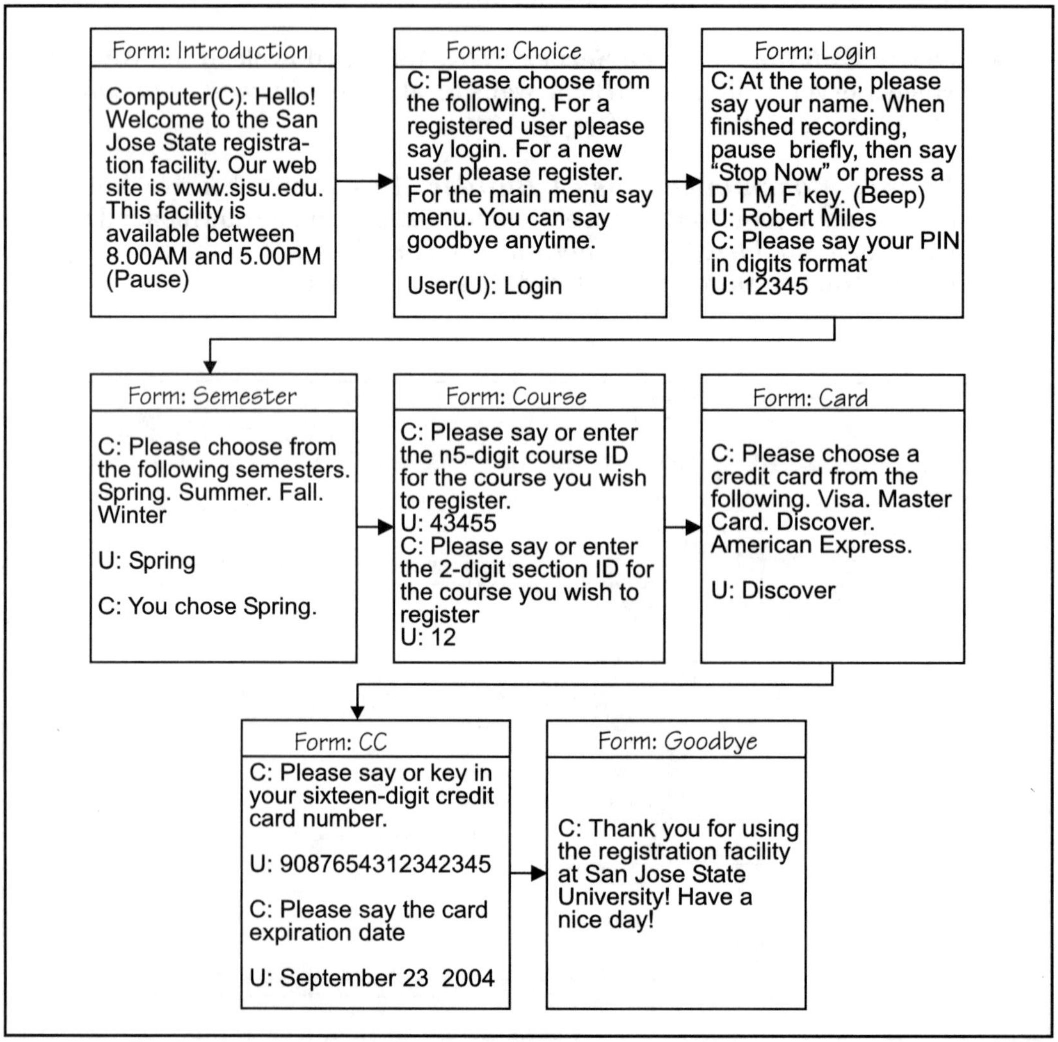

Figure 11-5. The Flow of the Registration VoiceXML Application.

VOICEXML TAGS

VoiceXML tags define the content that make up a VoiceXML document. This section describes VoiceXML tags defined in the DTD of the VoiceXML 1.0 version. VoiceXML contains numerous tags that roughly fall into the following categories:

- ***Root,*** the root for all VoiceXML tags.
- ***Dialog*** tags are used to organize a VoiceXML application and are the bread and butter of VoiceXML applications.
- ***Prompt*** tags are used to provide audio output.
- ***Field*** tags gather information from the VoiceXML user.
- ***Audio output*** and ***audio input*** tags are used in speech synthesis and speech recognition.
- ***Event*** tags to handle unexpected conditions.
- ***Control flow*** tags that aid in VoiceXML navigation.
- ***Call control*** tags for dialog navigation.
- ***Miscellaneous*** tags for the more esoteric VoiceXML functions.

Let's begin our trek through the VoiceXML tags. We will present a description of each category, show a high-level DTD diagram of each of the VoiceXML tag categories, then present a brief description of the individual tags. For the individual tag descriptions we'll present the syntax of the tag with its attributes and then follow up with an example.

Root

The root tag for VoiceXML is <vxml> (see Figure 11-6). The syntax for the <vxml> tag is as follows:

```
<vxml application="URI" base="string" xml:lang="string"
        version = "string">
  child tags
</vxml>
```

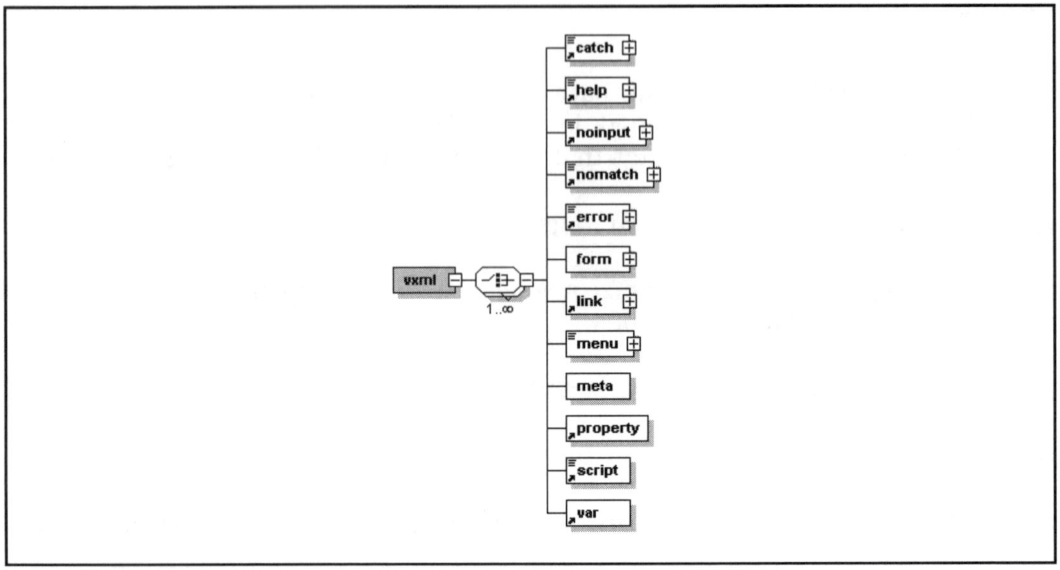

Figure 11-6. The Root VoiceXML Tag.

The *application* attribute points to the root document (which will be executed first by the interpreter) to which the current document is a part.The *base* attribute points to the base URI of the current document. The *xml: lang* attribute is used to indicate the language, but is implementation specific. The *version* attribute indicates the version of the VoiceXML DTD to be used.

As with HTML, a *<meta>* tag is available in VoiceXML. Here is the syntax for the <meta> tag:

```
<meta content="string" http-equiv="string" name="string" />
```

The <meta> tag can be used to provide any information about the VoiceXML document.

VoiceXML is an XML application, so when using characters for relational operators like less than or greater than, you must use the appropriate escape sequences. For example, you need to use "&lt" for "<".

Since the default character encoding for XML documents is UTF8, and the character codes in VoiceXML can be greater than 127 (i.e., greater than the bits in UTF8), there might be a possibility for misinterpretation. So, in some cases you need to force the XML parser to use a different encoding such as such as "ISO 8859" (as given below):

```
<?xml version="1.0" encoding="iso-8859-1"?>
```

VoiceXML Root Documents

Details

All VoiceXML applications have the concept of a root document. The root document defines constructs that are common across many documents in a VoiceXML application—like error handlers, variables, links, and form-level grammar. The voice interpreter loads the root document and can switch context between a document and the root document as needed—for example, to refer a global error handler.

An *application* is a set of documents sharing the same *root document*. Whenever the user interacts with a document in an application, its root document is also loaded. The root document remains loaded while the user is transitioning between other documents in the same application, and it is unloaded when the user transitions to a document that is not in the application. While it is loaded, the root document's variables are available to the other documents as *application variables*, and its grammars can also be set to remain active for the duration of the application (see the following example).

```
<?xml version="1.0" encoding="iso-8859-1"?>
<!--Tie this to your application by pointing to root document -->
<vxml version="1.0" application="your_app_root.vxml">
. . .
</vxml>
```

❑

Dialogs

VoiceXML has two ways of interacting with users—one is through forms, using the <form> tag, and the other is through menus, using the <menu> tag. We will generically refer to these as dialogs. The <form> tag—like any HTML form—gathers inputs for fields from users, while <menu> presents the choice values. Each dialog determines the next dialog (or another document) to transition to using a URI. The conversation is terminated when a dialog does not specify the next dialog to transition to, or if it has a tag that explicitly exits the conversation. See Figure 11-7 for an overview of the VoiceXML dialog tags.

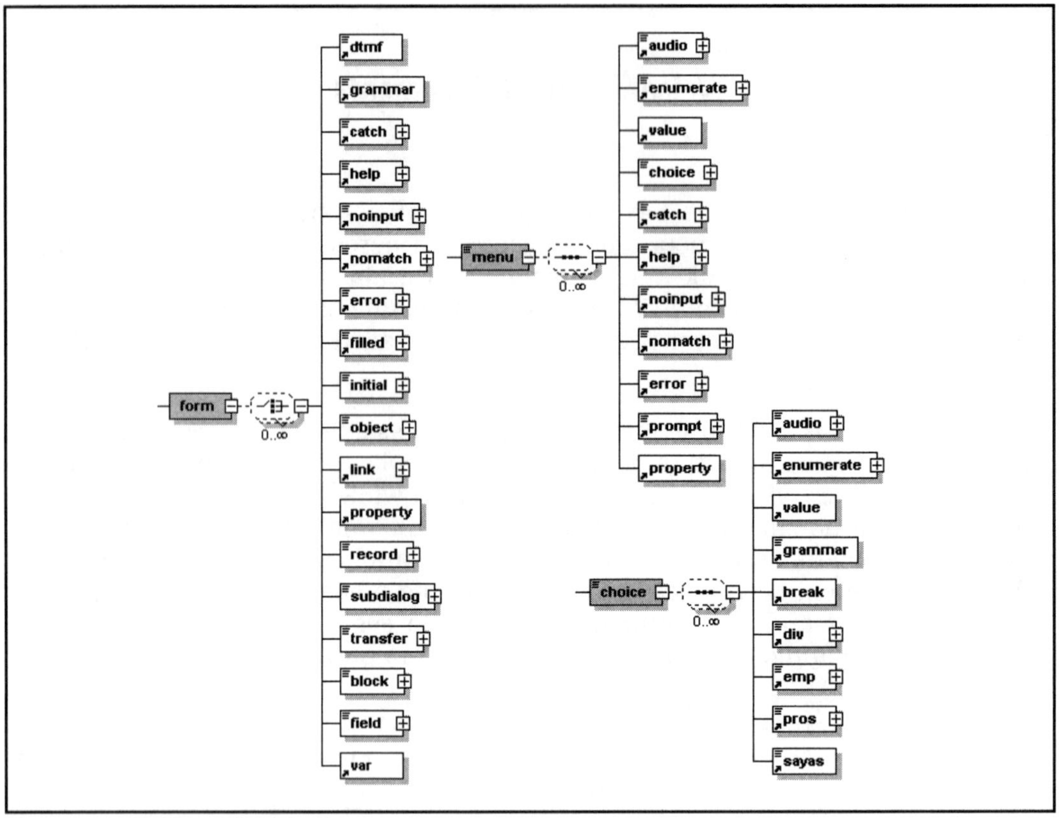

Figure 11-7. The VoiceXML Dialog Tags.

The <form> Tag

In forms, values for a set of fields can be obtained and collected in specified variables. The grammar to specify the allowed inputs can be specified either at the field-level (specifically for each field), or at the form-level (collectively for all fields). We describe grammar in more detail in the audio input section. The following is the syntax for the <form> tag:

```
<form id="string" scope="level">
  child tags
</form>
```

The *id* attribute of the <form> tag specifies an ID for the form so that a form can be a target of other dialogs within the same VoiceXML document or another document. The *scope* attribute details the scope of the form's grammars, which is either the current dialog or the entire document.

<pre><form id = "Semester">
  <field name ="sem">
    <prompt>
      Please choose from the following
           semesters.
           Spring.
           Summer.
           Fall.
           Winter.
    </prompt>
    <grammar>
      Spring | Summer | Fall | Winter
    </grammar>
    <filled>
      <prompt>
        You chose <value expr="sem"/>.
      </prompt>
      <goto next ="#Course"/>
    </filled>
  </field>
</form></pre>
	C: Please choose from the following semesters. Spring. Summer. Fall. Winter.
	
	U: Spring
	
	C: You chose Spring

The <menu> Tag

A <menu> tag presents a user with a list of options and then, based on the user's choice, proceeds to a URI corresponding to that choice. The following is the syntax for the <menu> tag:

```
<menu id="string" scope="level" dtmf="boolean">
  child tags
</menu>
```

The *id* attribute of the <menu> tag specifies an ID for the menu so that it can be a target of another dialog within the same document or outside the document. The *dtmf* attribute, when set to *true*, provides implicit values for choices that do not have explicit DTMF keys specified (e.g., 1, 2, etc). The *scope* attribute, like the <form> tag, specifies the scope of the menu's grammars, which is either at a dialog or a document level.

The <choice> Tag

Each of the menu items is described by a <choice> tag. The syntax for the <choice> tag is provided below:

```
<choice next="URI" dtmf="key" event="string" caching="mode"
     expr="'expression'" fetchhint="mode" fetchaudio="URL"
     fetchtimeout="time" >
  grammar
</choice>
```

The *next* attribute specifies the URI of a document or a dialog that is invoked when a choice is selected. The *dtmf* attribute specifies the DTMF key corresponding to the choice (e.g., 1, 2, etc.). The *event* attribute specifies the event that is thrown when a choice is selected. The *caching* attribute defines whether to look for a newer version no matter what (*safe*), or look for a newer version only if expired or missing (*fast*). The *fetchint* attribute defines when the voice browser should use a file (audio, grammar, etc.) to download. This attribute accepts two values; *prefetch* and *safe*. With *prefetch* the file is downloaded when the VoiceXML document loads. With *safe* the file is downloaded only when required. The *fetchaudio* attribute defines the location of an audio file that is to be played while a fetch is being performed. The *fetchtimeout* attribute specifies the time (in seconds or milliseconds) to wait for a file to be retrieved (e.g., “4s” or “50ms”).

The grammar for menu choices is automatically generated. By default, the generated grammar is an exact match of the text value of the choice. However, if you choose to provide a grammar, you could choose to use the <grammar> tag; or if you want to assign a digit for touch-tone menu, you can use the *dtmf* attribute.

The following is an example showing how a series of choice tags is used to present a list to the user. The <enumerate> tag is used to generate the list of spoken choices. The <enumerate> tag will be discussed in more detail in a later section.

<table>
<tr>
<td><pre><menu>
  <prompt>
    Welcome to S J S U. Say one of:
    <enumerate/>
  </prompt>
  <choice
      next="/wirelessjava/classes.vxml">
    Classes
  </choice>
  <choice</pre></td>
<td>C: Welcome to SJSU. Say one of: Classes. Events. Grades.

U: Classes

(control transfers to classes.vxml)</td>
</tr>
</table>

(continued)

```
      next="/wirelessjava/events.vxml">
    Events
  </choice>
  <choice
      next="/wirelessjava/grades.vxml">
    Grades
  </choice>
  <noinput>Please say one of
    <enumerate/>
  </noinput>
</menu>
```

FYI

Directed and Mixed-initiative Dialogs

Briefing

User interaction with dialogs is of two types: directed and mixed initiative.

In directed dialogs, the computer (or the server-side application) prompts the user for input values and the user supplies the values following the flow. The computer (or the application) decides the order in which the user has to respond and the user can provide the inputs for the specific prompts only. These are the machine-directed conversations. In this case the grammar is active only when user input is expected.

However, mixed-initiative dialogs are interactions between a user and a computer where the dialog is driven by either the user or the computer, based on user's input. In this case, logic is built in the form itself so that the machine can also decide the flow if the user has not provided input values. This is achieved by using the <initial> tag and the form-level grammar. Typically in this type of dialog, a user may fill more than one field (with a single response) and the fields may be filled in any order. A text example of a mixed-initiative dialog is provided below:

Computer: For what city and state would you like the weather?
User: Uh, California.

Computer: Please say the city in California for which you want the weather.
User: San Francisco, please.

Computer: Do you want to hear the weather for San Francisco, California?
User: No.

Computer: What city?
User: Los Angeles.

Computer: Do you want to hear the weather for Los Angeles, California?
User: Yes.

Computer: Mostly sunny today with highs in the 80s. Lows tonight from the low 60s.
❑

Prompts

Prompts control the output of synthesized speech and recorded audio. Based on the interpretation of the prompts, the user says or keys in information accordingly. In this section we will discuss three tags that aid in the prompting of speech to users: <prompt>, <enumerate> and <reprompt>. See Figure 11-8 for an overview of the Prompt tags.

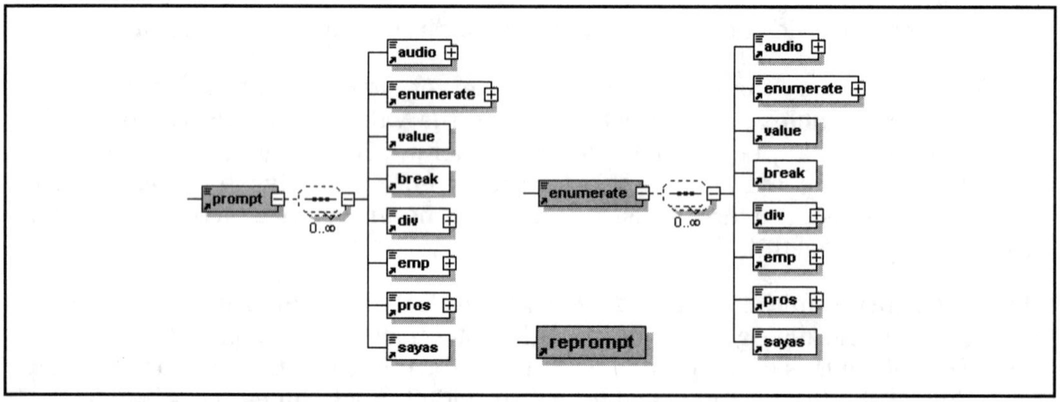

Figure 11-8. The Prompt VoiceXML Tags.

The <prompt> Tag

The syntax for the <prompt> tag is as follows:

```
<prompt bargein="boolean" cond="'expression'" count="integer"
        timeout="time">
  child tags
</prompt>
```

The attribute *bargein* specifies whether a user can interrupt a prompt; the default value is *true*. The attribute *cond* is an expression that specifies if the prompt has to be spoken; the default is *true*. The *count* attribute is used to provide different prompts for each try by the user. The *timeout* attribute is used to specify the time to wait for user input. If there is no input within this time interval, the platform will throw a *noinput* event.

`<prompt>`
`  Please say or enter the 5-digit course I D`
`  for the course you wish to register.`
`</prompt>`
	C: Please say or enter the five digit course I D for the course you wish to register

The <enumerate> Tag

The content within the <enumerate> tag is used as a template to generate additional spoken text that goes with any spoken choice values as defined within a <choice> tag that is part of a menu dialog.

```
<enumerate>
  child tags
</enumerate>
```

`<menu id="example" scope="dialog" >`
`  <prompt> You have chosen student`
`          assistance.`
`    <enumerate>`
`      Please say <value expr="_prompt"/> or`
`      press <value expr="_dtmf"/> for`
`      <value expr="_prompt"/>`
`    </enumerate>`
`  </prompt>`
`    <choice dtmf="1" next="register.vxml">`
`      Registration`
`    </choice>`
`    <choice dtmf="2" next="grades.vxml">`
`      Grades`
`    </choice>`
`</menu>`
	C: You have chosen student assistance.
	
	Please say registration or press 1 for registration.
	
	Please say grades or press 2 for grades.
	
	U: Grades
	
	(Control transfers to grades.vxml)

The <reprompt> Tag

Using the <reprompt> tag results in the playing of a previously played prompt.

```
<reprompt/>
```

<prompt count="1"> Please say or enter the 5-digit course I D for the course you wish to register. </prompt> ... <noinput> <reprompt/> </noinput>	C: Please say or enter the 5-digit course I D for the course you wish to register. (When no input from user, the prompt is replayed)

Fields

Spoken input and other data are stored using the <field> and <var> tags. In addition to these tags, we will discuss data referencing and data manipulation tags such as the <value>, <clear>, <assign>, and <option> tags. Figure 11-9 illustrates the Field tags.

The <field> Tag

In a <form>, any inputs to be obtained from users are usually specified using the <field> tag and the input values are obtained via ASR or DTMF grammars. The following is the syntax for the <field> tag:

```
<field name="string" expr="'expression'" cond="'expression'"
       type="grammar" slot="string" modal="boolean" >
  child tags
</field>
```

The *name* attribute defines a variable that holds the result for the field. The *type* attribute defines the internal (built-in) grammar that the field represents. The valid

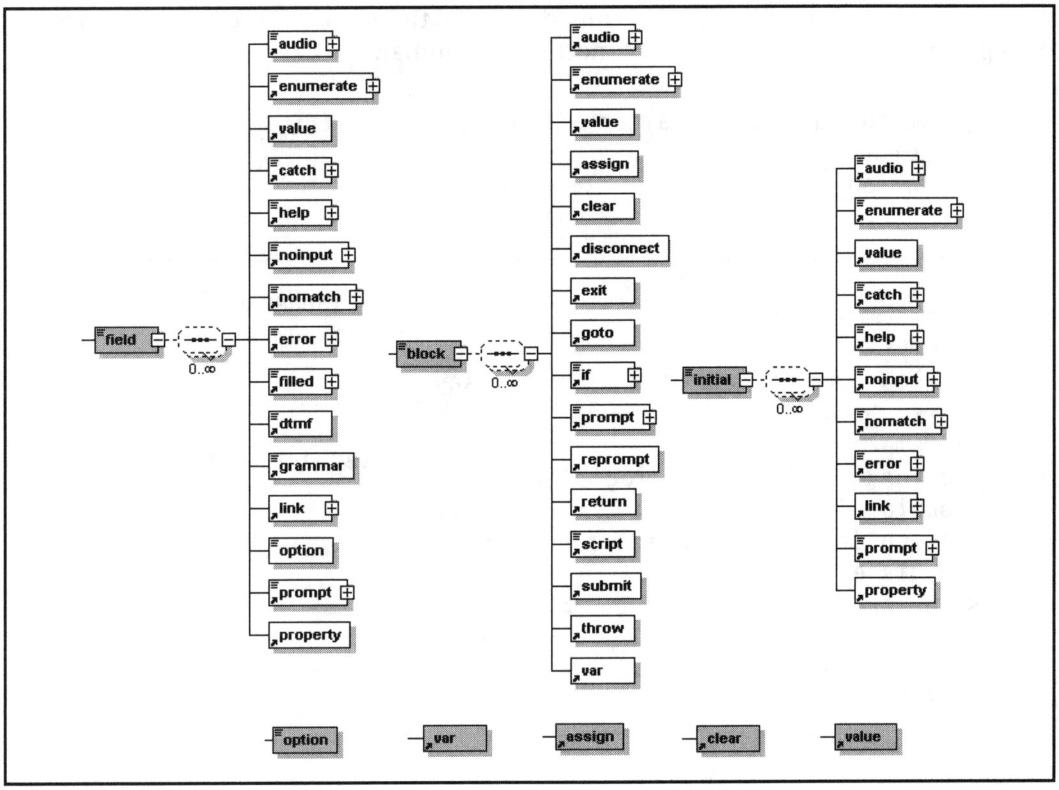

Figure 11-9. The Field VoiceXML Tags.

internal grammar types are: *digit*, *boolean*, *date*, *phone*, *currency*, *number*, and *time*. We will discuss internal grammar in more detail in the audio input sections.

```
<field name ="cid" type="digits">
  <prompt>
    Please say or enter the 5-digit
    course I D for the course you wish
    to register.
  </prompt>
</field>
```

C: Please say or enter the 5-digit course I D for the course you wish to register

The <option> Tag

To present a simple set of alternatives as input for a field, the <option> tag can be used. This can be used as an alternative to the <grammar> tag. It is represented by a set of <option> tags contained in a <field> tag. Each <option> tag encapsulates text that is used to generate a grammar.

```
<option dtmf="key" value="string">
  text
</option>
```

```
<form>
  <field name="maincourse">
    <prompt>
       Please select an option. You can say
       <enumerate/>
    </prompt>
    <option dtmf="1" value="login">login
    </option>
    <option dtmf="2" value="register">
      register
    </option>
    <filled>
      ......
    </filled>
  </field>
</form>
```

C: Please select an option. You can say:

Login.

Register.

The <var> Tag

Variables in VoiceXML are in all respects the same as ECMAScript variables. However, variable names starting with the underscore character (_) are reserved for internal use. A variable may be declared using the <var> tag. An important note is that a variable has to be declared before using it. Here is the syntax for <var> tag:

```
<var name="string" expr="'expression'"/>
```

The *name* provides the name of the variable and the *expr* provides an expression to be evaluated to obtain the value for this variable. If the expression is not supplied, the variable is assigned the value *undefined* as defined in ECMAScript.

The <field> tag can also be used to declare variables. In this case, the *name* attribute specifies the name of the variable and the *type* specifies the internal grammar type (e.g. *digit*, *boolean*, *date*, etc).

```
<var name="home_phone"/>
<var name="pi" expr="3.14159"/>
<var name="city" expr="'San Jose'"/>

<field name="num_tickets" type="number">
      <prompt>How many tickets do you wish to purchase?</prompt>
</field>
```

Details

Variable Scoping

Variables can have session, application, document, dialog, and anonymous scoping. The following describes these in more detail:

- ***Session variables*** are read-only variables that are not declared in VoiceXML documents but are declared and set by the voice browser. These variables are available for the duration of the user session.
- ***Application variables*** are declared with <var> tags, which are children of the root document's <vxml> tag. These variables are initialized when the root document is loaded and are visible to all the documents belonging to the same application.
- ***Document variables*** are declared with <var> tags, which are children of a VoiceXML document's <vxml> tag. A variable with document scope can only be accessed from within the same document.
- ***Dialog variables*** are declared within dialogs. Dialog variables are declared by <var> tags from within the <form> tag, or by <var> tags inside executable content (such as <block> content or <catch> tag content), or by the various form item tags. The child <var> tags of <form> are initialized when the form is first visited. The <var> tags inside executable content are initialized when the executable content is executed. The form item variables are initialized when the form item is collected.
- ***Anonymous variables*** are declared within <block>, <filled>, and <catch> tags.

Figure 11-10 shows the scope hierarchy.

If there is a variable X that is declared and available in every block of the diagram, then to refer to the variable X in the document scope, you can use the syntax document.X in your VoiceXML document. This will distinguish the variable in the document scope from those in the anonymous, dialog, and document scopes. ❑

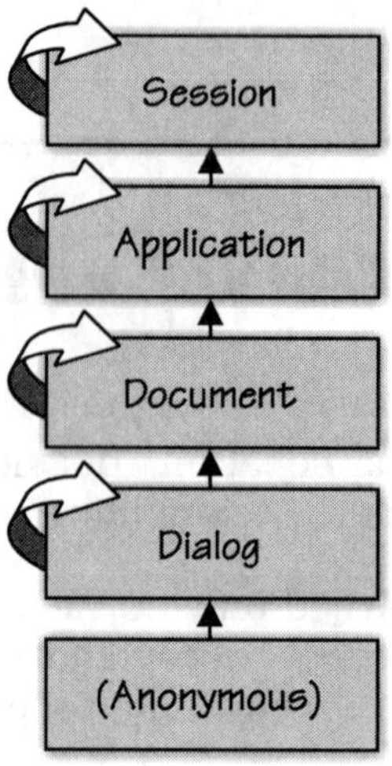

Figure 11-10. The Variable Scope Hierarchy.

The <initial> Tag

The <initial> tag is used to set up <field> items for mixed-initiative forms and matches user inputs with form-level grammar. In mixed-initiative forms, the speaker is prompted to answer the questions either all at once or in whichever order they choose. The following is the syntax for the <initial> tag:

```
<initial name="string" expr="expression" cond="expression">
  child tags
</initial>
```

```
<initial name="start">
  <prompt> For what department and course would you
    want the information?
  </prompt>
  . . .
</initial>
```

The <block> Tag

The <block> tag encapsulates a sequence of procedural statements used to prompt or compute, but is not used to get user inputs, as is the case with the <prompt> tag. The following is the syntax for the <block> tag:

```
<block name="string" expr="'expression'" cond="'expression'">
  child tags
</block>
```

`<block>`
`  You have chosen main menu.`
`  Wait Please.`
`  <goto next="#choice"/>`
`</block>`
	C: You have chosen main menu. Wait please. (control transfers to the 'choice' dialog)

The <assign> Tag

To assign a value to a variable, the <assign> tag is used. The *expr* attribute is used to specify an ECMAScript expression that has to be evaluated to obtain the value to be assigned to that variable. The following is the syntax for the <assign> tag:

```
<assign name="string" expr="'expression'">
</assign>
```

```
<assign name="sem" expr="'spring'"/>
<assign name="var1" expr="var1 + 1"/>
```

The <clear> Tag

Variables can be cleared or reset to an undefined value (as specified in ECMAScript) using the <clear> tag. If more than one variable has to be reset, then they are specified as a comma-separated list. If no variable name is provided then all the variables in the current form are reset. The following is the syntax for the <clear> tag:

```
<clear namelist="variable list"/>
```

The <value> Tag

The <value> tag is used to provide variable values as speech to users. The ECMAScript expression specified by the *expr* attribute is evaluated and spoken as output. Since text-to-speech is used in this case, the value for the *mode* attribute is set as *tts*. It is also possible to play a pre-recorded audio sample (such as from a *.wav* file) by setting the mode attribute to *recorded* and specifying the URI of the wave file using the *recsec* attribute. The following is the syntax for the <value> tag:

```
<value expr="'expression'" mode="tts or recorded" class="string"
        recsrc="URI"
/>
```

```
<prompt>
  The cost of your pizzas is <value expr="cost"/>
<prompt>
```

Audio Output

To further enrich the clarity of interaction with the user, VoiceXML provides five tags that allow better presentation and fine-tuning of text-synthesized speech in a VoiceXML document. These tags are <break>, <emp>, <sayas>, <div>, and <pros> (see Figure 11-11). The following section briefly discusses these audio output tags.

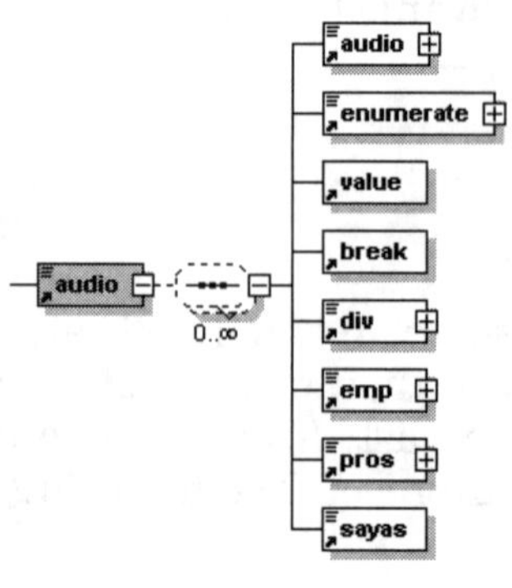

Figure 11-11. The Audio Output VoiceXML Tags.

The <break> Tag

The <break> tag is used to specify a break in speech output. The duration in milliseconds is specified using the *msecs* attribute and the duration of the break is specified using *size* attribute:

```
<break msecs="time" size="string"/>
```

The *msecs* attribute specifies the break time in milliseconds. The *size* attribute defines the alternative of the *msecs* attribute in terms of the magnitude of the break (*none*, *small*, *medium*, or *large)*.

The <emp> tag

The <emp> tag is used to specify emphasis on a portion of the output:

```
<emp level="string">
  text
</emp>
```

The levels of emphasis can be *strong*, *moderate (default)*, *none*, or *reduced.*

The <sayas> tag

The <sayas> tag is used to specify exactly how a word has to be pronounced. The attribute *class* can be specified with one of the following values: *phone*, *date*, *digits*, *literal*, *currency*, *number*, and *time:*

```
<sayas phon="string" sub="string" class="string">
  text
</sayas>
```

The <div> tag

The <div> tag is used to group output speech as either a sentence or paragraph (and is specified using the *type* attribute):

```
<div type="sentence" or "paragraph">
  text
</div>
```

The <pros> tag

Prosody refers to the pitch, pitch range, how fast and how loud words have to be spoken when speaking a prompt to the user. This can be managed by using the <pros> tag:

```
<pros rate="number" vol="number" pitch="number" range="number">
  text
</pros>
```

<table>
<tr>
<td>

```
<form>
  <block>
    Hello !!!!.
    <div type="sentence">
      <emp level="strong"> Welcome
      </emp> to San Jose State University
    </div>
    <break msecs="100"/>
    <pros rate="200" vol="0.2">
      The facilities are available between
      <sayas class="time">8:00 AM and 5:00 PM
      </sayas>
    </pros>
      Our telephone number is
      <sayas>
        class="phone">408-924-4288
      </sayas>
  </block>
</form>
```

</td>
<td>

C: Hello.

Welcome to San Jose State University.

(pause)

(fast)

The facilities are available between eight AM and five PM (normal) Our telephone number is four oh eight nine two four four two eight eight.

</td>
</tr>
</table>

The <audio> Tag

The <audio> tag is used when you want to play a sound file rather than synthesized speech. If the source of the specified audio file cannot be found, the alternate text within the audio tag is spoken out using the text-to-speech engine.

The following is the syntax for the <audio> tag:

```
<audio src="URI" caching="mode" fetchtimeout="time"
       fetchhint="mode"
</audio>
```

The following is an example of how the <audio> tag is used:

```
<prompt>
  <audio src="welcome.wav"><emp>Welcome</emp> to SJSU Campus Portal
  </audio>
   . . . .
</prompt>
```

Audio Input

Speech input in VoiceXML can be captured in three ways. A grammar can be specified (using the <grammar> tag) to contain words a user can say to perform an action, and to assign a corresponding string value or attribute value pairs when those words are uttered. A grammar can also be specified (using the <dtmf> tag) to contain DTMF key inputs that a user might press during the course of interaction. Finally, using the <record> tag results in the recording of user input as an audio clip (see Figure 11-12). This section briefly describes each of the audio input tags.

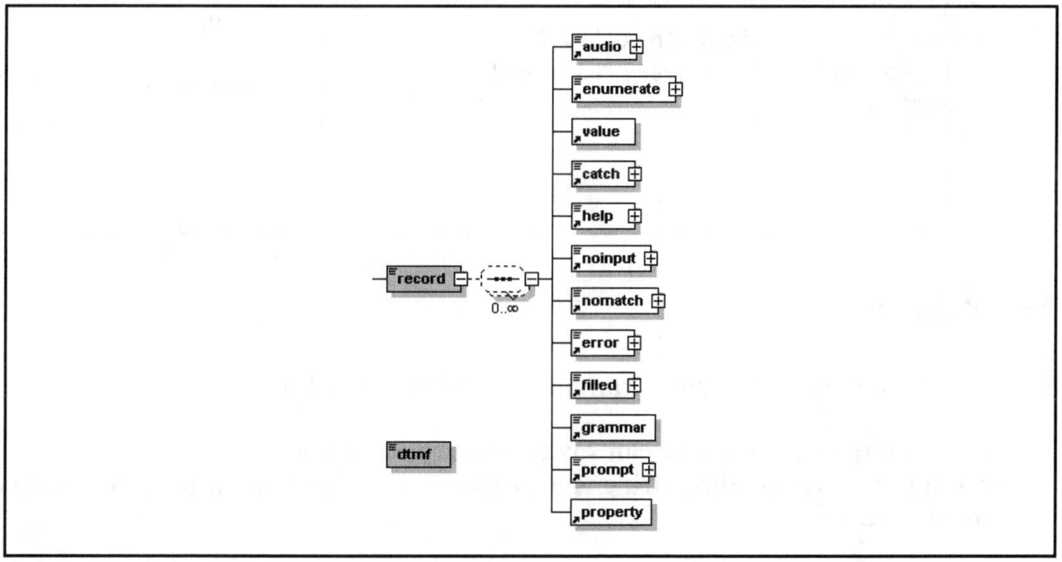

Figure 11-12. The Audio Input VoiceXML Tags.

The <dtmf> Tag

When specifying DTMF key inputs, the tag <dtmf> is used to specify the grammar. This defines the keys that the user can press.

The following is the syntax for specifying dtmf grammar:

```
<dtmf src="URI" scope="level" type="MIME_type" caching="mode"
      fetchhint="mode" fetchtimeout="time">
   #PCDATA
</dtmf>
```

<table>
<tr>
<td>
<pre>
<form name="newForm">
  <field name="sem_type">
    <prompt>
      Please key in the semester for which you
      want to register. Press 1 for Spring.
      Press2 for summer. Press 3 for fall.
    </prompt>
    <dtmf type="application/x-jsgf">
      1 {spring} | 2 {summer} |3 {fall}
    </dtmf>
  </field>
  . . .
</form>
</pre>
</td>
<td>C: Please key in the semester for which you want to register. Press 1 for Spring. Press 2 for summer. Press 3 for fall.

U: (keys in ‘2’)</td>
</tr>
</table>

The <grammar> Tag

In a VoiceXML document, the <grammar> tag is used as follows:

- To specify the words a user can say to perform an action.
- To assign a corresponding string value or attribute value pairs when those words are uttered.

The following is the syntax for specifying grammar:

```
<grammar src="URI" scope="level" type="MIME_type" caching="mode"
         fetchhint="mode" fetchtimeout="time">
  #PCDATA
</grammar>
```

The example below shows the usage of the grammar element.

`<form name="newForm">`
`<grammar src="http://wirelessjava.engr.`
`               sjsu.edu/Project/campus.gram"`
`scope="dialog" type"="application/x-jsgf"/>`
`  <field name="sem_type">`
`  <prompt>`
`    Please say the semester for which`
`    you want to register.`
`  </prompt>`
`  <grammar type"="application/x-jsgf">`
`     Spring {spring} | Summer {summer}`
`                     | Fall {fall}`
`  </grammar>`
`  </dtmf>`
`  </field>`
`...`
`</form>`
	C: Please say the semester for which you want to register.
	
	U: Spring

In both the <dtmf> and <grammar> tags, the grammar could be inline grammar—which means that the grammar is embedded in the VoiceXML document within CDATA section—or an external grammar—which means that the grammar can be placed in a separate file and a reference (URI) to that file is specified using the *src* attribute (shown in the <grammar> tag listing).

If the grammar is a field-level grammar, then the user input becomes the only values that the field can accept. If it is form grammar, the grammar is valid for all the fields within the form. Also, the grammar (both speech and dtmf)—inline or external—can be just a fragment or a complete grammars. Typically, fragmented grammars are used to specify choice options.

The *type* attribute is an optional attribute that is used to specify the MIME type of a grammar format. These MIME types are listed in Table 11-1.

The attribute *scope* is used to specify the scope for the grammar. The scope can be dialog, document, or determined by the containing tag. If the scope is dialog then the grammar is active for the current dialog, which could be a form or menu. If the scope is document then it is valid for the entire document. In the grammar example shown above the scope of the campus.gram is active for the form dialog.

Details

The World of VoiceXML Grammar

During the making of the VoiceXML 1.0 specification, there was no standard defined for writing grammars. Consequently, different platforms supported adopted different standards. Two of the most popular formats for writing grammar are the JSGF, which follows the Java syntax, and GSL, a proprietary format from Nuance Communications. The working draft of VoiceXML 2.0 published by the W3C mandates that VoiceXML browsers support the XML form of the W3C Speech Recognition Grammar format (SRGF). Here's a quick summary of the VoiceXML grammars you'll likely encounter.

Java Speech Grammar Format (JSGF)

JSGF is a platform-independent and vendor-independent representation of grammars for use in speech recognition. This format was developed by Sun Microsystems. JSGF is a rule-based grammar that specifies the utterances that a user might say. IBM's ViaVoice and HeyAnita use JSGF. The basic syntax of JSGF is as follows:

```
JSGF V 1.0
Grammar Header;
<rule1> = rule_expansion
```

where the rule name is enclosed within "<" and ">" and the input words being referenced by the rule name are placed on the right hand side. JSGF allows specifying inline and external grammars. In addition, the grammar specified could be complete grammar or just fragments (which are used to specify alternate ways of saying something with no formal rule definitions).

The listing below is an example of a JSGF-based grammar file that can be accessed externally from within a VoiceXML document. The options that a user has while prompting a city's name are Sydney, Brisbane, or Melbourne.

```
JSGF V 1.0
Grammar mycity;
<city> = Sydney | Brisbane |Melbourne
```

Grammar Specification Language (GSL)

GSL is a proprietary grammar language developed by Nuance Communications, the leading voice solutions provider. Like JSGF, GSL too is a rule-based grammar. The format is similar to JSGF but the syntax is slightly different. Tellme, BeVocal, and Voxeo, all use the Nuance GSL format. The basic syntax of the GSL format is:

```
.grammar_rule1 grammar_definition1
```

where the rule name is preceded by a "." and the rule definition follows the rule name. The GSL equivalent of the "mycity.gram" defined in JSGF section is shown in the following example.

```
.city  [Sydney Brisbane Melbourne]
```

Speech Recognition Grammar Format (SRGF)

SRGF is derived from JSGF. SGRF has two forms—one is based on XML and the other Augmented Backus-Naur Form (ABNF) form. VoiceXML 2.0 compliant browsers are mandated to support the XML form of SRGF. The MIME types for SRGF are application/grammar+xml for the XML form and application/grammar for ABNF form. The SRGF listing of the familiar city example is shown below.

```
<?xml version="1.0"?>
<grammar xml:lang="en" version="1.0">
  <rule id="city" scope="public">
    <one-of>
      <item> Sydney </item>
      <item> Brisbane </item>
      <item> Melbourne </item>
  </rule>
</grammar>
```

Built-in Grammar in VoiceXML

VoiceXML provides what is known as built-in grammar for capturing input of some simple types—like boolean, date, digits (for strings of individual digits), currency, number, phone, and time. These can be used by specifying *builtin: grammar/builtin-type* as the value for *src* attribute of <grammar> or <dtmf> tag. In case of a <field> tag, the value of the *type* attribute can be set to one of these built-in types to use the built-in grammar.

```
<grammar src=builtin:grammar/boolean/>

<field name="homePhone" type="phone">
  ...
</field>
```

```
<field name=:"ssnNo" type="digits">
  ...
</field>
```

Grammar Precedence

If more than one grammar is matched for user input, then the following precedence rules are used. (However, if more than one grammar with the same precedence is matched, then the grammar that occurs earlier in the document is chosen.) The following precedence rules are listed in order beginning with the one to be chosen first:

1. Grammar defined in the current form items including links and fields.
2. Grammar defined in the current form.
3. Grammar defined in the current page and grammar for other menus and forms in the current page that have document scope.
4. Grammar defined in the root document (which has document scope).

Universal grammars for the default events *help*, *exit*, and *cancel* are defined by the voice browser and usually have proprietary implementation based on the browser used. ❑

The <record> Tag

If the user input has to be obtained as an audio clip (i.e., without performing any conversion or interpretation) then the <record> tag is used. The following is the syntax for the <record> tag:

```
<record name="string" expr="expression" cond="expression"
        modal="boolean" beep="boolean" maxtime="time"
        finalsilence="time" dtmfterm="boolean" type="audio_type">
  child tags
</record>
```

The value for the *name* attribute is a variable that will hold the recording. The *modal* attribute indicates that if set to *false* all the grammars (including DTMF) pertaining to the dialog, document, and application are turned on. If *true* these grammars are turned off. The *maxtime* attribute indicates the amount of recording time (in seconds or milliseconds). The *finalsilence* attribute indicates the interval of silence that indicates the termination of speech. The *dtmfterm* attribute when set *true* allows the termination of a recording using a DTMF key. The *type* attribute indicates the format of the audio file type (e.g. *wav, etc*).

```
<form id = "login">
  <record beep="true" dtmfterm="true"
          name="Fname">
    <prompt>
      At the tone, please say your name.
      When finished recording, pause briefly,
      then say "Stop Now" or press a D T M F
      key.
    </prompt>
    <grammar>stop now</grammar>
  </record>
  <filled>
    <prompt>
      Hello <value expr="Fname"/> !
    </prompt>
    <goto next="#Semester"/>
  </filled>
  ...
</form>
```

C: At the tone, please say your name. When Finished recording, pause briefly, then say stop now or press a DTMF key.

U: John Doe

(pause)

U: Stop now

C: Hello John Doe! (transfers control to the semester dialog)

Events

VoiceXML supports an event mechanism to handle unexpected conditions similar to the exception-handling feature available in most languages like Java. The platform throws events under a variety of conditions, such as when the user does not respond, or when the response is unintelligible, or when user requests help, etc. The VoiceXML interpreter also throws events if it finds a semantic error in a VoiceXML document. One can also define their own events using the <throw> and <catch> tags. See Figure 11-13 for an overview of the event VoiceXML tags.

The <help> Tag

When a user requests help in the process of a conversation, a *help* event is thrown and this is caught by the <help> tag defined in the application. The *count* attribute allows you to specify a different behavior for the first occurrence of an error from subsequent occurrences.

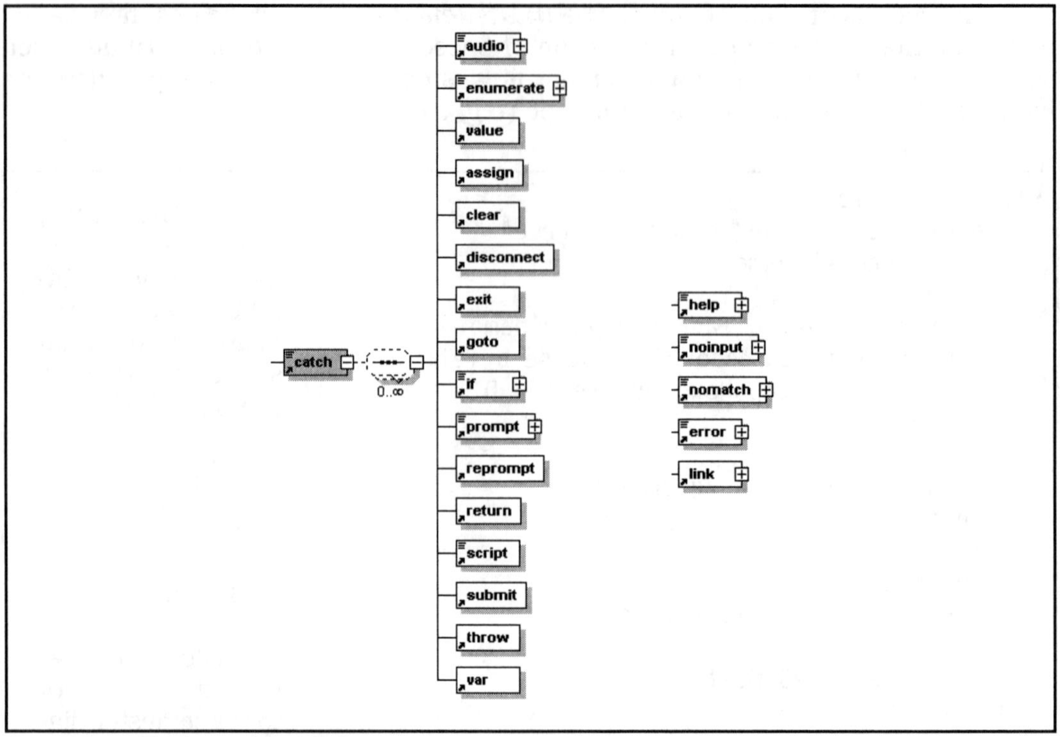

Figure 11-13. The Events VoiceXML Tags.

```
<help count="integer" cond="expression">
  child tags
</help>
```

The <nomatch> and <noinput> Tags

During a conversation, if the user's input does not match the grammar provided, or if there is no input for a period of time, the *nomatch* and *noinput* events respectively are thrown by the <nomatch> and <noinput> tags.

```
<nomatch count="integer" cond="expression">
  child tags
</nomatch>

<noinput count="integer" cond="expression">
  child tags
</noinput>
```

```
<?xml version="1.0"?>
<vxml version="1.0">
  <form>
  <prompt> Welcome to the SJSU registration system. What services
         would you like to use?
  </prompt>
  <grammar>...</grammar>
  <noinput>Please say something </noinput>
  <nomatch>I did not get it. Please say again </nomatch>
  <help>Please say a service name that you want to use</help>
  ...
</form>
</vxml>
```

The <error> Tag

The <error> tag is used to contain content to be executed when an error event occurs. It is the shorthand for <catch event="error">. We will discuss the <catch> tag in the next section. The syntax for the <error> tag is:

```
<error count="integer" cond="'expression'">
  child tags
</error>
```

```
<form>
  <block>
    ...
  </block>
  <error>
    An error has occurred. Exiting now.
    <exit/>
  </error>
</form>
```

The <throw> and <catch> Tags

You can define your own events using the <throw> and <catch> tags. In addition, one can throw and catch pre-defined events mentioned before like the *noinput*, *help*, etc. You can also catch more than one event:

```
<throw event="string" count="integer" cond="'expression'" />
<catch event="string" count="integer" cond="'expression'">
  child tags
</catch>
```

```
<form>
<prompt> Welcome to the SJSU registration system. What services
         would you like to use?
  </prompt>
  <grammar> . . .</grammar>
    <catch event ="invalid_entry">
       <prompt>
         The input is in valid
       </prompt>
   <filled>
       <if cond="invalid">
           <throw event="invalid_entry"/>
       </if>
   ...
  </filled>
</form>
```

The <link> Tag

The <link> tag contains a grammar entry which, when uttered by the user, transfers to a new dialog based on the value of the link's *next*, *expr*, or *event* attribute. The grammar in the <link> tag cannot specify the scope. Links are mostly used in mixed-initiative conversations. A <link> tag can also be used to throw an event instead of providing direction to a destination URI:

```
<link next="URI" event="string" caching="mode" fetchaudio="URL"
fetchtimeout="time" fetchhint="mode" expr="'expression'">
  child tags
</link>
```

The following listing shows how the <link> tag can be used to specify grammar for direction to an URI and to generate an event.

```
<link next="register.vxml">
  <grammar type="application/x-jsgf"> register</grammar>
</link>
...
<link event=help>
  <grammar type="application/x-jsgf">help | please help</grammar>
</link>
```

Control Flow

VoiceXML supports statements that may be used to alter the control flow in a VoiceXML application. This section deals with the control flow tags that are permitted by the VoiceXML specification. See Figure 11-14 for an overview of the control flow VoiceXML tags.

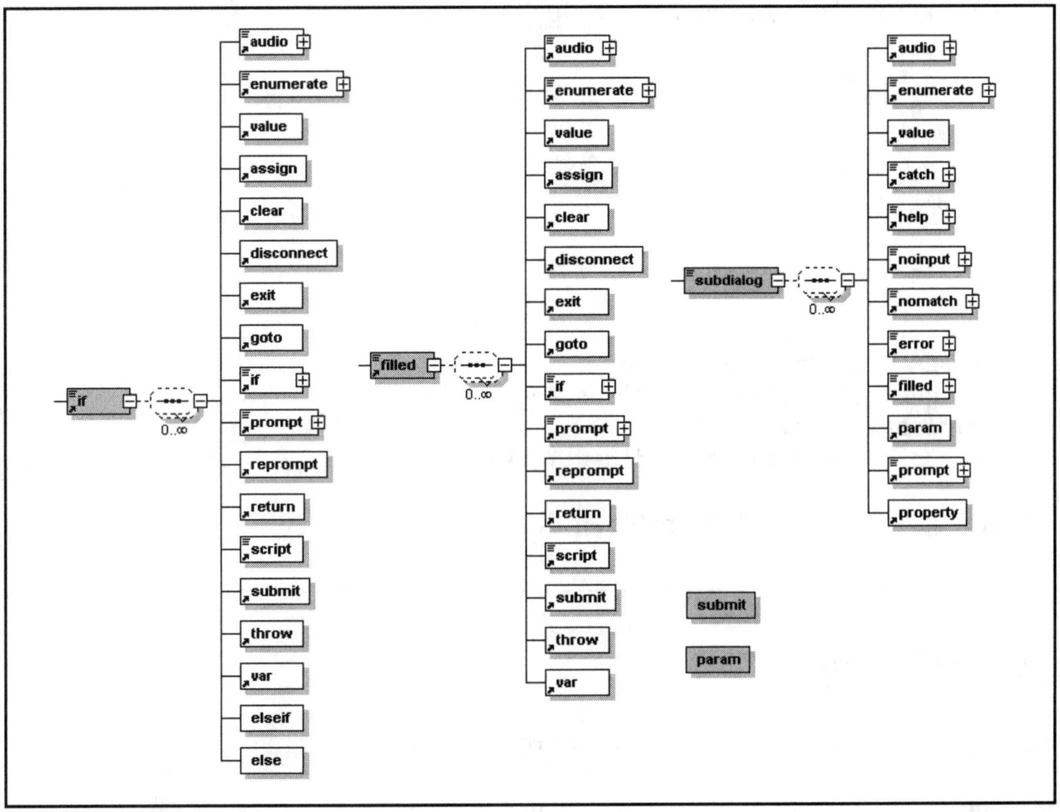

Figure 11-14. The Control Flow VoiceXML Tags.

The <param> Tag

The <param> tag can be used to pass a parameter to an object or a subdialog just like passing a parameter to a function. The *name* attribute should point to the variable name that has to be passed as the parameter. The *valuetype* attribute is used to indicate if the *value* is an URI (in which case the *valuetype* is ref) or if it itself is a value (in which case the *valuetype* is data). The following is the syntax of <param> tag:

```
<param name="string" expr="expression" value="string"
       valuetype="string" type="MIME_type"/>
```

The <if>, <else>, and <elseif> Tags

The <if>, <else>, and <elseif> tags may be used to create conditional logic in VoiceXML as in any other languages like Java or C. The following is the syntax for if, else, and the elseif statements in VoiceXML.

```
<if cond="'expression'">
  child tags
</if>

<else/>

<elseif cond="'expression'"/>
```

```
<if cond="city == 'Seattle'">
  <assign name="city" expr="'Los Angeles'"/>
<elseif cond="city == 'NY'"/>
  <assign name="city" expr="'New York'"/>
<else/>
  <assign name="city" expr="'San Jose'">
</if>
```

The <goto> tag

The <goto> tag is used to transfer control to:

- Another form item in the current form.
- Another dialog in the current document.
- Another document.

The following explains the syntax of the <goto> tag:

```
<goto next="URI" expr="'expression'" nextitem="string"
      expritem="'expression'" caching="mode" fetchhint="mode"
      fetchtimeout="time" fetchaudio="URL"  />
```

To transfer control to another form item, you can use the *nextitem* attribute, as shown below:

```
<goto nextitem="ssn_confirm"/>
```

The *expritem* attribute may be used to transfer control, if the form item name is computed using an ECMAScript expression as below:

```
<goto expritem="(type==12)? 'ssn_confirm' : 'reject'"/>
```

To go to another dialog in the same document, use *next* with only a URI fragment:

```
<goto next="#another_dialog"/>
```

The *expr* attribute may be used to compute the dialog to go to as shown below:

```
<goto expr="'#' + 'another_dialog'"/>
```

To transition to another document, use *next* with a URI:

```
<goto next="http://target/document"/>
```

Or an *expr* attribute may be used to compute the document to go to as shown below:

```
<goto expr="' http://' + 'another_document'"/>
```

<form id = "menu"> <block> You have chosen main menu. Wait please. <goto next="#choice"/> </block> </form>	C: You have chosen main menu. Wait please. (transfers control to the choice dialog)

The <submit> Tag

The <submit> tag is used to submit a list of variables to a document server via HTTP GET or HTTP POST (using the *method* attribute). As with other tags, the target document can either be specified as a URI by use of the *next* attribute, or as a dynamic value by use of the *expr* attribute.

```
<submit next="URI" expr="'expression'" namelist="string"
        method="GET|POST" caching="mode" fetchaudio="URL"
       fetchtimeout="time" fetchhint="mode" enctype="MIME_type" />
```

By default, all the field item variables are submitted. However, you can use the *namelist* attribute to specify a list of variables to be submitted.

```
<form id="rates">
  <field name="service">
   . . .
  </field>
  <field name="ccno" type="number">
   . . .
  </field>
  <filled>
  <submit next="http://wirelessjava.engr.sjsu.edu/examples/
              servlet/register"/>
  </filled>
</form>
```

The <return> Tag

The <return> tag marks the end of a subdialog and returns variables and control back to the calling dialog. The *namelist* attribute is used to specify the variables to be returned. The following explains the syntax for the <return> tag:

```
<return event="string" namelist="string"/>
```

It is also possible to specify an event to be thrown at the invocation of the return.

The <exit> tag

The usage of the <exit> tag results in the termination of all loaded VoiceXML documents and transfer of control to the context (of the voice browser) that originally loaded the documents. The following is the syntax for the <exit> tag:

```
<exit expr="'expression'" namelist="string"/>
```

```
<if cond="choose=='yes'">
   <exit />
</if>
```

Once control is returned to the context, it is up to the context to decide if the call has to be dropped, or transferred to an operator or to a menu.

The <subdialog> Tag

Subdialogs allow control to be transferred to another VoiceXML document or dialog (much like a function call in programming languages shifts control flow) to perform some user interaction and then return control back to the calling form. Subdialogs have a separate execution context and so the calling document's local data, grammar, and state are preserved when a subdialog is called and are restored upon return of control from the subdialog.

Subdialogs are implemented using the <subdialog> tag. Variables can be passed from the subdialog to the main dialog as ECMAScript objects by using the <return> tag from within the subdialog VoiceXML document:

```
<subdialog name="string" src="URI"
          expr="expression" cond="expression"
        modal="boolean" namelist="string" method="GET|POST"
        caching="mode" fetchaudio="URL" fetchtimeout="time"
        fetchhint="mode" enctype="MIME_type">
  child tags
</subdialog>
```

The *src* attribute is used to point to the dialog to which control is transferred. The form or menu ID is provided along with the document's name with a number sign (#) separating these two—for example documentURI#form. The *caching* attribute defines whether to look for a newer version no matter what (*safe*), or look for a newer version only if expired or missing (*fast*). The *fetchint* attribute defines when the browser should use a file to download. With *prefetch* the file is downloaded when the page loads. With *safe* the file is downloaded when needed.

<table><tr><td>

```
<form id = "rate">
  <block> You are in the class service</block>
    <subdialog name="result" src="#subDialog">
    <filled>
      You just said <value expr =
                       "result.subDgVar"/>
    <submit namelist="result.subDgVar"
    next="http://wirelessjava.engr.sjsu.edu
          /examples/servlet/register"/>
    </filled>
    </subdialog>
</form>

  <!--Sub dialog that is called from the form -->
  <form id="subDialog">
    <field name="subDgVar" type="digits">
      <prompt>Enter your course number. </prompt>
       <nomatch>
          . . .
       </nomatch>
       <filled>
         <return namelist="subDgVar"/>
       </filled>
    </field>
  </form>
```

</td><td>
C: You are in the class service.

C: Enter your course number.

U: 2347

C: You just said 2347
</td></tr></table>

The <filled> Tag

A <filled> tag defines an action that occurs just after a field item variable in a form is filled. The following is the syntax for the <filled> tag:

```
<filled mode="all|any" namelist="string">
  child tags
</filled>
```

Call Control

This section deals with the <disconnect> and <transfer> tags that enable handling of calls within a VoiceXML application. See Figure 11-15 for an overview of the call control VoiceXML tags.

```
<form id = "rate">
  ...
  <filled>
   You just said <value expr = "testvar"/>
   <submit namelist="testvar"
  next="http://wirelessjava.engr.sjsu.edu/examples/servlet/register"/>
  </filled>
  ...
</form>
```

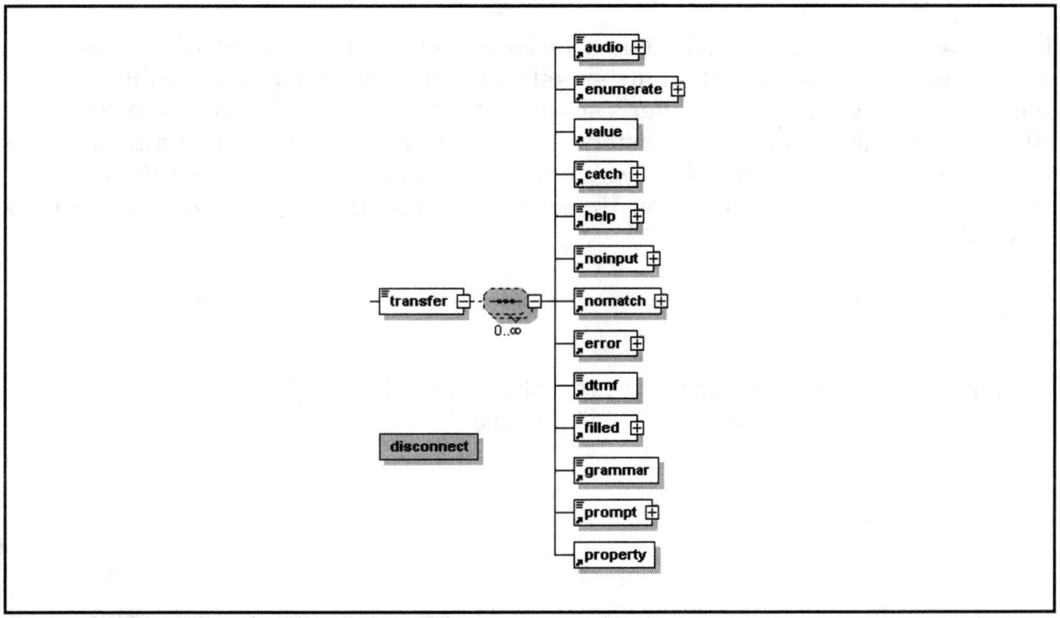

Figure 11-15. The Call Control VoiceXML Tags.

The <disconnect> tag

The <disconnect> tag disconnects the call and hangs up the browser's session. This is an empty tag with no attributes. The following is the syntax for the <disconnect> tag:

```
<disconnect/>
```

A <disconnect> differs from an <exit> tag in that it forces the interpreter context to drop the call.

The < transfer> tag

If the user has to be transferred to another phone number, this can be specified using the <transfer> tag. The following is the syntax for the <transfer> tag:

```
<transfer name="string" expr="'expression'" cond="'expression'"
dest="URI" destexpr="'expression'" bridge="boolean"
connecttimeout="time" maxtime="time">
  child tags
</transfer>
```

The value for the *name* attribute is a variable that will hold the result/status of the transfer attempt. The *dest* attribute specifies the destination phone number. *Bridge* specifies whether or not the caller's session with the VoiceXML interpreter resumes after the call initiated by the transfer ends. The *connecttimeout* attribute specifies the time (in *seconds* or *milliseconds*) to wait while trying to connect the call. The *maxtime* attribute specifies the time (in *seconds* or *milliseconds*) the call is allowed to last.

```
<form>
  ...
  <transfer name="testCall" dest="phone://4083430768"
            connecttimeout="10s" bridge="true">
    <filled>
      ...
    </filled>
  </transfer>
</form>
```

Miscellaneous

Figure 11-16 illustrates some of the VoiceXML tags that don't fit into other categories.

The <property> tag

The <property> tag is used to set named values that affect platform behavior, such as the recognition process, timeouts, caching policy, etc. Properties may be defined for the entire application (i.e., in the root document), in a specific VoiceXML

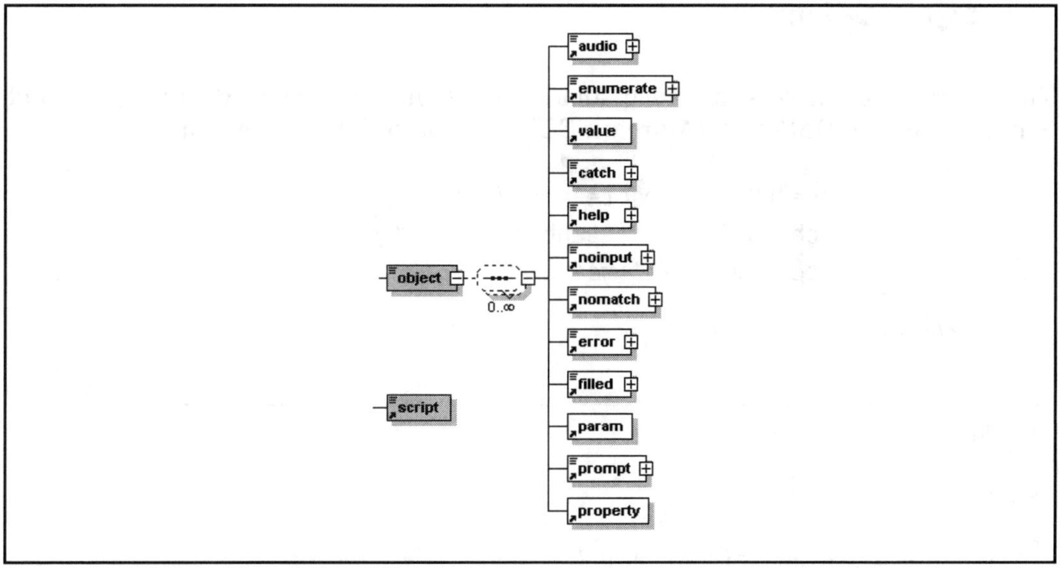

Figure 11-16. The Miscellaneous VoiceXML Tags.

document, in a dialog (in the form or menu), or in form items. The following is the syntax for the <property> tag:

```
<property name="property name" value="new property value"/>
```

There may be voice browser-specific properties, too. If the voice browser encounters an unsupported property it is ignored and no *error.unsupported.property* event is thrown.

```
<form id="no_bargein_form">
  <property name="bargein" value="false"/>
   . . .
</form>
```

The <object> Tag

The <object> tag is used to invoke platform-specific functionality that a VoiceXML application wants to use, such as speaker verification, native components, and additional telephony functionality. It is analogous to the HTML <OBJECT> tag. Discussion on the <object> tag is beyond the scope of this chapter.

The <script> Tag

The <script> tag allows the encapsulation of client-side scripting language, which is in this case ECMAScript. (More on ECMAScript in the next section.)

```
<script src="URI" charset="encoding"
        caching="mode" fetchhint="mode"
        fetchtimeout="time">
  #PCDATA
</script>
```

```
<vxml>
 . . .
<script>
  <![CDATA[
  function courseCost(courses) {
  return ( courses*course_cost );
  }
  ]]>
</script>
 . . .
</vxml>
```

FYI

VoiceXML 2.0: The Next Generation

Briefing

On October 23, 2001, the World Wide Web Consortium released the first working draft of VoiceXML 2.0. VoiceXML 2.0 improves upon VoiceXML 1.0 by increasing the portability of applications between vendors.

VoiceXML 2.0 features two new XML standards for writing voice-recognition grammars and speech synthesis markups:

- Speech Recognition Grammar Format (SRGF) (discussed earlier).
- Speech Synthesis Markup Language (SSML).

The SRGF replaces the predominantly used JSGF and GSL grammar formats of VoiceXML 1.0 and enforces standardization throughout the industry. The inclusion of this standard reduces the application portability issues that arise due to the use of proprietary formats for writing speech grammars.

SSML improves synthesized speech quality of voice-driven applications through various tags, including "break" tags which create pauses in the speech and "voice" tags which define the age and/or gender of the speech interface (speaker). Other SSML tags are helpful in clarifying inflection, pronunciation, pitch, range, and volume.

VoiceXML 2.0 also includes a new <log> tag that aids in specifying debug messages. In addition, VoiceXML 2.0 provides substantial clarifications for VoiceXML 1.0 ambiguities, a list of VoiceXML 1.0 tags rendered obsolete, and a list of modified VoiceXML 1.0 tags.

The details of these changes can be obtained at the W3C site *http://www.w3.org/TR/2001/WD-voicexml20-20011023.* ❑

The VoiceXML Registration Listing

Listing 11-1 shows the complete listing for the Registration VoiceXML document.

Listing 11-1. The Registration VoiceXML Document (Registration.vxml).

```
<?xml version="1.0"?>
<vxml version="1.0">
<form id="Introduction">
  <block>
    Hello!
   <prompt><emp level ="strong">Welcome to the San Jose State registration
    facility.</emp>
    Our web site is
    <sayas sub="W W W dot s j s u dot edu">
      www.sjsu.edu
    </sayas>
    This facility is available between
    <sayas class="time">8:00 AM and 5:00 PM
    </sayas>
    </prompt>
    <break msecs="40"/>
    <goto next="#choice"/>
  </block>
</form>

<form id = "choice">
  <field name ="choose">
    <prompt>
```

```
        Please choose from the following:
        For a registered user please say <emp level="strong">login</emp>.
        For a new user please say <emp level="strong">register</emp>.
        For main menu say menu.
        You can say goodbye anytime.
      </prompt>
      <grammar>
        login | register | menu | goodbye
      </grammar>
      <noinput>
        Please say something. <reprompt/>
      </noinput>
      <nomatch>
        Sorry.I did not get that. Please say again. <reprompt/>
      </nomatch>
      <filled>
        <if cond="choose=='login'">
          <goto next="#login"/>
        <elseif cond="choose=='register'"/>
          <goto next="#events"/>
        <elseif cond="choose=='menu'"/>
          <goto next="#menu"/>
        <elseif cond="choose=='goodbye'"/>
          <goto next="#goodbye"/>
        </if>
      </filled>
    </field>
</form>

<form id = "login">
  <record beep="true" dtmfterm="true" name="Fname">
    <prompt>
      At the tone, please say your name.
      When finished recording, pause briefly, then
      say "Stop Now" or press a D T M F key.
    </prompt>
    <grammar>stop now</grammar>
  </record>
  <field name ="id" type="digits">
  <prompt>
    Please say your PIN in digits format
  </prompt>
  <filled>
  <prompt>
    Hello <value expr="Fname"/> !
```

```
      Your pin is <value expr="id"/>
    </prompt>
    <goto next="#Semester"/>
    </filled>
  </field>
</form>

<form id = "Semester">
  <field name ="sem">
    <prompt>
      Please choose from the following semesters.
      Spring.
      Summer.
      Fall.
      Winter.
  </prompt>
  <grammar>
    Spring | Summer | Fall  | Winter
  </grammar>
  <filled>
    <prompt>
      You chose <value expr="sem"/>.
    </prompt>
    <goto next ="#Course"/>
    </filled>
  </field>
</form>

<form id = "Course">
  <field name ="cid" type="digits">
    <prompt>
      Please say or enter the 5-digit course I D for the course you wish to
      register.
    </prompt>
  </field>
  <field name ="sid" type="digits">
    <prompt>
      Please say or enter the 2-digit section I D for the course you wish
      to register.
    </prompt>
  </field>
  <filled>
    You have  chosen the course i d <value expr="cid"/> and section i d
    <value expr="sid"/>
    <goto next ="#Card"/>
```

```
  </filled>
</form>

<form id = "Card">
  <field name ="card">
    <prompt>
      Please choose a credti card from the following :   .
      Visa.
      Master Card.
      Discover.
    </prompt>
    <grammar>
      Visa | Master Card |American Express | Discover
    </grammar>
    <filled>
      <prompt>
        You have chosen the <value expr="card"/> as your credit card.
      </prompt>
      <goto next="#cc"/>
    </filled>
  </field>
</form>

<form id = "cc">
  <field name="ccnum" type="digits?length=16">
    <prompt bargein="true">Please say your sixteen
    digit credit card number now.
    </prompt>
  </field>
  <field name="expirydate" type="date">
    <prompt>What is the expiration date of this card? </prompt>
  </field>
  <filled>
    Your card number is <value expr="ccnum"/> and the expiration date is
    <value expr="expirydate"/>
    <goto next="#end"/>
  </filled>
</form>
<form id ="end">
  <block>
    Thank you for using the registration facility at San Jose State
    University!
  Have  a nice Day!
  </block>
</form>
```

```
<form id = "menu">
  <block>
    You have chosen the main menu.
    Wait Please.
    <goto next="#choice"/>
  </block>
</form>

<form id = "goodbye">
  <block>
    Thank you for using San Jose State University Portal.
    GoodBye.
    <exit/>
  </block>
</form>
</vxml>
```

INTRODUCTION TO ECMASCRIPT

VoiceXML is a dialog-oriented language that facilitates a conversation with an end user. However it does not provide features for the following:

- Validating user inputs.
- Accessing the browser file system, or a database on the same server as the voice browser.
- Generating alerts, confirmations, and dialogs locally so that a network trip all the way to the back-end Internet service could be avoided.

To address these without modifying VoiceXML, the VoiceXML Forum supports a scripting language called ECMAScript. ECMA stands for European Computer Manufacturer Association and ECMAScript is based on JavaScript 1.1. Brendan Eich of Netscape invented ECMAScript. ECMAScript is a very powerful scripting language. It uses the syntax and programming language concepts of Java, C++, and C. As with any programming language, ECMAScript has so many features that it would take an entire book to explain.

So how do VoiceXML and ECMAScript go together? All variables in VoiceXML are completely equivalent to ECMAScript variables. We also recall that the *expr* attribute that we encountered in many tags defines an ECMAScript expression. The following is the familiar validation ECMAScript example that we encountered in the WML chapter. The example below shows how ECMAScript is used within a

VoiceXML document. The ECMAScript is buried in the same VoiceXML document in which it is used, and is placed within the <script></script> tags. In the following section we will discuss in brief the features of ECMAScript. One important point to remember is that the ECMAScript contents should be within the CDATA section.

```
<?xml version="1.0" encoding="iso-8859-1"?>
<vxml version="1.0">
<var name="number"/>
  <form>
    <field name="number" type="digits">
      <prompt>Please say a number between 25 and
              50</prompt>
    </field>
    <filled>
      <prompt>
        One moment while I validate your number.
      </prompt>
      <assign name="document.number"
              expr="number"/>
    </filled>
    <block>
     Your number is <value
       expr="validate(number)"/>
    </block>
  </form>
  <script>
   <![CDATA[function validate(number){
     var max = 50;
     var min = 25;
     if ((number < min)||(number > max))
       return "invalid";
     else
       return "valid";
     }
    ]]>
  </script>
</vxml>
```

C: Please say number between 25 and 50

U: 36

C: One moment while I validate your number

C: Your number is valid

Variables and Data Types

ECMAScript supports both global and local variables. Local variables are those that are declared inside functions or passed as function parameters. Variable declarations are not compulsory in ECMAScript. This means that variables need not be declared before they can be used in ECMAScript. Variables are declared by placing *var* before unique variable names within a function. ECMAScript supports six standard variable types. These are enumerated in Table 11-3 below:

Table 11-3. The ECMAScript Variable Types.

Variable Type	Example	Comments
Undefined	<var name="semester"/> <field name="ccno">	Default value of a variable before it is assigned a value.
Null	<assign name="sem" expr="null"/>	Assigning a null value to a variable.
String	<var name = "name" expr="'Mike'"/>	Strings are delimited by double quotes (" "). Character strings are referenced by using single quotes (' ').
Boolean	<var name=has_num expr="false"/>	Boolean variable take on values true or false.
Number	<var name="fee_cost" expr="736.65"/>	Any positive or negative integer or decimal value.

Built-in Objects

ECMAScript supports many built-in objects like **Array**, **Date**, **Math**, **Function**, **Boolean**, **Number**, **String**, and **RegExp**.

- **Arrays** are created using the *new Array( )* statement. The **Array** constructor can take either the items to be placed in the array or the array index. The array referencing is the same as in Java.
- The **Date** object allows the VoiceXML programmer to work with dates and times. The statement *mydate= new Date( )* will return the current date.
- The **Math** object allows operations like *abs( )*, *Random( )*, and *Exp( )*.
- The **String** object allows string manipulations like *CharAt( )* and *ToUpperCase( )*.

- **Functions** in ECMAScript can be declared in two ways. One is by using the **Function** object or by declaring a function as in Java.

Operators and Expressions

ECMAScript supports most of the expressions and operators available in many programming languages. The simplest of the expressions are constants and variable names that would evaluate to their values. Simple expressions can be combined together with operators to make complex expressions. Invoking functions simply as expressions and ignoring their return values is allowed, too. Table 11-4 enumerates the ECMAScript operators:

Table 11-4. The ECMAScript Operators.

Operator Type	Operator	Description
Assignment	=	assign
	+=	add (numbers)/concatenate (strings) and assign
	-=	subtract and assign
	*=	multiply and assign
	/=	divide and assign
	div=	divide (integer division) and assign
	%=	remainder or modulus and assign
	<<=	bitwise left shift and assign
	>>=	bitwise right shift with sign and assign
	>>>=	bitwise right shift zero fill and assign
	&=	bitwise AND and assign
	|=	bitwise OR and assign
	^=	bitwise XOR and assign
Arithmetic	+	add (numbers)/concatenate strings
(Basic binary)	-	subtract
	*	multiply
	/	divide
	div	integer division
Arithmetic	%	remainder or modulus
(Complex binary)	<<	bitwise left shift
	>>	bitwise right shift
	>>>	bitwise right shift zero fill
	&	bitwise AND
	|	bitwise OR
	^	bitwise XOR

Table 11-4. The ECMAScript Operators. (Continued)

Operator Type	Operator	Description
Arithmetic	+	plus
(Basic unary)	-	minus
	--	pre-or-post decrement
	++	pre-or-post increment
	~	bitwise NOT
Logical	&&	logical AND
	||	logical OR
	!	logical NOT (unary)
Comparison	<	less than
	<=	less than or equal
	==	equal
	>=	greater than or equal
	>	greater than
	!=	inequality
Comma	,	Combines multiple evaluations into one expression.
Conditional	x ? a : b	Assigns a value based on Boolean result of initial value
Typeof	typeof	Returns an integer value representing the type of the expression
Invalid	invalid	Returns false if the type of an expression is invalid

Functions

As with other programming languages, functions, group expressions and statements return a value. If the return statement is not placed in a function then it will not return a value (and will behave like a subroutine). Also, a function name cannot be overloaded. As mentioned above functions can be created in two ways:

- `fnObj = new Function(optionalArg1, optionalArg2, optionalArgN, function_body)`

In this case the arguments are optional, and the function_body would be placed in double quotes. For example:

```
mySum = new Function(a, b, "return (a+b)");
```

- The same function can be declared as:

```
Function mySum (a, b) {
```

```
    return (a+b);
  }
```

In both cases the function can be called as 'mySum(5, 6);'

Statements

ECMAScript supports statements that define expressions, variable declaration, and execution control. Table 11-5 below enumerates the ECMAScript statements:

Table 11-5. ECMAScript Statements.

Statement	Example	Comments
Empty	`while (!test(bit));`	Used when statement is needed but no operation.

Expression	`var hello = Hello World;`
	`alert(Invalid);`
	`area = pi * r * r;`
		Expressions assign values, calculates values, calls functions, and so on.

Block	`{ var j = 2;`
	`j =  j + 4;`
	`Console.print(j = + j);`
	`}`
		A group of statements contained inside curly braces ({}). Used to form a compound statement.

Variable	`var result = 0;`	Declares and initializes variables.

If	`if (happy)`
	`{ smile = true`
	`} else`
	`  {smile = false;`
	`}`
		Conditional branching execution depending on Boolean test value.

While	`while (i < 3)`
	`{ total++;`
	`}`
		Conditional looping execution depending on Boolean test value.

For	`for (var i=0; i<10; i++)`
	`{ total++;`
	`}`
		Conditional looping execution depending on Boolean test value.

Break	`if (i == 3) break;`	Terminates a for or while loop.

Table 11-5. ECMAScript Statements. (Continued)

Statement	Example	Comments
Continue	`if (i == 3) continue;`	Terminates execution of a for or while loop and resumes execution with the next iteration.
Return	`return Invalid Number;`	Returns a value for a function call.

VOICEXML PROGRAMMING TIPS

In this section we will highlight some of the issues to keep in mind while developing VoiceXML application. Though conventional software engineering methodologies still hold well, the approach has to be shifted to focus on user interaction and usability factors.

General Programming Tips

When designing VoiceXML applications keep the focus on the end user's need and the user experience. Since these applications will be run from phones, the user experience, and usability factors determine the usefulness of the application and whether the application provides just the right amount of the right information.

Users find it confusing to navigate in deep menus, so try not to use more than three levels of menus. Place the most frequently used items at the top dialog. Always allow users to enter data via DTMF keys in addition to voice inputs because user pronunciation might differ. Thus, if you rely solely on voice inputs the user might not be able to access your application with a great deal of success.

Try to build your application with mixed-initiative dialogs such that a first time user and an experienced user could use it with ease. Try to use *tapered* prompting (i.e., shortened prompts if the same prompt has to be repeated) whenever appropriate. Also try to use *incremental* prompts (the inverse of tapered prompts where you start with an initial prompt and if the user fails to provide any input or the right input, you provide more prompting information and increase the prompt size). Also try and use confirmations wherever appropriate.

If possible, provide a way to remember and reuse user inputs from previous visits. If you are providing a suite of applications, like a portal, then maintain consistency throughout. Also, embedding the <value> tag within <choice> is not very useful and avoidance of this construct is recommended—even though VoiceXML DTD allows such usage. The reason being that, the grammar for the <choice> tag will be created when the document is initially loaded, and the variable used in the

<value> tag should also be defined before the document is loaded, that is, defined with application scope. That way, the value held in the variable will not be reevaluated when reading out choices for the user.

Grouping Dialogs

Dialogs may be grouped based on the resources they require and functions they perform. They may also be grouped based on the frequency of expected use and the number of VoiceXML documents the voice browser should download from the application server. Also, breaking down complex dialogs into sub-dialogs would help create reusable components.

Where to Define Grammar

A grammar may be defined inline or with an external file. This decision will impact the interactions between the voice browser and the application server and the response time from an end user's perspective. For quicker responses the grammar can be made inline. However, if the grammar is huge and resides in an external file, then *fetching* and *caching* may be used to improve performance.

Fetching and Caching Resources to Improve Performance

Any VoiceXML resource (like an audio clip or a grammar file) can be "prefetched" and cached to improve performance. In the root document you could define a dialog (say, a form) and set the *fetchint* and *caching* attributes of the resources to appropriate values. This would cause all the resources to be loaded when any document in that application is loaded. This dialog should not be referred from any other documents, but merely serves to fetch and cache resource files. Listing 11-2 is a sample root document that illustrates this:

Listing 11-2. Fetching and Caching to Improve Performance.

```
<!-- This form should never be visited. It is defined the fetch and caching
attributes of the resources -->
<!–all resources that you might need to be cached–like grammar  -->
<form id="prefetchGrammar">
 <grammar src="sjsu.gram" type="application/x-jsgf" fetchhint="prefetch"
caching="safe"/>
</form>
```

SUMMARY

In this chapter, we have covered the programming aspects of VoiceXML 1.0 and ECMAScript and how they can be used together to build voice-enabled applications. Like WML, VoiceXML has certain features of a programming language in it—like providing support for variables, control statements and events—that makes it really powerful. VoiceXML is better suited for use when both eyes and hands are preoccupied performing other tasks (like driving), and it does not impose any resource requirements on client devices. The ease and simplicity of ECMAScript enhances VoiceXML programming too. All these explain why VoiceXML is becoming popular among enterprise application developers.

Chapter 12

Your First Java Card Application

Java Cards, if you recall from our previous discussion about Java technologies on handheld devices, are a specialized form of Smart Cards that contain a mini Java Virtual Machine. In this chapter, we're going to take a closer look at Java Cards and how to write Java programs that run on them. We'll revisit *cardlets* (which were briefly introduced in Chapter 7), discuss the cardlet life cycle, take a look at the Java Card API, and write a simple cardlet of our own. We'll also discuss some of the more bread-and-butter aspects of Java Card programming, like how to download your cardlets to your card and activate them. We'll wrap up by creating a simulated host application that interacts with our simple cardlet, illustrating a complete end-to-end example of Java Card programming.

If you are coming from a J2EE or J2SE background, you may have felt cramped by the memory limitations and size restrictions posed by J2ME, as discussed in the previous chapters. Prepare to feel even more cramped, as a typical Java Card virtual machine environment makes even MIDP seem spacious! Whereas the MIDP specification recommends a few hundred KB for MIDlets, a typical Java Card will only provide from 8KB-32KB of storage space for all the cardlets installed on the entire card!

The tiny space available for cardlets may seem incredibly small at first. As you'll see throughout this chapter, although the memory limitation is something that we have to live with, the types of applications suited for Java Cards don't really require much memory in the first place.

FYI Briefing

Cardlets, Applets, and Java Card Applications

So, is it a cardlet, an applet, or a Java Card application?

We've seen what we call "cardlets" described as all three terms. The official name of a small application residing on a Java Card is an applet. However, to avoid confusion with Web browser-based applets, these small applications are oftentimes called cardlets or Java Card applications. Although there is a slight difference between a Java Card application and a Java Card applet, we won't go into the detail here.

We'll use the three terms interchangeably in this chapter. ❑

APPLICATIONS SUITED FOR JAVA CARDS

As the processor and memory resources of a typical Java Card are minimal, it doesn't make much sense to try and create a database server, XML parser, or other heavyweight system as a Java Card application. Instead, Java Cards are suited for smaller, computationally easy tasks. Some of these tasks include:

- ***Wallet or Credit Cards:*** Wallet cards manage a virtual cash balance on a credit card. Typically, a wallet card would be "charged" with a certain amount of cash, the amount being registered with the wallet cardlet living on the card. As the card is used, the cardlet is sent messages to deduct or increase the value in the simulated "wallet". Credit cards can also be augmented with features to track spending habits, store information about the card holder, etc.

- ***Gift Cards:*** A variation on the wallet theme described above, a gift card could be used by a business to replace typical paper gift certificates. A Java enabled gift card could be activated with a certain amount of money by the purchaser, given as a gift, and then used by the recipient to purchase goods or services at the business. When the card is exhausted, the business would either recycle the card for the next purchaser, or perhaps convert the card into a wallet card, allowing the holder to continually recharge the balance on the card.

- ***Security IDs or other security-conscious applications:*** As cardlets are inherently tamper-proof, a security-aware business could use Java Cards for identification or security clearance purposes. When combined with a cryptographic or encrypted identity key, a formidable security framework can be established. (See the next Details box.)

- ***Personal Information Storage:*** Trade show and conference exhibitors use attendee identification cards to retrieve information about the individuals who visit their exhibit. By having such personal information managed by a cardlet, much more information about the visitor can be retrieved, which can lead to a more focused presentation being given to prospective customers. Further, at recent shows, systems have been created to manage the attendee's entire agenda or schedule inside the cardlet on the identification card. Personalization applications such as these can be even more useful when paired with GSM-enabled cellular phones, since the phone's personalization information (like address books, call logs, etc.) can also be stored with identity information on the same chip.

- ***Single Sign-On:*** Since a Java Card can be used to establish identity, a set of applications can be configured to accept the identity established by a Java Card while only requiring the user to sign on a single time, at the beginning of the transaction. A Java Card "sign on" may be nothing more then the act of putting the card into the reader.

Java Cards and Cryptography

Details

Although cryptography and encryption are computationally expensive processes, some Java Cards include a chip to perform encryption, key generation, and other cryptographic functions in hardware, thus greatly speeding up the process on an otherwise slow Java Card CPU. ❑

These are the most popular uses for Java Card technology, but given that the Java Card VM is a full-featured Java Virtual Machine, there are many more possible applications we haven't discussed.

THE JAVA CARD VM

A typical Java Card has only 24-32KB of ROM, which has to include the Java VM executable, all class libraries, and the card's OS. Such devices typically include on the order of 8KB-16KB of EEPROM for storing cardlets. RAM constraints on a typical Java Card are even smaller, with 1KB-2KB being the norm. In an environment this small, there is not enough room to include code to support all of the J2SE language constructs. Hence, the Java Card VM only supports a subset of the standard Java specification.

Unsupported Constructs

When designing the Java Card specification to run on resource-constrained devices, the architects of the specification chose to omit certain items from the standard Java language:

- ***Primitive Types:*** Java Card does not support the **char**, **float**, **double**, or **long**. In addition, arrays of more than one dimension are not supported.
- ***Dynamic Class Loading:*** Java Card applications cannot load classes dynamically at run time. Class definitions are either burned into ROM at the factory when the card is produced, or are downloaded into EEPROM at the time of deployment. Programs executing on the card need to be self-contained—all referenced classes need to be installed on the card at the time the application is run.
- ***Security Managers:*** In Java Card applications, the VM itself is responsible for enforcing security policies. The security model of Java Card applications does not depend on a **SecurityManager** class to make security policy decisions.
- ***Object Finalization:*** Object finalization is not supported. An object's *finalize* method will never be called.
- ***Object Cloning:*** Objects cannot be cloned in a Java Card application. Java Card's **Object** class contains no clone method, and the Java Card API contains no **Cloneable** interface.
- ***Threads:*** The Java Card specification does not support multiple threads of control. The **Thread** class and any thread-related keywords (such as *synchronized*) cannot be used in a Java Card application.
- ***Native Methods:*** Native methods are not supported in the Java Card API, as both the *native* keyword and the **java.lang.System** class are both missing from the standard.
- ***Core API Classes:*** Most of the standard system classes in *java.lang* are not present, with the exception of **Object**, **Throwable**, and various exception classes which exist in limited form with most of their "standard" methods missing.

Optionally Supported Constructs

It is not mandatory for a Java Card VM implementation to support *integer types*. Although the *int* keyword is part of the specification, the decision to implement it is up to the vendor. If *int* is unsupported, the VM must reject the application using it.

Other Differences

There are other restrictions placed on the number of classes that can be in a package, the number of interfaces implemented by a class, the number of methods in a class, and so forth. While this may limit some more complicated applications, the number that is placed as the ceiling is high, so it shouldn't affect simple applications such as the ones we'll be discussing in this chapter. If you are planning on writing large Java Card applications, you may want to think carefully how deep you nest your classes, packages, and methods.

The bytecodes interpreted by the VM itself are also reduced from their J2SE counterparts. Some bytecodes from standard Java are supported, some others are supported in a limited or restricted fashion, and yet others are unsupported altogether. A thorough discussion of which bytecodes fall into which category is beyond the scope of this chapter. A complete listing is contained in the Java Card VM specification whitepaper, available as part of Sun's Java Card SDK (see *http://java.sun.com/products/javacard*).

With all these restrictions, what kinds of applications can we write? Indeed, if support for something as simple as integers is optional, what types of applications can be produced? We'll show an example in a bit. First, we need to lay some groundwork to cover how cardlets work and how we interact with them from our host applications and card readers.

Cardlet Life Cycle

Before we can look at how to write our Java Card applications, we need to understand the application life cycle imposed on our applications by the Java Card specification.

A Java Card obtains its electrical power from the card reader into which it is inserted. When the card receives power, the CPU on the card continues executing from where it left off the last time power was removed. When power is removed, the card goes into "suspended animation," in effect it is stopped at the place where it was executing when the card was removed from the reader. The card's VM is activated when the card is first initialized (in other words, when the card is first programmed and personalized for the user), and effectively runs forever thanks to the stop/restart behavior described above.

The way in which your applications actually get downloaded into the card is implementation-specific, but a few general rules hold:

1. You compile your *.java* files into *.class* files using a Java compiler.
2. You convert the *.class* files into a CAP file (more on this later).
3. You download the CAP file onto the card as a program file (a type of elementary file in the card's EEPROM).

The Converted APplet (CAP) file is a special type of binary file which is built for the VM on the card. It contains all the bytecodes from the original *.class* file, but may also include any other information needed by the VM to successfully execute the code in the card's environment. Your SDK vendor or provider will usually provide a tool to convert the *.class* files into CAP files suitable for their particular environment.

Once the CAP file is resident on the card, the VM must be instructed to install the application that resides in the CAP file. You can think of the CAP file as the equivalent of a regular *.class* file from standard Java. By itself, it does not constitute an application until an instance of the class contained in it is instantiated and brought to life. This is the same way standard Java does things—you have *.class* files sitting on your disk that are "blueprints" of objects that can be instantiated from them. A CAP file works the same way in the Java Card environment.

So how does this instantiation occur? In Java Card terminology, instantiating a applet (cardlet) is known as *installing*. It works like this:

1. From some external application (typically included with the SDK software provided by your card reader's vendor), you instruct the VM running on the card to install a new instance of the cardlet defined in the CAP file you previously downloaded.
2. The VM, upon receiving this instruction, calls the static *install* method of your cardlet, defined in your CAP file. The *install* method accepts a parameter called 'AID'—the Applet ID. The AID is a sequence of bytes, and is used to distinguish one cardlet instance from another.
3. Your static *install* method creates a new instance of your cardlet by using the Java *new* operator.
4. The new instance calls its *register* method to inform the VM that it is ready to process incoming commands.

These steps allow for multiple instances of a given cardlet to be running at the same time, since the procedure allows one to install any number of copies of the same cardlet by following the same steps. Since (as we stated earlier), the VM is always running, this means that our cardlets are effectively always running as well.

We've been mentioning the notion of "sending commands" or "sending instructions" to the card, but have not expanded on this idea. In the next section, we'll look at APDUs, *Application Protocol Data Units*, which we briefly discussed in Chapter 7. APDUs are the commands we'll be using to send instructions to our cardlets.

Details

Java Card Security Model

The Java Card VM security model is not unlike the Web browser applet security model. We've mentioned that multiple instances of multiple cardlets exist on the card, selectable one at a time. These multiple cardlet instances have no idea that any others exist, and have no way of communicating between one another. ❑

APDUs

APDUs are messages sent to your cardlet from some source outside the cardlet. These messages can come from an operator console (which could include tools that were included with your card reader) or host applications. An APDU is a message that can convey a number of meanings, depending on your cardlet. For example, an APDU can embody the message "Increase the balance on this wallet card by $20", or "List the courses the student holding this card is enrolled in". Since the types of messages conveyed by APDUs can be literally anything, their structure is fairly loose and left open to interpretation by each individual developer and application.

APDUs follow a request/response model. Some bytes are sent to your cardlet as a request. Some bytes are generated by your cardlet as a response and sent back to the caller.

An APDU, if you recall from Chapter 7, consists of a set of hexadecimal bytes. The APDU format is defined by the ISO 7816-4 standard, which means that APDUs are going to be handled the same on every card reader supporting the standard (nearly all do).

The APDU format includes the elements shown in Table 12-1. Let's look at what each component means in more detail.

- **CLA**: The CLA, or CLASS byte, indicates the class of application that this message is intended for. Typically, a cardlet supports one value of CLASS byte. Your cardlet should check to make sure that, for each APDU received, that the CLA byte matches what is expected. If not, your cardlet should throw an exception indicating that a mismatched APDU is detected. (We'll see how this is done in the sample cardlet in a later section.)

- **INS**: The INS byte specifies what instruction is being sent to the card. The INS byte is cardlet-specific. In other words, the same INS value sent to two different types of cardlets may result in different actions being taken. (See the next briefing box.)

- **P1, P2**: The P1 and P2 bytes specify any input parameters related to the command specified in the INS instruction. The value of these parameters obviously depends on the instruction being sent.

- **P3**: The P3 byte specifies the length of any optional data being passed into the card as extra parameters.

Table 12-1. APDU Structure.

Message Component	Description
CLA	Command Class.
INS	Command instruction identifier.
P1, P2	Input parameters related to the command.
P3	Length parameter that specifies the number of data bytes to transfer to or from the card as part of the command.

FYI Briefing

Cardlet Instruction Mapping

Each value of the INS byte being sent to your cardlet logically corresponds to some operation that is specific to your cardlet. Let's say we were developing a cardlet that managed student information at a university. We might store the student's name, student ID, and enrolled courses inside our cardlet. Thinking of a situation like this, we logically have six operations that our cardlet should support:

- *getName/setName*
- *getID/setID*
- *getCourses/setCourses*

Since there are six operations, we would probably have six methods defined in our cardlet to perform these operations. We need to associate INS byte values with each of these operations. *The mapping of INS byte values to cardlet operations is entirely up to the developer.* We could use a mapping such as:

- getName = 0x10
- setName = 0x20
- getID = 0x30
- setID = 0x40
- getCourses = 0x50
- setCourses = 0x60

The mapping is completely up to us. Further, the mapping rules are completely cardlet-specific. We could have a different cardlet that accepted the 0x10 instruction but performed a completely different application.

We'll see how to decode APDUs and INS bytes and route the call to the appropriate method in our sample cardlet, coming in a later section. ❑

APDUs have to model every type of card interaction, from status messages, changes in the host application, method invocations, and the like.

APDU Dispatching and Applet Selection

How does the VM dispatch APDUs to cardlets? How does the VM even know what cardlet should receive a particular APDU in the first place?

In order to answer these questions, we need to introduce the concept of *Applet Selection*. Although more than one cardlet instance can be resident on the card, only one can be selected at a time, which means that only one cardlet can receive commands at a time. (See the Briefing box entitled "What's New In Java Card 2.2?" for a peek at how Java Card 2.2 addresses this limitation.)

So, how does an applet become selected, and therefore capable of receiving APDUs from the VM? The applet selection is generally performed from a tool that ships with your Java Card SDK, but can actually be performed programmatically as well. There is a special SELECT FILE APDU which, when sent to the card, selects the applet whose AID (or file ID) is supplied as a parameter to the APDU. Since the SELECT

FILE APDU can be constructed and dispatched from inside an applet (or host application), the currently selected applet can be changed as needed during the lifetime of a Java Card application.

The VM dispatches incoming APDUs to the currently selected applet, which, as we mentioned earlier, should inspect the APDU to make sure that it is the right recipient of the APDU being issued (by checking the CLA and INS bytes).

We now have enough background to start looking at the Java Card API, and the classes and interfaces it provides. Once we understand how the Java Card API is organized, we'll write a simple cardlet and host application to illustrate the whole enchilada in operation.

The Java Card API

As mentioned earlier, the Java Card class library is severely truncated from its J2SE counterpart due to the ROM size limitation present on most cards. In this section, we'll take a look at the most widely-used classes in the API and conclude with a look from a programming standpoint at how a typical cardlet is organized. The next section will build upon what we've learned here and walk through the construction of a simple cardlet to handle Java Card-based authentication for our Campus Portal.

The Java Card API is organized into several packages, akin to J2SE:

- *java.lang:* Contains only **Object**, **Throwable**, and various exception classes. Everything else from J2SE's **java.lang** package is deleted.
- *javacard.framework:* Contains classes to model AIDs, APDUs, Applets, and various card-related system exceptions.
- *javacardx.framework:* Contains classes to handle lower-level file I/O directly on the card.
- *javacardx.crypto:* Contains classes to perform on-the-card encryption and cryptography.

The Applet Class

The **javacard.framework.Applet** class is the base class from which all cardlets are derived. The **Applet** class contains the following methods that are typically overridden by your cardlets:

- *install*: The *install* method is called when the VM detects a request from an operator (or another applet) to install a new instance of a particular applet class. The *install* method is passed an APDU that contains information about the new instance, such as its AID and other information.

- *register*: After creating a new instance of your applet class, the *install* method calls the *register* method which taps into the VM to signal that the instance is ready to receive commands. (The *register* method is not usually overridden by your applets, but is included in this list since it is an important method.)

- *process*: The VM will call your applet's *process* method when an APDU is to be delivered. The *process* method is passed the APDU that embodies the message being sent.

- *select*: The *select* method is called by the VM when your applet is selected by the card reader or another application. You can treat the *select* method like an "awaken" method, since its invocation signals that the applet will now be able to receive instructions again.

- *deselect:* The *deselect* method is called by the VM when your applet is de-selected (in other words, another applet is selected to be the active applet). Just before your applet is de-selected, the *deselect* method is called, allowing you to take care of any cleanup that needs to be performed. Your applet will still be "alive", but won't receive any messages.

The **Applet** class has a protected constructor—that means you can't create a new instance of an applet from outside the class itself. This restriction is in place to enforce the use of the *install* method to create new applet instances.

The APDU Class

The **APDU** class encapsulates the information contained in an APDU message being sent to your applet. Since the **APDU** class models a real APDU, it contains a buffer that records the bytes that comprise the primary fields of the APDU (CLA, INS, P1, P2, and P3). The **APDU** class also contains an extended buffer that is used to record the "extra data" passed to or from a card during an instruction dispatch.

The **APDU** class provides the following methods:

- *getBuffer*: The *getBuffer* method is used to access the bytes contained in the APDU. It returns an array of bytes that contains the major components of the APDU (CLA, INS, P1, P2, and P3) and any extra bytes that are passed along as extra data.

- *receiveBytes*: The *receiveBytes* method is used to receive more bytes that may have not been present in the original buffer. The *receiveBytes* method requires the programmer to manage the buffer manually.

- *sendBytes*: The *sendBytes* method causes the reply bytes that are currently in the APDU buffer to be returned to the VM as the reply to the instruction. Your applet populates the buffer with the reply to the instruction and uses the *sendBytes* method to cause the reply to be returned to the originator.

- *sendBytesLong:* The *sendBytesLong* method allows the program to supply more bytes into the APDU buffer when responding to an incoming message.

The *sendBytes*, *sendBytesLong*, and *receiveBytes* methods are low-level routines that are somewhat cumbersome to use. In order to address this concern, the Java Card specification introduces two “convenience” methods that take care of some of the work for us:

- *setOutgoingAndSend*: The *setOutgoingAndSend* method is the easiest way to send a short reply from your applet. You set the reply bytes inside the APDU buffer, and then call *setOutgoingAndSend*, passing the offset of the start of your reply and the number of bytes.

- *setIncomingAndReceive:* This is the most common receive method. It configures the APDU object for receive mode, retrieves the input bytes from the APDU's originator, stores the bytes in the APDU buffer, and returns the number of bytes successfully received.

These convenience methods are used when your input data (in the case of receive), or reply data (in the case of send) is sufficient to fit entirely within the APDU buffer. If not, you need to use one of the lower level methods described earlier.

You use the **APDU** class to decode an incoming message and to create a reply to the message. The APDU object can be in either “incoming” mode or “outgoing” mode, indicating the direction of data flow. When receiving data from an external source (a host application, the card reader, or other source), the APDU object is placed in incoming mode. Your applet uses the *setIncomingAndReceive* method to set the APDU object's mode to incoming, and to receive the extra input bytes. Similarly, when replying to the message, the reply bytes are set inside the APDU buffer and the mode is changed to outgoing. These actions result in the VM returning the result bytes to the original sender of the APDU.

An **APDUException** class is provided to model APDU-related exceptions. The **APDUException** class contains a field which indicates the type of error generated:

- ILLEGAL_USE: This type of APDUException is thrown when trying to read data when the APDU is in outgoing mode, or vice-versa

- BUFFER_BOUNDS: This type of APDUException is thrown when the size of the APDU buffer is insufficient to accommodate all bytes requested during a send or receive operation.

- BAD_LENGTH: This type of APDUException is thrown when the supplied length differs from the actual data length during a send or receive.

- IO_ERROR: This type of APDUException is thrown when the runtime detects an I/O error of some type (miscommunication with the card, host application, or the like).

File Management Classes

The Java Card API provides several classes to work with files stored on the card. Most cards support the ISO 7816-4 file system format, which is the format supported by the classes supplied in the Java Card API.

Working with files requires intimate knowledge of the card layout of each card you're working with. Since multiple card styles exist, we are going to only cover the high-level classes. If your cardlets need to read and write on-card files, consult the documentation that came with your card reader SDK. We talked about the different types of files in Chapter 7. You should review the file types if you need a refresher.

The file management classes are part of the *javacardx.framework* package and include:

- **File:** The base class for all other file system and file classes.

- **FileSystem:** The **FileSystem** class represents the file system structure on the card and represents the containment hierarchy of the files contained "inside" it.

- **CyclicFile:** Represents a file whose structure resembles a ring of fixed-length records.

- **DedicatedFile:** Represents a simple file container that can hold other dedicated files or elementary files.

- **ElementaryFile:** Represents an elementary file. This class is the base class for all linear files and transparent files.

- **LinearFixedFile:** Represents a file with fixed length records, organized in a linear fashion (as opposed to the cyclic file described above).
- **LinearVariableFile:** Represents a file with variable length records, organized in a linear fashion.
- **TransparentFile:** Represents a linear file with no fixed record structure at all.

Since certain cards may or may not support these types of files, you are again urged to consult your card reader's SDK for more information.

Other Classes

Aside from the classes mentioned above, there are other classes you may find useful during your Java Card development. A detailed description of all the classes in the Java Card API is beyond the scope of this chapter—if you intend to create more sophisticated cardlets than those described in this chapter, you may want to review the Java Card API specification in more detail.

Some of the other interesting classes include:

- **javacard.framework.AID:** Models the AID of a particular applet instance. Your applet can query the system to find its own AID; the system returns the AID in an AID object. The class contains methods to query the AID for the bytes it comprises.
- **javacard.framework.System:** Contains methods that manage system-related constructs like transactions. Generally, this class is not needed, but if your applets are going to support transactions, you may need to use the transaction management functions in the System class. We won't talk any more about Java Card transactions. If you are interested in this topic, you should review the full specification.

FYI Briefing

What's New In Java Card 2.2?

Until now, we've been discussing the API classes provided in the Java Card 2.0 (and to some extent 2.1) specification level. As of June 2002, the Java Card 2.2 API has been made available, offering several new improvements and enhancements over the older versions.

The biggest news is that Java Card 2.2 now supports a primitive form of RMI. This is excellent news for developers, as it means we don't have to encode our operations as APDU requests—we can simply code our interactions between our host applications and applets as method calls. Although the form of RMI supported by the new specification is limited (it does not support on-card registries or TCP/IP), it does simplify the task of encoding and decoding APDUs.

Typical J2SE-style RMI is encoded over TCP/IP. Method invocations are encoded into JRMP packets which are sent to the target VM for dispatching. Since cards only support APDU communication, Java Card RMI is based on an APDU transport. Method calls are translated into specially-formatted APDUs instead of JRMP. This means that Java Card RMI applications can't participate in regular RMI applications without some help. The RMI exchange that takes place occurs only between the host application and the cardlet.

Java Card 2.2 also includes a more robust memory management system. One of the hallmarks of standard Java is the fact that we don't have to manage deletion of objects inside our programs since the garbage collector handles it for us. This has a drawback (or benefit, depending on how you look at it): since we no longer manage our own memory, we no longer have anyway of accessing particular bytes in memory.

While these features make sense in a standard Java environment, a card's restricted memory environment makes it desirable to have a little more fine-grained control over when our objects are deleted. The new 2.2 standard specifies several classes that our cardlets can use to dictate to the VM exactly when objects should be deleted from memory. The onus is on us to make sure that we're managing memory correctly—if the new technique is used, the specification says there are no guarantees that the cardlet will remain in a stable condition if objects are surreptitiously deleted while still in use. ❑

YOUR FIRST CARDLET

We're now ready to create our own cardlet. For this example, we've chosen a simple cardlet that manages a PIN code stored on the card. The cardlet supports two operations:

- GETPIN: To retrieve the current PIN value.
- SETPIN: To set a new PIN value.

Listing 12-1 shows the source code for our cardlet. In our cardlet, we define some constants for the INS and CLA bytes we recognize. We decided to use 0x10 for the

INS value for the SETPIN operation and 0x20 for the INS value for the GETPIN operation. We could have used other values if desired.

Our cardlet maintains a byte array to store the PIN. This byte array is created inside the constructor. Our cardlet supplies an *install*, *process*, and *select* method that override the default methods in the **Applet** class. Our cardlet's *install* method simply creates a new instance of the cardlet class, which results in the constructor being called in the new instance. The constructor creates the byte array to store the PIN, populates the PIN array with the default PIN 'ABCD', and then registers itself by calling the VM's *register* method.

The interesting method in our cardlet is the process method. Our process method performs the following tasks:

1. It retrieves the APDU buffer by calling *getBuffer* on the supplied APDU object.
2. It checks the instruction in the APDU to see if it is a SELECT instruction, indicating that the VM has selected this instance to be the active cardlet. If so, it returns control to the VM by throwing the "exception" called SW_NO_ERROR. The VM interprets this return as success.
3. It checks the CLA byte inside the APDU to ensure that it matches the expected value for our applet.
4. It extracts the INS byte and calls either *GetPIN* or *SetPIN*, depending on the value of the INS byte. If an invalid INS byte was supplied, the SW_INS_NOT_SUPPORTED exception is thrown to indicate that an INS value was supplied that the cardlet did not understand.

The *GetPIN* method copies the contents of the local PIN buffer into the APDU buffer, sets the APDU mode to outgoing, and calls *sendBytes* to cause the bytes in the buffer to be returned to the instruction originator.

Similarly, the *SetPIN* method copies the contents of the incoming APDU buffer into the local PIN buffer. It returns a copy of the new PIN to the originator.

Listing 12-1. The StorePIN Cardlet.

```
import javacard.framework.*;
import javacardx.framework.*;

public class StorePIN extends javacard.framework.Applet {

    final static byte StorePIN_CLA = (byte)0x02;

    //  APDU instructions recognized by our applet
```

```
final static byte SETPIN = (byte)0x10;
final static byte GETPIN = (byte)0x20;
final static byte SELECT = (byte)0xA4;
final static byte PIN_LENGTH = (byte)4;

byte buffer[];
byte MyPIN[];

private StorePIN() {

MyPIN = new byte[4];
MyPIN[0]='A';
MyPIN[1]='B';
MyPIN[2]='C';
MyPIN[3]='D';

//  Register the instance with the VM
register();
}

public static void install(APDU apdu) {

new StorePIN();
}

public boolean select() {
return true;
}

public void process(APDU apdu) throws ISOException{

   buffer = apdu.getBuffer();

   // we might receive a SELECT message. If so,
   // just return.
   if ((buffer[ISO.OFFSET_CLA] == ISO.CLA_ISO) &&
       (buffer[ISO.OFFSET_INS] == ISO.INS_SELECT)){
       ISOException.throwIt(ISO.SW_NO_ERROR);
   }

   // Check the APDU to ensure it's a class
   // of message intended for us.
   if (buffer[ISO.OFFSET_CLA] != StorePIN_CLA)
       ISOException.throwIt(ISO.SW_CLA_NOT_SUPPORTED);
```

```
      // Check the instruction inside the APDU to
      // see what to do.
      switch (buffer[ISO.OFFSET_INS])
          {
          case SETPIN:
             SetPIN(apdu);
             break;
          case GETPIN:
             GetPIN(apdu);
             break;
          case ISO.INS_SELECT:
             ISOException.throwIt(ISO.SW_NO_ERROR);
             break;
          default:
             ISOException.throwIt(ISO.SW_INS_NOT_SUPPORTED);
          }

   }

   private void SetPIN(APDU apdu) {

      byte index, size;

      buffer = apdu.getBuffer();
      size = (byte)(apdu.setIncomingAndReceive());

      if(size>PIN_LENGTH)
          size=PIN_LENGTH;

      for (index = 0; index < size; index++)
          MyPIN[index] = buffer[ISO.OFFSET_CDATA + index];

      for(index = 0; index< PIN_LENGTH; index++)
          buffer[index]=MyPIN[index];

      apdu.setOutgoing();
      apdu.setOutgoingLength(PIN_LENGTH);
      apdu.sendBytes((short)0, (short)PIN_LENGTH);
   }
private void GetPIN(APDU apdu) {

      byte index;

      buffer = apdu.getBuffer();
```

```
        apdu.setOutgoing();
        apdu.setOutgoingLength(PIN_LENGTH);

        for (index = 0; index < PIN_LENGTH; index++)
            buffer[index] = MyPIN[index];

        apdu.sendBytes( (short)0 ,(short)PIN_LENGTH);
    }
}
```

Preparing Your Cards and Activating Cardlets

Now that we've reviewed our sample cardlet, we're going to walk through how to compile it and install it on a real Java Card. For this example, we will be using the Cyberflex Access SDK from Schlumberger. This SDK provides the class libraries needed to compile our cardlet, and also contains the local libraries we'll need later when we build our host application. If you will be using a different SDK and card reader, the steps needed to build and upload your cardlets may differ—consult your SDK documentation for more information.

Our Schlumberger SDK was supplied with a Reflex USB card reader, so we'll be using these pieces when building our cardlet. After installing the Cyberflex Access SDK, we compile our cardlet using the standard javac compiler:

```
set CLASSPATH=path to Schlumberger class libraries
javac StorePIN.java
```

After these two steps, we have a StorePIN.class file that contains our cardlet's byte codes. The next step is to convert this *.class* file into a CAP file and upload it to our card. The Cyberflex Access SDK ships with a tool called the Cyberflex Access Toolkit that makes it easy to build the CAP file and upload it to the card in just a few steps:

1. Plug in the card reader and insert a blank card.
2. Start the Cyberflex Access Toolkit.
3. From the Tools menu, select 'Make Cardlet' (see Figure 12-1).
4. Select the *.class* file just created from the file dialog.

At this point, we have a CAP file on the disk (Schlumberger calls its CAP files *.bin* files), but no instances of our cardlet have been created yet. The Cyberflex Access Toolkit provides another easy-to-use procedure to create an instance of our applet:

1. From the Tools menu, select 'Cardlet Manager'.

2. From the File menu, select 'New Program'. The 'Create Program' dialog will appear (Figure 12-2).
3. Enter the path to the *.bin* file we just created in the 'Program Name' text field. (You can use the '…' button next to the text field to bring up a file browser.) By default, the *.bin* file is placed in the same directory as the original *.class* file.
4. Click the '…' button next to the 'Key' text field and select the 'Default Signed Applet Key'. This is the key used to sign our CAP file when uploading to the card.
5. Enter '1008' for the file ID. You can use any value, but be careful you don't use a file ID that is already used by another file on the card.
6. We need to verify the signed applet key before we can sign the applet. Select 'Verify Keys' to bring up the 'Verify Key' dialog. Click 'Select Key' and choose the 'Default AUTO'. These actions are shown in Figure 12-3 and Figure 12-4.
7. From the 'Cardlet Manager' dialog, right click on the new '1008' file in the window (or whatever file ID you chose). Select New->Instance.
8. In the 'Create Instance' window, enter '090504030201' for the AID (we'll explain the significance of this ID in a minute). Enter '1009' for the Instance Directory ID. Enter '972' for 'Instance Directory Size' and '500' for 'Data Container Size' (shown in Figure 12-5). Click the 'Create' button and your instance will be created (972 bytes for our program and data, of which 500 bytes is dedicated for data).

At this point, we now have a running instance of our applet with AID 0x090504030201. What's the significance of this number? Nothing, really, except that's the AID that our host program is expecting.

More on Application (AIDs)

Details

We're using an invented '0x090504030201' hex number for our AID. This number is invented—we could have used anything. However, as stated, our host application is expecting the same ID to represent a running applet on our card.

Locating AIDs is SDK-dependent—there may or may not be APIs available for your host applications to enumerate which AIDs are available. Because of this, our AIDs are hardcoded into the host application.❑

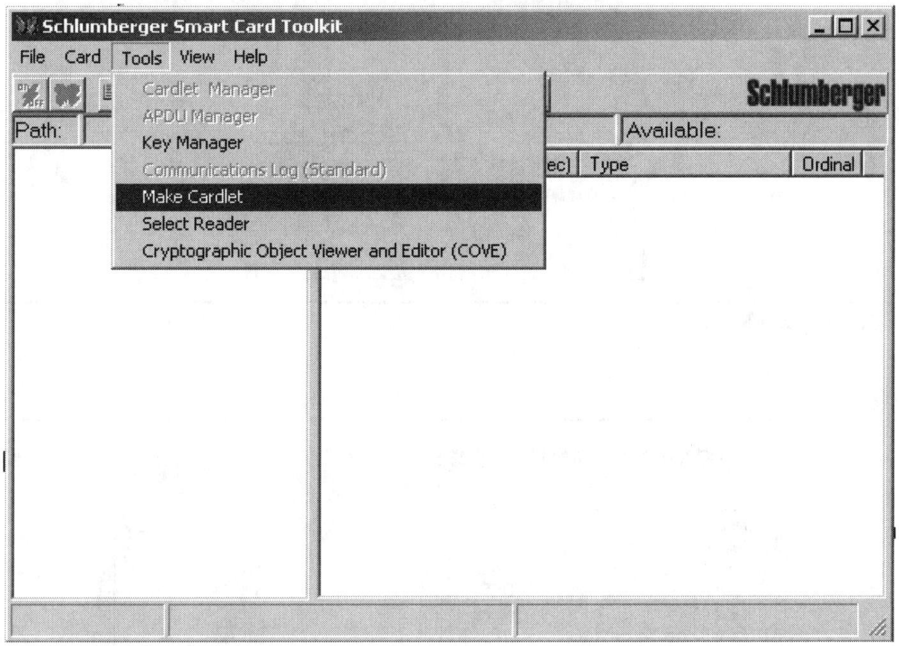

Figure 12-1. Using Cyberflex Access Toolkit to Make a Cardlet

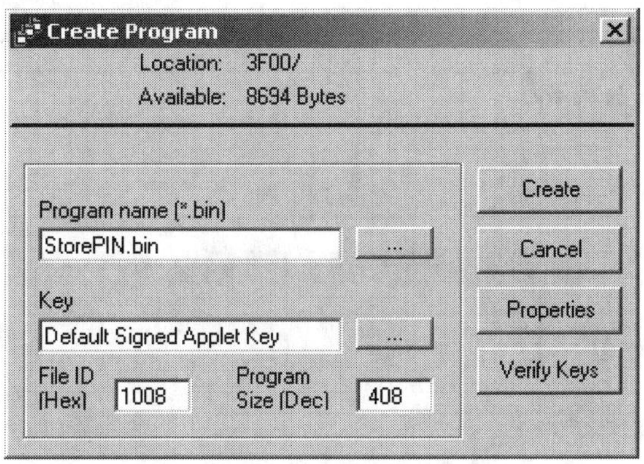

Figure 12-2. Uploading a CAP (.bin) File.

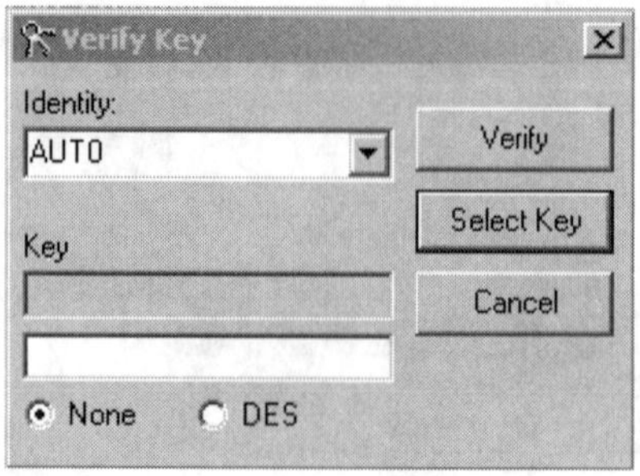

Figure 12-3. Verify Key Dialog.

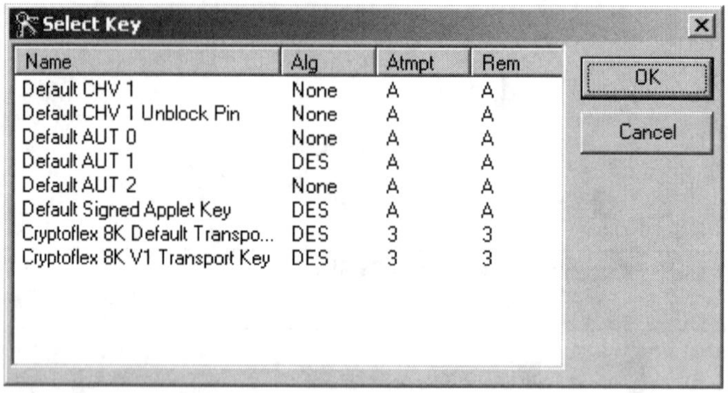

Figure 12-4. Selecting a Key to Verify.

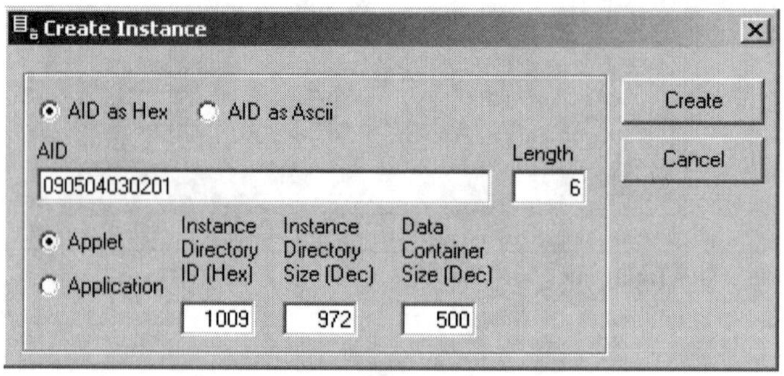

Figure 12-5. Creating an Instance.

We now have an instance of our cardlet running on the card. At this point, we're ready to look at how to communicate with our cardlet by creating a host application.

Details

The APDU Manager

Schlumberger's Cyberflex Access Toolkit also provides an APDU manager that you can use to manually construct APDUs and send them to your card. This could be useful for testing purposes to make sure your cardlet is working properly before you start creating your host application. For more information on how to use the APDU manager, consult the Schlumberger documentation. ❑

THE HOST APPLICATION

Our host application (whose listing is shown in Listing 12-2) will be Swing-based and provide a simple GUI that interacts with our cardlet. It provides a text field into which we can enter a PIN, and has two options to set the PIN to the value entered in the text field and to validate the PIN entered against the value stored in the card. We're going to be using the Schlumberger libraries in our host application, since we need a way to access the card reader from our application. If you are using a different SDK, you may need to adjust the code accordingly.

Host applications can be standalone Java applications (like ours), J2EE components, or even Web browser-based applets. In all cases, the card reader vendor's proprietary libraries and classes need to be installed on the machine that has the card reader connected to it.

Warning

Samples Not Secure!

The sample cardlet and host application described in this chapter are not designed to be secure. You would typically not provide ways to retrieve the PIN from the card by a simple APDU call. A more secure method would be to provide a *ValidatePIN* method that checked the supplied PIN against the recorded PIN, but did not divulge the recorded PIN to non-trusted sources.

Our examples are intended to show how basic cardlets and host applications are built, not to illustrate exhaustive Java Card security (which would also take advantage of the cryptographic functions provided by the card).

In general, your cardlets should try to hide as much personal or sensitive information as possible. If your system needs to verify things like PINs or passwords, it's a good idea to have the card verify these, rather than permitting the host application to read them and perform the checking itself. This "don't trust the host application" idea is a good one, as it prohibits a malicious host application from reading data from your cardlets without need or permission.❑

We won't go into the GUI part of our host application (shown in Figure 12-6)—the layout and GUI-generating code should be self explanatory. When either the "Change PIN" or "Check PIN" button is clicked, control transfers to the *actionPerformed* method (since the host application object is registered as an action listener for each button).

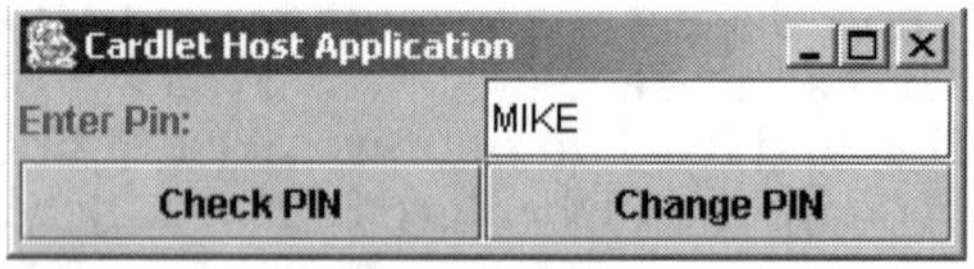

Figure 12-6. The Host Application GUI.

Our *actionPerformed* method provides most of the functions of our host application. The code uses the Schlumberger classes (**slb.iop.SmartCard** and **slb.iop.IOP**) to locate the reader that is plugged into the machine. Schlumberger uses the **slb.iop.SmartCard** class to model the card that is inserted into a particular reader. Once we find the **SmartCard** object that corresponds to the card we're using (we take a naïve approach and just assume that the card is inserted into the first reader that is found), we call its *SelectCardlet* method to select our cardlet based on its AID.

At this point, the action taken depends on what button was clicked. If "Check PIN" was clicked, we use the **SmartCard** object's *SendCardAPDU* method to send an APDU that represents the GETPIN method. The result value is the byte array that is constructed inside our cardlet. If you recall, our cardlet's GETPIN operation constructs a byte array that contains the bytes of the PIN stored inside the cardlet. After we receive the byte array, we convert it to a string, and compare this string to the string entered in the text field of our host application. Depending on the outcome of the comparison, an appropriate message is printed to the screen.

If "Change PIN" was clicked, we again use the **SmartCard** object's *SendCardAPDU* method to send an APDU that represents the SETPIN method. We first read the text entered in the host application's text field and convert this into an array of integers that we pass in as the extra bytes inside our APDU. If you recall, our cardlet's SETPIN method receives the extra bytes and copies them to the local PIN buffer inside the cardlet.

Note that in all cases, our PIN is limited to PIN_LENGTH characters, which is set to 4 bytes in both the cardlet and the host application. Even if we enter more text in the text field, only the first PIN_LENGTH bytes will be sent to the cardlet.

Listing 12-2. The Host Application.

```
import java.awt.*;
import java.awt.event.*;
import java.security.*;
import javax.swing.*;
import slb.iop.*;

public class HostApplication extends JFrame implements ActionListener
{   private JButton CheckPINButton, ChangePINButton;
    private JTextField PINTextField;
    private JLabel EnterPINLabel;
    private Container contentPane;

    final static int iAID[] = {0x09,0x05,0x04,0x03,0x02,0x01};
    final static byte CLA = (byte)0x02;
    final static byte SETPIN = (byte)0x10;
    final static byte GETPIN = (byte)0x20;
    final static byte PIN_LENGTH = (byte)4;
    final static byte P1 = (byte)0;
    final static byte P2 = (byte)0;

    public HostApplication()
    {  tk = Toolkit.getDefaultToolkit();
      Dimension d = tk.getScreenSize();
      setTitle("Cardlet Host Application");
      setSize(300, 80);
      setLocation(d.height/3,d.width/3);
      addWindowListener(
                      new WindowAdapter()
                      { void windowClosing(WindowEvent e)
                          {.exit(0);
                          }
                      }
```

```
                );

    contentPane = getContentPane();
    contentPane.setLayout(new GridLayout(2,2));

    CheckPINButton = new JButton("Check PIN");
    ChangePINButton = new JButton("Change PIN");
    PINTextField = new JTextField(10);
    EnterPINLabel = new JLabel("Enter Pin: ");
    contentPane.add(EnterPINLabel);
    contentPane.add(PINTextField);
    contentPane.add(CheckPINButton);
    contentPane.add(ChangePINButton);
    CheckPINButton.addActionListener(this);
    ChangePINButton.addActionListener(this);

    setVisible(true);
}

public void actionPerformed(ActionEvent evt)
{  Object source = evt.getSource();
    SmartCard sCard;
    IOP sIOP;
    String s;
    byte INS;
    short sArray[]=null;
    int iArray[]=null;
    byte bArray[]=null;
    String strPin=null;
    boolean bRet;
    int numReaders, index;
    String rs[] = null;

    try
        {  sCard = new SmartCard();
           sIOP = new IOP();

           rs = sIOP.ListReaders();

           if (!sIOP.Connect(sCard, rs[0],false))
               {  JOptionPane.showMessageDialog(
                  this,"Unable to connect to card.");
                  return;
               }
```

```
bRet = sCard.SelectCardlet(iAID);
if (!bRet)
    {  s=sCard.GetErrorMessage();
       return;
    }

    if (source == CheckPINButton)
    {
       INS = GETPIN;

       // GETPIN takes no extra bytes
       iArray = new int[0];

       sArray=sCard.SendCardAPDU(
            CLA, INS, P1, P2, iArray, PIN_LENGTH);
       if (sArray != null)
           {  byte sBytes[] = new byte[sArray.length];

              for (int i = 0; i < sArray.length; i++)
                  { sBytes[i] = (byte)sArray[i];
                  }
              strPin = new String(sBytes);

    if(PINTextField.getText().trim().equals(strPin.trim()))
                  { JOptionPane.showMessageDialog(
                      this, "Pin Number Entered is Equal.");
                      return;
                  }
              else
                  { JOptionPane.showMessageDialog(
                      this,"Pin Not Valid. Try Again!");
                      return;
                  }
           }
       else
           {  s=sCard.GetErrorMessage();
           }

    }
else if(source == ChangePINButton)
    { INS = SETPIN;

       bArray = PINTextField.getText().trim().getBytes();
       for(index=0;
            index<bArray.length;
```

```
                    index++, iArray[index]=(int)bArray[index]);
                sArray=sCard.SendCardAPDU(
                    CLA, INS, P1, P2, iArray, PIN_LENGTH);

                JOptionPane.showMessageDialog(this, "Pin Changed.");
            }
        }
    catch(Throwable Exc)
        { JOptionPane.showMessageDialog(
          this, "Exception",
          "Exception Occured", JOptionPane.ERROR_MESSAGE);
        }
    }

    public static void main(String[] args)
    {
        HostApplication hostapp = new HostApplication();
    }
}
```

SUMMARY

We've taken a quick look at the Java Card specification, including the cardlet life-cycle and cardlet development. We looked at the messaging model used by Java Card applications, the exchange of APDUs, and how we can use these messages inside our cardlets. We developed a simple cardlet, tested it, and then developed a host application that runs on our desktop PC and talks to our cardlet through the card reader.

In Part 3, we'll shift gears and talk about the server side of our wireless application development environment.

Part 3
Wireless Device Programming Meets the Enterprise

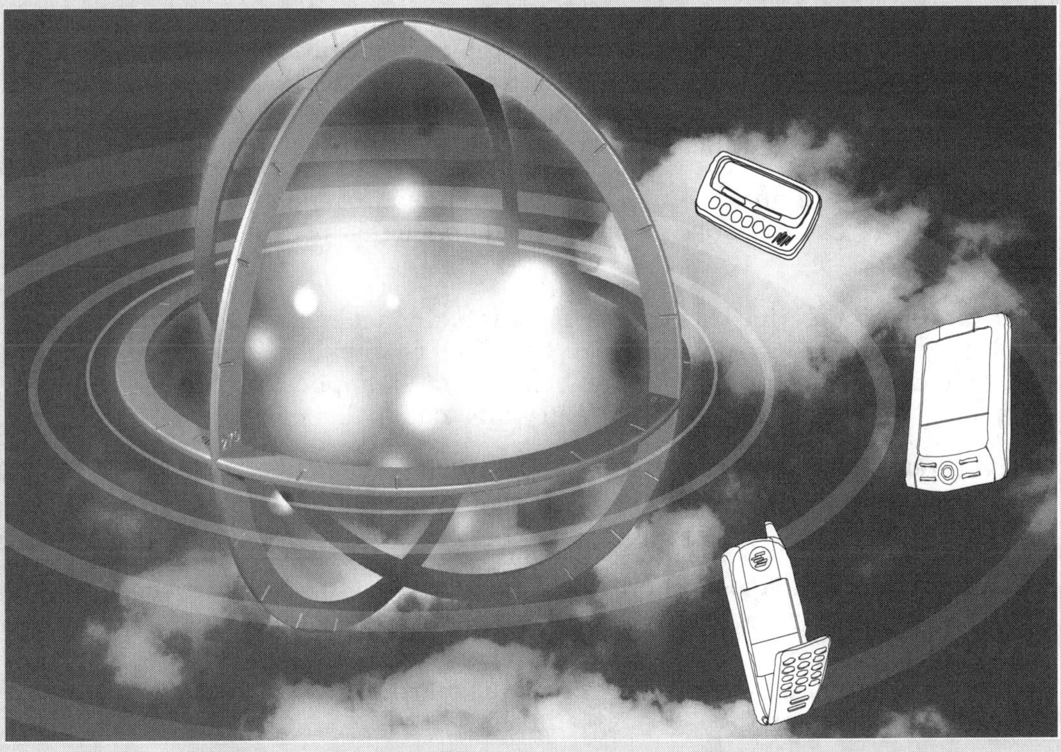

An Introduction to Part 3

With Part 3 we switch gears and explore combining applications that run on a handheld device with enterprise level applications that are accessed via the Internet. Enterprise applications augment handheld devices by off-loading compute intensive function from the device to enterprise servers and by providing collaboration between users. The enterprise infrastructure also provides a way to isolate an enterprise level service from the variety of handheld device clients by providing transcoding and personalization, and to integrate a number of services using the Web Services technologies.

We begin the part by giving an overview of the *Java 2 Enterprise Edition (J2EE)* as a platform for developing a service that can be accessed from a handheld device. We then explore the Web Services technologies and how they can used to expose a J2EE application to the Internet and ultimately be accessed from a handheld device. Next we describe the enterprise technologies that can be used to support numerous handheld device types and for implementing personalization. The final two chapters of Part 3 explore the use of XML as a means of exchanging data between handheld and enterprise applications, and also cover best practices for *Java 2 Micro Edition (J2ME)* programming on a handheld device.

Here's a brief description of the chapters in Part 3:

- ***Chapter 13*** describes the J2EE architecture and technologies. It also uses one of the many development tools, Borland's JBuilder, to provide a step-by-step implementation of a J2EE service using Enterprise JavaBeans, Servlets, JavaServer Pages (JSPs), and JavaBeans. The chapter ends with the implementation of a J2ME MIDlet that acts as a client to the enterprise service. The MIDlet is developed using Borland's MobileSet extension to the JBuilder development environment.
- ***Chapter 14*** describes the Web Services architecture and technologies and how a J2ME MIDlet can invoke an enterprise service that is exposed to the Internet as a Web Service.
- ***Chapter 15*** describes the details of transcoding technologies for transforming display content for a specific device type. The chapter also looks at personalization technologies for tailoring the behavior of an application to a specific user and to lighten the burden of text entry on a handheld device.
- ***Chapter 16*** explores the use of XML for data exchange between devices and enterprise applications. This includes an overview of XML parsers for MIDP.
- ***Chapter 17*** explores the strategies and best practices for writing J2ME applications. This strategies will provide a guide for you to write efficient programs for a handheld device and to take advantage of enterprise services.

With these techniques in hand, you will be well on your way to writing wireless Java enterprise applications.

Chapter 13

J2EE and Wireless Applications

This chapter is the beginning of our journey to understand the server side of our wireless applications. Servers, with their rich set of computing resources, have architectures that can subsume much of the function of a wireless enterprise application. Given the limited capabilities of our handheld devices, it is desirable that the server side of our wireless application be the portion of our "application iceberg" that is under water.

Server-side architectures provide frameworks that allow us to build our applications piecemeal with components. Some components are written and deployed as extensions to Web servers; others are designed to be encapsulations of application business logic. We will continue our adherence to Java as a programming language and architecture for our server-side components and use the Java 2 Enterprise Edition (J2EE) as a platform for developing components for the server side of our wireless applications.

These components also allow us to build an infrastructure for our wireless applications. It is useful to revisit the architecture diagram we presented in Chapter 1 (see Figure 13-1). Note that the portion of the diagram labeled Wireless Portal, as well as the end user applications (the puzzle pieces in the diagram), are built using J2EE components. This gives a modular system that allows the incremental development and maintenance of our server wireless infrastructure as well as our applications.

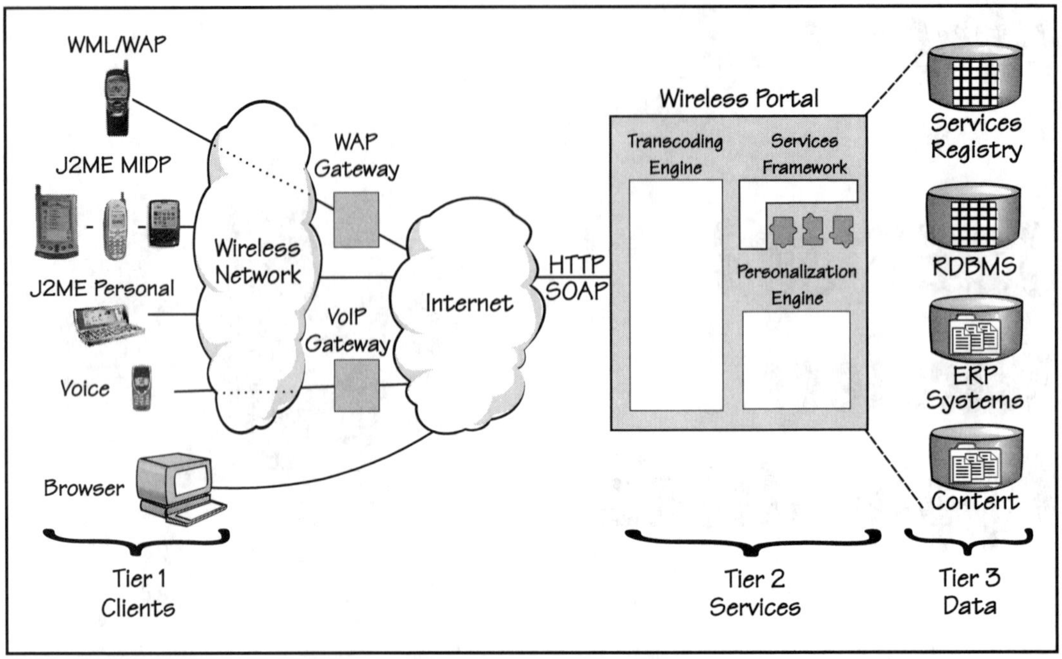

Figure 13-1. An Architecture for Enterprise Application Delivery to Wireless Devices.

J2EE ARCHITECTURE: AN OVERVIEW

This section gives an overview of the J2EE version 1.3 architecture and the components that are used for development. The architecture is very extensive and beyond the scope of this book to describe in detail. Thus, we will focus on an overview of the architecture then reinforce the description with a simple programming example in the next section. The programming example will use the same J2EE components that we will use to develop the wireless infrastructure for the Campus Portal and its applications. However, this discussion will not give a complete view of the J2EE architecture, and we encourage you to read some of the books specific to this topic that are listed in the list of references at the end of this book.

Figure 13-2 gives a high-level overview of the J2EE architecture. The central feature of J2EE is three *containers*:

- An ***Application Client container*** for Java applications.
- A ***Web container*** for components that extend a Web server.
- An ***Enterprise JavaBean (EJB) container*** that provide components for business logic and data abstraction.

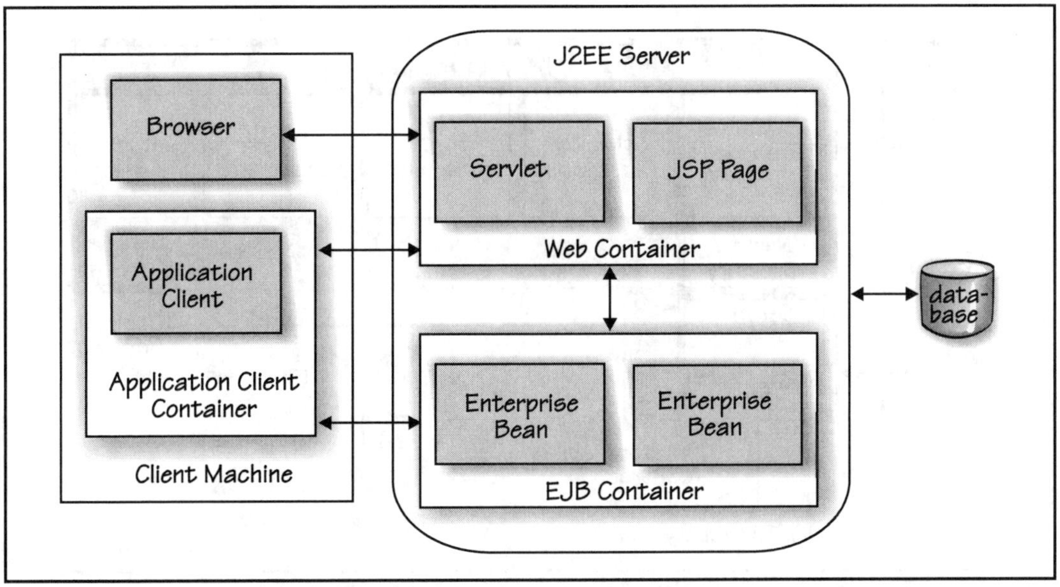

Figure 13-2. An Overview of the J2EE Architecture.

The Web and EJB containers provide the middle-tier in the J2EE 3-tier architecture. The first tier contains clients. In the J2EE architecture this typically includes a browser, J2EE application clients, and standalone J2EE applications. We will extend the list of clients to include connected handheld devices.

The third tier in the J2EE architecture contains persistent data sources. These are typically relational databases but can also include object databases, Enterprise Information Systems (EIS), and legacy database and transaction systems (such as IMS databases and CICS and Tuxedo transaction systems). We will focus on relational databases as our datastore.

Containers

Containers provide a framework for the deployment and execution of components. A container's interaction with components, other containers, and system level function such as component deployment is defined by *services* and *contracts* (see Figure 13-3). The J2EE architecture defines a number of Java technologies that a container uses to provide these services.

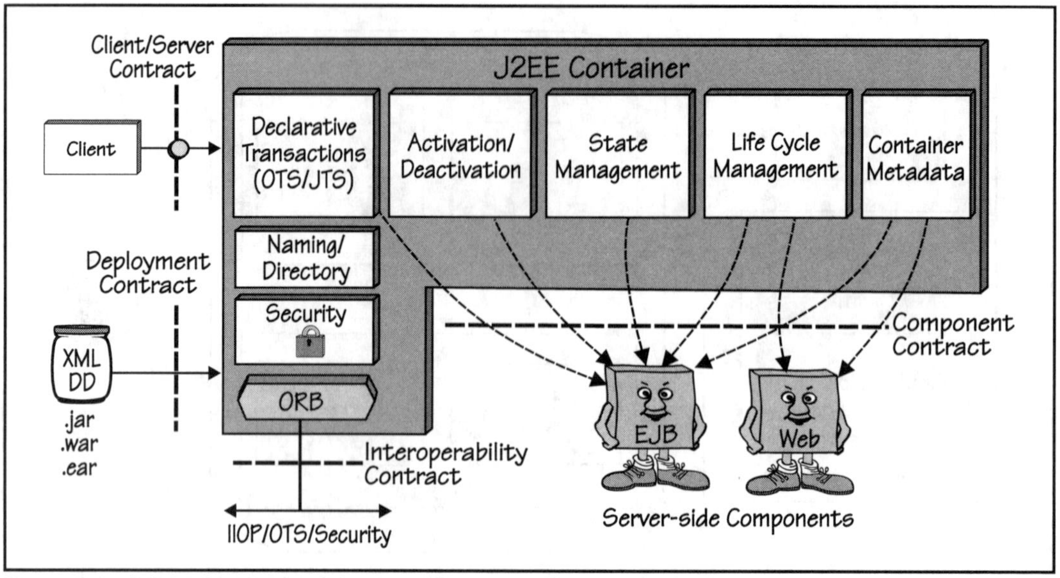

Figure 13-3. A J2EE Container with Services, Contracts, and Java Technologies.

Like an operating system, containers provide a number of services for the components they execute. The services a container provides include:

- ***Life cycle control:*** The container controls the initial creation of a component instance and its removal when no longer needed. The container may also temporarily passivate components during periods of high activity, then activate the component again when accessed by a client.

- ***State management***: If the container controls persistent state, then the container controls when a component's persistent state is loaded and stored to the back end data storage.

- ***Transaction Management:*** The container implements declarative transactions and provides transaction interfaces to clients and components. The attributes for declarative transactions are specified in a component's *deployment descriptor*. A deployment descriptor is an XML document that describes deployment and runtime attributes for a component.

- ***Security:*** The container provides a secure execution environment for a component. This includes authentication, authorization, encryption, non-tampering of messages, auditing, and other security features. The container also provides some information and APIs to a component so it can make business logic decisions based on caller identity and roles.

- ***Naming and Directory Services:*** The container provides lookup services so a client or component can locate other components, resources (like database and queue resources), and environment entries.

- ***Component Metadata:*** The container maintains deployment and runtime metadata about each component type and instance. Components and clients can access this metadata through APIs provided by the container.

Contracts define the interactions between other parts of a J2EE system. Contracts can include programming interfaces (such as those between a client and a container or between a component and a container), on-the-wire protocols, descriptive information (such as deployment descriptors), and packaging for components. The following describes the contracts in a J2EE system:

- The ***Client/Server Contract*** defines the interface between a client and a container. For EJBs, this contract is defined by two Java interfaces, a home interface (that defines factory and finder methods) and a component interface (that defines business methods).

- The ***Component Contract*** defines the interface between the container and the component. This interface includes callback methods that the container uses to manipulate the component during its life cycle and business methods.

- The ***Deployment Contract*** defines the descriptive data, packaging, and procedures for deploying a component. The descriptive data is an XML document called a *Deployment Descriptor (DD)*. A *Document Type Definition (DTD)* for each component defines the required and allowable deployment descriptor information. Each component is packaged into a JAR file and each JAR file has information and a specific extension unique to each component type.

- The ***Interoperability Contract*** defines the protocols and semantics required for interoperability between containers. The definitions include on-the-wire formats (such as IIOP) and transaction and security semantics.

Technologies for Container Services

Containers provide services for components by integrating a number of Java, Internet, and *CORBA (Common Object Request Broker Architecture)* technologies. The following briefly describes each of these technologies:

- ***Java Naming and Directory Services (JNDI)*** provides a standard mechanism for locating components, environment data, and resource connections.

- ***Java Transaction API (JTA)*** and ***Java Transaction Service (JTS)*** provides a standard mechanism for initiating and completing transactions.
- ***Java Database Connectivity (JDBC)*** provides Java access to relational databases.
- ***Java Messaging Service (JMS)*** provides asynchronous messaging using both the point-to-point model and the publish-and-subscribe model.
- ***Java Mail*** enables a component to initiate email notifications.
- ***Java API for XML Processing (JAXP)*** provides for XML parsing using the SAX and DOM APIs.
- ***Java Connector Architecture (JCA)*** allows resource adapters to be added to a container to extend its data access capability beyond relational databases.
- ***Java Authentication and Authorization Service (JAAS)*** enables the authentication and authorization of users of a J2EE system.
- ***Communication protocol*** standards such as HTTP (HyperText Transfer Protocol), IIOP (Internet Inter-ORB Protocol), and TLS (Transport Level Security) are used to provide client access and interoperability.

J2EE Roles

The J2EE architecture also defines roles for companies and individuals. These roles are defined below:

- A ***J2EE Product Provider*** implements the J2EE specification and sells the implementation as a commercial product.
- An ***Application Component Provider*** writes components and clients for use in an enterprise application.
- An ***Application Assembler*** composes enterprise applications using components provided by an application component provider.
- A ***Deployer*** takes a complete enterprise application and tailors its deployment descriptor information to a specific execution environment.

- A ***System Administrator*** monitors the runtime execution of an application and makes changes to the system and application configuration to meet the anticipated load.
- ***A Tool Provider*** provides development, debugging, assembly, deployment, and runtime monitoring tools for each of the roles described above.

Web Components

Web components run in a container that is an extension of a Web server. Web components, like Common Gateway Interface (CGI) scripts, provide a way to generate dynamic content from a Web server. Web components provide an efficient alternative to CGI scripts. The basic Web component type is a *Java Servlet*. The J2EE architecture also defines an alternative to Servlets called *JavaServer Pages (JSPs)*. JSPs are HTML pages with embedded Java *scriptlets* and macros and are compiled into Servlets prior to execution.

Enterprise JavaBeans

The J2EE architecture also defines Enterprise JavaBean (EJB) components to implement the business logic of an enterprise application. There are three types of EJB components:

- ***Session EJBs*** provide a functional extension of a J2EE client. A J2EE client invokes the session bean by calling its methods. A session bean is not shared among clients and does not contain persistent data. When the client or the J2EE server terminates, the state of the session bean is lost.

  Two types of session beans are defined in the J2EE architecture: *stateless* and *stateful*. Stateless session beans do not retain *conversational* state for their clients between method calls. A stateful session bean does maintain conversational state.

- ***Entity EJBs*** provide an abstraction of data in persistent storage. Typically the persistent store is a relational database. Entity EJB instances are shared by their clients. Entity beans have a primary key that uniquely identifies each entity bean instance for a given type. Entity EJBs also define state and relationships to other entity EJBs (similar to foreign keys in a relational database). The J2EE architecture defines a query language called EJB QL for locating entity EJB instances based on primary key, state, and relationship values.

  The J2EE architecture defines two types of Entity beans, based on the management of their persistent state and relationships: *Container Managed Persistence (CMP)* and *Bean Managed Persistence (BMP)*. The state for a CMP entity bean, as the name implies, has its state synchronized with the underlying data store by the containers. BMP entity beans, on the other hand, manage their own state synchronization during the execution of call-back methods from the container.

- ***Message-driven EJBs*** receive asynchronous invocations from a JMS message queue. Messages on the queue could have originated from a client, a Web component, or another EJB. Message-driven EJBs are similar to stateless session EJBs except they receive their invocations from the JMS queue. This call could be from multiple clients.

Component Development, Packaging, and Deployment

Components are packaged into JAR files. A JAR file provides a way to collect the files that compose a component into one place. This includes Java *.class* files, resource files, icons, bitmaps, help files, and deployment descriptors. Each component type has specific deployment descriptor data. The data is represented in XML according to a DTD for the component type. The different types of component packaging files are:

- ***Application client component (.jar)*** for client-side components.
- ***Web component (.war)*** for Web components (JavaServer Pages and Servlets).
- ***EJB (.jar)*** for EJBs.
- ***Enterprise Application (.ear)*** for enterprise applications.

Enterprise applications package the other components into one file for delivery as a complete application.

Figure 13-4 illustrates the J2EE development and deployment process. The developer performs these steps:

1. ***Writes*** the source and deployment descriptor for each of the components that will comprise the application. This includes *.java* files for the client, JSP JavaBeans, Servlets, and EJBs. Web pages may also be written using *.jsp* files for JavaServer Pages. An XML deployment must also be written for each component.
2. ***Compiles*** the Java files using javac. The Java *.class* files are placed into an archive file along with the deployment descriptor XML files. Each component type is placed into a component-specific archive file type.

3. ***Assembles*** the components that will form the application into an Enterprise ARchive (*.ear*) file. This includes writing an enterprise application deployment descriptor that will describe the assembly of components. This step may be performed from a library of components by someone other than a developer.

4. ***Deploys*** the enterprise archive to an application server for testing and debugging. The deployment step allows deployment descriptor parameters to be updated, stub and skeleton files for the underlying distributed objects to be generated, and application verification. Once the application reaches a production environment, deployment may be performed by someone other than a developer.

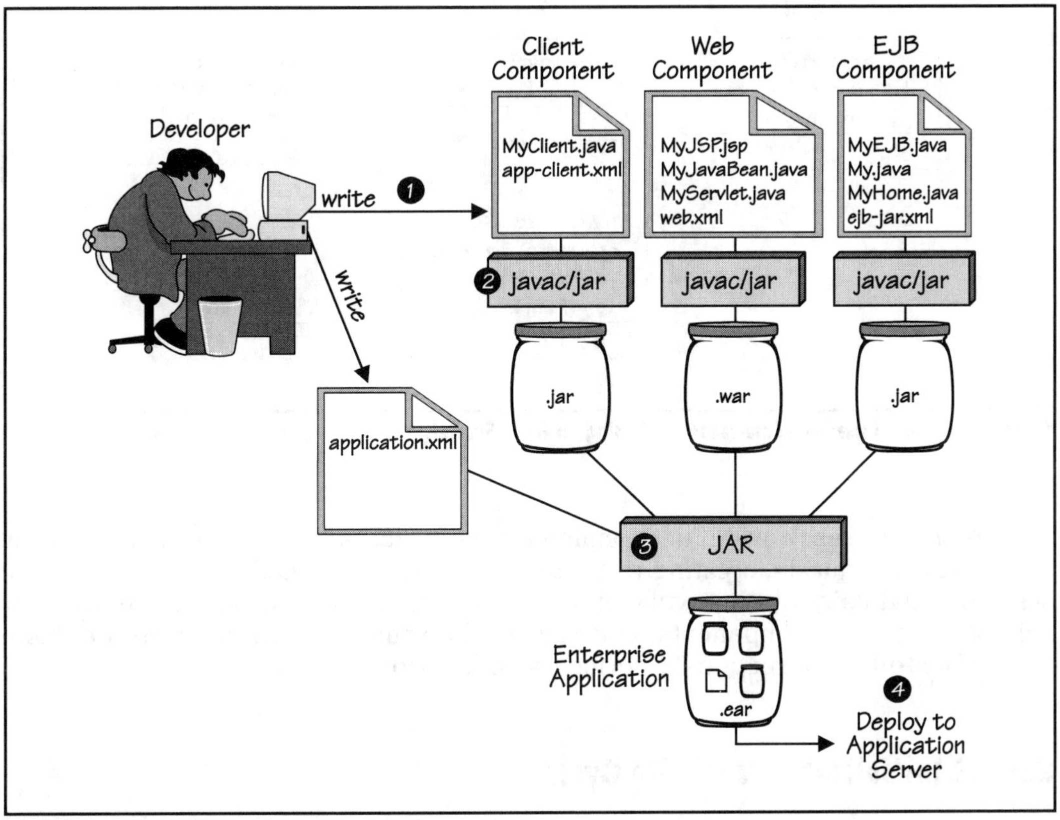

Figure 13-4. The J2EE Development and Deployment Process.

PROGRAMMING SERVLETS

Servlets are Java classes that implement the *javax.servlet.Servlet* interface either directly or indirectly. Servlets that process HTTP requests (the typical case) extend the **HttpServlet** class. Servlets that process other types of requests extend the **GenericServlet** class. A number of supporting classes provide a complete environment for programming Servlets. Figure 13-5 shows some of the Servlet interfaces and classes for writing an HTTP Servlet.

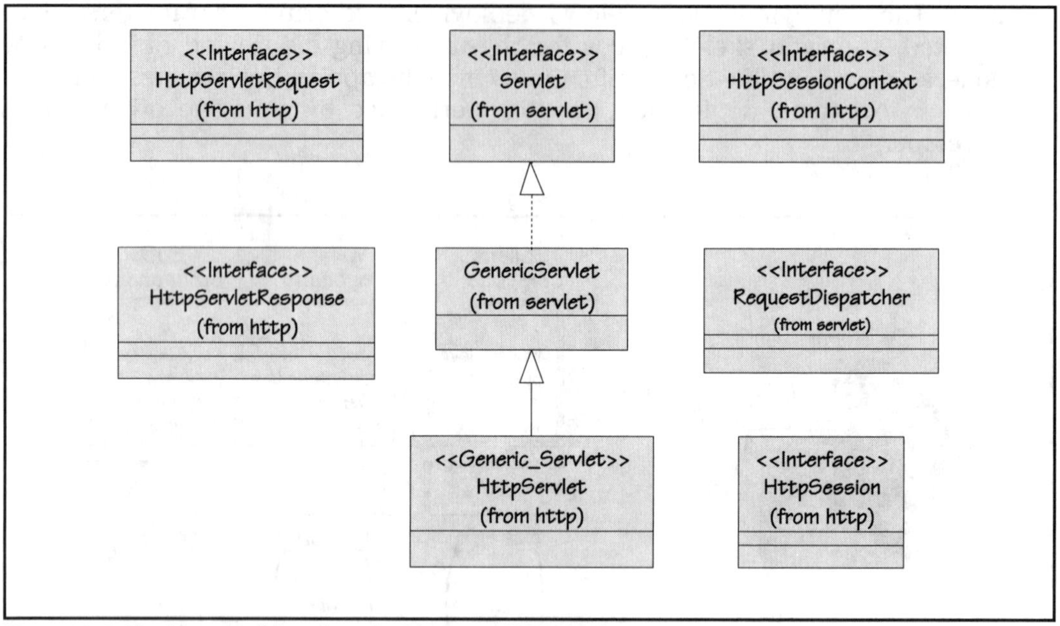

Figure 13-5. The Classes and Interfaces for Writing an HTTP Servlet.

The Servlet classes provide functionality for initialization and life cycle, processing and handling requests, creating responses to requests, maintaining client state via sessions, maintaining contextual information, filtering requests and responses, calling other Web components, and sharing information with other Web components. The following sections describe these functions briefly.

Servlet Initialization and Life Cycle

Servlets instances are created when a request from a client is received and a Servlet instance does not already exist. When a Servlet is created, its class is loaded, an instance of the Servlet is created, the Servlet's *init* method is called, then a service method such as *doGet* or *doPost* is called.

When the container decides to remove a Servlet instance, it calls the destroy method of the Servlet.

You can also define Servlet listener classes to capture life cycle events that occur during the execution of the Servlet. Life cycle events can include the creation and destruction of Servlets, the manipulation of attributes, and the creation, invalidation, and timeout of sessions.

Writing Servlet Application Functions

The container passes requests to a Servlet by invoking a service method that is part of the Servlet. For HTTP Servlets, this is typically the *doGet* or *doPost* methods. For a generic Servlet, the service method is called. Service methods define request and response objects as parameters and the container passes these objects to the Servlet when the method is invoked. The following shows the method signature for the *doGet* service method call for an HTTP Servlet:

```
public void doGet (HttpServletRequest request,
                   HttpServletResponse response)
                   throws ServletException, IOException
```

Servlet Request and Response Objects

The Servlet Request object passes information from the client and the container to the Servlet. The information can include:

- ***Parameters*** passed from the client program. Parameter values can be obtained by name by calling the *getParameter* method.

- ***Attributes*** that are maintained by the container as part of context or session information. Attribute values can be obtained by name by calling the *getAttribute* method for the appropriate context.

- ***Protocol*** and ***localization*** information.

The request object can also be used to obtain an input stream so that input data can be processed manually.

Response objects allow the Servlet to return information to the Servlet's client. The information can include:

- ***Character and binary data:*** For character data the Servlet uses the response object's *getWriter* method to get a *PrintWriter* object.

- ***The content type.*** For example, the content type might be set to text/html.

- ***The type of buffering:*** This controls when the response data is returned to the client.
- ***Localization information:*** Language, cultural and other requirements of a specific locale.
- ***Status Codes:*** These indicate the success (or failure) of an HTTP request.
- ***Cookies:*** These track client-specific data across multiple HTTP requests.

Client State

Web clients often need to maintain state across multiple HTTP calls to a Servlet. HTTP is inherently stateless, so Servlets use an **HttpSession** object to maintain the client state. The session object for each client is maintained by passing an identifier between the client and the server. This can be done through cookies or through a technique called URL rewriting where the Servlet inserts a session identifier as a parameter string in the URL when the client has cookies disabled. The *encodeURL* method of the response object should be used to create all URLs returned by a Servlet so the session ID is returned correctly if cookies are disabled.

A new session can be created or an existing session retrieved using the *getSession* method of the request object. Once the session is created or retrieved, the *getAttribute* and *setAttribute* methods can be called to access and modify client session state.

Servlet Context

The container maintains a context for every Web component instance. The context contains:

- Initialization parameters
- Resources (such as JDBC connections) associated with the Web context
- Attributes
- Logging capabilities

The Servlet context can be obtained by calling the *getServletContext* method.

Servlet Filters

Filters are used to modify the header or content of a Web component's request or response. Requests are typically processed by a filter before the request is passed to a Web component and the response is processed by a filter prior to the response being returned to the client. Filters provide generic functions that can be associated with any Web component. Filters can perform such functions as image conversion, data compression, encryption, tokenizing streams, and XML transformations.

Filters are Java classes that implement the **Filter** interface. The filtering function is performed in a *doFilter* method. The filter also includes the life cycle methods, *init* and *destroy.* Filters are associated with Web components at deployment time. There can be zero or more filters associated with a Web component that form a filter chain.

Calling Other Web Components

Web components can invoke other Web components as part of processing a client's request. The invocation can be either indirect or direct. Indirect invocations occur when a Servlet returns a URL to a client, then the client invokes the Web component with the URL.

Direct invocation occurs when a Web component calls another Web component using the **RequestDispatcher** object from the Servlet context. The *include* method call can be made to insert content generated by another Servlet into the response or the *forward* method can be called to transfer control to another Web component. Request forwarding is used when some preliminary processing is done by a Web component, then the task of generating the response for the client is passed to another Web component.

WRITING JAVASERVER PAGES

JavaServer Pages (JSPs) combine text-based static content with dynamic content generated by JSP elements. The static content can be markup languages such as HTML, WML, and XML. The JSP elements are inserted within the static elements to generate the dynamic content and to control how a JSP page executes. JSP elements interface with Java objects to generate dynamic content. The JSP architecture defines some *implicit objects* similar to those defined for Servlets. JSP elements include:

- ***Directives*** provide information to the Web Container for executing the JSP.
- ***Scripting Elements*** provide programming language constructs for generating dynamic content. The programming language is typically Java. Scripting elements include declarations, scriptlets, and expressions.

- ***Actions*** generate content for return to the JSP client. Actions follow the form of XML statements and can have a body or be empty elements. New action elements can be defined with a Tag extension.

JavaServer Pages are compiled to create a Servlet prior to execution.

JSP Objects

JavaServer Pages can create, access, and modify objects during the processing of a request to a JSP. These objects are created by scriptlets, actions, or are provided to the JSP by the container as implicit objects. Objects are accessible to scripting elements through a script language variable. Created objects have a unique name and a scope (such as *page*, *request*, *session*, and *application*) that define where the created object is accessible and when it is destroyed.

Table 13-1 describes the implicit objects available during the execution of a JSP.

Table 13-1. The JSP Implicit Objects.

Variable	Scope	Description
request	request	The Servlet request object.
response	page	The Servlet response object.
pageContext	page	The context object for this JSP page.
session	session	The session object for the requesting client.
application	application	The Servlet context object.
out	page	Object for writing to the output stream.
config	page	The ServletConfig for this JSP page.
page	page	The instance of the Servlet class processing the current request.
exception	page	The uncaught Throwable object that caused an error page to be generated.

JSP Elements

JSP elements generate the dynamic content for a JSP. The following sections describe the JSP elements in more detail.

Directives

Directives provide JSP execution instructions to the container. Directives have the syntax:

```
%@directive {attr="value " }}*%
```

The attributes define the information that each directive provides to the Web container. Table 13-2 summarizes the JSP directives.

Table 13-2. The JSP Directives.

Directive	Attribute	Description
page		Defines page-level attributes.
	language	Defines the scripting language. Java is a required language for JSP 1.2.
	extends	A superclass that the JSP class will extend.
	import	The list of import classes.
	session	Defines participation in an HTTP session.
	buffer	Defines the buffering model for content output.
	autoFlush	Defines if the buffer should be flushed when full.
	isThreadSafe	Defines if the container should serialize requests to the JSP.
	info	Adds an arbitrary string to the JSP translated page.
	isErrorPage	Indicates if JSP is the target of another JSPs error page.
	errorPage	Defines a URL to a page where uncaught exceptions will be forwarded for processing.
	contentType	Defines the character encoding for the JSP page, the JSP response, and the MIME type for the JSP reponse.
	pageEncoding	Defines the character encoding for the JSP page.
taglib		Defines custom tag extensions to the standard JSP tag set.
	uri	An absolute or relative URI that identifies the tag library descriptor.
	tagPrefix	Defines a prefix for custom tag extension usage.
include		Allows for the substitution of text or code into a JSP page.
	file	The name of the file to be included.

Scripting Elements

Scripting elements perform computations and manipulate objects to generate dynamic content for JSP pages. There are three types of scripting elements:

- ***Declarations*** define scripting element constructs for use by other scripting elements and have the syntax:

```
<%! declaration(s) %>
```

Declarations do not produce any output.

- ***Scriptlets*** contain code fragments that may perform computations or produce output. Scriptlets have the syntax:

```
<% scriptlet %>
```

- ***Expressions*** are scripting language expressions that are evaluated and converted to strings for insertion into the JSP page output. Expressions have the syntax:

```
<%= expression %>
```

Actions

Actions produce output for the JSP page or manipulate objects. Table 13-3 describes the standard set of actions defined by the JSP specification.

Table 13-3. The JSP Standard Actions.

Action	Description
jsp:useBean	Associates a scripting variable with a Java class instance. The Java class typically follows the JavaBean definition rules.
jsp:setProperty	Sets the value of the property in a JavaBean.
jsp:getProperty	Retrieves the value of a JavaBean property, converts it to a string, then writes it to the JSP output.
jsp:include	Includes static and dynamic content from the identified page into the output stream.
jsp:forward	Forwards the current request to a Web component.

Table 13-3. The JSP Standard Actions. (Continued)

Action	Description
jsp:param	Provides key/value information for the jsp:include, jsp:forward, and jsp:params elements.
jsp:plugin	Enables the inclusion of an applet into the JSP page output.
jsp:params	Always a child of the jsp:plugin action that defines parameter values for an applet.
jsp:fallback	Always a child of the jsp:plugin action that defines the error handling if the applet cannot be started.

Tag Extensions

Tag extensions allow a developer to make new actions available to the authors of JSP pages. Three steps are required to create and use a new tag extension:

1. Write a *tag handler* that implements the function of a new action tag. The Tag handler is a Java class.
2. Define the new tag in a *Tag Library Descriptor (TLD)*. The TLD is defined using an XML document that uses a JSP taglib XML DTD.
3. Declare the usage of the tag library in a JSP page using the taglib directive. The following is an example of a taglib directive:

```
<%@ taglib uri="/WEB-INF/myTagLib.tld" prefix="myTag" %>
```

Custom tags are used within a JSP page following XML syntax. Tags can have a start and end tag and possibly a body, or have no body. Attributes can also be defined for tags.

DEVELOPING J2EE APPLICATIONS

In this section we will illustrate J2EE application development by creating a simple application that uses EJB and JSP components. While developing the example, we will use a high-level development tool, Borland's JBuilder 6.0 Enterprise Edition, to create the components we need for our application. JBuilder allows us to:

- ***Define*** our components using a UML-like symbology.
- ***Write*** code to define the functionality of our components.
- ***Make*** all of our application code as part of a *project*.
- ***Create*** deployment descriptors and package our components into *.jar*, *.war*, and *.ear* files.

- ***Deploy*** our components to one of several commercial application servers.
- ***Run*** and ***debug*** the components.

JBuilder is not the only tool with such capabilities. Others include WebGain's WebGain Studio, IBM's WebSphere Studio Application Developer, and Sun's Forte for Java.

Let's return to our university theme for a sample J2EE application.

The application, which we call RoomFinder, will help a student locate the room for a university course. The RoomFinder application, when given a Course ID, will return the room where the course will be held. This simple J2EE application includes four EJBs (one remote Session EJB and three local entity EJBs) and one JSP Web component. One entity EJB, called Course, will represent all of the details about the course, including relationships to other entity EJBs that represent the professor for the course and the room where it will be taught. The RoomFinder JSP produces the HTML page that allows the user to input the Course ID and view the room where the course will be taught. Figure 13-6 shows a screen capture of the Web page generated by the RoomFinder JSP. Figure 13-7 shows the page after a Course ID has been entered and the corresponding room has been retrieved.

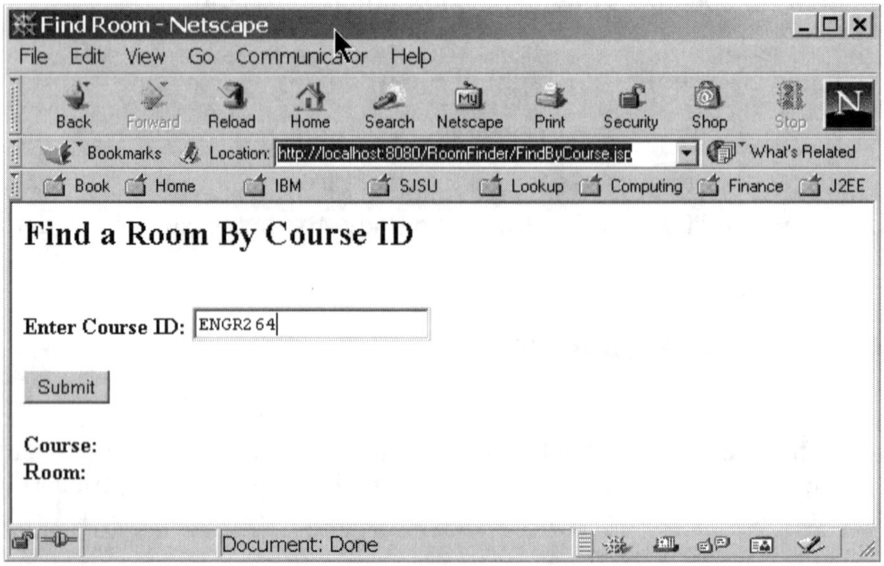

Figure 13-6. The RoomFinder Application HTML Page Generated from the FindRoomByCourse JSP.

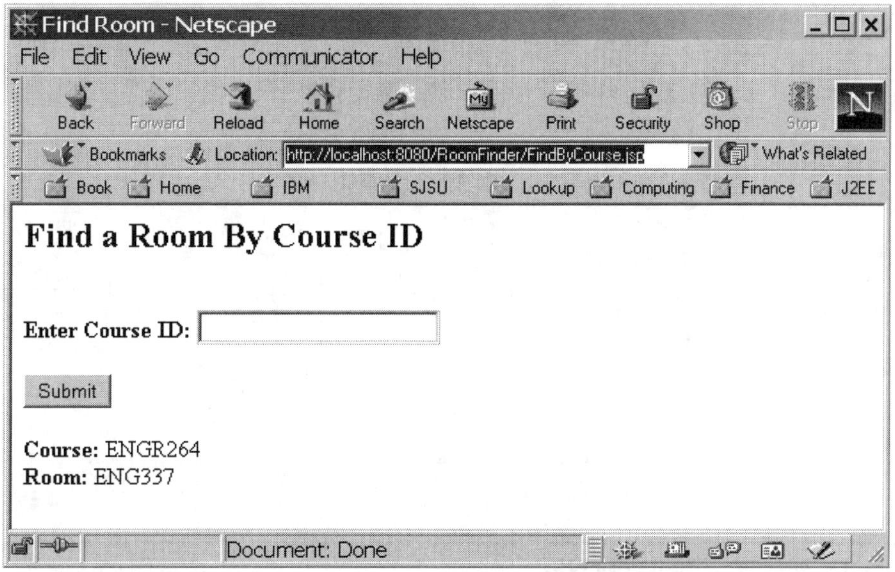

Figure 13-7. The RoomFinder Application after Finding the Room for the ENGR264 Course.

The RoomFinder EJBs

JBuilder 6 provides a graphical Designer for EJB 2.0-compliant entity beans that use Container Managed Persistence (CMP). This designer allows you to visually accomplish several tasks in EJB development:

- Define individual EJBs by type (session, entity, or message-driven).
- Define the type of interfaces they support (remote or local).
- Define fields, business methods, finder methods, select methods, and relationships for each EJB.
- Associate fields and relationships with a DataSource.
- Define the DataSource schema.

Figure 13-8 shows a screen shot of the RoomFinder EJBs as represented in the JBuilder EJB designer. The EJBs are shown in a UML-like format. Three entity EJBs (**Professor**, **Course**, and **Room**) are shown on the right. Each entity bean is subdivided into four areas: title; fields; business and callback methods; and finder and select methods. A different icon for each type indicates the method types. The session EJB, called **RoomFinder**, is on the left. It is subdivided into three areas: title, fields, and business and callback methods. The lines drawn between the EJBs represent relationships between the EJBs. For instance, the **Course** EJB has relationships that indicate the professor and room for the course.

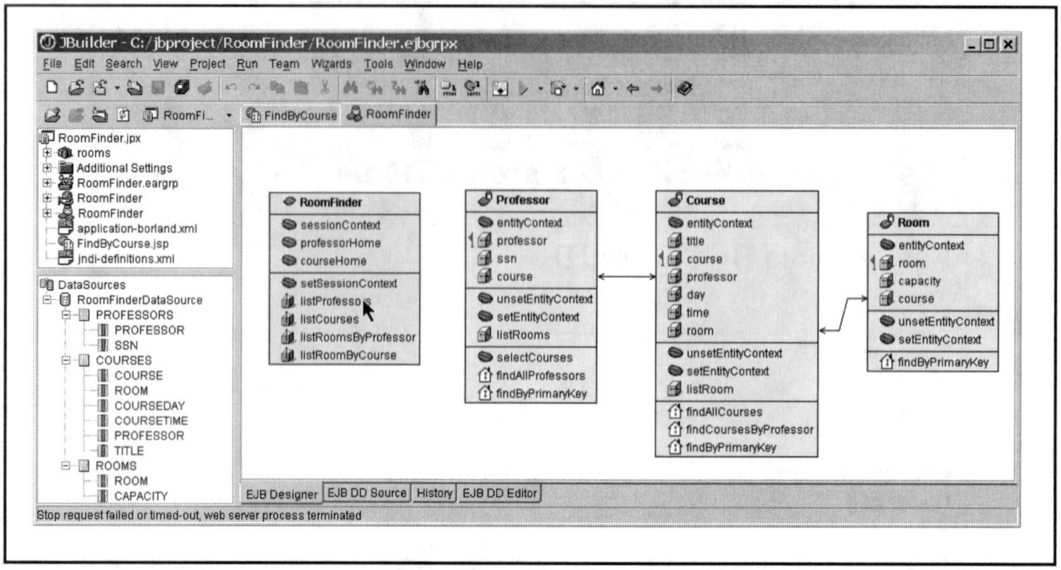

Figure 13-8. The RoomFinder Application EJBs.

Each EJB consists of two interfaces and one Java class. The client/server contract for an EJB is defined by the interfaces. One interface, called the *home interface*, defines methods for creating, removing, and locating a bean instance. The second interface, called a *component interface*, defines the business methods for an EJB. The home and component interfaces can be local (meaning the client executes within the same Java VM) or remote (meaning the client can also execute on another Java VM or machine). Each interface extends interfaces appropriate to the EJB type and whether or not it is local or remote. The container generates implementations for the home and component interfaces during deployment.

All EJBs have a Java class that implements the component contract and the business methods for the EJB. Each EJB type—session, entity, and message-driven—implements interfaces appropriate to its type.

Message-driven EJBs do not have home and component interfaces. Instead the container provides access for clients of message-driven EJBs through a JMS queue.

Developing the RoomFinder Application with JBuilder

So lets begin our journey in J2EE application development. We'll use a number of screen shots to illustrate the significant steps required in JBuilder to define a project, add our J2EE components, and run/debug the project. After all, many pictures are worth several thousand words.

Creating a JBuilder Project

Figure 13-9 shows the wizard that helps you define a project. The wizard has three panels that let you define the project name, derive the project from a template that has been preconfigured, set up project paths, and enter descriptive information.

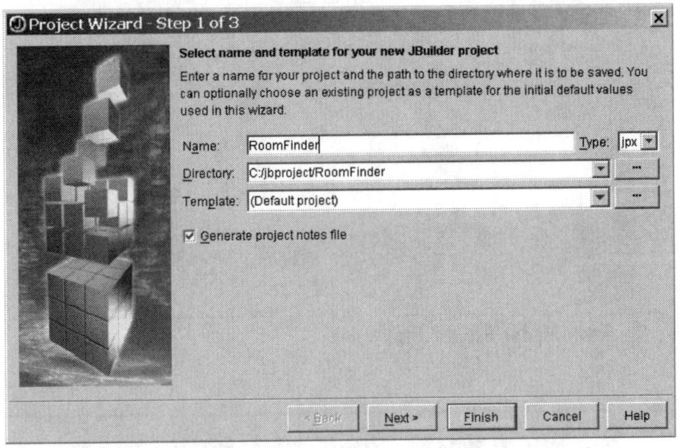

Figure 13-9. Creating the RoomFinder JBuilder Project (RoomFinder.jpx).

Once a project has been defined, it may be necessary to set other project properties. This can include properties for running and debugging the application, defining build options, code and editing style, and which application server will be your target for execution. Figure 13-10 shows the first of the wizard pages that help you set up your project properties. It is invoked from the Project->Properties menu bar.

Adding Components to the Project

Our next step is to define the components for our project. Figure 13-11 shows the wizard that can be invoked from the File->New menu bar. The Web tabbed page of the wizard allows you to define a Web application (with a *.war* file) and include JSPs and Servlets that will be part of the Web application. The Enterprise page allows you to create a designer for your EJB 2.0 components (including an EJB archive file) and an Enterprise ARchive (*.ear*) file. The Enterprise ARchive file is used to assemble and package the Web, EJB archive, and other files we create for our RoomFinder application.

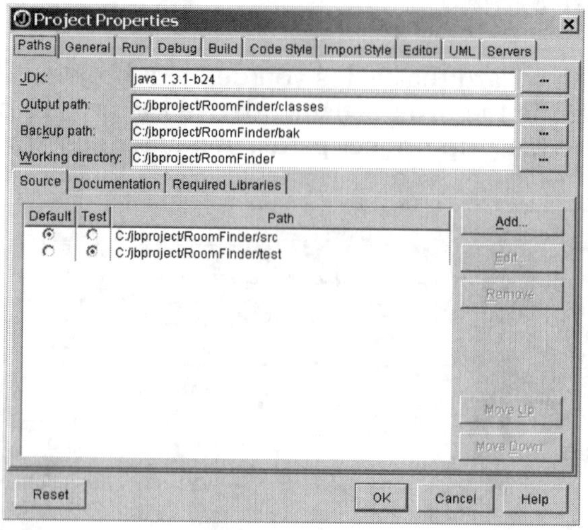

Figure 13-10. Changing the RoomFinder Project Properties.

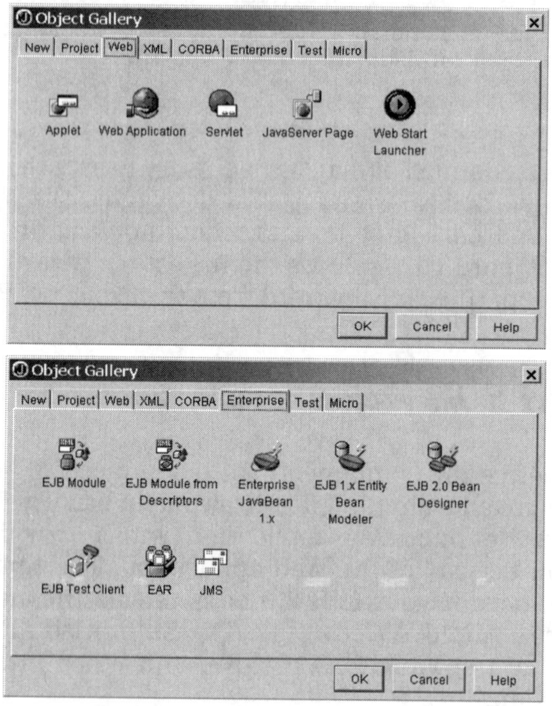

Figure 13-11. Adding Enterprise JavaBean, Web, and Enterprise Components to the RoomFinder Project.

Using the JBuilder EJB Designer

When using the EJB Designer, the lower left-hand pane in JBuilder displays the DataSource that represents the relational database tables that the entity EJB field will be mapped to. Importing a schema from an existing database can create a DataSource or you can add a DataSource then define tables and columns to create a new schema. Once a schema is defined, an EJB can be associated with a table and the fields in the EJB can be associated with a column in the table. Figure 13-12 shows the DataSource and its schema for the RoomFinder application.

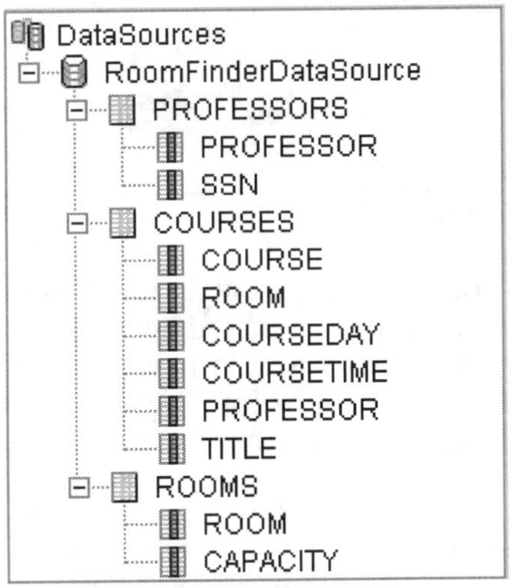

Figure 13-12. The RoomFinder DataSource.

EJBs can be added to the designer and modified as you go through your design process. Pop-up dialogs help you enter descriptive information for your EJB as a whole (such as name, schema name, whether the EJB is local, remote, or both, table name, and table references for relationships). Other pop-ups allow you add fields and methods and then define the properties for the fields and methods. Figure 13-13 shows the pop-up dialog for defining the EJB name and other EJB related information. This dialog is presented when you left-button click on the EJB name.

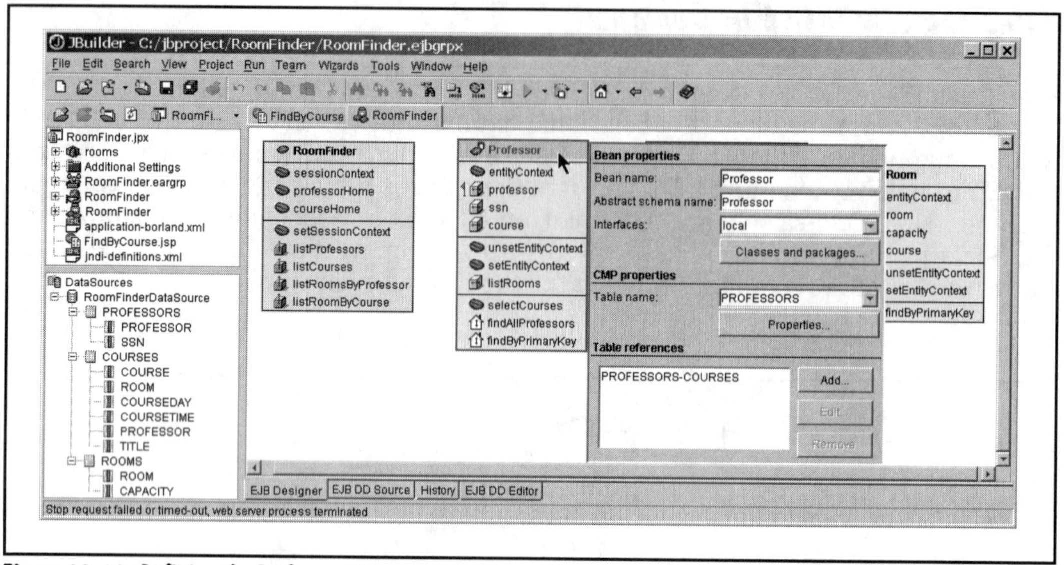

Figure 13-13. Defining the Professor EJB Properties.

Figure 13-14 shows the pop-up dialog that is used to add *fields*, *methods*, and *relationships* to an EJB. The dialog is invoked by clicking on the EJB with the right mouse button. This dialog also lets you add new EJBs, remove the selected EJB, view the bean source, and open an editor for the deployment descriptor. If the right mouse button is clicked over a field of the EJB, the dialog will also allow you to remove the field.

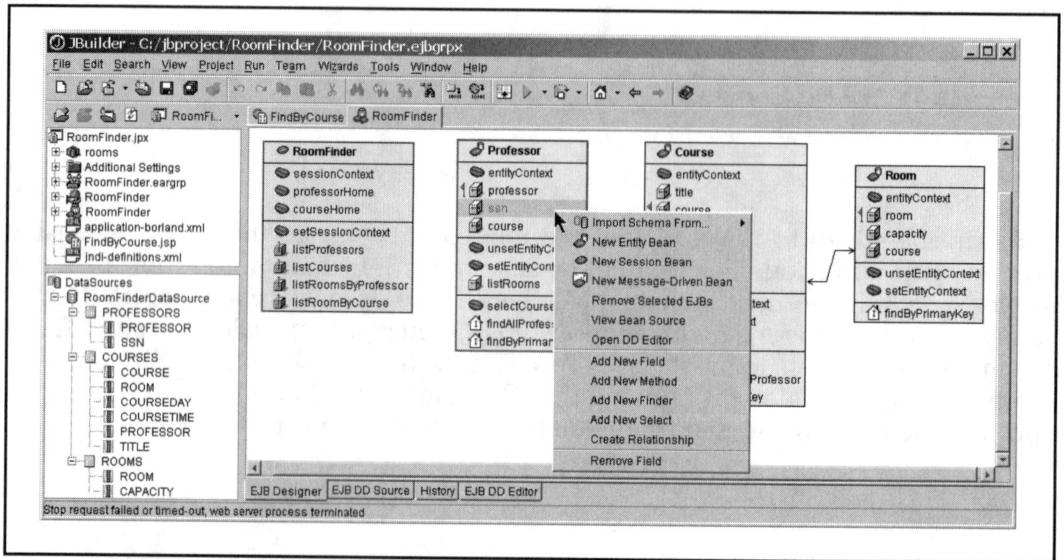

Figure 13-14. Defining the RoomFinder EJB Fields and Methods.

If the left mouse button is clicked on an EJB field or method, a dialog is presented to modify the various parameters for the items. Figure 13-15 shows the dialog presented for the primary key field for the Professor entity bean. Figure 13-16 shows the dialogs presented for the *listRooms* method, the *selectCourses* Select method, the *findAllProfessors* Finder method, and the *course* Relationship field of the **Professor** EJB. Note that for Select and Finder methods, EJB-QL is used to specify the Select or Finder search criteria.

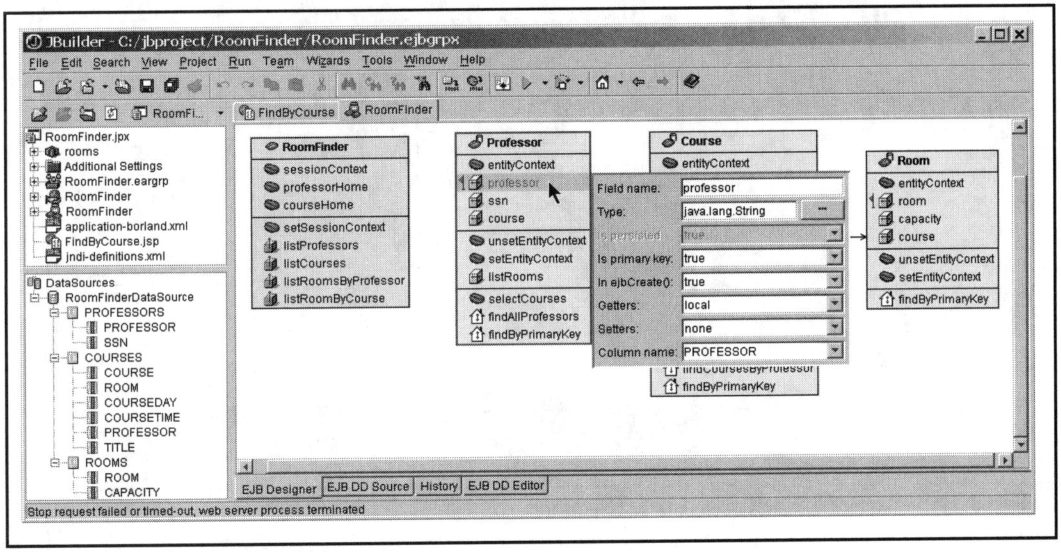

Figure 13-15. Defining EJB Field Properties.

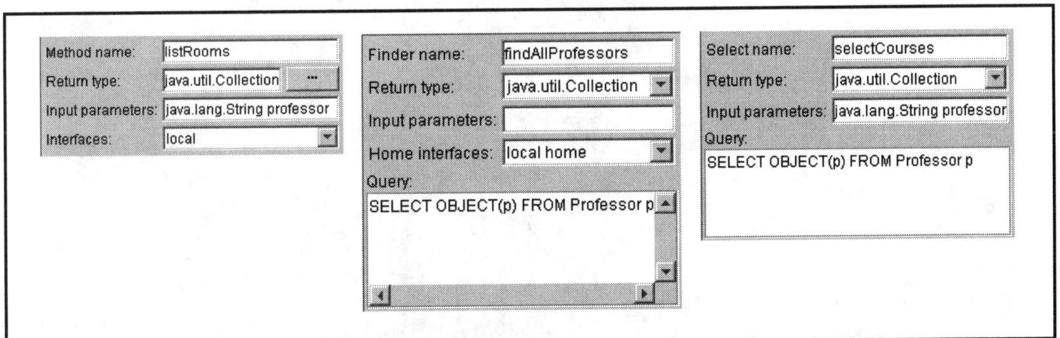

Figure 13-16. Defining Properties for EJB Business Methods, Finder Methods, and Select Methods.

Relationships can also be established between entity beans using the Designer. If you right-button click on an entity EJB and select Create Relationship, a relationship field will be created and a rubber-band arrow like that shown in Figure 13-17 will appear. You use the arrow to select the target EJB of the relationship. After the relationship is created, the relationship dialog is used to set the details of the relationship (see Figure 13-18). This includes defining the Table References for the relationship. This is done through another dialog (also shown in Figure 13-18) that allows an arrow to be drawn graphically between the two database tables that represent the EJBs that are part of the relationship.

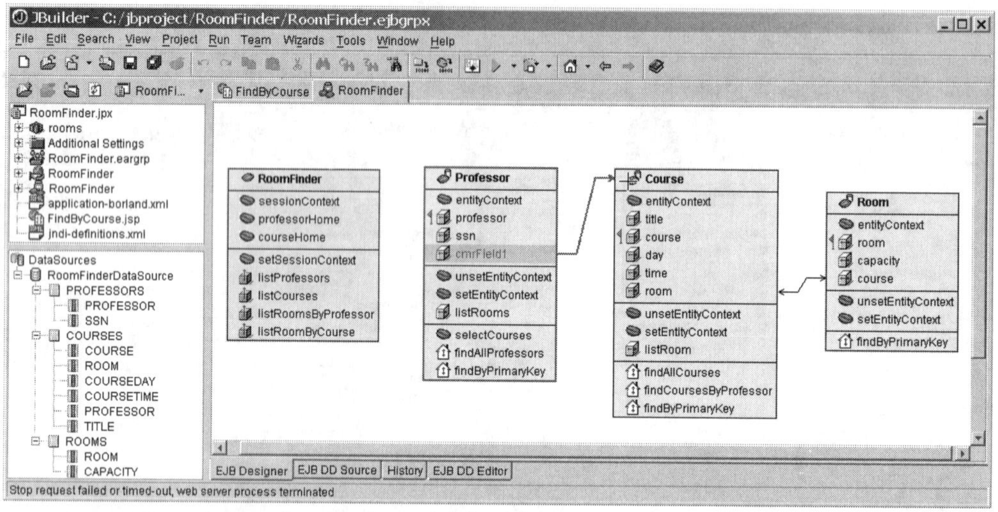

Figure 13-17. Defining the RoomFinder EJB Relationships.

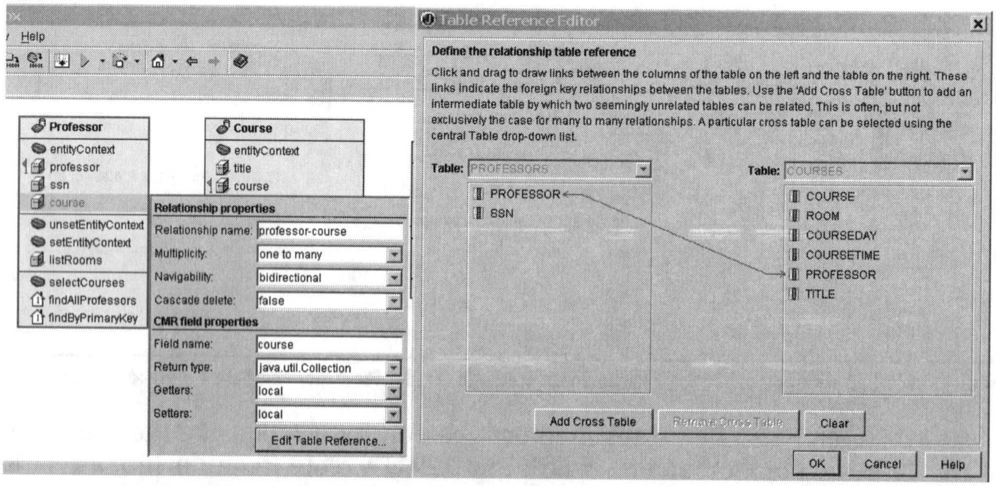

Figure 13-18. Defining the Professor-Course EJB Relationship Properties.

Adding the Business Logic to the EJB Design

At this point, we have completed the design of all of the EJBs for our RoomFinder application. The EJBs for the application have been created and all of the fields and methods have been defined and their properties have been chosen. The next step is to complete the implementation for the EJBs. This is the first time we will have to write Java code for our EJBs. Fortunately, all of our effort to this point has produced the home and component interfaces for the EJBs and a skeleton for the EJB class. All we have left to do is add the behavioral code to the methods in the EJB class.

JBuilder provides an Explorer-like pane in the upper left to navigate all of the pieces of your project, including the Java source files that it creates or other files that you add to the project. Figure 13-19 shows the explorer pane. The source for the **RoomFinder** Session EJB (RoomFinderEJB.java) has been selected by double clicking on the file name in the explorer window. When a file is selected, it appears in the window on the right as a tabbed window. This allows you to easily navigate from one source file to another as well as to other tabbed windows that represent other parts of the project.

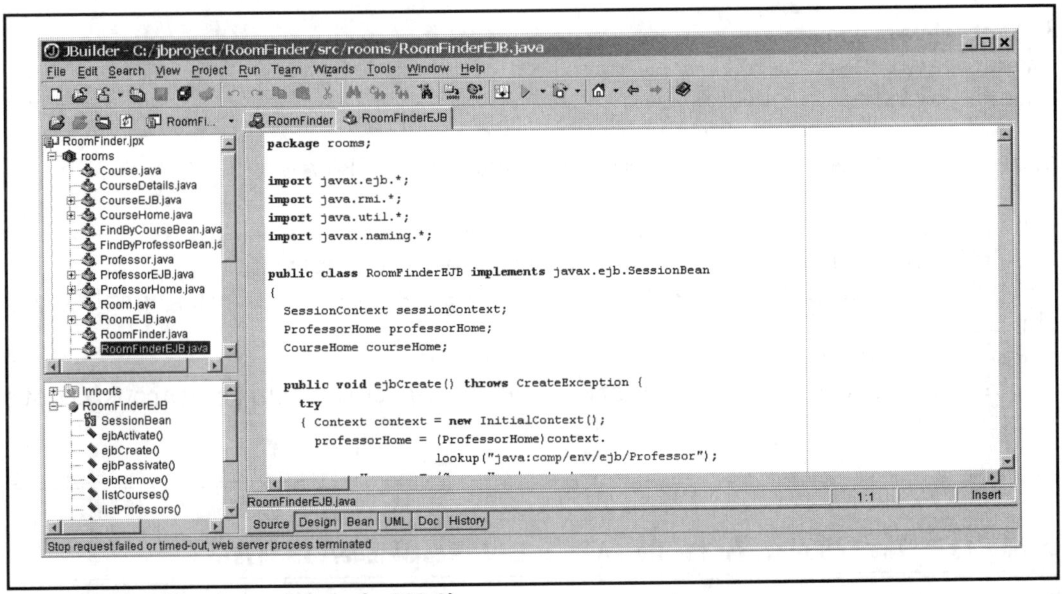

Figure 13-19. Adding Source Code to the EJB Class.

The source code for the Home and Component interfaces that is produced by JBuilder for the **RoomFinder** session EJB and the **Course** entity EJB is presented in Listings 13-1 through 13-6. Note that the **Course** entity EJB has only one business method that needs to be implemented. The remainder of the EJB's source code is callback methods and some abstract methods for fields that represent the entity's data and relationships. For brevity, the source code for the **Professor** and **Room** EJBs is not shown, but is available at *http://wirelessjava.engr.sjsu.edu.*

Listing 13-1. The Home Interface for the RoomFinder EJB (RoomFinderHome.java).

```
package rooms;

import javax.ejb.*;
import java.util.*;
import java.rmi.*;

public interface RoomFinderHome extends javax.ejb.EJBHome
{
  public RoomFinder create() throws CreateException, RemoteException;
}
```

Listing 13-2. The Component Interface for the RoomFinder EJB (RoomFinder.java).

```
package rooms;

import javax.ejb.*;
import java.util.*;
import java.rmi.*;

public interface RoomFinder extends javax.ejb.EJBObject
{
  public String[ ] listProfessors() throws RemoteException;
  public String[ ] listCourses() throws RemoteException;
    public  java.util.Collection  listRoomsByProfessor(String  professor)
throws
  RemoteException;
  public rooms.CourseDetails listRoomByCourse(String course) throws Remote
  Exception;
}
```

Listing 13-3. The EJB Class for the RoomFinder EJB (RoomFinderEJB.java).

```
package rooms;

import javax.ejb.*;
import java.rmi.*;
import java.util.*;
import javax.naming.*;
```

```
public class RoomFinderEJB implements javax.ejb.SessionBean
{
  SessionContext sessionContext;
  ProfessorHome professorHome;
  CourseHome courseHome;

  public void ejbCreate() throws CreateException
  { try
    { Context context = new InitialContext();
      professorHome = (ProfessorHome)context.
                      lookup("java:comp/env/ejb/Professor");
      courseHome    = (CourseHome)context.
                      lookup("java:comp/env/ejb/Course");
    } catch (NamingException ex)
    { throw new CreateException(ex.getMessage());
    }
  }

  public void ejbRemove() {}
  public void ejbActivate() {}
  public void ejbPassivate() {}
  public void setSessionContext(SessionContext sessionContext)
  { this.sessionContext = sessionContext;
  }

  public String[ ] listProfessors()
  { try
    { Collection professors = professorHome.findAllProfessors();
      ArrayList names = new ArrayList();
      Iterator i = professors.iterator();

      while (i.hasNext())
      { Professor prof = (Professor) i.next();
        names.add(prof.getProfessor());
      }
      String[ ] myProfs = new String[names.size()];
      names.toArray(myProfs);
      return myProfs;
    } catch (Exception e)
    { throw new EJBException(e.getMessage());
    }
  }

  public String[ ] listCourses()
  { try
```

```
    { Collection courses = courseHome.findAllCourses();
      ArrayList courseIDs = new ArrayList();
      Iterator i = courses.iterator();

      while (i.hasNext())
      { Course course = (Course) i.next();
        courseIDs.add(course.getCourse());
      }
      String[ ] myCourses = new String[courseIDs.size()];
      courseIDs.toArray(myCourses);
      return myCourses;
    } catch (Exception e)
    { throw new EJBException(e.getMessage());
    }
  }

  public java.util.Collection listRoomsByProfessor(String professor)
  { try
    { Professor prof = professorHome.findByPrimaryKey(professor);
      return prof.listRooms(professor);
    } catch (Exception e)
    { throw new EJBException(e.getMessage());
    }
  }

  public rooms.CourseDetails listRoomByCourse(String course)
  { try
    { Course cour = courseHome.findByPrimaryKey(course);
      return cour.listRoom();
    } catch (Exception e)
    { throw new EJBException(e.getMessage());
    }
  }
}
```

Listing 13-4. The Local Home Interface for the Course EJB (CourseHome.java).

```
package rooms;

import javax.ejb.*;
import java.util.*;

public interface CourseHome extends javax.ejb.EJBLocalHome
{ public Course create(java.lang.String title, java.lang.String course,
```

```
  java.lang.String day, java.lang.String time) throws CreateException;
  public java.util.Collection findAllCourses() throws FinderException;
 public java.util.Collection findCoursesByProfessor(java.lang.String pro
  fessor)throws FinderException;
 public Course findByPrimaryKey(java.lang.String course) throws FinderEx
  ception;
}
```

Listing 13-5. The Local Component Interface for the Course EJB (Course.java).

```
package rooms;

import javax.ejb.*;
import java.util.*;

public interface Course extends javax.ejb.EJBLocalObject
{ public java.lang.String getTitle();
  public void setTitle(java.lang.String title);
  public void setRoom(rooms.Room room);
  public rooms.Room getRoom();
  public void setProfessor(rooms.Professor professor);
  public rooms.Professor getProfessor();
  public void setDay(java.lang.String day);
  public java.lang.String getDay();
  public void setTime(java.lang.String time);
  public java.lang.String getTime();

  public java.lang.String getCourse();

  public rooms.CourseDetails listRoom();
}
```

Listing 13-6.The EJB Class for the Course EJB (CourseEJB.java).

```
package rooms;

import javax.ejb.*;

public abstract class CourseEJB implements EntityBean
{
  EntityContext entityContext;

  public java.lang.String ejbCreate(java.lang.String title,
 java.lang.String course, java.lang.String day, java.lang.String time)
```

```
throws CreateException { setTitle(title);
  setCourse(course);
  setDay(day);
  setDay(day);
  setTime(time);
  return null;
}

public void ejbPostCreate(java.lang.String title, java.lang.String
            course, java.lang.String day, java.lang.String time)
            throws CreateException {}

public void ejbRemove() throws RemoveException {}

public abstract java.lang.String getTitle();
public abstract void setTitle(java.lang.String title);
public abstract void setCourse(java.lang.String course);
public abstract java.lang.String getCourse();
public abstract java.lang.String getDay();
public abstract void setDay(java.lang.String day);
public abstract void setTime(java.lang.String time);
public abstract java.lang.String getTime();
public abstract void setRoom(rooms.Room room);
public abstract void setProfessor(rooms.Professor professor);
public abstract rooms.Professor getProfessor();
public abstract rooms.Room getRoom();

public void ejbLoad() {}
public void ejbStore() {}
public void ejbActivate() {}
public void ejbPassivate() {}

public void unsetEntityContext()
{ this.entityContext = null;
}
public void setEntityContext(EntityContext entityContext)
{
  this.entityContext = entityContext;
}

public rooms.CourseDetails listRoom()
{ try
  { String professor = getProfessor().getProfessor();
    String room = getRoom().getRoom();
    return new CourseDetails(getCourse(), getTitle(),
```

```
                                  professor, room,
                                  getDay(), getTime());
    } catch (Exception e)
    { throw new EJBException(e.getMessage());
    }
  }
}
```

The EJB Deployment Descriptor Details

Now that the EJBs have been defined and the code written for each EJB, the next task is to provide deployment descriptors for the EJBs. All of the EJBs that were defined in the EJB Designer are part of one JBuilder *EJB Module* and are packaged in one EJB JAR file. The EJB JAR file contains a deployment descriptor that describes the JAR file as a whole and also has segments that describe each EJB individually. The deployment descriptor information is captured using XML and JBuilder provides wizards to enter the XML information.

The deployment descriptor wizards are invoked from the project explorer pane. Double clicking on the **RoomFinder** EJB Module in the Project Explorer pane displays a wizard for entering the global information for the EJB JAR (see Figure 13-20).

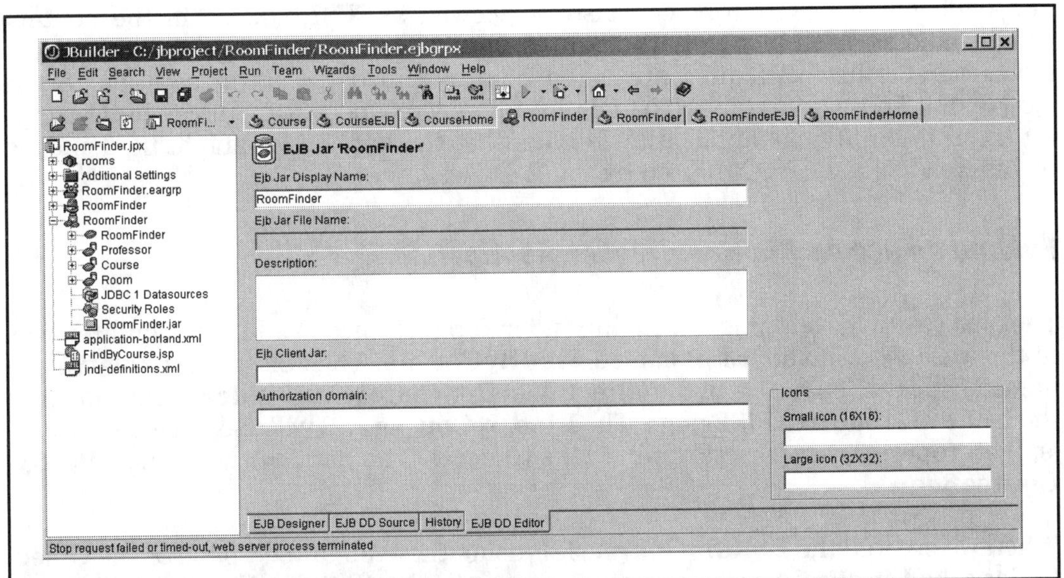

Figure 13-20. The EJB JAR Deployment Descriptor Wizard.

Figure 13-21 shows the deployment descriptor wizard for the **Course** EJB. Note that the wizard presents several tabbed panels for setting general, environment, reference, property, and security information. Most of the information is preset from information entered in the EJB Designer.

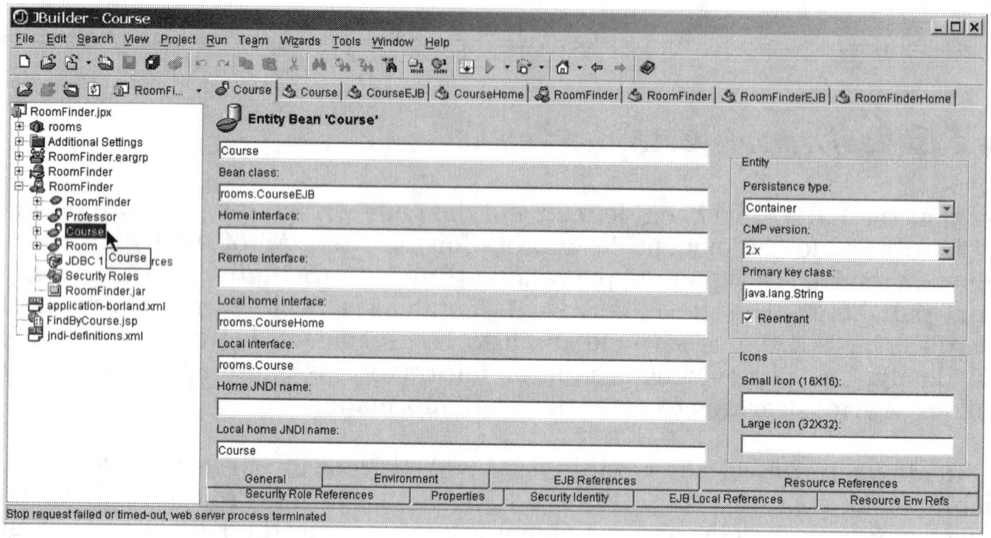

Figure 13-21. The Course EJB Deployment Descriptor Wizard.

The deployment descriptors can also be viewed as XML source in the content window and as an XML tree in the pane at the lower left (see Figure 13-22). Note that several other XML files are listed in the tabbed panels in addition to the ejb-jar.xml. These other tabbed panels are for *server-specific* Deployment Descriptor information for the application servers that the application might be deployed on, as well as modeler XML used by the EJB Designer.

Putting a Face on the RoomFinder Application

The next step in developing our RoomFinder application is to add our user interface. In this case we will add a JSP named **FindByCourse**. The JSP will use a JavaBean named **FindByCourseBean** to retrieve data from our **RoomFinder** EJBs and make the data available to the JSP page. The **FindByCourse** JSP will use the jsp:useBean, jsp:getProperty, and jsp:setProperty tags to access the data stored in the **FindByCourseBean**.

When we created the **FindByCourse** JSP using the Web tabbed page of the Object Gallery, the **FindByCourseBean** was also created by JBuilder. Figure 13-23 shows an expanded view of the *Rooms* Web Application in the JBuilder explorer pane. Note

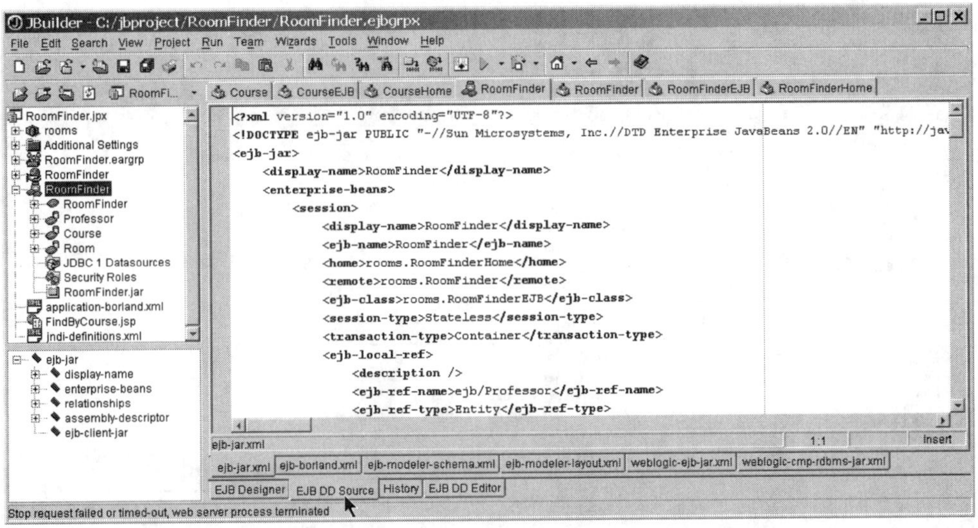

Figure 13-22. The EJB Deployment Descriptors.

that the JSPs, Web component deployment descriptors, and *.war* file are listed. A Web page view of the JSP is shown in the content pane. If the Source tab is clicked in the content pane, then the source for the JSP is displayed in the content pane and a hierarchical view of the JSP source is shown in the pane in the lower left of the JBuilder window (see Figure 13-24).

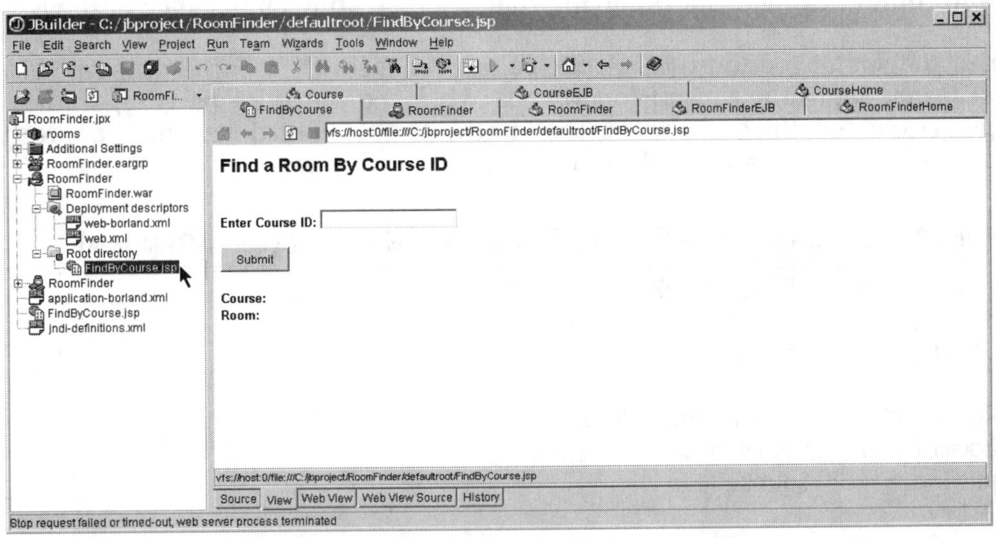

Figure 13-23. The RoomFinder Web Application and JBuilder View of the "Find by Course" HTML Page.

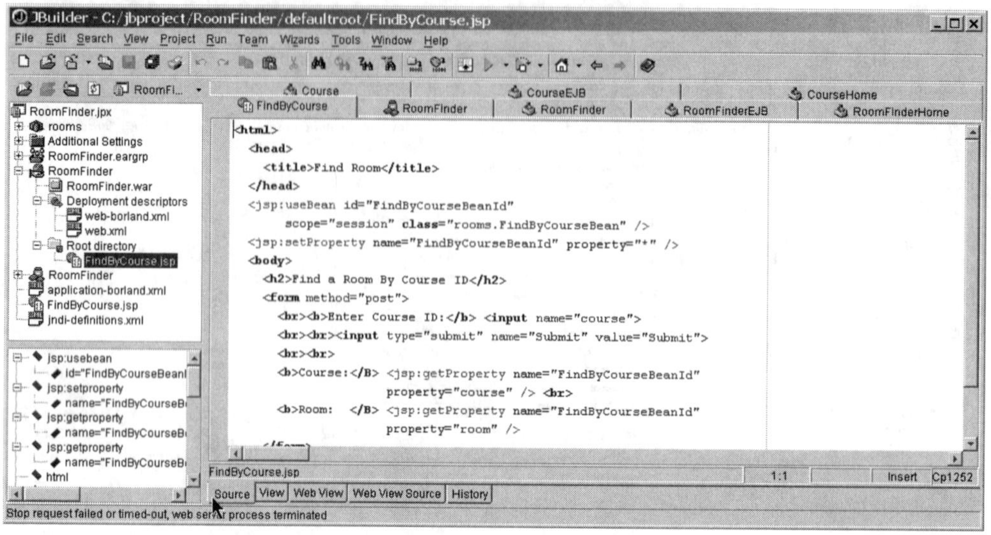

Figure 13-24. Writing the JSP Source for the "Find by Course" HTML Page.

The JBuilder Project pane allows you to select a Java source file and edit the file. Listing 13-7 shows the completed source for the FindByCourseBean.java that supports the **FindByCourse** JSP page. The JavaBean provides a constructor and two properties: *course* and *room*. With the jsp:usebean tags, the JSP is able to get and set property values in the bean. The constructor prepares the bean for accessing data from the suite of EJBs that we have created by doing a JNDI lookup of the **RoomFinder** session bean Home interface then creating an instance of the EJB.

When the submit button on the JSP page is clicked, the course property in the JavaBean is set by the jsp:setProperty tag. When the jsp:getProperty tag in the JSP page is executed, the *getRoom* method in the JavaBean retrieves the room for the specified course and returns it to the JSP page for display.

Listing 13-7. The Source for the "Find By Course" JavaBean (FindByCourseBean.java).

```
public class FindByCourseBean
{ String room = null;
  String course = null;

  RoomFinderHome roomFinderHome;
  RoomFinder roomFinder;
  public FindByCourseBean()
  { try
    { Context context = new InitialContext();
      roomFinderHome = (RoomFinderHome)PortableRemoteObject.narrow(
```

```
                        context.lookup("java:comp/env/ejb/RoomFinder"),
                        RoomFinderHome.class);
        roomFinder = roomFinderHome.create();
      } catch (Exception e)
      { e.printStackTrace();
      }
  }

  // Access course property
  public String getCourse()
  { return course;
  }

  // Set course property
  public void setCourse(String newValue)
  { if (newValue!=null)
    { course = newValue;
    }
  }

  // Access room property
  public String getRoom()
  { if (course!=null)
    { try
      { CourseDetails courseDetails = roomFinder.listRoomByCourse(course);
          this.room = courseDetails.getRoom();
        } catch (Exception e)
        { e.printStackTrace();
        }
    }
    return room;
  }
  // Set room property
  public void setRoom(String newValue)
  { if (newValue!=null)
    { room = newValue;
    }
  }
}
```

Making and Testing the RoomFinder Application

We now have a complete application. The next steps are to *make* and *test* the execution of the application. Making the project is initiated by clicking the *make* button on the toolbar (see Figure 13-25), or invoking it from the Project menu bar selection. Note that following the completion of the make step, a compiler tab appears in the lower portion of the JBuilder window. This compiler window contains any informational, warning, and error messages that were produced during the make step. Double-clicking on compile error messages will open the related source file and place the cursor on the offending line of source code.

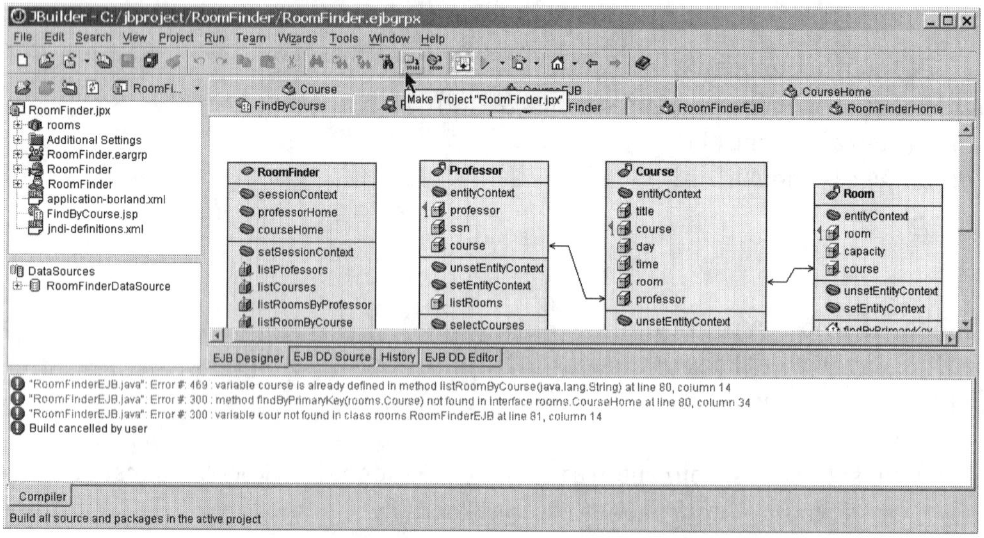

Figure 13-25. Making the RoomFinder Application.

JBuilder can deploy a J2EE application to one of several commercial application servers. The server can be selected from the Project Properties->Server tabbed dialog page (see Figure 13-26).

Running and *debugging* the application is a simple push button task that can be initiated from the tool bar (see Figure 13-27). When debugging the project, breakpoints can be set in any of the project's Java source code. When a break point is reached during execution, a debug panel is presented that allows the examination of variables and performing other debugging tasks (see Figure 13-28).

Running the project allows the application to execute without breakpoints.

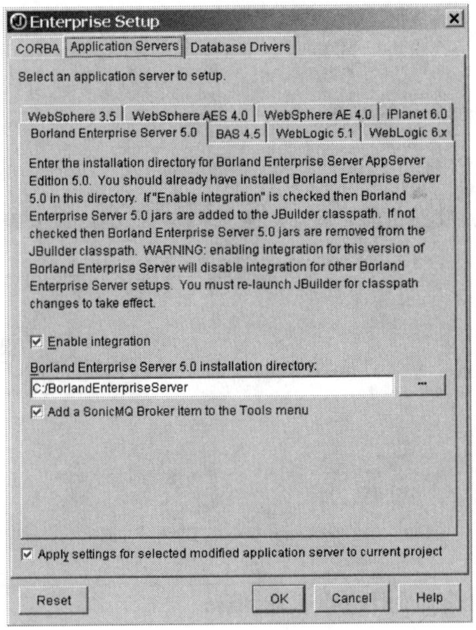

Figure 13-26. Configuring the RoomFinder Project for a Specific Application Server.

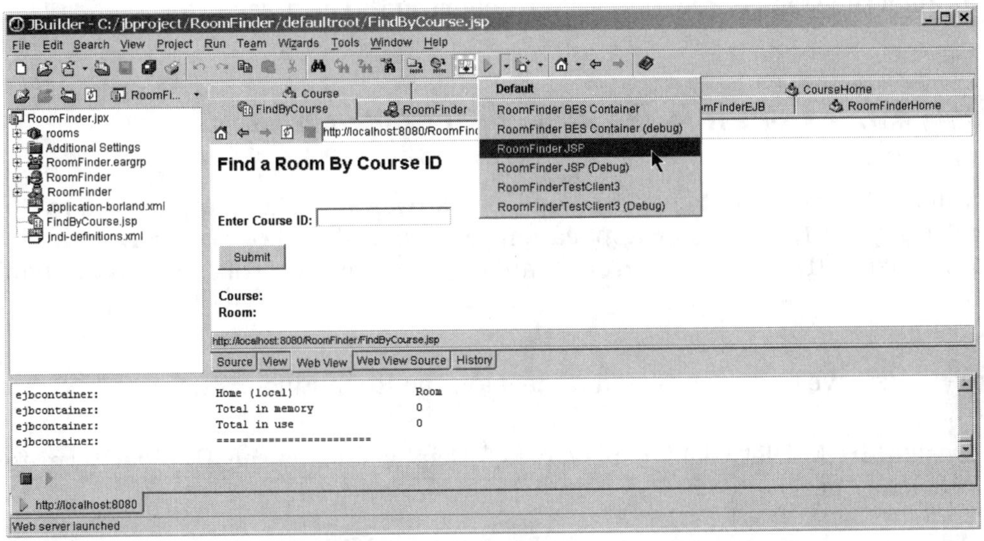

Figure 13-27. Running and Debugging the RoomFinder Application.

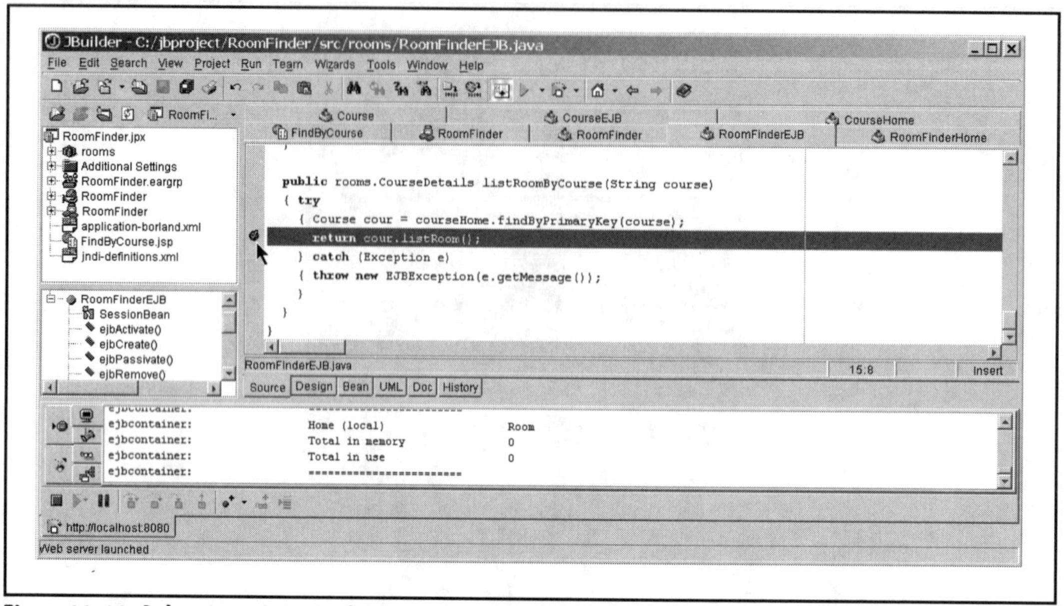

Figure 13-28. Debugging a J2EE Application within JBuilder.

This concludes a brief overview of creating a J2EE application using JBuilder Enterprise. So what does this have to do with wireless applications, you might ask? JBuilder also includes a J2ME development environment called MobileSet. We will go through the steps of creating a J2ME client in the next section.

Adding a J2ME Client to the RoomFinder Application

We will use the MobileSet environment to create a MIDP client application to find a room using the RoomFinder application. Figure 13-29 shows screenshots of the RoomFinder MIDlet running in an emulator for the Motorola i85s Java-enabled phone.

Three tasks have to be completed to develop the RoomFinderMIDlet:

1. Design the MIDlet GUI "displayables." Displayables is the JBuilder term for a MIDlet screen.
2. Add the code required to traverse between the MIDlet screens.
3. Add the code to invoke services from the J2EE application.

Figure 13-29. The RoomFinderMIDlet J2ME Client Application.

JBuilder MobileSet allows a MIDP application to be developed much as a Swing or AWT application would be developed. MobileSet provides a design panel with a palette of widgets that are specific to MIDP.

Creating the JBuilder MobileSet Project

JBuilder creates projects based on the type of JDK that is being used. Because the J2ME application we will create uses the CLDC/MIDP development kit, we have to create our J2ME application in a project separate from the one that was used to develop the JSPs, Servlets, and EJBs. Figure 13-30 shows defining the project in the first step using the Project then selecting the Java CLDC-10 MIDP-1.0 JDK in the second step.

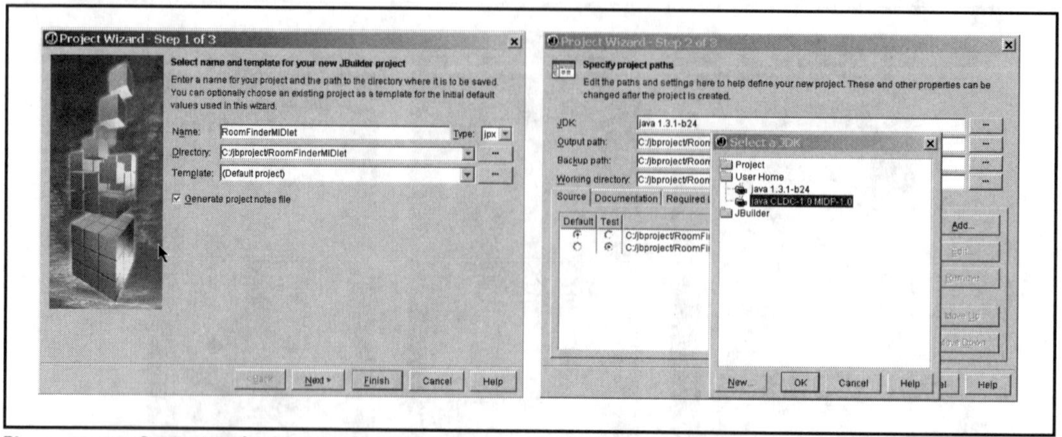

Figure 13-30. Designing the RoomFinderMIDlet "Displayables" Using JBuilder's MobileSet.

After creating the J2ME project, the next step is to add a MIDlet and MIDP Displayables to the project. This is done using the Micro tab on the Object Gallery wizard (see Figure 13-31).

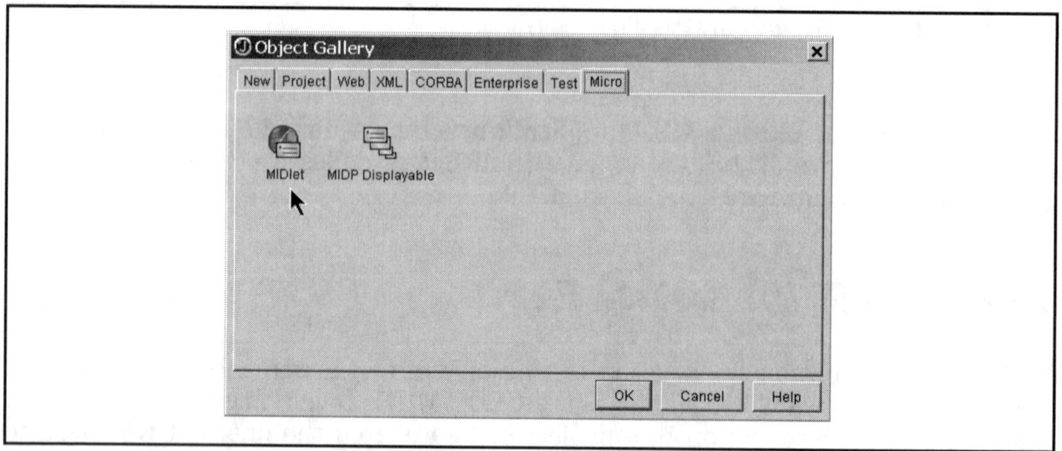

Figure 13-31. Choosing MIDlet and MIDP Objects from the Object Gallery.

When the MIDlet is added, a wizard appears for choosing Package and Class names and to add the first Displayable class (see Figure 13-32). The second wizard page has fields for entering the Class name, the Title for the Displayable class, the displayable type, and the type of command handling. The **RoomFinderMIDlet** will have two Displayables: **CourseSelect** and **RoomDisplay**.

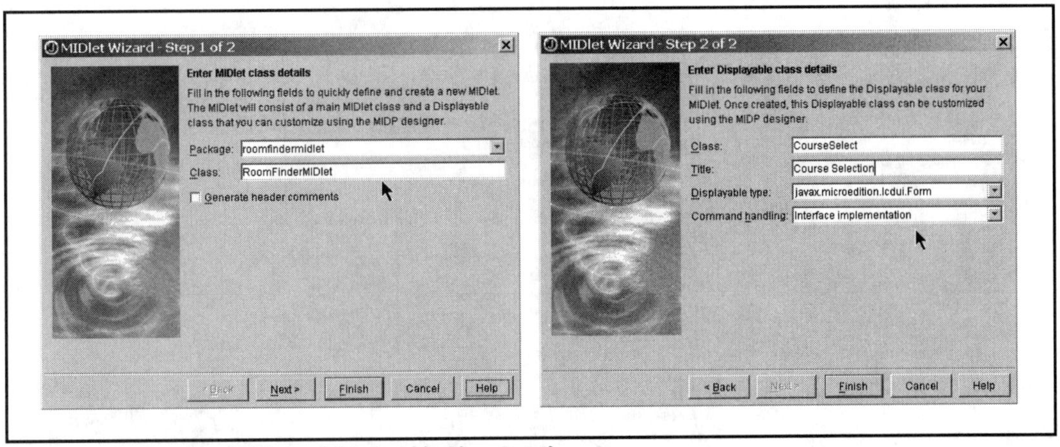

Figure 13-32. Entering MIDlet and Displayable Class Details.

When additional Displayable classes are added to the MIDlet, a wizard appears to enter the details (see Figure 13-33).

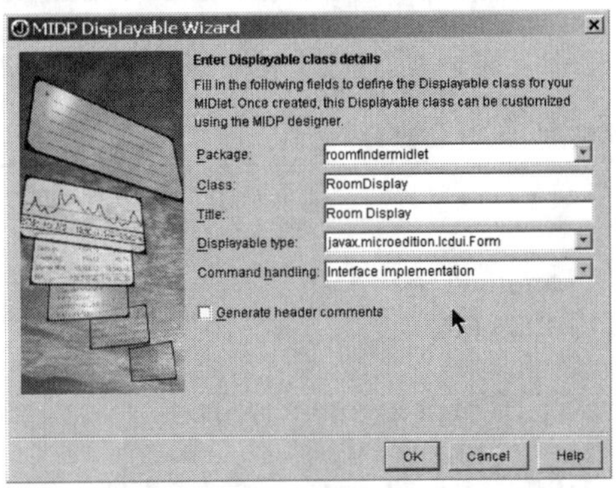

Figure 13-33. Adding the RoomDisplay Displayable to the RoomFinder MIDlet.

With the **RoomFinderMIDlet** and the **CourseSelect** and **RoomDisplay** Displayable classes defined, we are ready to complete the code for the J2ME application. The J2ME project also allows us to navigate through the project files using an explorer pane at the upper left, a method navigation pane at the lower left, and the source code on the right (see Figure 13-34).

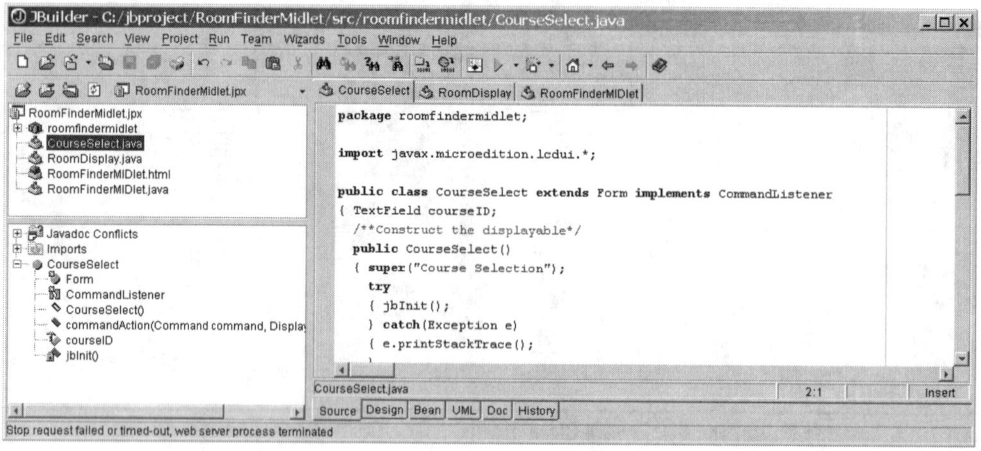

Figure 13-34. Completing the J2ME RoomFinder MIDlet and Displayable Code.

Listings 13-8 through 13-10 present the RoomFinderMIDlet source code that was generated by the JBuilder MobileSet environment, then modified to add the Room Finder function. You might want to review the MIDP classes in Chapters 9 and 10 if you need a refresher.

Listing 13-8. The RoomFinder MIDlet Class.

```
package roomfindermidlet;

import javax.microedition.midlet.*;
import javax.microedition.lcdui.*;

public class RoomFinderMIDlet extends MIDlet
{ static RoomFinderMIDlet instance;
  CourseSelect displayable = new CourseSelect();

  /** Constructor */
  public RoomFinderMIDlet()
  { instance = this;
  }

  /** Main method */
  public void startApp()
  { Display.getDisplay(this).setCurrent(displayable);
  }

  /** Handle pausing the MIDlet */
```

```
  public void pauseApp()
  { }

  /** Handle destroying the MIDlet */
  public void destroyApp(boolean unconditional)
  { }

  /** Quit the MIDlet */
  public static void quitApp()
  { instance.destroyApp(true);
    instance.notifyDestroyed();
    instance = null;
  }
}
```

Listing 13-9. The CourseSelect Displayable Class.

```
package roomfindermidlet;

import javax.microedition.lcdui.*;

public class CourseSelect extends Form implements CommandListener
{ TextField courseID;
  /**Construct the displayable*/
  public CourseSelect()
  { super("Course Selection");
    try
    { jbInit();
    } catch(Exception e)
    { e.printStackTrace();
    }
  }

  /**Component initialization*/
  private void jbInit() throws Exception
  {
    // set up this Displayable to listen to command events
    courseID = new TextField("", "", 15,TextField.ANY);
    courseID.setLabel("Enter Course ID:");
    setCommandListener(this);

    // add the commands
    addCommand(new Command("Exit", Command.EXIT, 1));
    addCommand(new Command("OK", Command.OK, 1));
```

```
    this.append(courseID);
  }

  /**Handle command events*/
  public void commandAction(Command command, Displayable displayable)
  { if (command.getCommandType() == Command.EXIT)
    { RoomFinderMIDlet.quitApp();
    }

    if (command.getCommandType() == Command.OK)
    { RoomDisplay display = new RoomDisplay(courseID.getString());
      Display.getDisplay(RoomFinderMIDlet.instance).setCurrent(display);
    }
  }
}
```

Listing 13-10. The RoomDisplay Displayable Class.

```
package roomfindermidlet;

import javax.microedition.lcdui.*;
import java.util.*;
import javax.microedition.io.*;
import java.io.*;

public class RoomDisplay extends Form implements CommandListener
{ StringItem courseID;
  StringItem room;

  /**Construct the displayable*/
  public RoomDisplay(String courseID)
  { super("Room Display");
    try
    { jbInit();
      displayRoom(courseID);
    } catch(Exception e)
    { e.printStackTrace();
    }
  }

  private void displayRoom(String course)
                           throws IOException, EOFException
  { courseID.setText(course);
    HttpConnection conn = null;
```

```
  DataInputStream dis = null;
  try
  { String roomFinderURL =
           "http://localhost:8080/RoomFinder/RoomFinderServlet";
    String urlString = roomFinderURL + "?" +  "CourseID=" + course;
    conn = (HttpConnection)Connector.open(urlString,
                                Connector.READ_WRITE);
    dis = new DataInputStream(conn.openDataInputStream());
    room.setText(dis.readUTF());
 } finally
 { if (conn != null) conn.close();
   if (dis != null) dis.close();
 }
}

/**Component initialization*/
private void jbInit() throws Exception
{ // create StringItems
  courseID = new StringItem("", "");
  room = new StringItem("", "");
  room.setLabel("Room:  ");
  courseID.setLabel("Course ID:  ");

  // set up this Displayable to listen to command events
  setCommandListener(this);

  // add commands
  addCommand(new Command("Exit", Command.EXIT, 1));
  addCommand(new Command("Back", Command.BACK, 1));

  // add StringItems to form
  this.append(courseID);
  this.append(room);
}

/**Handle command events*/
public void commandAction(Command command, Displayable displayable)
{ if (command.getCommandType() == Command.EXIT)
  { RoomFinderMIDlet.quitApp();
  }

  if (command.getCommandType() == Command.BACK)
  { CourseSelect select = new CourseSelect();
    Display.getDisplay(RoomFinderMIDlet.instance).setCurrent(select);
  }
```

```
    }
}
```

SUMMARY

This completes our overview of the J2EE programming environment. We also showed how to add a wireless client that can access our J2EE application. Tools were a big part of the story. They relieve much of the drudgery of various programming tasks required for our J2EE application and also helped add a MIDP client to our application.

So what's next? We want to extend the reach of our J2EE applications and also make them easy to provide as a service to our wireless clients. Web Services is a developing technology that opens up new possibilities to make applications accessible from the Internet. It also lets us define and compose services using a registry. We will explore Web Services in the next chapter.

Chapter 14

Web Services for Wireless Devices

In this chapter we will introduce the next revolution in distributed computing—Web Services. Web Services are self-describing, lightweight, inter-galactic components that comprise a *Service-Oriented Development (SOD)* approach to crafting distributed applications. The creation and assembly of these distributed services is expected to foster the emergence of a new breed of developers who will be more focused on the business process flow within an application. These developers can leverage new code with pre-existing logic in building distributed composite applications.

In this new model, XML is expected to play an important role and will be used to bind diverse services together and match the impedance differences amongst them. The difference between the traditional distributed computing approach and this SOD methodology will be its increased flexibility and the ability for enterprises to reconfigure business process logic much more quickly than in the past.

WEB SERVICES—EVERYONE'S FAVORITE BUZZWORD

It seems that everywhere you look, someone is talking about how Web Services are the "be all, end all" of software development. While the Web Services design model does lend itself well to certain applications, it certainly is not applicable to all scenarios. To make things even worse, the Web Services landscape is littered with a veritable alphabet soup of acronyms. Every week, a new three- or four-letter standard is being announced, making it difficult to keep track of which standards are viable and which are dead ends.

Fortunately, the three main standards seem to be well entrenched in the industry. We will concern ourselves with these three main standards: SOAP, WSDL, and UDDI, leaving research into the more esoteric standards as an exercise for the reader.

What Web Services Are...and Aren't

Before jumping into our discussion on Web Services standards, let's digress for a moment and examine the similarities between Web Services and traditional distributed object systems (such as CORBA and Java RMI).

- Both traditional distributed object systems and Web Services attempt to provide solutions for distributed computing over a network, typically involving software components executing in different processes on different systems.
- Both provide mechanisms for locating components. In traditional systems, this is typically accomplished by using a naming or directory service of some type. In the Web Services model, a software program called a *service registry* is utilized in much the same fashion.
- Both provide mechanisms for suitably encoding method or function invocations between participants.
- Both provide mechanisms for suitably marshalling parameters passed as arguments in function calls or method invocations.
- Both provide rules detailing on-the-wire bits and bytes that are sent over the network.
- Both provide mechanisms for determining if faults have occurred during a method call or function invocation.
- Both models provide *introspection* and *reflection* capabilities. This means that all the information needed to interact with a distributed component can be discovered at runtime.

From the above descriptions, it sounds as if Web Services fill much the same niche as traditional distributed object systems. However, this is not an entirely accurate view. There are certain fundamental differences between the two models that make

Web Services more suited for particular tasks, and traditional distributed object systems more suited for other tasks. Some of the key differences between the two approaches include:

- ***Object Orientation:*** Web Services are not object oriented. Yes, you heard that right. Web service invocations are much closer to traditional RPC calls. When you invoke a function provided by a Web Service, the invocation is stateless—a different service provider could handle your request each time it was issued. This is in sharp contrast to traditional distributed object systems, which are by nature object-oriented. You can be sure that when you hold a reference to a particular object in a traditional distributed object system, you are always assured that subsequent method invocations on that object will be operating against a set of data encapsulated by that object. Web Services have no encapsulated data—you either have to pass along the data you want to operate on, or provide a means for the service to locate the data.

  Since Web Services are stateless service providers, the same function code is always executed when a particular message is received. Object-oriented systems provide *polymorphism*, which means the same message may be interpreted and handled differently by different objects.

- ***Protocols:*** Web Services may have an upper hand here, as the three core standards (SOAP, WSDL, and UDDI) all have their roots in XML and are well defined and accepted by most of the key players in the industry. This is in contrast to traditional systems whose protocols run the gamut from JRMP to GIOP to DCOM, etc.

- ***Architecture and language neutrality:*** With the exception of CORBA, traditional distributed object systems are typically tied to one programming language (for example, Java in the case of the EJB system), or architecture (DCOM or COM+ are examples). Web Services, on the other hand, by virtue of their exchange of XML messages for most interactions, can coexist and interact with any platform or programming language.

So what applications are Web Services well suited for? Some examples include:

- ***Data Mining:*** A collection of data is returned based on a few criteria supplied as parameters to a Web Service.

- ***Weather Display:*** Some location parameters specify a location on the globe and the weather related to that location is returned.

- ***Real Estate Listing:*** Specify a zip code and other similar search criteria. A listing for houses for sale in the location is then returned.

As you can see, these applications are browsing applications that fit the stateless Web Service model.

WEB SERVICES—A BIRD'S EYE VIEW

Now that we've seen how Web Services compare to traditional systems and how they fit into the overall distributed computing picture, we can take a closer look at the subsystems that comprise a typical Web Services system.

We mentioned earlier how Web Services are platform-neutral and language-neutral. What this means is that you, as an application developer, need not care about which language or platform upon which a particular service has been implemented. Further, because Web Services support run-time discovery and configuration, you really only need to know the service name, and perhaps some metadata distinguishing the service from other similar services. All other information about the service, including its parameters and calling conventions, can be discovered. This approach is analogous to searching for a service in the yellow pages. There could be hundreds of doctors, lawyers, and car repair shops listed. All we need to know is the name of the service we are interested in using, and the yellow pages directory contains all the information we need to locate the service (the phone number, address, business hours, etc).

The Web Services model contains a directory that can be used by your applications just like we humans use the phone company's directory. In fact, the directory, or *service registry,* explicitly provides services for "yellow pages-style" lookups, "white pages-style" lookups, and "green pages-style" lookups. We'll see more of this in a subsequent section. The directory, known as the *Universal Description Discovery and Integration (UDDI)* registry, records information about deployed Web Services in the same fashion as the white pages and yellow pages telephone directories record information about people and businesses.

Web Services are loosely-coupled software modules; they do not have to be bound together. This implies that Web Services do not have to know about each other during build time or development time as they can discover each other during runtime.

Each deployed service has an associated *service description* that describes to the world its capabilities. Since this description has to be sufficiently flexible and extensible so as to represent the service's parameters (and their data types), a standard way of communicating this information has been agreed upon. This standard, the *Web Services Description Language (WSDL)*, is an XML application that is designed to fully describe everything needed to bind to and invoke a service. The UDDI service registry described above can point parties interested in a particular service to that service's associated WSDL document.

Web Services address the need for interoperability by defining a standard XML-based protocol called *Simple Object Access Protocol (SOAP)*. Using SOAP, XML messages are used to model interactions between Web Services, which typically involve invocations of methods. SOAP messages can be synchronous (i.e., every request would wait for a response before proceeding) or asynchronous (i.e., after making a request, the execution continues without waiting for a response.

Web Services can be modeled by considering the interactions between three actors, each of which is typically a software application:

- The ***service provider***, which publishes information about the service (the service description) to a service registry.
- The ***service registry***, which maintains a catalog of deployed (published) services and makes this catalog available to new service providers and service requesters.
- The ***service requester***, which finds the service by using a lookup against the service registry, and then binds to the service provider. At this point, the provider and requestor can interact in a meaningful way.

These concepts are illustrated in Figure 14-1.

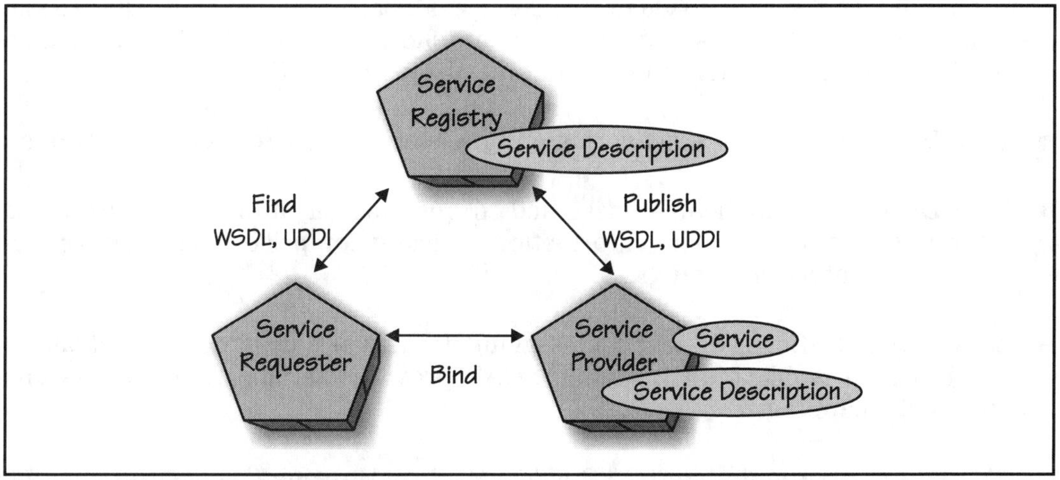

Figure 14-1. Bird's Eye View of Web Services.

TECHNOLOGY BUILDING BLOCKS FOR WEB SERVICES

To understand how these pieces fit together, let's take a look at the technological building blocks used when designing a Web Services-based software system. These building blocks are shown in Figure 14-2 below.

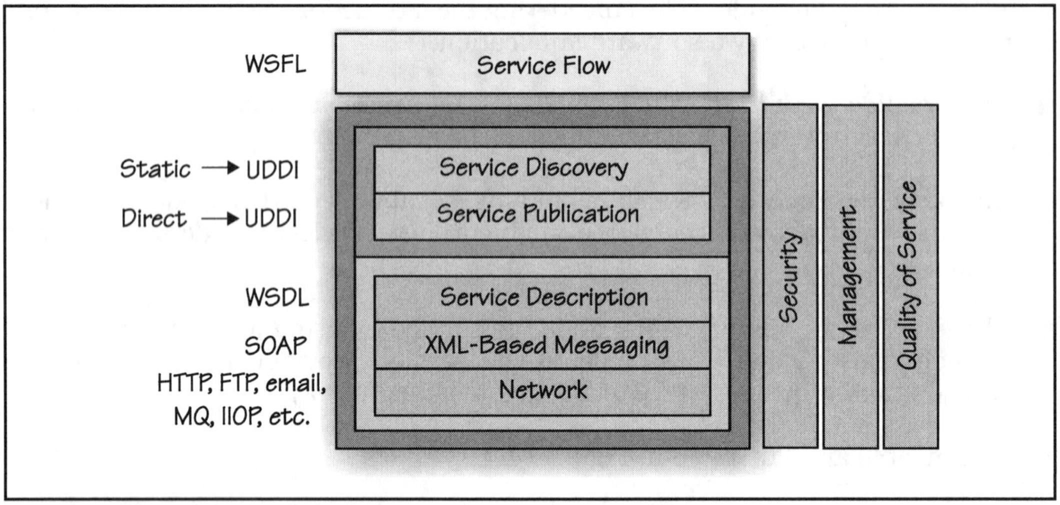

Figure 14-2. The Web Services Technologies.

Some building blocks, such as Quality of Service, Management, and Security are still under development. The bulk of the components shown in Figure 14-2 are explained below. We will discuss each in more detail in subsequent sections.

- ***UDDI:*** A repository for discovering and registering Web Service descriptions.
- ***WSDL:*** An XML application designed to describe all information needed to interact with a service. This information typically comprises messages, operations, and protocol mappings.
- ***SOAP:*** A specific XML-based messaging framework designed to act as an application-to-application protocol. SOAP provides serialization formats and RPC-style mapping.
- ***Transport protocols*** like HTTP, FTP, SMTP, or Message Queues for carrying SOAP messages.
- ***Flow languages*** describe the ordering of interactions between services. This is important, as there typically would be no centralized server to control business processes. For describing the flow of interactions between services,

IBM has proposed Web Services Flow Language (WSFL) and Microsoft has proposed XLANG.

In subsequent sections, we will provide detailed information about a few of these fundamental building blocks. Before we do so, however, we need to understand how the pieces fit together to form an overall architecture.

FYI

Web Services' Ancillary Building Blocks

Briefing

You'll notice that in the discussion above, we only covered the major Web Services standards. There are other standards in various stages of industry acceptance and approval.

Standards governing security, transaction management, and flow among Web Services seem to be gathering the most attention lately as standards organizations push the Web Services model toward the same level of functionality that current distributed object systems provide.

When considering security issues posed in the Web Services model, standards such as SAML, the *Security Assertions Markup Language* could be employed. Standards governing security must provide the ability to resolve security credentials between different security domains. This basically means that any security framework used for Web Services, if it is to gain any reasonable amount of industry acceptance, must account for the various security models used by each participating service, whatever they may be. Furthermore, these security models should address *non-repudiation* concerns (being able to prove an activity took place) and provide the means to audit service interactions should the need arise.

Transactional characteristics in the scope of Web Services could be similarly represented using *XA Markup Language (XAML)* or *Business Transaction Processing (BTP)*. These new proposals are designed to extend traditional transactional capabilities present in existing applications to the Web Services arena. The proposals address coordination of invocations on multiple services, compensating transactions to undo any undesired side-effects, and the ability to commit or cancel the transaction at any point during its life.

We're going to take a wait-and-see approach as the key players in the industry jockey for positions, trying to get their standards ratified and accepted. Security and transaction support seem to be necessities, but it's unclear right now what standards are going to beat out the others. ❑

WEB SERVICES ARCHITECTURE

From an application's perspective, a Web Service can be defined as a self-describing, self-contained, modular unit of application logic that provides some business functionality through an Internet connection. Applications access Web Services via ubiquitous Web protocols and data formats, such as HTTP and XML, and due to their self-describing nature, the services can be located and invoked with no prior knowledge of how they are implemented. Web Service execution/interactions can be ordered, linked, and chained together with other Web Services to execute a larger workflow or business transaction.

Figure 14-3 illustrates the overall architecture of a Web Service.

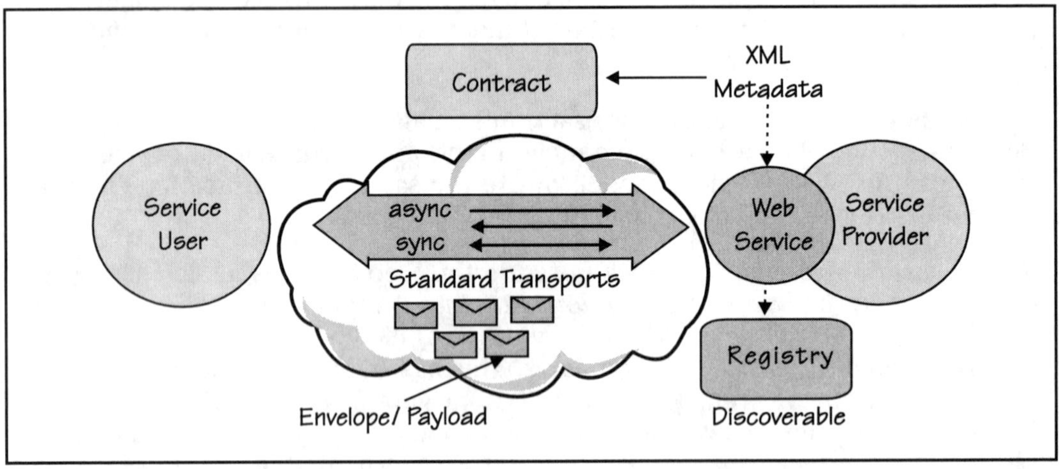

Figure 14-3. Architecture of Web Services.

The interactions of the service user and service provider are represented in XML messages (typically SOAP messages) and passed between the parties as synchronous or asynchronous messages. Although we haven't talked about SOAP in detail yet (we will shortly), SOAP messages can be exchanged using a variety of protocols. Some of these protocols, like HTTP, are synchronous in nature; a message is sent, and a response is received shortly thereafter. However, SOAP messages can also be exchanged using asynchronous protocols like SMTP (e-mail). The SOAP messages exchanged by the parties contain an envelope, inside which the payload of the message (usually a method invocation or a response to one) is placed.

The service contracts (which include interface definitions containing the location and technical information about each service) are also represented using XML and stored in the service or business registry to enable them to be discoverable as needed at run-time. The standard typically used to represent the service contract is WSDL, and the registry used is a UDDI registry, although this configuration is not mandatory.

Any Web Service has four distinct development life cycle phases:

- The ***build phase***, in which the Web Service would be developed and tested, and the service contracts would be created.
- The ***deployment phase***, in which the service contracts would be published (typically in a UDDI registry) and the run-time code would be deployed on a server (typically a SOAP server).
- The ***run phase***, when the service is available for a client to discover it, locate it (or bind to it), and invoke all defined operations.
- The ***management phase***, where scheduled management, maintenance, and administration should be addressed.

Once a service has been developed, it must be deployed on a server and its service contract must be published to a UDDI registry. Applications use SOAP when interacting with the registry to publish services. We stated earlier that the service contract is typically modeled using WSDL, but do be aware that there is no formal relationship between UDDI and WSDL (although the technologies are complementary). This means it is completely legal for a service to register its service contract using some other language besides WSDL, although in practice this is rarely done.

Once a service contract has been published, service requesters can find it by browsing the registry in which it is published. Using the information in the service contract, the requester can generate a proxy object or class capable of interacting with the service. Since the service contract is platform- and language-neutral, service requesters can create these proxy objects or classes in whatever language needed. Subsequently, the service proxy can be compiled and used to invoke the service. The act of creating a link between a service requester and a service provider is called *binding*. Binding to a service can be either static or dynamic. When a proxy class is used to access the service, this is known as static binding. When the information in the service contract is used at run time to construct an appropriate service invocation, this method is known as dynamic binding. We'll talk more about both types of binding shortly.

Web Services Operations

We briefly touched upon several concepts in the previous section. Let's take a closer look at the four important operations that constitute what most ordinary Web Services-based applications perform:

- ***Publish:*** A service's contract is published to a service or business registry.

- ***Find:*** A service requester can query the registry for services matching certain criteria. The requester receives all the information needed to interact with the services.

- ***Bind:*** A link is created between the service requester and service provider for service invocations. The service requester uses this link to invoke the service.

- ***Invoke:*** The service requester asks the service provider to perform the service, optionally passing any needed parameters or arguments to the service.

Publishing a Web Service

Publishing a Web Service is a three-step process. The first step is to create a set of WSDL documents describing a service's interface. The second step involves publishing the service in a UDDI registry and attaching the location of the generated service documents to the registration information so interested parties can locate the service. The final step involves actually taking the code that provides the service and deploying it on an application server. Figure 14-4 illustrates what needs to occur when publishing a service:

1. A set of WSDL service description documents needs to be created or generated. When viewed in combination, these documents define the characteristics of an individual service.
2. The service is published to the UDDI registry using the UDDI API. UDDI implementations often include helpers called WSDL Proxies. These WSDL proxies make it easier for the programmer to publish business information and WSDL documents relating to services in one interaction, reducing the amount of boilerplate code needed to perform these common operations.
3. The service provider code is published on an application server.

Finding a Web Service

A service requester uses the find operation to locate an actual service. First, the UDDI registry would be searched to find a particular service. Depending on the UDDI implementation being used, the service requester may be able to make use of the same WSDL proxies described in the earlier section. Furthermore, the service requester may even be provided with automatically generated proxy classes they can use to interact directly with the service. This is illustrated in Figure 14-5.

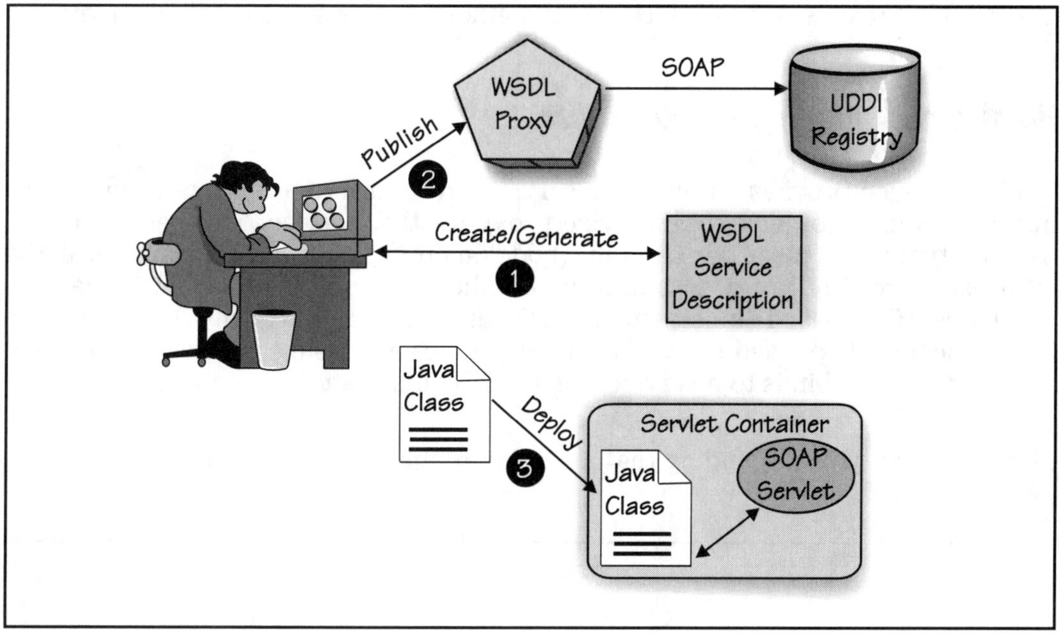

Figure 14-4. Publishing a Web Service.

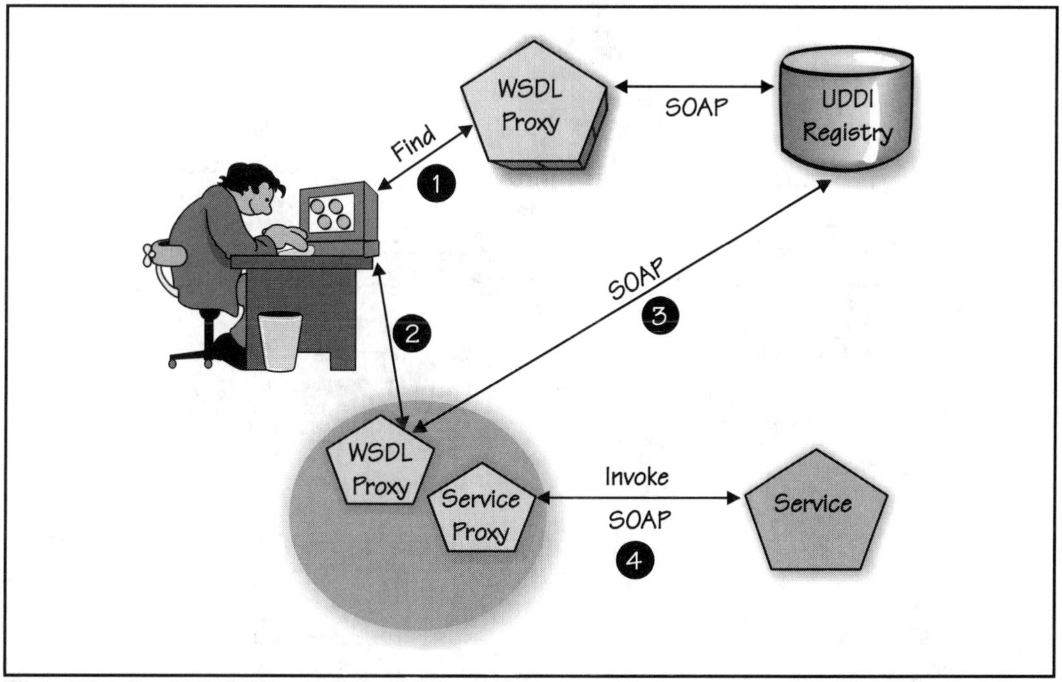

Figure 14-5. Finding a Web Service.

The service proxy is then used to invoke methods on the service implementation.

Binding and Invoking a Web Service

When a service provider publishes a service (in the service registry), the service interface is used to generate a service proxy for that service. This service proxy, which is typically a piece of executable code (such as a Java class), contains all the information required to access and invoke the service. For example, if the Web Service is SOAP-based (most usually are), the service proxy will contain all the SOAP client code needed to invoke the SOAP service. Using this service proxy, a service requester binds to a service. Figure 14-6 illustrates this concept.

There are two types of binding that a service requester can use when binding to a Web Service:

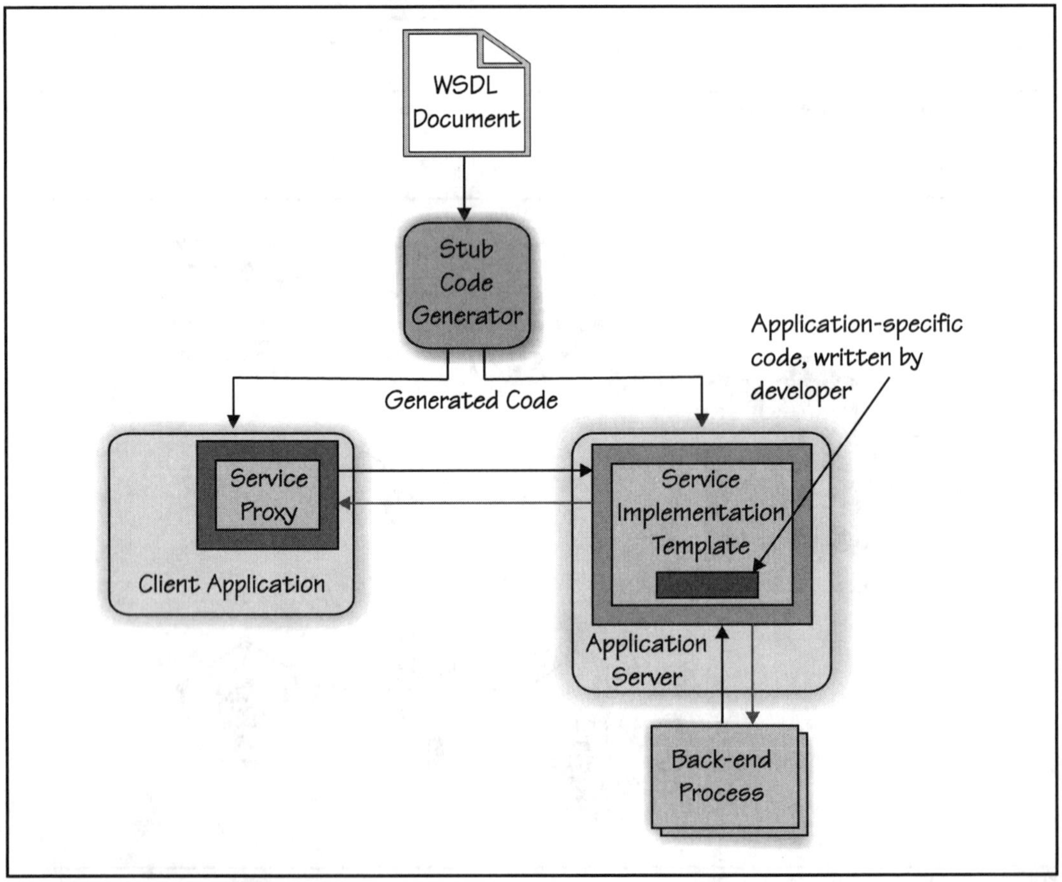

Figure 14-6. Invoking a Web Service.

- ***Static binding:*** In this model the service implementation definition is found during build-time. This implementation would contain the reference to the service interface definition and the location where the actual service can be found. Then, using this service implementation definition and the service location, the service interface definition is located and the service proxy is generated. All these would happen during build-time. A service requester using this proxy will always try to access the service at that same location every time the service is invoked.

- ***Dynamic binding:*** This method is used when the requester knows the type of service to use and when the actual implementation is unavailable or could change at run-time. In this method the service interface is located and the proxy is generated during build-time. However, during run-time, the requester has to identify the particular service implementation before the proxy can invoke the desired service.

Web Services Standards—The Devil Is in the Details

Now that we understand how Web Services are organized, and have a basic understanding of their architecture, let's examine each major Web Service standard in more detail.

Simple Object Access Protocol (SOAP)

SOAP is a lightweight protocol for exchanging structured and typed information between peers in a decentralized, distributed environment using XML as the encoding language. Due to its simplicity, flexibility, and power, SOAP has become the cornerstone technology for Web Services.

The SOAP standard defines rules for expressing some common concepts in the context of an XML message:

- ***Message formatting rules*** that describe the message, so that all participants know how the message should be processed.
- ***Encoding rules*** to describe how application-defined data types are encoded.
- ***Elements*** that describe the desired remote procedure calls and responses.

SOAP evolved as a standardized way of doing XML-RPC. It provides standard encoding for the most commonly used data types, such as strings, numbers, and arrays.

The SOAP specification does not mandate a transport protocol, so SOAP could be used with messaging protocols like SMTP, or even in an asynchronous system like a message queue. SOAP can work in conjunction with any protocol; typical implementations are based on protocol bindings that describe how a SOAP message can be carried in an HTTP payload. As such, HTTP is the only binding defined for SOAP as part of the standard.

SOAP messages are one-way messages (blocking calls) and when encoded using HTTP, the SOAP responses are delivered as HTTP responses using the same connection that was used for the request. Figure 14-7 illustrates this.

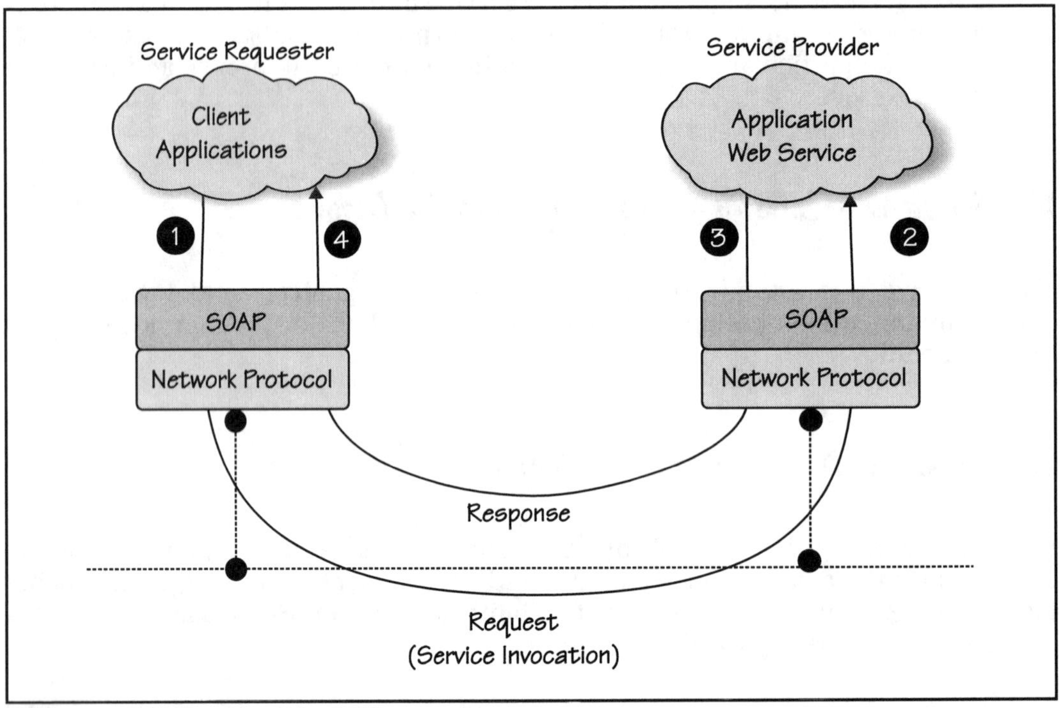

Figure 14-7. XML Messaging Using SOAP.

SOAP Message Format

SOAP messages consist of:

- An ***Envelope***, which contains an optional header element and a mandatory body element. The envelope is the outermost element in the SOAP document.
- A ***body***, which contains the SOAP message being delivered.

- An ***optional header***, which may contain information relating to routing, rules that are to be followed when processing the message, etc.

- An ***optional fault element***, which typically contains information about an error or exception that occurred when processing a previous message.

The SOAP message format, although adhering to well-defined organizational rules, provides flexibility by allowing certain extensions:

- ***Message syntax:*** The Header element can be used to add any type of extension to a message.

- ***Data Payload:*** You can use your own custom serializers and send any data using a SOAP message.

- ***Purpose:*** Although typically associated with remote procedure calls, SOAP messages can serve a number of other purposes. SOAP can even be used as a general-purpose messaging system.

- ***Transport protocol:*** You can use any protocol that you want—The SOAP specification does not mandate a protocol.

Figure 14-8 shows a sample SOAP document and marks the header and body. Figure 14-9 illustrates two SOAP documents. The first document, the "request", contains a message called *GetLastTradePrice*. The second document, the "response," con-

```
<SOAP-ENV:Envelope
   xmlns:SOAP-ENV="SoapEnvelopeURI"
   SOAP-ENV:encodingStyle="SoapEncodingURI">
  <SOAP-ENV:Header>
<t:Transaction xmlns:t="TxURI">
 SOAP-ENV:mustUnderstand="1">
<tid>123456</tid>
</t:Transaction>
  <SOAP-ENV:Header>

  <SOAP-ENV:Body>
<m:GetLastTradePrice xmlns:m="ServiceURI">
<tickerSymbol>ACME</ticketSymbol>
</m:GetLastTradePrice>
  </SOAP-ENV:Body>
</SOAP-ENV:Envelope>
```

HEAD

BODY

Figure 14-8. Example of a SOAP Envelope.

REQUEST

```
<SOAP-ENV:Body>
  <m:GetLastTradePrice
    xmlns:m="ServiceURI">
    <tickerSymbol>ACME</tickerSymbol>
  </m:GetLastTradePrice>
</SOAP-ENV:Body>
```

RESPONSE

```
<SOAP-ENV:Body>
  <m:GetLastTradePriceResponse
    xmlns:m="ServiceURI">
    <result>45.25</result>
  </m:GetLastTradePriceResponse>
</SOAP-ENV:Body>
```

Figure 14-9. The Body of a SOAP Request and Response.

tains a message called *GetLastTradePriceResponse*. These two documents show how SOAP can be used to deliver a request and a response to the request.

In this request document, we are modeling an RPC message being sent to the party receiving the request. The method name, *GetLastTradePrice*, becomes the element name and the parameter(s) become child elements of the method element.

In the response message, the word "Response" is appended to the name of the service method to indicate that this is the response for the invocation of the service method. Hence, in this example, we see *GetLastTradePriceResponse*.

Custom SOAP Serializers

Details

When using SOAP to exchange parameters, XML is used as the underlying format for marshalling and unmarshalling an object. However, the SOAP standard only defines encoding rules for some well-defined, common types. What do you do if the data you're transferring doesn't adhere to these simple types?

There are a few options open to you. You could:

- Encode everything as a string using Java's toString(..) methods and pass serialized representations of your data as SOAP parameters. This method is cumbersome and adds a significant amount of overhead to your SOAP message. Furthermore, you have to ensure that your service provider and service requestor are in lockstep with their Java versions. Finally, this method is completely useless unless you're using a 100% Java environment on both ends.
- Rewrite your code to use only standard SOAP types. This method might not be possible if you don't have control over the code.
- Use a custom serializer. This technique allows you to define a Java class that will marshal an object or objects of your choice into a SOAP-compatible representation, behind the scenes. A similar de-serializer class reconstitutes your object on the remote side.

Both types of custom serializers define how the object has to be encoded when sending it over the network. Most SOAP servers (such as Apache Axis and its older variety, Apache SOAP) provide a bean serializer class, which you can use to exchange objects between your client and server, provided your objects exhibit bean-like characteristics.

The bean serializer class imposes some constraints on your objects. First and foremost, the objects that you choose to exchange need to follow standard JavaBean naming conventions for attribute accessors. You also need to make sure that a parameter-less constructor is provided. You also need to tell the SOAP toolkit when to use the bean serializer (e.g., for what types should the bean serializer be utilized). When the SOAP toolkit encounters an object being passed as a parameter to some SOAP service (or as a return value from some SOAP service) whose type matches one of the types for which a bean serializer was specified, the SOAP toolkit inspects the state of the object by invoking its get(..) methods. The results are serialized into XML and passed as part of the SOAP message.

The corresponding deserializer would then be able to instantiate an instance of the same type and populate it using the setter methods provided by the object.

It is also possible to write your own "from-scratch" custom serializer. You would want to write your own serializer if you wanted to perform custom actions before sending your object, or if you are attempting to serialize an object for which the bean serializer is unsuitable—perhaps the object does not possess bean-like get(..) and set(..) methods. ❑

Web Services Description Language (WSDL)

WSDL allows Web Services to describe a service's interfaces and bindings. When developers began implementing SOAP-based Web Services, they needed to somehow explain how to interact with their services. Early Web Services developers solved this problem by providing a sample SOAP document that illustrated the method invocations their Web Services supported, typically including dummy or sample parameters to illustrate the correct calling convention. However, this method did not provide reflection-like capabilities and was not robust enough to provide any reasonable amount of type checking. Without a way to specify what the parameter types were (which the sample SOAP document did not provide), no type checking could be performed. So, to allow Web Services to be self-describing, WSDL was introduced.

Figure 14-10 shows the various components of a WSDL document. The following describes the components.

- ***Message:*** This is an abstract definition of a message between two parties. Typically, the message descriptions provided in a WSDL document reflect the formatting and naming conventions of the associated SOAP request/response messages they describe.

- ***Types:*** These define data types to be used in the messages, typically in the form of XML Schemas, although WSDL allows other mechanisms. These are the elements or the data structures that will be contained in a message.

- ***Operation:*** This is an abstract definition of the operation for a message, such as naming a method, message queue, or business process that will accept and process the message. You can think of an operation as an activity that will be performed by a remote service provider.

- ***Port Type:*** A port type represents a set of operations that are implemented by some service provider residing at some location. Port types do not represent a particular service provider. Rather, they define the possible operations for any number of service providers who provide a common service. Port types also don't contain any information relating to a particular protocol or binding. The closest analogy to a port type in the traditional distributed object world would be an *abstract interface*. Thanks to its abstract nature, once a port type is defined, it can be mapped to multiple transports through different bindings.

- ***Binding:*** This defines how an operation is bound to a protocol. The combination of a protocol and data format for a port type defines how the port type is mapped on to a particular transport. Any protocol can be used to bind messages.

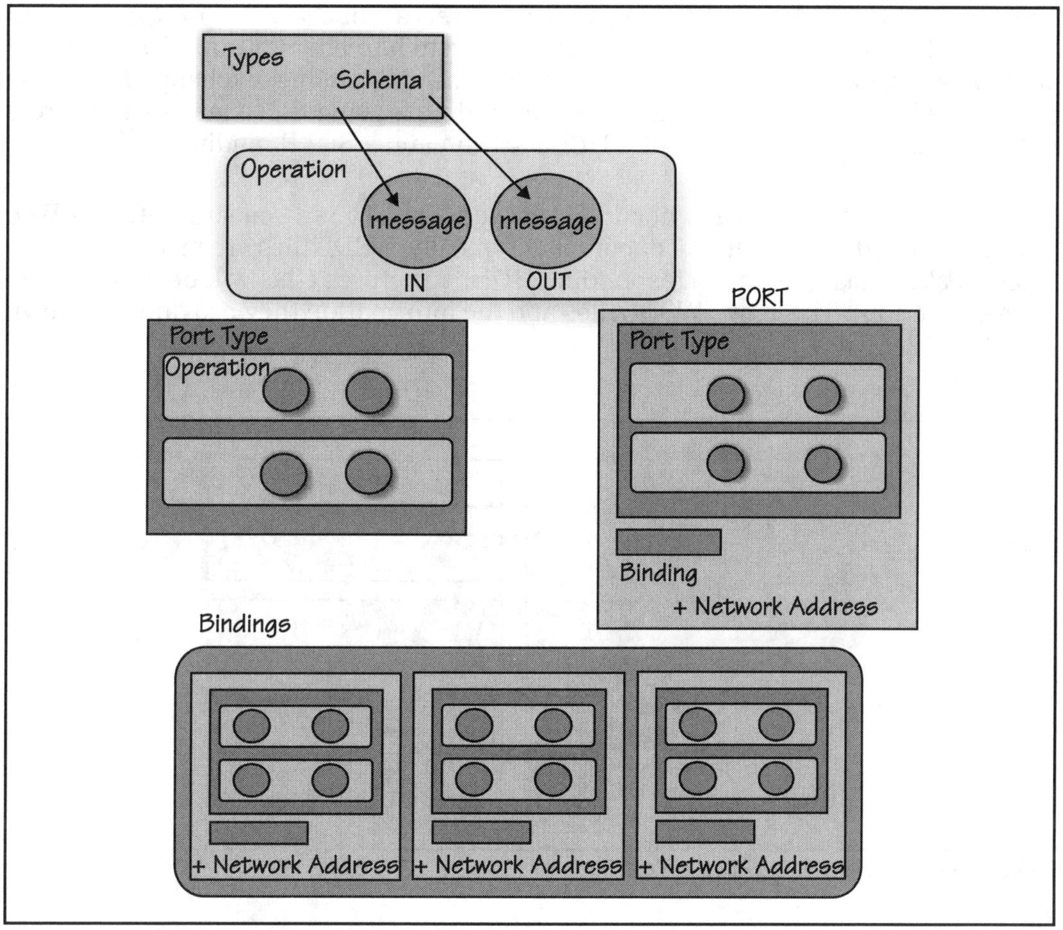

Figure 14-10. The Components of WSDL.

The WSDL specification defines SOAP, HTTP GET/POST, and MIME message binding formats.

- ***Port:*** A combination of a binding and a network address, that provides the target address of the service communication. The ports define the actual network end-point that associates a binding with a protocol-specific address.

- ***Service:*** This is a collection of related endpoints encompassing the service definitions in the file; services map the binding to the port (in short, a service is a collection of ports).

The components of a WSDL document fully describe how to invoke a service. The document conveys information about the data being exchanged, the sequence of messages for an operation, protocol bindings, and the location of the service.

Typically, a WSDL document defines services as a collection of *endpoints*, but separates the abstract definition from the concrete implementation. *Messages* and *port types* provide abstract definitions for the data being exchanged and the operations being performed by a service. A binding is provided to map to a concrete set of ports, usually consisting of a URL location and a SOAP binding.

The root element of a WSDL document is `<definitions>`—meaning that a Web device is being described in the document. Typically, WSDL files are broken into two pieces that are mapped to corresponding UDDI structures (this will be explained in the next section). These two WSDL files and the information they provide are shown in Figure 14-11.

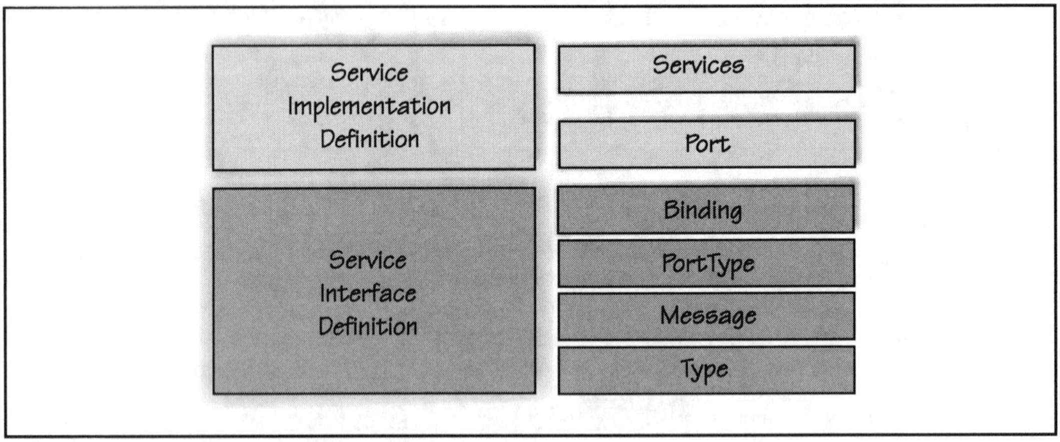

Figure 14-11. WSDL Basic Service Description.

A given set of services may share common protocol bindings, port types, or message formats. Such common information is generally defined in the service interface definition WSDL file, which can be reused for each service in the set.

The information specific to a particular service endpoint is included in the service implementation definition WSDL document, one document for each service in the set.

Universal Description, Discovery, and Integration (UDDI)

UDDI stands for Universal Description, Discovery, and Integration. The UDDI Project is an industry initiative that is working to enable businesses to easily and dynamically find and interact with each other. As the name suggests, UDDI enables a business to:

- ***Describe*** its business and its services.
- ***Discover*** other businesses that offer desired services.
- ***Integrate*** with these other businesses.

One of the important building blocks of the Web Service technology stack is the UDDI Business Registry, usually referred to simply as a "UDDI registry." UDDI is a cross-industry open specification that is built on top of existing standards like TCP/IP, XML, HTTP, DNS, and SOAP. The UDDI registry provides a place for a company to register its business and the services that it offers. People or businesses that need a service can use this registry to find a business that provides the service.

The registry is created as a group of multiple operator nodes. Although each operator node is managed separately, information contained within each registry is synchronized across all nodes.

The UDDI specification defines APIs for programmers to interact with UDDI registries:

- Inquiry APIs are provided to locate businesses, or information about services provided by businesses.
- Publishing APIs are included for creating and deleting UDDI data in the registry.

UDDI Data Structures

Like telephone directories, the information stored in a UDDI registry is grouped into *White Pages*, *Yellow Pages*, and *Green Pages*. In this section, we will explain these categories and the data structures UDDI uses to store them.

Here's a description of the categories:

- ***White Pages:*** The information in the UDDI white pages contains the business name, description of the company, and the contact information.

- ***Yellow Pages:*** The information in the UDDI yellow pages is classified according to industry standard classification schemes. These schemes include the *North American Industry Classification System (NAICS)*, the *Universal Standard Products and Services Classification (UNSPSC)*, and another unnamed geographic taxonomy. A fourth category is also defined—named "ETC"—for general-purpose keyword type classification. In the yellow pages grouping, a business can appear in more than one category.

- ***Green Pages:*** The information contained in the UDDI green pages represents technical information about the services.

The information in the white, yellow, and green pages is logically organized as shown in Figure 14-12.

- ***Business Entity:*** A business entity represents information about a business. Each business entity contains a unique identifier, the business name, a short description of the business, some basic contact information, a list of categories and identifiers that describe the business, and a URL pointing to more information about the business.

- ***Business Service:*** Associated with the business entity is a list of business services offered by the business entity. Each business service entry contains a business description of the service, a list of categories that describe the service, and a list of *specification pointers* to references and information related to the service.

- ***Binding Template:*** Associated with each business service entry is a list of *binding templates* that point to specifications and other technical information about the service. For example, a binding template might point to a URL that supplies information on how to invoke the service. The specification pointers also associate the service with a *service type*.

- ***tModel:*** A service type is defined by what is known as a *tModel* in UDDI. Multiple businesses can offer the same type of service, as defined by the tModel. This provides the ability for a group of businesses to share a set of tModel specifications. A tModel is used to find out how to interact with a service of a particular type. The tModel structure contains the name of the organization that published the tModel, a list of categories that describe the service type, and

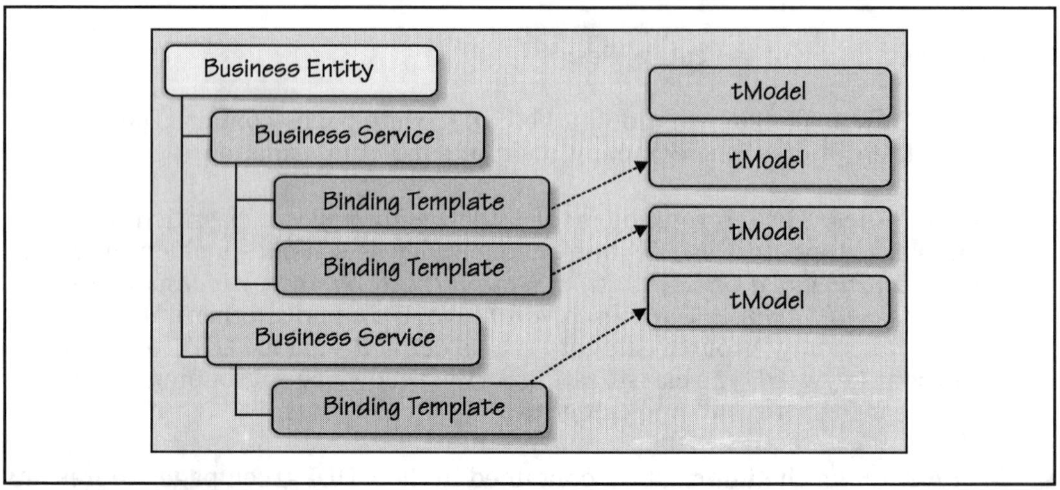

Figure 14-12. UDDI Data Types.

pointers to technical specifications for the service type such as interface definitions, message formats, message protocols, and security protocols.

Users can search the registry for businesses, services, or service types. The UDDI Business Registry offers both a Web-based user interface and a programmatic interface. The UDDI Business Registry uses standard industry taxonomies, or classification schemes, to categorize businesses, services, and service types. Each business entity, business service, and tModel can be associated with a variety of identifiers and categories, enabling users to search the registry by industry, product category, and geographical location.

Any kind of service can be registered in the UDDI Business Registry but the primary intent behind UDDI is to provide a global registry for Web Services.

In the following sections, we describe each UDDI data structure in some detail. A thorough treatment of each structure's nuances is not provided. The UDDI standard fills several hundred pages and an entire book could be written only covering UDDI.

Figure 14-13 illustrates the various elements in the major UDDI data structures.

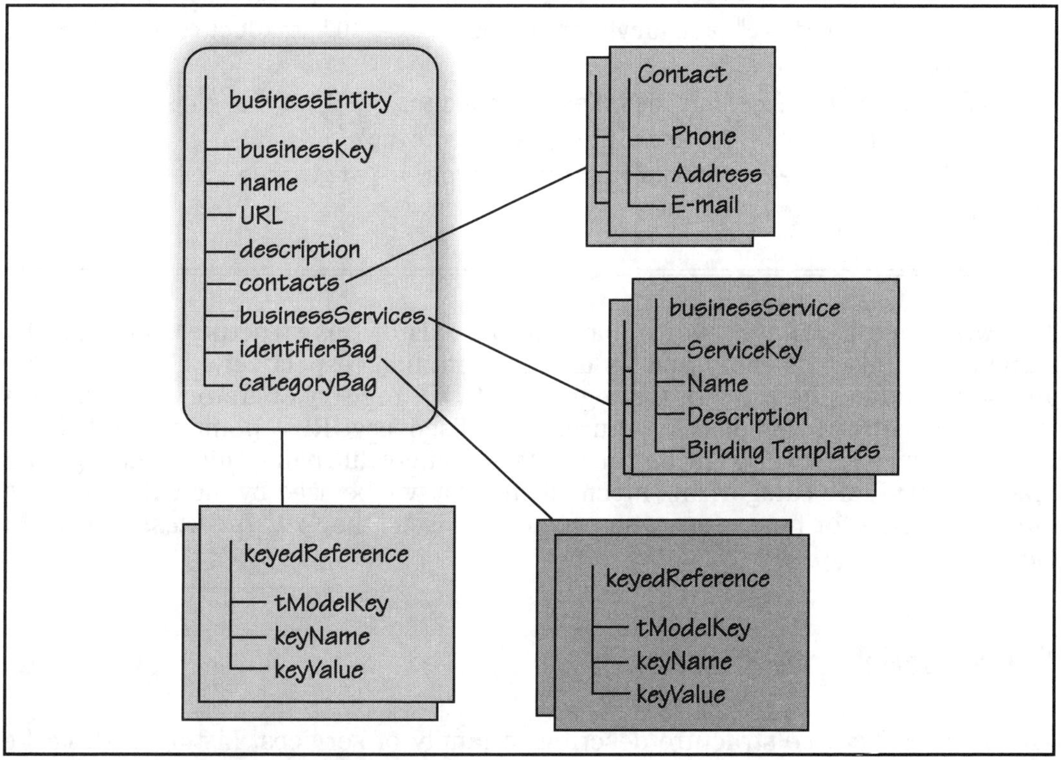

Figure 14-13. UDDI Structures and Their Elements.

The following sections describe the WSDL structures in more detail.

The businessEntity Structure

The *businessEntity* structure provides all published information about a business entity. Listing 14-1 is the specification for the businessEntity structure:

Listing 14-1. Structure Specification for businessEntity Structure.

```
<element name = "businessEntity">
  <type content = "elementOnly">
    <group order = "seq">
      <element ref = "discoveryURLs" minOccurs = "0" maxOccurs = "1"/>
      <element ref = "name"/>
      <element ref = "description" minOccurs = "0" maxOccurs = "*"/>
      <element ref = "contacts" minOccurs = "0" maxOccurs = "1"/>
      <element ref = "businessServices" minOccurs = "0" maxOccurs = "1"/>
      <element ref = "identifierBag" minOccurs = "0" maxOccurs = "1"/>
      <element ref = "categoryBag" minOccurs = "0" maxOccurs = "1"/>
    </group>
    <attribute name = "businessKey" minOccurs = "1" type = "string"/>
    <attribute name = "operator" type = "string"/>
    <attribute name = "authorizedName" type = "string"/>
  </type>
</element>
```

The *name* and *businessKey* are mandatory fields of this structure, denoting the name of the business entity and its unique identifier, respectively. The "operator" attribute contains the name of the registered UDDI registry operator who manages the master copy of this data structure. The "discoveryURLs" point to the URL of files for service discovery. The optional lists of name-value pairs, "identifierBag" and "categoryBag", are categorizing mechanisms that will be used by the UDDI registry during searches for businesses when customers issue the *find_business* UDDI API call.

The businessService Structure

The *businessService* structure describes a family of services. Listing 14-2 is the specification for the businessService structure:

Listing 14-2. Structure Specification for businessService Structure.

```
<element name = "businessService">
  <type content = "elementOnly">
    <group order = "seq">
      <element ref = "name"/>
      <element ref = "description" minOccurs = "0" maxOccurs = "*"/>
      <element ref = "bindingTemplates"/>
      <element ref = "categoryBag" minOccurs = "0" maxOccurs = "1"/>
    </group>
    <attribute name = "serviceKey" minOccurs = "1" type = "string"/>
    <attribute name = "businessKey" type = "string"/>
  </type>
</element>
```

The *name*, *serviceKey*, and *bindingTemplate* are the mandatory fields in this structure. The “name” attribute holds the human readable name for the family of services described by the businessService. The “serviceKey” is generated when that family is initially created.

The bindingTemplate Structure

The *bindingTemplate* structure holds all the technical details about a service. This usually contains information about the technology used, and the settings and parameters needed for an application. Listing 14-3 is the specification for the bindingTemplate structure:

Listing 14-3. Structure Specification for bindingTemplate Structure.

```
<element name = "bindingTemplate">
  <type content = "elementOnly">
    <group order = "seq">
      <element ref = "description" minOccurs = "0" maxOccurs = "*"/>
      <group order = "choice">
        <element ref = "accessPoint" minOccurs = "0" maxOccurs = "1"/>
        <element ref = "hostingRedirector" minOccurs = "0"
                 maxOccurs = "1"/>
      </group>
      <element ref = "tModelInstanceDetails"/>
    </group>
    <attribute name = "bindingKey" minOccurs = "1" type = "string"/>
    <attribute name = "serviceKey" type = "string"/>
  </type>
```

```
</element>
```

The *bindingKey, accessPoint, hostingRedirector,* and *tModelInstanceDetails* are the mandatory fields in this structure. The "bindingKey" is a unique ID for each bindingTemplate structure and generated when that structure is initially created. The "accessPoint" provides the Web address/entry point for the Web Service. This can be any address—not necessarily a URL. The "hostingRedirector" is used to indicate that the current bindingTemplate structure cannot be used and the bindingTemplate structure pointed to by this field has to be obtained and used instead. "tModelInstanceDetails" is a list of structures that contain information about tModels.

The tModel Structure

The three structures explained above contain information that is typically used when searching the registry for candidate services. Not much technical information is stored in them. However, a tModel structure is designed to provide information that would facilitate interaction with a service.

A tModel can define anything, but currently tModels are used as sources for determining compatibility and as keyed namespace references (by way of categorizing and grouping services). Listing 14-4 is the specification for the tModel structure:

Listing 14-4. Structure Specification for tModel Structure.

```
<element name = "tModel">
  <type content = "elementOnly">
    <group order = "seq">
      <element ref = "name"/>
      <element ref = "description" minOccurs = "0" maxOccurs = "*"/>
      <element ref = "overviewDoc" minOccurs = "0" maxOccurs = "1"/>
      <element ref = "identifierBag" minOccurs = "0" maxOccurs = "1"/>
      <element ref = "categoryBag" minOccurs = "0" maxOccurs = "1"/>
    </group>
    <attribute name = "tModelKey" minOccurs = "1" type = "string"/>
    <attribute name = "operator" type = "string"/>
    <attribute name = "authorizedName" type = "string"/>
  </type>
</element>
```

The *tModelKey* and *name* attributes are the mandatory fields in this structure. The "tModelKey" is a unique key for each tModel structure and is generated when the structure is created for the first time. The "name" is the name provided for this tModel.

How UDDI Structures Relate to WSDL Documents

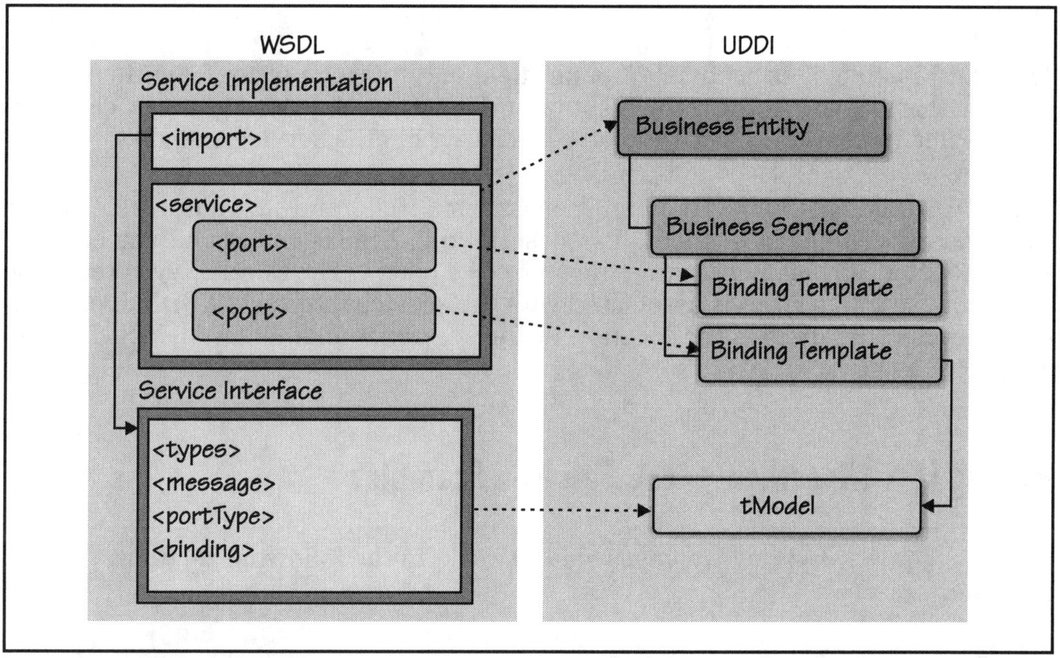

Figure 14-14. WSDL to UDDI Mapping.

The service description information found in WSDL is complementary to the information found in the UDDI registry. Some elements described in tModels are constructed from the Service Interface document. When publishing a WSDL service description, a service interface must be available as a tModel before a service implementation is published as a businessService.

We will now very briefly discuss how and from where the various UDDI structure elements/fields are populated with values.

The UDDI tModel has a field "name" and its value is set to the target namespace from the service interface WSDL document. There is another field called "overviewURL" in the tModel structure and its value points to the service interface WSDL document itself. If there is more than one service binding described in the WSDL document pointed to by the "overviewURL" element, those bindings have to be explicitly referenced in the URL. If there is only one service binding then it is specified after the URL separated by a pound symbol as "URL#service-binding". The other significant field in the tModel structure is "categoryBag", and its value is a reference to the uddi-org:types tModel. This element specifies what UDDI categories this tModel is associated with. The "keyedReferences" fields associated with each "categoryBag" indicate the intended business use for this service interface description. Finally, the tModel's description is set to the value of the <documentation> element that is associated with the definitions element of the WSDL document.

The UDDI businessService has a "name" field and its value is set from the service name in the WSDL service implementation document. Its "description" field is set to the value in the <documentation> element within the service element.

The UDDI bindingTemplate has an "accessPoint" field and its value is set to the soap:address extension element. Its "description" field is set to the value of the <documentation> element within the <port> element. The tModelKey is set to the UUID for the tModel that is associated with the service interface WSDL document. Its "overviewURL" field is the location of the service implementation WSDL document.

Putting It All Together—Web Services Examples

It's time to put everything we've learned to work. In the following sections, you will learn how to:

- Expose an EJB as a Web Service.
- Interact with Web Services from MIDlets.

Let's get started!

EJBS AS WEB SERVICES

EJBs make excellent Web Service providers. In this section, we'll expose our RoomFinder session EJB from Chapter 13 as a Web Service, and illustrate how a client can interact with it.

We will be using Apache Axis as our SOAP provider. Axis provides a straightforward deployment mechanism and allows us to package our Web Service as a WAR file that can be deployed in any J2EE-compatible Web container.

In addition to creating the server-side WAR file, we'll also be creating a WSDL document that describes our new service. From this WSDL file, we'll generate a client proxy that can be employed by users of our service.

The good news is that Axis can generate all these files for us—we don't need to write a single line of code!

Here are the steps we need to follow to create and deploy our EJB Web Service (see Figure 14-15):

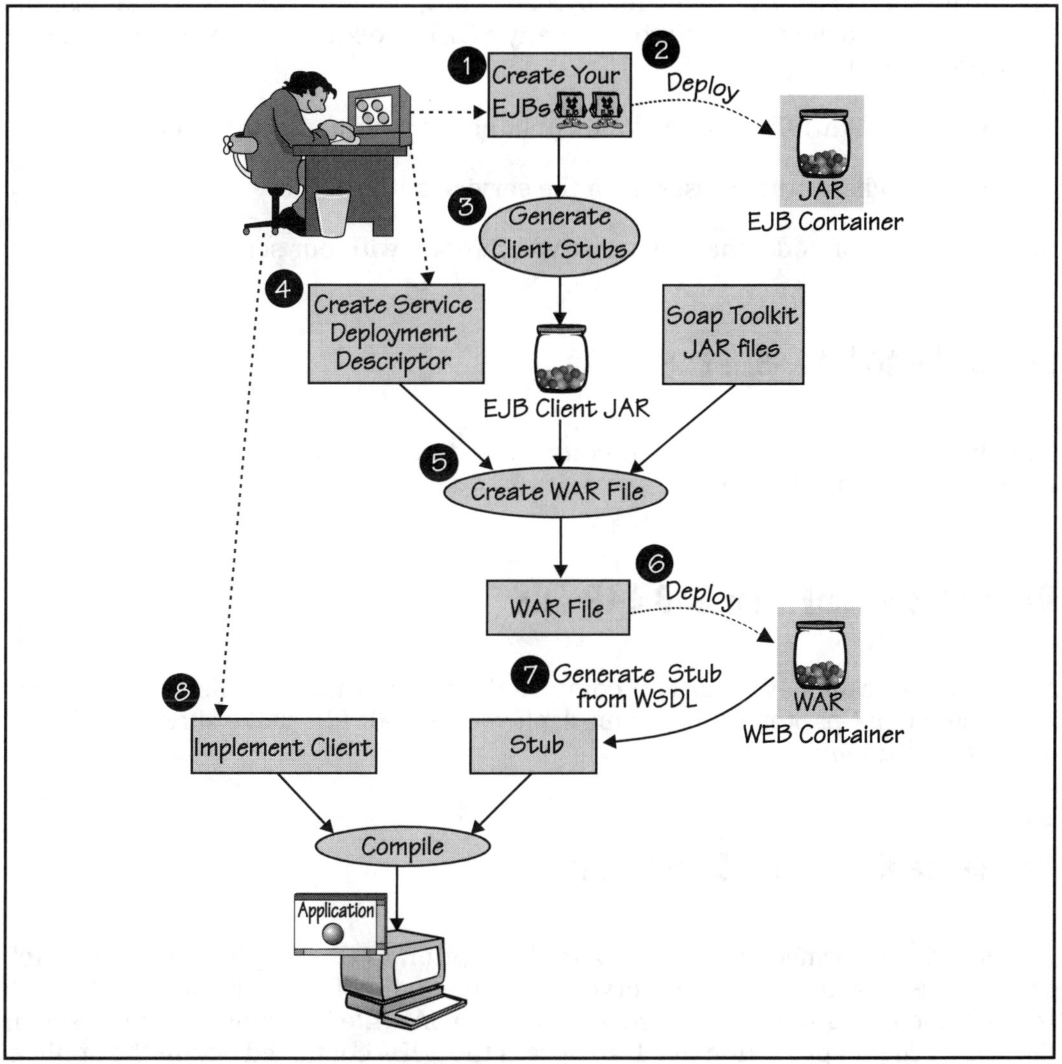

Figure 14-15. Creating and Deploying an EJB Web Service.

1. Create the EJBs that will be exposed as a service. We'll be using the RoomFinder EJBs from Chapter 13.
2. Deploy the EJBs in our EJB container. If you ran the RoomFinder example, you have already performed this step.
3. Create the EJB client stub classes. These stub classes are typically generated by your application server via an included tool or wizard.
4. Create the Axis service deployment descriptor for our service.
5. Create the server-side WAR file by packaging the EJB client stub classes, the service deployment descriptor, and any SOAP toolkit JAR files together into a WAR archive.
6. Deploy the WAR file into the Web container of your application server.
7. Generate client stub classes from the service's WSDL file
8. Implement, or code, the client that will interact with our service.

Create the Web Service EJBs

You should already have this step completed if you compiled and ran the RoomFinder example in Chapter 13.

Deploy the RoomFinder EJB JAR File

You should already have this step completed if you left the JAR file deployed from our example in Chapter 13. If you un-deployed the JAR file, make sure to re-deploy it before continuing.

Create the EJB Client Stub Classes

This step is performed by invoking a tool or option in your application server. You should consult your application server's documentation for details on how this task is performed. The output of this step will be a JAR file containing any classes or libraries a client application needs to access the EJBs contained within the application server's EJB container.

Create the Service Deployment Descriptor

Axis requires a special deployment descriptor called “server-config.wsdd” to be present in your WAR file. Fortunately, the folks at Apache provide a tool that generates this file for us, based on a simpler XML input file called deploy.wsdd:

Listing 14-5. Axis deployment descriptor (deploy.wsdd).

```
<deployment
    xmlns="http://xml.apache.org/axis/wsdd/"
    xmlns:ns="http://borland.com"
    xmlns:java="http://xml.apache.org/axis/wsdd/providers/java">
  <service name="SOAPRoomFinder" provider="java:EJB">
    <parameter name="beanJndiName" value="RoomFinder"/>
    <parameter name="homeInterfaceName" value="rooms.RoomFinderHome"/>
   <parameter name="allowedMethods" value="listProfessors listCourses"/>
  </service>
</deployment>
```

The deploy.wsdd file shown in Listing 14-5 contains the information Axis needs to deploy the service described in the file. You provide the following information:

- The service name (in our case, “RoomFinder”).
- The EJB’s JNDI name (in our case, again “RoomFinder”). This name should match the JNDI name in the EJB deployment descriptor we created in Chapter 13 and deployed in our EJB container.
- The fully package-qualified home interface name of the EJB we’re exposing as a service.
- A list of allowed methods—what methods we want to expose in our service.

You can create the deploy.wsdd file in any text editor. Once you’ve created it, we can use the Axis-provided translation tool to translate our deploy.wsdd file into a more descriptive server-config.wsdd file:

```
java org.apache.axis.utils.Admin server deploy.wsdd
```

The command shown above will produce the server-config.wsdd file as output, which we include in our WAR file:

Listing 14-6. Axis server config (server-config.wsdd).

```
<?xml version="1.0" encoding="UTF-8"?>
<deployment xmlns="http://xml.apache.org/axis/wsdd/"
      xmlns:java="http://xml.apache.org/axis/wsdd/providers/java">
 <globalConfiguration>
  <parameter name="adminPassword" value="admin"/>
  <parameter name="sendMultiRefs" value="true"/>
  <parameter name="sendXsiTypes" value="true"/>
  <parameter name="sendXMLDeclaration" value="true"/>
  <requestFlow>
   <handler type="java:org.apache.axis.handlers.JWSHandler"/>
  </requestFlow>
 </globalConfiguration>
 <handler name="URLMapper"
      type="java:org.apache.axis.handlers.http.URLMapper"/>
 <handler name="LocalResponder"
      type="java:org.apache.axis.transport.local.LocalResponder"/>
 <handler name="Authenticate"
      type="java:org.apache.axis.handlers.SimpleAuthenticationHandler"/>
 <handler name="RPCDispatcher"
      type="java:org.apache.axis.providers.java.RPCProvider"/>
 <handler name="MsgDispatcher"
      type="java:org.apache.axis.providers.java.MsgProvider"/>
 <service name="RoomFinder" provider="java:EJB">
  <parameter name="allowedMethods" value="listProfessors,listCourses"/>
  <parameter name="beanJndiName" value="RoomFinder"/>
  <parameter name="homeInterfaceName" value="rooms.RoomFinder"/>
 </service>
 <service name="SOAPRoomFinder" provider="java:EJB">
  <parameter name="allowedMethods" value="listProfessors listCourses"/>
  <parameter name="beanJndiName" value="RoomFinder"/>
  <parameter name="homeInterfaceName" value="rooms.RoomFinderHome"/>
 </service>
 <service name="AdminService" provider="java:MSG">
  <parameter name="allowedMethods" value="AdminService"/>
  <parameter name="enableRemoteAdmin" value="false"/>
  <parameter name="className" value="org.apache.axis.utils.Admin"/>
 </service>
 <service name="http://xml.apache.org/axis/wsdd/" provider="java:MSG">
  <parameter name="allowedMethods" value="AdminService"/>
  <parameter name="enableRemoteAdmin" value="false"/>
  <parameter name="className" value="org.apache.axis.utils.Admin"/>
 </service>
 <transport name="http">
```

```
  <requestFlow>
   <handler type="URLMapper"/>
   <handler type="java:org.apache.axis.handlers.http.HTTPAuthHandler"/>
  </requestFlow>
 </transport>
 <transport name="local">
  <responseFlow>
   <handler type="java:org.apache.axis.transport.local.LocalResponder"/>
  </responseFlow>
 </transport>
</deployment>
```

Note that the server-config.wsdd file shown in Listing 14-6 contains more descriptive information about our service. You generally don't edit this file—it's something that you should let Axis automatically generate for you.

Since we want to create a WAR file that can be used with any J2EE-compliant application server, we need to place the server-config.wsdd file inside the WAR file's WEB-INF directory. We also place the EJB client JAR file along with the Axis JAR files inside the WEB-INF\lib directory. By including the Axis JAR files inside our WAR file, we can use our application on servers that don't already have Axis installed.

The WAR file that comprises our Web Service will consist of the following parts:

- An EJB client JAR file containing client-side classes needed for interacting with our RoomFinder session bean. Your development tool, such as JBuilder, or your J2EE container will provide a tool to generate these client-side classes from your EJBs.
- An Axis-specific deployment descriptor.

When you've created the WAR file, deploy it in your application server's Web container.

Use the WDSL File to Generate Client Stub Classes

Axis organizes services under the "/services/" URL, prefaced by the URL of your application server's Web container. The service name is appended to the end of the URL. Hence, to invoke our SOAPRoomFinder service, we would route our SOAP messages to:

http://localhost:8040/SOAPRoomFinder/services/SOAPRoomFinder

At this point, we have a service deployed in our application server, but without a WSDL file to describe it, no clients will be able to use it. Axis provides another utility we can use to obtain the WSDL for our deployed service. By appending the string "?wsdl" to the URL above, Axis will generate the WSDL for our service on demand:

http://localhost:8040/SOAPRoomFinder/services/SOAPRoomFinder?wsdl

Figure 14-16 shows the WSDL file as seen via a standard Web browser.

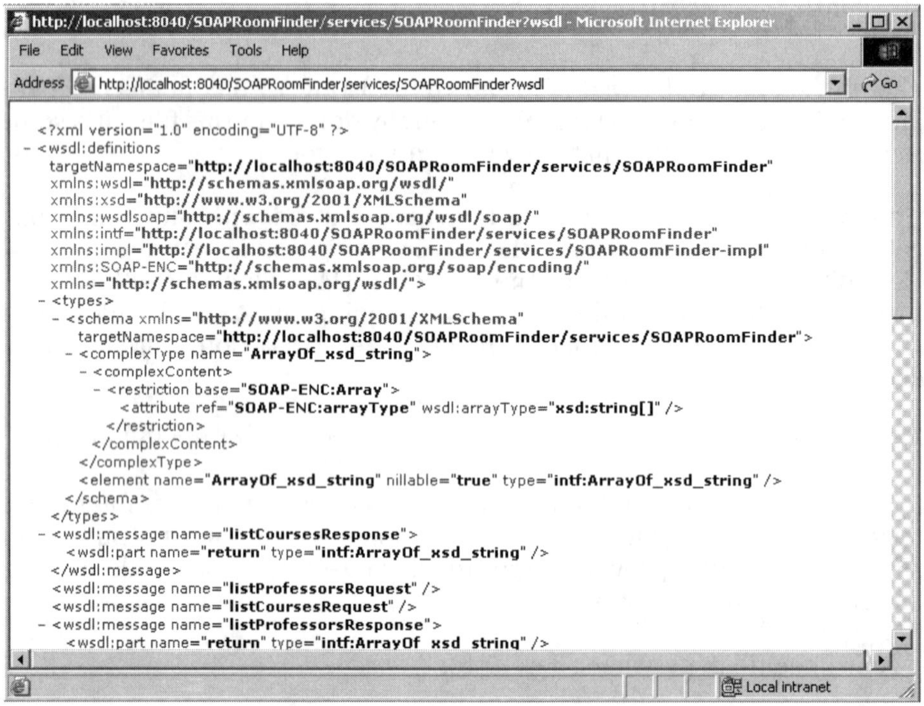

Figure 14-16. Browsing Axis-generated WSDL.

Listing 14-7 shows the entire WSDL document.

Listing 14-7. SOAPRoomFinder.wsdl.

```
<?xml version="1.0" encoding="UTF-8"?>
<wsdl:definitions
 targetNamespace=
  "http://localhost:8040/SOAPRoomFinder/services/SOAPRoomFinder"
 xmlns:wsdl="http://schemas.xmlsoap.org/wsdl/"
 xmlns:xsd="http://www.w3.org/2001/XMLSchema"
```

```
 xmlns:wsdlsoap="http://schemas.xmlsoap.org/wsdl/soap/"
 xmlns:intf=
     "http://localhost:8040/SOAPRoomFinder/services/SOAPRoomFinder"
 xmlns:impl=
     "http://localhost:8040/SOAPRoomFinder/services/SOAPRoomFinder-impl"
 xmlns:SOAP-ENC="http://schemas.xmlsoap.org/soap/encoding/"
 xmlns="http://schemas.xmlsoap.org/wsdl/"><types><schema
 xmlns="http://www.w3.org/2001/XMLSchema"
 targetNamespace=
     "http://localhost:8040/SOAPRoomFinder/services/SOAPRoomFinder">

 <complexType name="ArrayOf_xsd_string">
  <complexContent>
   <restriction base="SOAP-ENC:Array">
    <attribute ref="SOAP-ENC:arrayType" wsdl:arrayType="xsd:string[]"/>
   </restriction>
  </complexContent>
 </complexType>

<element name="ArrayOf_xsd_string" nillable="true"
     type="intf:ArrayOf_xsd_string"/></schema></types>
  <wsdl:message name="listCoursesResponse">
    <wsdl:part name="return" type="intf:ArrayOf_xsd_string"/>
  </wsdl:message>
  <wsdl:message name="listProfessorsRequest">
  </wsdl:message>
  <wsdl:message name="listCoursesRequest">
  </wsdl:message>
  <wsdl:message name="listProfessorsResponse">
    <wsdl:part name="return" type="intf:ArrayOf_xsd_string"/>
  </wsdl:message>
  <wsdl:portType name="_RoomFinder_Stub">
    <wsdl:operation name="listProfessors">
      <wsdl:input message="intf:listProfessorsRequest"/>
      <wsdl:output message="intf:listProfessorsResponse"/>
    </wsdl:operation>
    <wsdl:operation name="listCourses">
      <wsdl:input message="intf:listCoursesRequest"/>
      <wsdl:output message="intf:listCoursesResponse"/>
    </wsdl:operation>
  </wsdl:portType>
  <wsdl:binding name="SOAPRoomFinderSoapBinding"
    type="intf:_RoomFinder_Stub">
    <wsdlsoap:binding style="rpc"
     transport="http://schemas.xmlsoap.org/soap/http"/>
```

```
    <wsdl:operation name="listProfessors">
      <wsdlsoap:operation soapAction="" style="rpc"/>
      <wsdl:input>
        <wsdlsoap:body use="encoded"
         encodingStyle="http://schemas.xmlsoap.org/soap/encoding/"
         namespace=
        "http://localhost:8040/SOAPRoomFinder/services/SOAPRoomFinder"/>
      </wsdl:input>
      <wsdl:output>
        <wsdlsoap:body use="encoded"
         encodingStyle="http://schemas.xmlsoap.org/soap/encoding/"
         namespace=
        "http://localhost:8040/SOAPRoomFinder/services/SOAPRoomFinder"/>
      </wsdl:output>
    </wsdl:operation>
    <wsdl:operation name="listCourses">
      <wsdlsoap:operation soapAction="" style="rpc"/>
      <wsdl:input>
        <wsdlsoap:body use="encoded"
         encodingStyle="http://schemas.xmlsoap.org/soap/encoding/"
         namespace=
        "http://localhost:8040/SOAPRoomFinder/services/SOAPRoomFinder"/>
      </wsdl:input>
      <wsdl:output>
        <wsdlsoap:body use="encoded"
         encodingStyle="http://schemas.xmlsoap.org/soap/encoding/"
         namespace=
        "http://localhost:8040/SOAPRoomFinder/services/SOAPRoomFinder"/>
      </wsdl:output>
    </wsdl:operation>
  </wsdl:binding>
  <wsdl:service name="SOAPRoomFinderService">
    <wsdl:port name="SOAPRoomFinder"
      binding="intf:SOAPRoomFinderSoapBinding">
      <wsdlsoap:address
       location=
        "http://localhost:8040/SOAPRoomFinder/services/SOAPRoomFinder"/>
    </wsdl:port>
  </wsdl:service>
</wsdl:definitions>
```

Once we've obtained the WSDL file describing our service by requesting it in the method shown above, we can use another Axis tool to generate the Java proxy classes our clients will use when interacting with the service. The tool, WSDL2Java, is invoked as follows:

```
java org.apache.axis.wsdl.WSDL2Java  SOAPRoomFinder.wsdl
```

The WSDL2Java tool creates a number of output files, based upon the information contained in the WSDL file supplied to it on the command line. For our SOAPRoom-Finder example, the tool generates two interfaces and two classes:

- **RoomFinderStub:** This is the stub interface our clients will use to invoke the exposed *listProfessors* and *listCourses* methods.
- **SOAPRoomFinderService:** This interface represents a unique instance of the SOAPRoomFinder service.
- **SOAPRoomFinderServiceLocator:** Clients use methods in this class to locate RoomFinder service objects. More precisely, this class locates a particular instance of the service using a given binding.
- **SOAPRoomFinderSoapBinding:** Contains the code that marshals the requests to a RoomFinder service. This class interfaces with the SOAP generation code contained inside the Axis libraries. Other bindings besides SOAP are available; the WSDL2Java tool generated this SOAP binding class since SOAP was the binding specified in the source WSDL file.

Design and Test the Web Service Client

Let's now look at a simple client that will use our RoomFinder service. This client will be truly minimalist—we'll just call each of the exposed methods and examine the results. You can easily extend the example if you want to test more functionality.

Listing 14-8 shows the JavaServer Page that generated the output shown in Figure 14-17. The scriptlet in the JSP uses the classes and interfaces that were created using WSDL2Java.

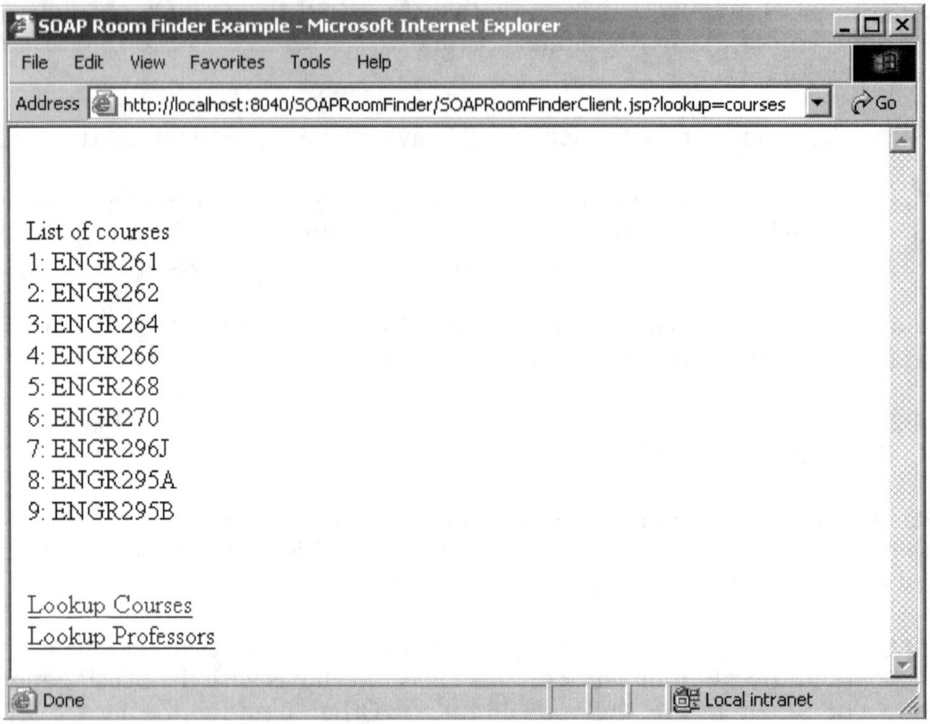

Figure 14-17. Output of SOAP Room Finder Client.

Listing 14-8. SOAPRoomFinderClient.jsp.

```
<%@ page import="localhost.*,rooms.RoomFinder" %>
<HTML>
 <HEAD>
  <TITLE>SOAP Room Finder Example</TITLE>
 </HEAD>
 <BODY>
  <%
   String lookup=request.getParameter("lookup");
   String Result[]=null;

   if(lookup!=null)
   {
    try
    {
     SOAPRoomFinderServiceLocator service =
```

```
      new SOAPRoomFinderServiceLocator();
     RoomFinderStub roomfinder = service.getSOAPRoomFinder();

     if(lookup.equals("courses"))
      Result=roomfinder.listCourses();

     if(lookup.equals("professors"))
      Result=roomfinder.listProfessors();

     out.println("<BR><BR>List of "+lookup+"<BR>");
     for(int i=0; i<Result.length; i++)
      out.println((i+1)+": "+Result[i]+"<BR>");
    }
    catch(Exception E)
    {
     out.println("Error invoking service: "+E);
    }
   }
  %>

  <BR><BR>
  <A HREF="SOAPRoomFinderClient.jsp?lookup=courses">Lookup Courses</A>
  <BR>
  <A HREF="SOAPRoomFinderClient.jsp?lookup=professors">Lookup
Professors</A>
 </BODY>
</HTML>
}
```

We've showed an end-to-end example using an EJB as a service provider. Thanks to the magic of Apache Axis, we don't have to write a single line of code to make it all happen. Of course, we still have to code our service's consumers (our clients).

INVOKING WEB SERVICES FROM YOUR MIDLETS

We've already shown how a JSP can act as a client for a Web Service. We can also turn our MIDlets into SOAP clients with a little work. This section illustrates how this is done.

In order to save time, we'll create a MIDlet that uses the service that was deployed in the previous example. Again, we're going to design a very minimal client—you can add additional functionality based on your handheld device's capabilities.

To invoke a Web Service from a MIDlet, we need to be able to compose a SOAP message and send it to the service. A version of SOAP from Enhydra (*www.enhydra.org*) called kSOAP is available for MIDP and we will use it for the example shown in this section.

To use kSOAP with the J2ME Wireless Toolkit, download kSOAP and kXML from Enhydra's Web site and place the files inside your MIDlet's "lib" directory when compiling using the J2MEWTK, or inside your project when compiling using JBuilder MobileSet. Other toolkits may have alternate means of including these libraries in your MIDlet.

The following MIDlet can be installed on a J2ME device. This MIDlet interacts with the Web Service we exposed from our RoomFinder EJB in the previous example. When compiled and executed in the emulator, the output shown in Figure 14-18 is displayed. Listing 14-9 shows the MIDlet code.

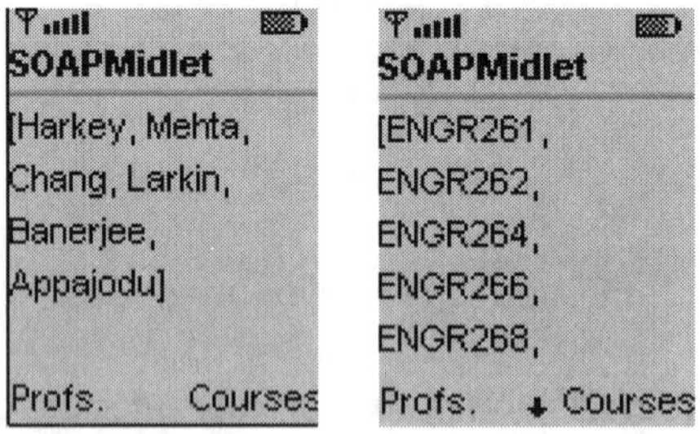

Figure 14-18. Invoking the RoomFinder Service from a MIDlet.

Listing 14-9. Invoking Web Service from your MIDlet (SOAPMidlet.java).

```
import javax.microedition.midlet.*;
import javax.microedition.lcdui.*;
import java.io.*;
import javax.microedition.io.*;
import org.ksoap.*;
import org.ksoap.transport.*;

public class SOAPMidlet extends MIDlet implements CommandListener
{ Form mainForm = new Form ("SOAPMidlet");
  StringItem resultItem = new StringItem ("", "");
  Command getCourses = new Command ("Courses", Command.SCREEN, 1);
  Command getProfessors = new Command("Profs.", Command.SCREEN, 1);
  SoapObject rpc;
```

```
  public SOAPMidlet ()
  { mainForm.append (resultItem);
    mainForm.addCommand (getCourses);
    mainForm.addCommand (getProfessors);
    mainForm.setCommandListener (this);
  }

  public void startApp ()
  { Display.getDisplay (this).setCurrent (mainForm);
  }

  public void pauseApp () { }
  public void destroyApp (boolean unconditional) { }
  public void commandAction (Command c, Displayable d)
  { try
    { if (c.equals(getCourses))
          rpc = new SoapObject("", "listCourses");
      else
          rpc = new SoapObject("", "listProfessors");

      resultItem.setText(""+new HttpTransport
       ("http://localhost:8040/SOAPRoomFinder/services/SOAPRoomFinder",
        "").call(rpc));
    }
    catch (Exception e) {
      e.printStackTrace ();
      resultItem.setLabel ("Error:");
      System.out.println(e);
    }
  }
}
```

Based upon the command selected by the user (either "Profs." or "Courses"), we construct a SoapObject which contains the method we want to invoke on our service (either "*listProfessors*" or "listCourses", respectively). We then create an **Http-Transport** object that connects to the URL supplied. The **HttpTransport** then marshals our **SoapObject** over the wire to the remote server, and receives the results.

WEB SERVICES TOOLS AND PLATFORMS

In the following sections we will briefly discuss the features of some of the available Web Services development tools. As the Web Services development model becomes more popular, the numbers of toolkits and development packages have grown exponentially. In the forthcoming sections, we focus on some of the more widely accepted products.

Apache Axis/Apache SOAP

Called Apache SOAP during its first few revisions, Apache Axis is a WSDL/SOAP Web Services provider that is capable of being dropped into any application server as a Web module. Because of this easy integration, it's no wonder that many commercial Web Services platforms have adopted Axis as their underlying Web Services technology.

In addition to providing a SOAP RPC router based on a Servlet that can be incorporated in a Web application, Axis also provides:

- Automatic WSDL generation from deployed services.
- Java to WSDL generation based on any arbitrary Java class.
- WSDL to Java stub/proxy class generation.

One other interesting feature that Axis provides is the ability for the user to place a plain Java file (non-compiled) into the Web application directory, and Axis will automatically turn the code contained in the file into a service. This is accomplished by renaming the *.java* file to *.jws*—for Java Web Service. Axis interprets any file ending in *.jws* in its Web application directory as a class comprising a Web Service.

We saw Axis in action when we developed our example RoomFinder Web Service earlier.

IBM's Web Services Tool Kit (WSTK)

This toolkit contains both design time and runtime components. Design time Web Services components include tools for discovering and publishing Web Services and for creating Web Services out of existing Java applications. WSTK's runtime components include components that are used on both the client side and server side. The server side component comprises a UDDI registry that can be deployed within an organization, removing the need to utilize the various public UDDI registries. Client runtime components provide various Java APIs allowing applications to:

- Publish and find Web Services by interfacing directly with UDDI (through the UDDI4J API).
- Publish and find Web Services via WSDL descriptions (WSDL Proxy).
- Access Web Services via SOAP (Service Proxy).

IBM's WSTK also includes several other features worth noting:

- A self-contained installation of WebSphere, IBM's application server product. WSTK applications can be tested and deployed within this installation, or moved to an existing WebSphere deployment.
- Preliminary support for the Web Services security model.

WSTK also includes an installation of Apache Axis, which we have already discussed.

HP's e-Speak

The e-Speak architecture also contains the fundamental Web Services components. The e-Speak product also provides a smart core to support ancillary operations such as accounting, auditing, or charging. e-Speak contains the following components:

- The core—known as the e-Speak Engine—that provides APIs for interactions between clients.
- An XML protocol definition (including asynchronous messaging) and other standards defined for interactions being defined in Service Specifications.
- A hosted directory of Web Services known as the "services village" that categorizes and organizes Web Services. This directory can be accessed at *http://www.hpespeak.com.*

The e-Speak platform software is released under GPL and LGPL open source licenses, and HP is the founding member of the e-Speak open source community. e-Speak uses Oracle as its database technology and Tomcat as its Servlet container.

Java Technology Web Services Architecture

The Java language specification, as released by Sun with the input from the various JSRs, has had increasing support for Web Services as new versions of Java have been released. You might have already heard of JAXM, JAXR, JAXB, or any of the other myriad JAX libraries. As a whole, the JAX packages provide much of the function needed to develop Web Services directly in Java, without the need for any external technology or tools. Figure 14-19 illustrates the architecture of Java's JAX technology family for Web Services.

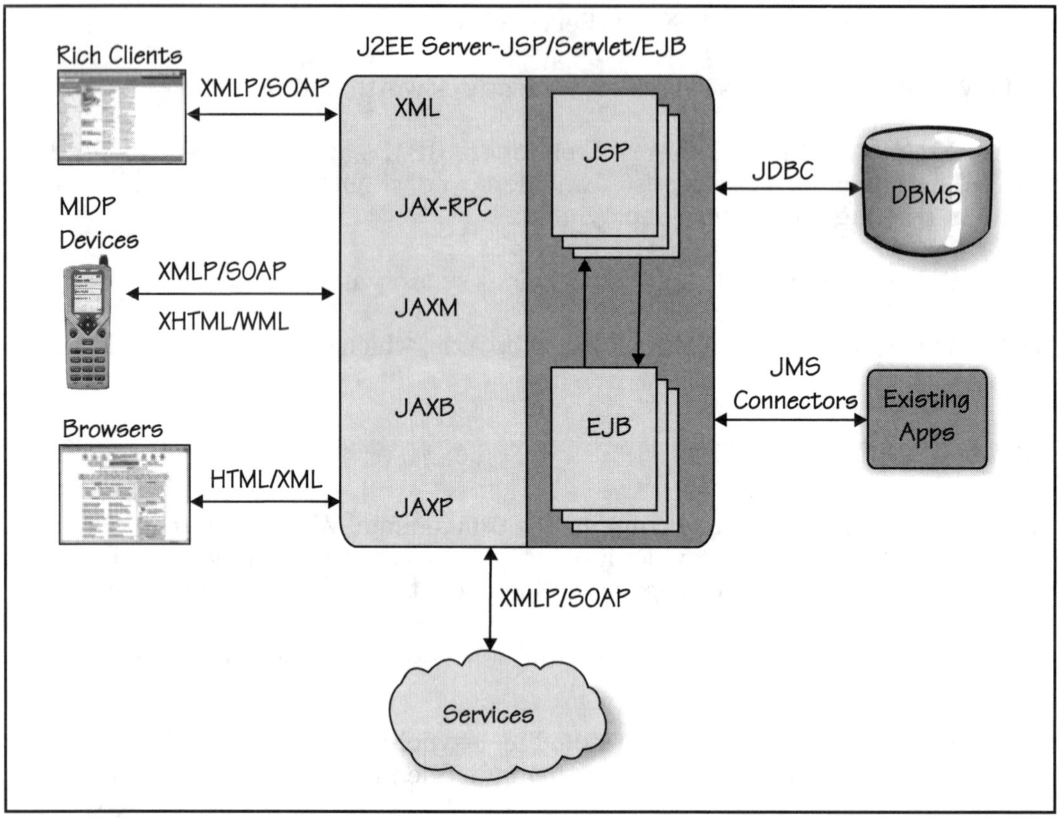

Figure 14-19. Java Technology Web Services Model.

So, what do the various JAX components do?

- **JAXP***:* The Java APIs for XML Processing provide XML parsing capabilities, marshalling and unmarshalling APIs, and support XML-based RPC call definitions.
- **JAXM***:* The Java APIs for XML Messaging support messaging constructs for XML messages. JAXM is built on SOAP 1.1, which means it has support for SOAP with attachments. The JAXM architects are also looking at providing the ability to expose an EJB as a Web Service end point, much like Axis already provides. JAXM is similar to JMS, the Java Messaging System, but not quite as exhaustive.
- **JAXB***:* The Java architecture for XML Binding provides an API for automatic two-way mapping between XML documents and Java objects.

- **JAXR:** The Java APIs for XML Registries provide a uniform API for accessing different types of registries (such as UDDI). JAXR also contains provisions for other types of registries, should the need arise.

- **Java APIs for WSDL:** Not to be confused with JAXW (that's another standard entirely), the Java APIs for WSDL support creating, parsing, and using definitions contained within WSDL documents inside your programs.

So, Sun and the JSRs are working tirelessly to crank out a full set of APIs that will hopefully be compatible across all Java VMs. Some of the technologies listed above are further along than others. Time will tell how well this set is received by the industry.

THE FUTURE DIRECTION OF WEB SERVICES

The current implementation of Web Services using SOAP, WSDL, and UDDI are supposed to be simple, flexible, and lightweight without being limited by language, platform, language, or object model. However there are a few important pieces of the puzzle still missing such as:

- Transactions
- Security
- Web Services management
- Quality of Service Policies
- Fault Tolerance

Most of these are available in robust architectures such as CORBA and J2EE. Hopefully, we will see these same concerns addressed in future extensions and enhancements to the Web Services design model.

FYI Briefing

The WS-I Organization

WS-I, the Web Services Interoperability Organization, is a group of companies who have joined together to address the growing problem of incompatible services. Their stated goal, to "promote consistent and reliable interoperability among Web Services across platforms, applications, and programming languages," would certainly go a long way toward quashing some of the doubts circulating about widespread Web Services adoption.

For certain, their stated goal is a noble one, and if reached, would alleviate many headaches when trying to integrate diverse services from many sources. Indeed, many companies (at the time of this writing, over 100) have joined WS-I to help out. Conspicuously missing is Sun, who is taking a "wait and see" attitude toward the organization.

Only time will tell whether or not WS-I can exert enough influence on service designers to ensure that services remain interoperable. ❑

SUMMARY

In this chapter we visited some of the important technologies and fundamentals of Web Services and walked you through the steps required to build and test a Web Service. Although some key players in the industry are whole-heartedly embracing the dawn of the Web Services age, others are taking a more cautious approach, preferring to wait and see how the competing technologies emerge. The Web Services paradigm is reaching a critical mass—once enough industry muscle is behind one set of standards or another, things should take a clearer shape.

We will revisit Web Services again in Part 4, when we look more closely at our Campus Portal.

Chapter 15

Transcoding and Personalization

This chapter will introduce two key features that any well-designed wireless enterprise application platform should provide. The first feature, *transcoding*, allows an application to tailor its presentation and interface for a particular device, based on the hardware and software capabilities of that device. The second feature, *personalization*, allows applications to customize their behavior for each and every user of the application, based on their identities and personal preferences.

In order to provide these services to clients, applications that utilize transcoding and personalization frameworks also need ways to profile the characteristics of the device interacting with the application. These frameworks allow the application to figure out what capabilities a user's device may have without previously knowing anything about the device.

This chapter will examine transcoding and personalization techniques, along with their support technologies (like device profiling) that are needed to support them. Although a thorough explanation of transcoding technologies is beyond the scope of this book, the discussion that follows will provide enough information for a skilled application developer to get started.

TRANSCODING: REALTIME CONTENT TRANSFORMATION

As Internet services evolved over the years, the primary interface used by most users remained the ubiquitous Web browser. Historically, this was not a problem; as long as the application developer ensured his or her application functioned properly when accessed through a Web browser, he or she could be reasonably certain that it would function properly for any potential user that might come along.

As the number of wireless and other mobile devices increased, the application developer was faced with a new challenge—how to continue to support the standard PC Web browser, while embracing the new generation of users accessing the application via a handheld device. Early solutions involved coding distinct support for each possible device in the application, but this approach quickly became infeasible as the number of different devices increased.

Since it became practically impossible to rewrite application logic to support many types of devices while simultaneously considering device-specific resource limitations, other alternatives were explored. One promising approach that was discovered was to have a single content-generating application while providing a separate software application to transform its output for specific devices. This approach embodies the key aspect of transcoding—the realtime transformation of content for a particular device, taking into consideration the device's characteristics and capabilities.

The following sections describe how content can be massaged so that users of different device types can use applications to the fullest extent supported by their hardware, while reducing the need to extensively rewire existing applications to support new devices. We'll start with an overview of transcoding technologies, including software frameworks that an application developer can use to easily incorporate transcoding into existing applications.

Generating Content for Specific Devices

Let's review the most popular techniques used for generating content suitable for mobile devices:

- ***Generating specialized content based on the device:*** In this approach, the content is generated from an Internet service in a particular markup language to suit the needs of a particular device. For example, an application might have one set of JSP pages that generate HTML output and another set of JSP pages that generate WML output. Even if these two sets of pages use the same back-end business logic, the application developer may need to rewrite and redesign the content source (by supplying new sets of JSP pages) to support each and every device desired. Having specialized content generation methods such as these would provide a richer user interface when accessing services from those devices. However, this approach has a downside; the overhead of maintaining existing content generators, and the future development effort required to add support for new devices could increase dramatically as the number of supported devices grows.

- ***Content generation for specific devices:*** This involves employing a software module or agent to fetch the contents of a Web page and rewrite the same contents using a different markup language. For example, when accessing an i-mode page from a WAP phone, a screen-scraping module could be employed that would "scrape off" the contents of the i-mode page and create a WML page dynamically for display on the WAP phone based upon the contents that were scraped. The advantage in using screen scraping is that we need only one source document and the contents can be rewritten to conform to different markup languages.

- ***Transcoding:*** In this method, an intelligent software agent is the interface between any client and the content generating server. Based on the client constraints, which may include a list of markup languages supported and/or hardware capabilities, the transcoding software converts the source content to the appropriate markup language suitable for the client.

What Is Transcoding?

As future wireless handheld devices are produced, content publishers, and application developers may need to handle future standards, different capabilities, new types of resource constraints and possibly different markup languages when tailoring content to each of these new devices. To support content tailoring to existing devices, and to easily support new devices, many enterprise application developers are taking the approach of developing a software system which transcodes, or translates content based on the characteristics of each type of device. This system, called a transcoder or transcoding engine, provides this ability. From a high-level viewpoint, transcoding may be thought of as simply converting content in one markup language to another other markup language using a program. Figure 15-1 shows how a transcoder works.

In Figure 15-1, the output of a service is transcoded into several output formats.

- ***Web Browser:*** The output of the service is translated into a form suitable for a Web browser. This would generally imply that the output is translated into HTML.

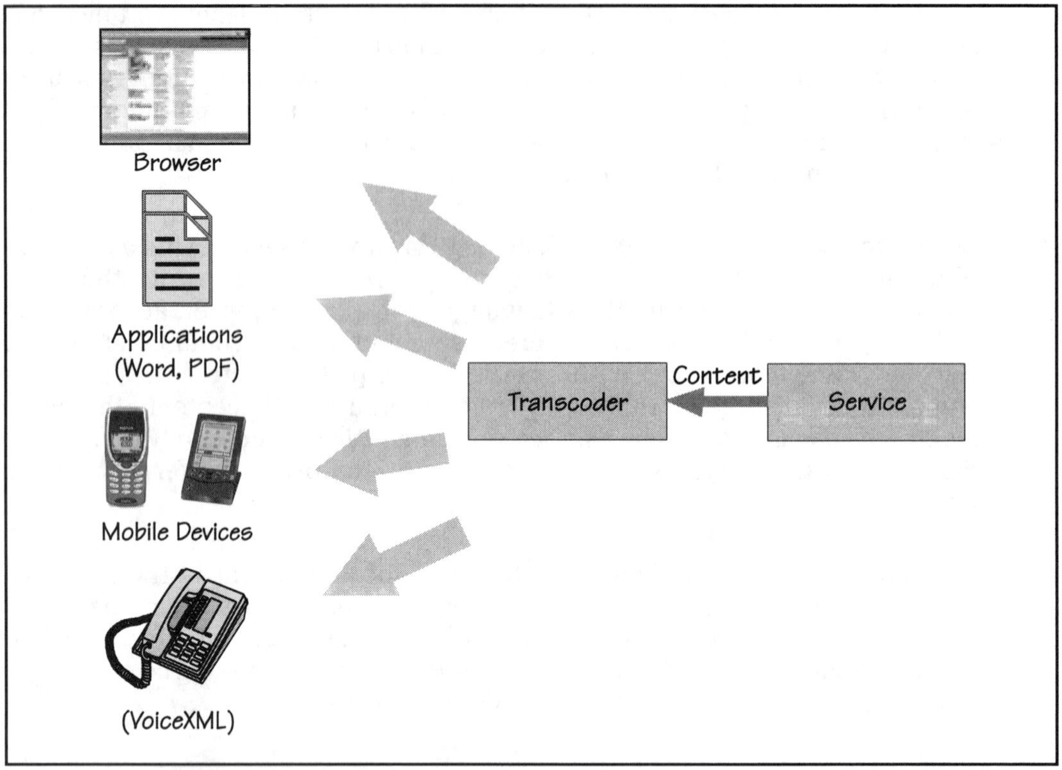

Figure 15-1. An Overview of Transcoding.

- ***Portable Document:*** The output of the service is translated into a portable document format such as Microsoft Word or Adobe PDF.
- ***VoiceXML:*** The output of the service is translated into a content format that can be converted to audio.
- ***Mobile Devices:*** The output of the service is translated into one or more formats suitable for mobile devices, such as WML.

The transcoder converts the content at request time, taking into consideration the characteristics of the device making the request. When converting the content, the transcoder follows a set of rules which tell it what transformations to apply to the content, and when. By following these rules, the transcoder can convert incoming content into any output format. The set of transformation rules for a given conversion are usually written in a format called XSL. We'll describe XSL in a subsequent section; for now, we have some background information we need to cover first.

TRANSCODING IMPLEMENTATIONS

In the following sections we'll examine what a transcoder needs to do in order for it to properly translate content. A transcoder must perform several tasks each time it converts content from one form to another.

It must:

- Discover the remote device's capabilities.
- Decide how much content to display on the device.
- Use rules specified in a stylesheet to convert one markup language to another.
- Translate UI (User Interface) elements (usually described in the initial content in a device-independent fashion) into UI elements suitable for the device.
- Provide the fully-translated content to the device, while considering its capabilities.

Describing, Exchanging, and Discovering Mobile Device Capabilities

In order for a transcoder to adapt content to a particular device, it must be made aware of the hardware and software characteristics of the device so that it may adapt the content appropriately. The *Composite Capability/Preference Profiles (CC/PP)* standard, whose overview is shown in Figure 15-2, is intended to provide the information necessary to adapt the content and the content delivery mechanism to best fit the capabilities of the device. The data about each device (which may also include the preferences of the user) is stored as a device profile. A library of these device profiles may be stored in a central location for later retrieval.

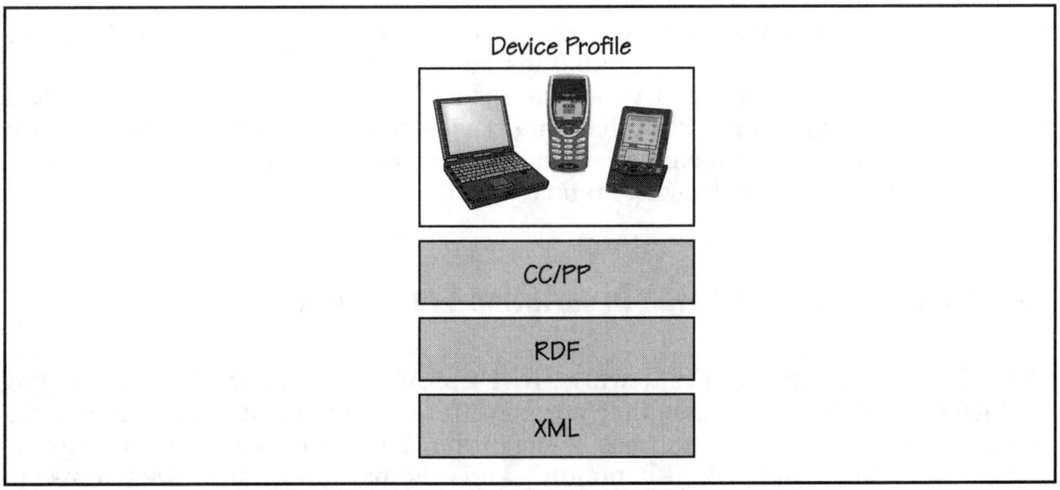

Figure 15-2. Composite Capability/Preference Profile with Device Profiling.

Since one important characteristic of the next Web generation is the proliferation of new types of client access devices, the potential need for CC/PP by the next generation of enterprise wireless applications is substantial. A single user may use several of these devices at the same time, and there must be a way for each type of device to share its capabilities with servers.

For example, some people access e-mail with a desktop computer, a laptop computer, and a handheld, as well as an Internet-enabled cell phone. Each of these devices has a different set of capabilities. To help application developers keep pace with heterogeneous clients, the CC/PP protocol provides a way for clients and servers to exchange information about the capabilities and preferences of each.

CC/PP is based on a lower-level standard called *Resource Description Framework (RDF)*. RDF provides the means for exchanging many different types of metadata between two parties. RDF is not limited to describing only mobile device characteristics—it is a generic framework that can be used to share any type of metadata.

An RDF document contains records describing properties about a Web resource, which is why it was chosen by the CC/PP working group to describe device capabilities. In this context, a device equipped with a browser is considered a Web resource; one which has properties like screen size and rendering capabilities.

Before clients and servers can exchange information about these capabilities and characteristics, there needs to be an agreement between the parties detailing the types of characteristics that may be shared, and how they are to be interpreted. At a technical level, this implies that there needs to be a sharing of metadata describing device capabilities. The shared metadata (or data about data) should fully describe the capabilities that the participants may see during an exchange.

If every device manufacturer was to describe device capabilities using their own proprietary metadata schema, capabilities could not be shared between devices from different manufacturers. To address this problem, CC/PP was developed as a standard way of encoding and exchanging device capabilities. To understand how CC/PP allows parties to exchange metadata about wireless device capabilities, we need to first look at its underlying standard, RDF.

An Overview of the Resource Description Framework

The Resource Description Framework (RDF) provides the means for exchanging modular and interoperable metadata across different resource description communities. Its main goal is to address a major hurdle in the resource description community, namely the lack of common standards that inhibit the effective use of metadata by software applications.

RDF addresses metadata exchange between two parties over the Web. It defines a standard way for metadata to be communicated (by using an XML-based dialogue), as well as defining a standard way for the parties to discover how to "understand" the metadata. RDF documents contain statements that describe properties about some entity, like a Web page, a book, a document, or anything else.

CC/PP as an HTTP Extension

Since RDF is a generic metadata description and exchange model, it is capable of being used in any number of scenarios. CC/PP, as mentioned earlier, uses RDF syntax to describe types of device characteristics.

CC/PP messages are communicated between parties via an extension of the HTTP protocol. A client obtains a resource from a server with an HTTP request GET and POST methods are supported) into which the client encodes its device capabilities and user preferences. In ordinary HTTP, such as what one might encounter in a Web browser scenario, the browser adds various HTTP headers to the GET/POST request. This would include such headers as the MIME type understood by the browser, the host name, and so on. CC/PP requests, being communicated via an HTTP extension, contain supplemental headers in addition to the basic ones recommended by the HTTP 1.1 specification.

CC/PP messages, though transmitted from one party to another inside HTTP headers, may contain XML documents due to CC/PP's RDF lineage. In Listing 15-1, a typical CC/PP message is shown. A mobile phone client could send such a message to a server when requesting documents.

Listing 15-1. Exchanging Mobile Phone Capabilities in CC/PP.

```
<?xml version="1.0"?>
<RDF xmlns="http://www.w3.org/1999/02/22-rdf-syntax-ns#"
     xmlns:rdf="http://www.w3.org/1999/02/22-rdf-syntax-ns#"
     xmlns:ccpp="http://www.w3.org/2000/07/04-ccpp#"
     xmlns:uaprof="http://www.wapforum.org/UAPROF/ccppschema-19991014#">
 <Description about="http://www.sjsu.edu/portal/MyProfile">
  <ccpp:component>
  <Description about="http://www.sjsu.edu/portal/Browser">
    <type resource="http://www.sjsu.edu/portal/Schema#BrowserUA" />
    <uaprof:BrowserName>Mozilla</uaprof:BrowserName>
    <uaprof:BrowserVersion>5.0</uaprof:BrowserVersion>
    <uaprof:CcppAccept>
     <Bag>
      <li>text/plain</li>
      <li>text/vnd.wap.wml</li>
```

```
    </Bag>
   </uaprof:CcppAccept>
  </Description>
  </ccpp:component>
 </Description>
</RDF>
```

Listing 15-1 shows an XML document containing a CC/PP description of the mobile phone's capabilities. This message conveys the following information:

- The client is a browser in the Mozilla family (via the "BrowserName" element).
- The client version is 5.0 (via the "BrowserVersion" element).
- The client accepts plain text or WML encoded documents (via the "CcppAccept" Bag of elements).

This is a minimal description. A more complete description would also contain the screen resolution, presence of a keyboard, voice recognition support, and so on. It might also contain user preferences, which for example could allow the application to determine whether or not the user prefers sound enabled or disabled, and what the user's favorite color is. A complete description of a device's capabilities is known as a *device profile*.

We've shown a typical CC/PP document above. The sender of this document (in this case, the mobile phone) has several ways of sending the contents to the server:

- ***The CC/PP capabilities are published on the Web and accessed via a URL.*** This method is typically used when supporting several devices. The application developer can publish the CC/PP documents that describe the capabilities of each type of device on the Web, and refer to them as needed. Using this method, clients do not have to send the entire document with each request; they simply send a URL pointing to the document containing the capabilities they support.

- ***The CC/PP capabilities can be embedded directly in an HTTP header.*** The clients compose a complete CC/PP description for each request, and send the document as one of the HTTP headers issued with the request, which is sent to the server.

- ***The CC/PP capabilities are published on the Web***. However, the client amends the descriptions stored there with any request-time differences.

Consider the example in Listing 15-2 that follows. In this example, a client is shown requesting the Web resource "/a-resource" by issuing an HTTP 1.1 "GET" request. The "Opt:" header, which is used to indicate the presence of an HTTP extension, indicates the client would like the server to process the request using the HTTP extension specified in the header's payload. In this case, the payload identifies the CC/PP exchange extension.

Listing 15-2. CC/PP Header Exchange.

```
GET /a-resource HTTP/1.1
Host: cslab35
Opt:  "http://www.w3.org/1999/06/24-CCPPexchange" ; ns=99
99-Profile: "1-uKhJE/AEeeMzFSejsYshHg=="
99-Profile-Diff-1: <?xml version="1.0"?>
    . . .
```

The client also adds the "Profile" and "Profile-Diff-1" headers to the HTTP request. The "Profile" header can contain one or more of the following:

- A URL pointing to a CC/PP capabilities document (such as the one shown in Listing 15-1).
- A string representing an MD5 hash of the payload of a subsequent "Profile-Diff" header. This string begins with a number, followed by a dash, followed by the MD5 hash. The number at the beginning of the string matches the number at the end of a subsequent "Profile-Diff" header.

An astute reader may be wondering what the significance of the number 99 is in the various HTTP headers shown in Listing 15-2. The "Opt" HTTP extension header that refers to the CC/PP extension is suffixed with an "ns=99"—this indicates that any subsequent headers that start with 99 are to be considered relevant to the CC/PP extension. This "namespace-qualification" of headers is used when multiple HTTP extensions are in use. This avoids name collision if different extensions specify the same header names. The header namespace, in this case "99", is randomly generated and should not be construed as having any particular significance in a given case.

In this example, the `99-Profile` header consists of a digit and a MD5 digest. The digit "1" indicates the digest represents the "`99-Profile-Diff-1`" header encoded with base64. The payload of the "Profile-Diff" header contains a CC/PP device profile, or a set of CC/PP elements that are used to override capabilities specified in another profile. URLs or MD5 hashes present in the "Profile" element are read in the order they are encountered—this means that rules contained in profiles or profile diffs override rules encountered in earlier profiles.

RDF in the Header

When a mobile device sends a request to a server for some content, the server either receives the device profile directly from the client (if the client directly encodes the device profile as a "Profile-Diff" header), or constructs it by examining the URLs provided in the "Profile" header and applying any changes noted in the "Profile-Diff" headers.

Hence, the origin server will acquire all the information needed about the device from the device profile and thus will be able to choose the right content for that device. In addition to the device profile, the document profile, which specifies the language capabilities needed by the client, is needed. The document profile can be thought of as a corollary to a device profile—a device profile indicates which languages a device supports. A document profile indicates what language support is needed by a client to correctly render the document.

As the server decides which document to deliver to the client, the document profile would be compared with the device profile, as shown in Figure 15-3. Subsequently, the best fit between the two would be discovered, and a suitable document would be generated or the best fitting variant would be selected.

CC/PP will be a key part of the coming pervasive Web as it provides a framework within which diverse clients can tell servers what they want or what they can do. It

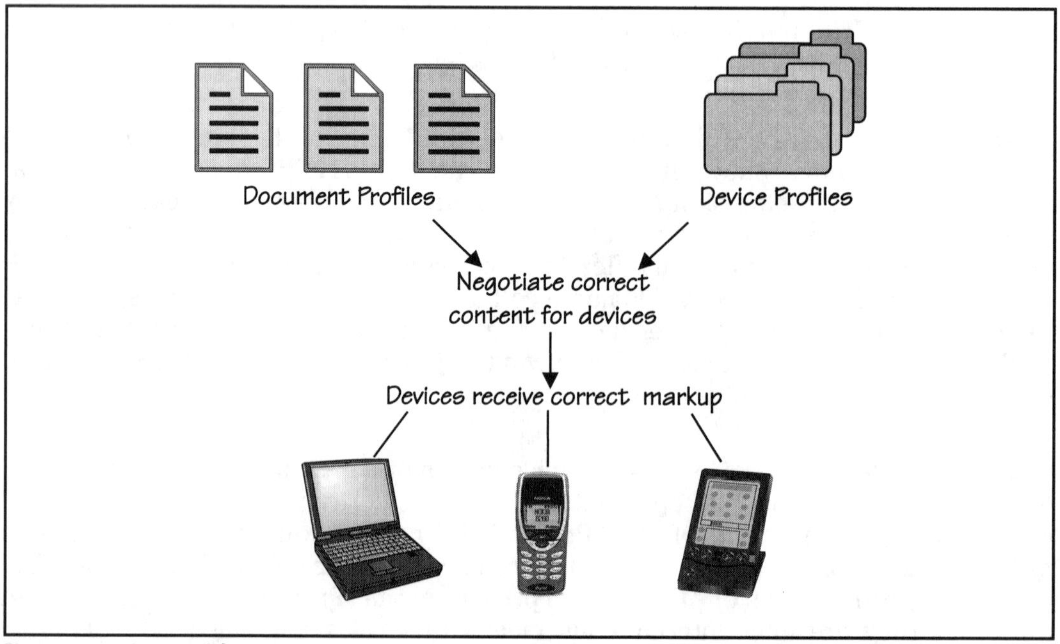

Figure 15-3. Content Adaptation.

will allow servers to adapt to a client's needs. When CC/PP becomes a W3C recommendation (it is currently still a W3C note), RDF will become as pervasive as the Web itself.

Transcoding Basics

Now that we understand how device capabilities are exchanged between clients and servers, we can refocus our attention on how the server actually adapts content to meet the client's needs. We mentioned earlier that a transcoding engine follows a set of rules when converting content to a different output format, and that these rules are usually encoded using a language called XSL. In the next section, we'll give a quick introduction to XSL and cover some of the standard transformation and matching rules it provides. XSL is a topic that could fill a book all by itself; our coverage is going to be at the introductory level only.

INTRODUCTION TO XSL

This section presents an introduction to Extensible Stylesheet Language (XSL) and Extensible Stylesheet Language Transformations (XSLT). We'll briefly discuss each and follow with a few short examples. XSL forms the core of transcoding technology—it's the workhorse behind the scenes doing the hard work for us.

What Is XSL?

XSL is a stylesheet language for XML documents. An XSL stylesheet specifies a filter that is applied to a source XML document to produce a different output XML document. Although XSL is generic enough to represent any type of transformation, some of its typical uses are to convert XML documents into a form understandable by some user agent, such as a Web browser. XSL consists of two parts:

- A mechanism for transforming XML documents (XSLT).
- An XML vocabulary for specifying formatting rules.

Note: Every XSL document is a well-formed XML document.

What Types of Translations Is XSL Capable of Performing?

With XSL, the stylesheet developer can:

- ***Add*** new XML elements to the output document, which were not present in the source document.
- ***Remove*** XML elements as the source document is processed, so that they are not copied to the output document.
- ***Rearrange*** and sort the source document's elements.
- ***Test*** and make decisions about which elements to copy to the output document.

The above rules, though very generally stated, are powerful enough to create any type of transformation imaginable. We'll see some examples on the next pages as we learn how a transcoding engine processes XSL stylesheets.

How Are XSL Stylesheets Processed?

In this section we will briefly walk you though how an XSL transformation will be applied. Let's first look at the steps that the stylesheet processor follows when processing your stylesheet (shown in Figure 15-4).

1. The XML parser reads the source document and XSL stylesheet.

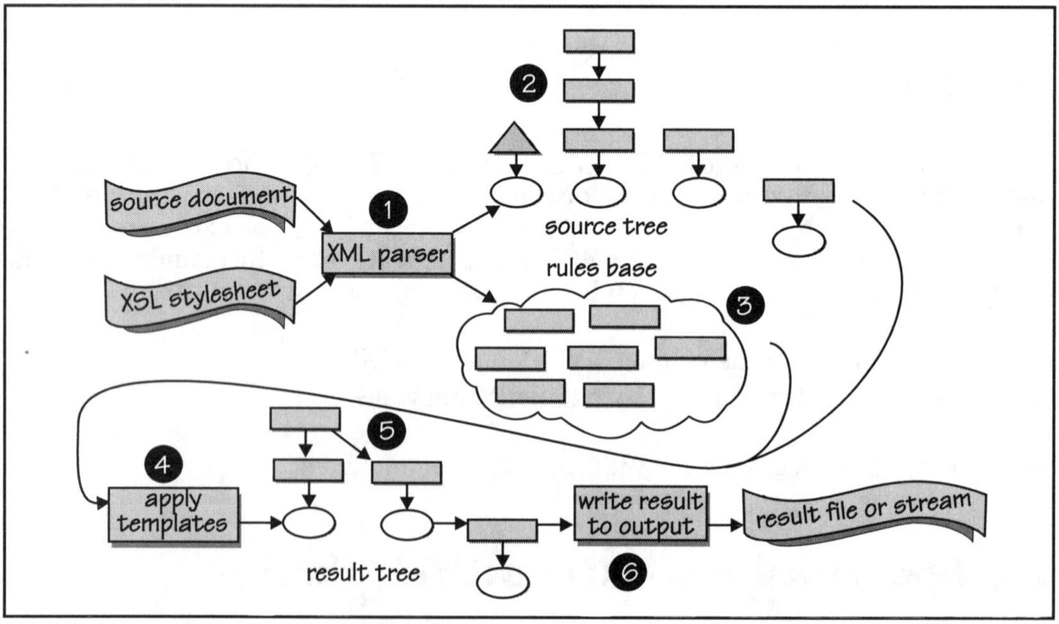

Figure 15-4. Sequences in Applying XSL to an XML Document.

2. An in-memory parse tree is created, reflecting the composition of the source document's elements.
3. A set of rules, called the rule base, is created from the transformation rules specified in the XSL stylesheet.
4. The parse tree is traversed, and any matching rules from the rule base are applied.
5. The resulting output elements are copied into a result tree.
6. The result tree is output as XML into an output file.

To illustrate these steps, let us consider the following simple XML document in Listing 15-3. Note that the element `<chapter>` is the root element of this document.

Listing 15-3. A Simple XML File that Represents a Book Chapter.

```
<chapter id="cmds">
  <chaptitle>FileCab</chaptitle>
  <para>
    This chapter describes the commands that
    manage the <tm>FileCab</tm>inet application.
  </para>
</chapter>
```

During the first stage in XSL stylesheet processing, the source document will be read into memory and stored as a parse tree. Figure 15-5 shows how the source tree for the Chapter XML document would be represented in memory.

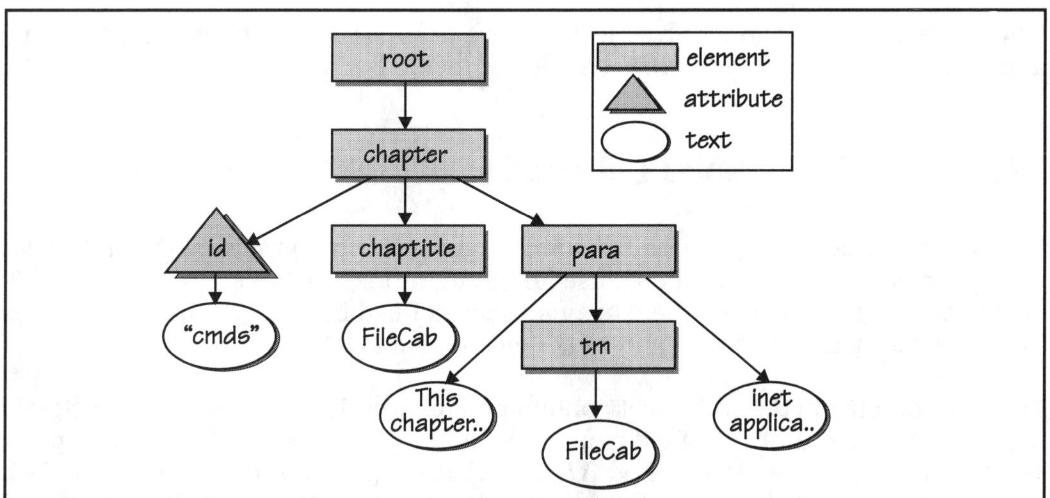

Figure 15-5. Possible Tree Representation for an XML Document.

The transcoding engine will use this in-memory parse tree when applying the XSL transformation.

After reading the source XML document and constructing the tree representation, the transcoding engine reads the XSL stylesheet and organizes the rules contained in it to provide efficient lookups as the stylesheet is applied. The transcoding engine then traverses the source tree starting from the root node, and attempts to match each node in the source tree to a corresponding rule defined in the stylesheet. If a match is found, the transcoding rule specified in the stylesheet is applied. The result of applying the transcoding rule (which could be a completely new element or set of elements, an alteration of the source element, or something else entirely) is copied into another parse tree that holds the intermediate results as the source document is processed. This matchmaking and results gathering continues with the source element's child nodes until the source tree has been completely traversed.

Once the source tree has been traversed completely, the XSL processor walks the result tree and copies what it finds into an output file or stream, using the syntax rules contained in the stylesheet, if such rules are present.

How Is an XSL Stylesheet Organized?

All XSL documents must be well-formed XML documents. Although any namespace can be used when qualifying XSL elements, the industry seems to have settled on the *xsl* namespace. All XSL stylesheets should use the root element <xsl:stylesheet>. Figure 15-6 illustrates an XSL document with its important elements and attributes annotated.

Now that you've seen how XSL stylesheets are processed and organized, let's delve a little deeper and learn what XSL can do for us.

XPath—Element Matching and Location

We mentioned earlier that the transcoding engine (or more precisely, the stylesheet processor contained inside the transcoding engine) scans the source document's elements and tries to match them against rules defined in our stylesheet. How does the transcoding engine determine which elements match?

The rules defined in the stylesheet contain path expressions that are used to clearly identify the element or set of elements to which the rule applies. The path expressions adhere to a standard called *XPath*. XPath represents the root of an XML document with a single slash (/) character, and also uses the same slash character to separate parent elements from child elements.

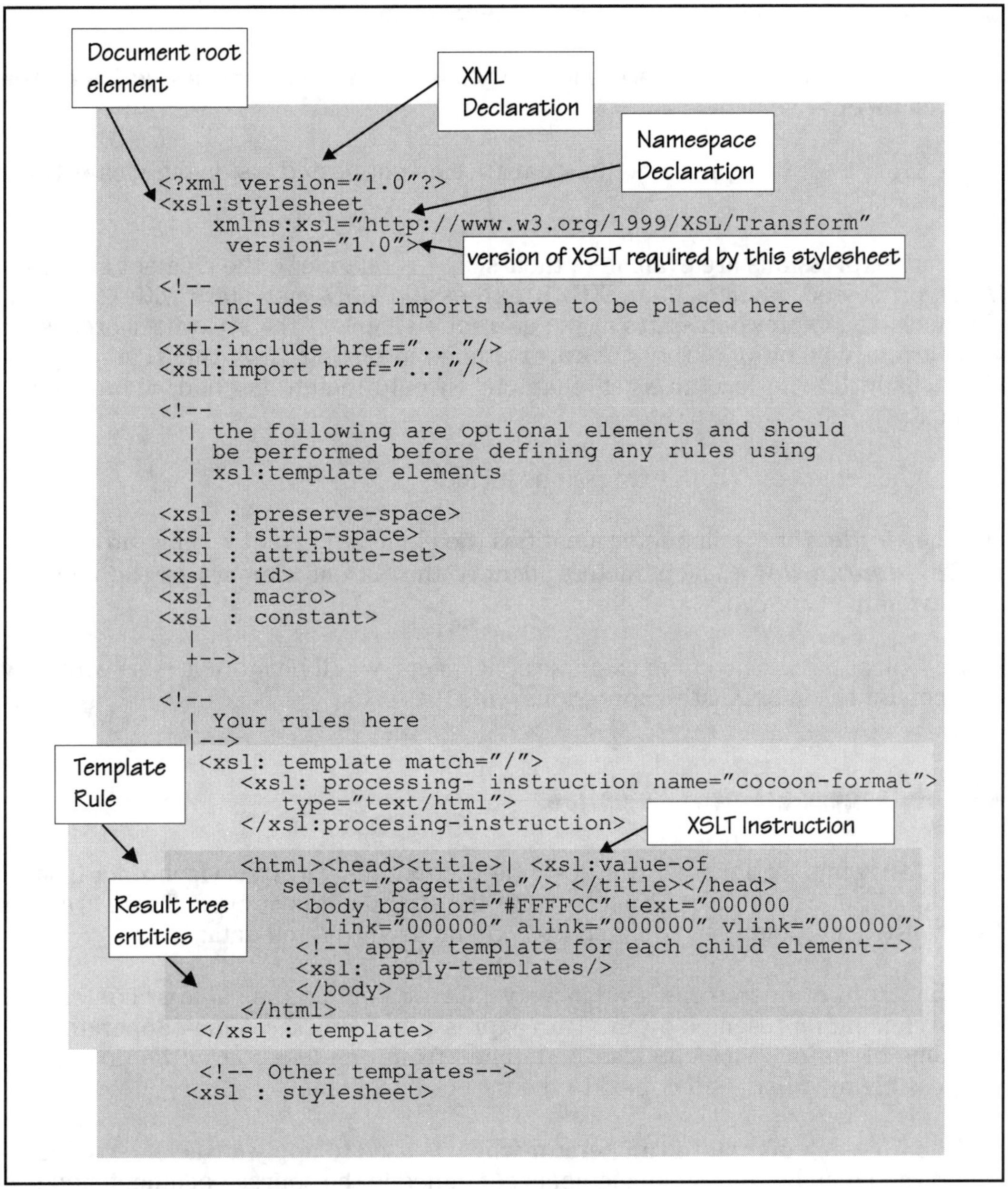

Figure 15-6. Components of a Stylesheet.

For example, when considering the XML document shown previously in Listing 15-3, all the following represent valid XPath expressions:

- `/chapter` represents the root **<chapter>** element
- `/chapter/chaptitle` represents the **<chaptitle>** element nested inside the **<chapter>** element
- `/chapter/para` represents the **<para>** element nested inside the **<chapter>** element

As XPath expressions are evaluated, their scope is relative to the element currently being processed. Hence, if an XPath expression does not start with the slash character, the expression starts at the current element. If the stylesheet processor was currently pointing at the **<chapter>** node in Listing 15-3, the XPath expression "chaptitle" (no leading slash character) would match its child **<chaptitle>** element.

Some other common XPath expressions include:

- The ***single dot (.)*** character identifies the element currently being processed
- The ***double dot (..)*** characters identify the parent element of the element currently being processed

As we look at the forthcoming example stylesheets, you'll be more at ease with how the scoping rules for XPath expressions work.

XSL Stylesheet Element Selection

In the following sections we will describe the various ways in which data from a source XML document can be accessed as the stylesheet is applied. There are four XSL constructs (operators) defined that can be used to fetch data:

- The **value-of** operator is used in a stylesheet to obtain the value or contents of an element represented by an XPath expression. Using the value-of operator, the value of various elements can be fetched from one location in the document, possibly modified, and copied to the output document.

- The **for-each** operator is an iterator which is used to apply a rule or set of rules on each member of a set of elements contained in the source document. You use the for-each operator to perform two common types of source document manipulation—sorting elements and filtering out unwanted elements.

- The **if (conditional if)** statement is used to apply a rule or set of rules if a condition is met. Typically the condition is used to test the existence of an element, or the value of an element.

- The **conditional choice** construct is used when you want to translate content one way if a certain match is made, and another way if not.

As we work through the following sections, we will provide an XML document and an XSL stylesheet using one of the above mechanisms when matching data in our source document. In these examples, XSL stylesheets were created that translate a simple XML document (a university course catalog) into various XHTML documents to show how these mechanisms can be used. As you look at the examples, refer to Listing 15-4 below—all the example stylesheets are processed against this XML file.

Listing 15-4. University Course Catalog (university.xml).

```
<?xml version="1.0" ?>
<?xml-stylesheet type="text/xsl" href="university.xsl"?>
<CATALOG>
 <COURSE>
  <DEPARTMENT>Engineering</DEPARTMENT>
  <COURSE>
   <CODE>ENGR262</CODE>
   <TITLE>Client/Server Overview</TITLE>
  </COURSE>
  <PROFESSOR>Harkey</PROFESSOR>
  <ROOM>ENGR 337</ROOM>
  <TIME>Tu 1800-1950</TIME>
 </COURSE>

 <COURSE>
  <DEPARTMENT>Engineering</DEPARTMENT>
  <COURSE>
   <CODE>ENGR264</CODE>
   <TITLE>Client/Server Distributed Objects</TITLE>
  </COURSE>
  <PROFESSOR>Larkin</PROFESSOR>
  <ROOM>ENGR 337</ROOM>
  <TIME>W 1800-1950</TIME>
 </COURSE>

 <COURSE>
  <DEPARTMENT>Engineering</DEPARTMENT>
  <COURSE>
   <CODE>ENGR296J</CODE>
   <TITLE>Mobile Technologies</TITLE>
  </COURSE>
  <PROFESSOR>Appajodu</PROFESSOR>
  <ROOM>MQH 324</ROOM>
```

```
  <TIME>M 1900-2050</TIME>
 </COURSE>

 <COURSE>
  <DEPARTMENT>Math/Computer Science</DEPARTMENT>
  <COURSE>
   <CODE>MATH226</CODE>
   <TITLE>Theory of Numbers</TITLE>
  </COURSE>
  <PROFESSOR>Smith</PROFESSOR>
  <ROOM>TBA</ROOM>
  <TIME>TBA</TIME>
 </COURSE>

 <COURSE>
  <DEPARTMENT>Math/Computer Science</DEPARTMENT>
  <COURSE>
   <CODE>MATH279</CODE>
   <TITLE>Graph Theory</TITLE>
  </COURSE>
  <PROFESSOR>Jones</PROFESSOR>
  <ROOM>TBA</ROOM>
  <TIME>TBA</TIME>
 </COURSE>

</CATALOG>
```

XSL value-of and XSL Sort

The *value-of* XSL operator is used to fetch the value represented by a given XPath expression. This is usually used to retrieve data from some element when constructing a new element.

Let's take the course catalog shown above and convert the contents to HTML, while simultaneously sorting the courses by department.

Figure 15-7 shows the output generated from the stylesheet in Listing 15-5. In the stylesheet, we used the value-of operator in addition to illustrating how the for-each operator can be used to sort elements. In the example, the sorting is ordered by department, via the inclusion of an *order-by* attribute in the *for-each* element:

```
<xsl:for-each select="CATALOG/COURSE" order-by="+ DEPARTMENT">
```

The order-by attribute consists of a plus (+) or minus (-) sign, indicating an ascending or descending sort order, followed by an element name that identifies the sort element.

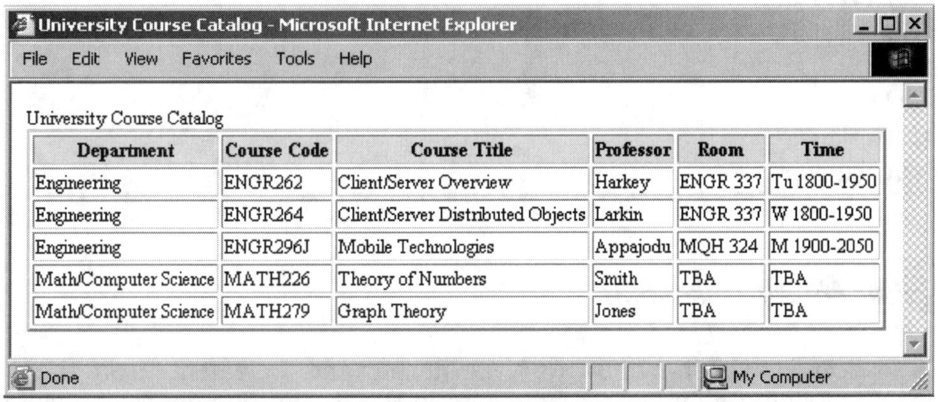

Department	Course Code	Course Title	Professor	Room	Time
Engineering	ENGR262	Client/Server Overview	Harkey	ENGR 337	Tu 1800-1950
Engineering	ENGR264	Client/Server Distributed Objects	Larkin	ENGR 337	W 1800-1950
Engineering	ENGR296J	Mobile Technologies	Appajodu	MQH 324	M 1900-2050
Math/Computer Science	MATH226	Theory of Numbers	Smith	TBA	TBA
Math/Computer Science	MATH279	Graph Theory	Jones	TBA	TBA

Figure 15-7. A University Course Catalog Sorted and Displayed as HTML.

Listing 15-5. Using XSL value-of and for-each Operators (university1.xsl).

```
<?xml version="1.0"?>
<xsl:stylesheet xmlns:xsl="http://www.w3.org/TR/WD-xsl" version="2.0">
<xsl:template match="/">
  <html>
  <title>University Course Catalog</title>
  <body>
    University Course Catalog
    <table border="2" >
      <tr>
        <th bgcolor="yellow">Department</th>
        <th bgcolor="yellow">Course Code</th>
        <th bgcolor="yellow">Course Title</th>
        <th bgcolor="yellow">Professor</th>
        <th bgcolor="yellow">Room</th>
        <th bgcolor="yellow">Time</th>
      </tr>
      <xsl:for-each select="CATALOG/COURSE" order-by="+ DEPARTMENT">
      <tr>
        <td>
         <xsl:value-of select="DEPARTMENT"/></td>
        <td><xsl:value-of select="COURSE/CODE"/></td>
        <td><xsl:value-of select="COURSE/TITLE"/></td>
```

```
        <td><xsl:value-of select="PROFESSOR"/></td>
        <td><xsl:value-of select="ROOM"/></td>
        <td><xsl:value-of select="TIME"/></td>
      </tr>
      </xsl:for-each>
    </table>
  </body>
  </html>
</xsl:template>
</xsl:stylesheet>
```

XSL Filter Query (for-each)

In this section we will show how to filter out unwanted elements from the source document, retaining only desired elements in the output document. We will again use the course catalog XML file, but will introduce a slight change in the XSL stylesheet.

The filter query provides the ability to match certain elements in the source document, while omitting others. The determinations as to which elements are matched and which are not are governed by the XPath expression contained in the select attribute. Simply construct an expression that identifies the elements you want to match, and supply the expression as the argument of the select attribute.

To illustrate this concept, we'll create another stylesheet, this time only matching Engineering courses:

Our new stylesheet is almost identical to our previous version. The only line that changed is the line that contains the for-each operator. We are only matching the elements that have "Engineering" for their DEPARTMENT element.

```
<xsl:for-each select="CATALOG/COURSE[DEPARTMENT='Engineering']">
```

Table 15-1 compares the available filter operators.

Table 15-1. XSL for-each filter operators.

Filter Operator	Used When
= (equal)	Testing for equality.
=!(not equal)	Testing for inequality. The rule matches if the operands are not equal.

Table 15-1. XSL for-each filter operators. (Continued)

Filter Operator	Used When
&LT& (less than)	Testing magnitude. The rule matches if the left operand is less than the right operand.
&GT& (greater than)	Testing magnitude. The rule matches if the left operand is greater than the right operand.

Figure 15-8 shows the output generated from the slightly modified XSL stylesheet in Listing 15-6.

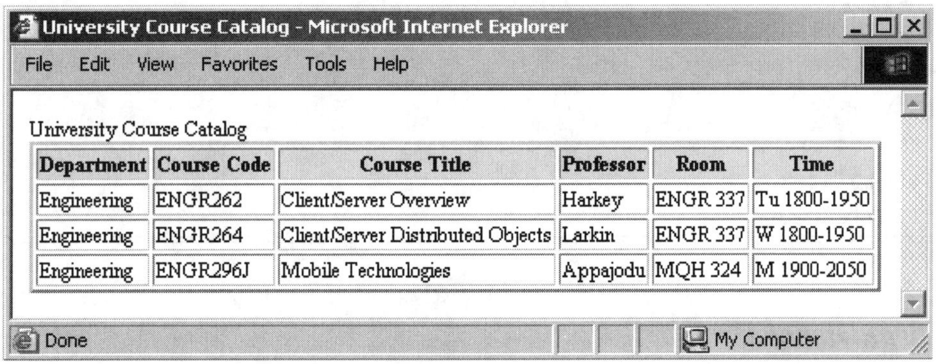

Figure 15-8. Catalog Displayed with Only Engineering Courses Matched.

Listing 15-6. Using XSL for-each Operator to Filter Elements (university2.xsl).

```
<?xml version="1.0"?>
<xsl:stylesheet xmlns:xsl="http://www.w3.org/TR/WD-xsl" version="2.0">
<xsl:template match="/">
  <html>
  <title>University Course Catalog</title>
  <body>
    University Course Catalog
    <table border="2" >
      <tr>
        <th bgcolor="yellow">Department</th>
        <th bgcolor="yellow">Course Code</th>
        <th bgcolor="yellow">Course Title</th>
```

```
          <th bgcolor="yellow">Professor</th>
          <th bgcolor="yellow">Room</th>
          <th bgcolor="yellow">Time</th>
        </tr>
        <xsl:for-each select="CATALOG/COURSE[DEPARTMENT='Engineering']">
        <tr>
          <td>
           <xsl:value-of select="DEPARTMENT"/></td>
          <td><xsl:value-of select="COURSE/CODE"/></td>
          <td><xsl:value-of select="COURSE/TITLE"/></td>
          <td><xsl:value-of select="PROFESSOR"/></td>
          <td><xsl:value-of select="ROOM"/></td>
          <td><xsl:value-of select="TIME"/></td>
        </tr>
        </xsl:for-each>
      </table>
    </body>
    </html>
</xsl:template>
</xsl:stylesheet>
```

XSL Conditional If

In this section we will describe how to use the *conditional if* test. We will again reference the course catalog document shown earlier.

To apply a conditional if test on the contents of an XML document, simply add an *xsl:if* element to your XSL document like this:

```
<xsl:if match=".[DEPARTMENT='Engineering']">
... some output ...
</xsl:if>
```

To illustrate this concept, let's add a column to our HTML output table. This column indicates whether or not the time for the course is "TBA", indicating that the student should see the instructor to find out the time the course is meeting.

Figure 15-9 shows the output for the modified stylesheet shown in Listing 15-7.

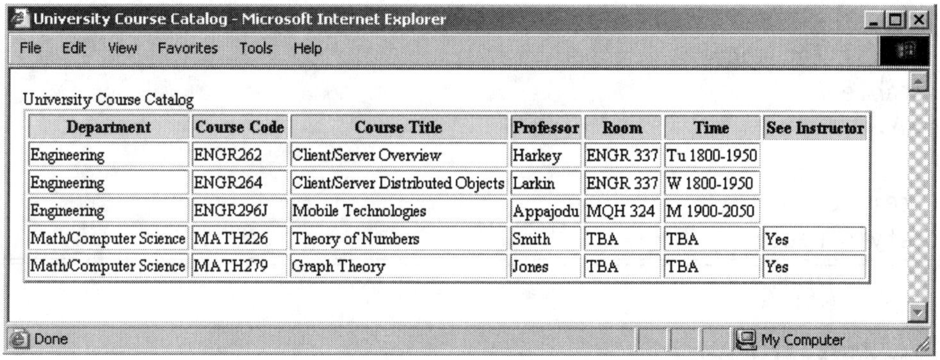

University Course Catalog

Department	Course Code	Course Title	Professor	Room	Time	See Instructor
Engineering	ENGR262	Client/Server Overview	Harkey	ENGR 337	Tu 1800-1950	
Engineering	ENGR264	Client/Server Distributed Objects	Larkin	ENGR 337	W 1800-1950	
Engineering	ENGR296J	Mobile Technologies	Appajodu	MQH 324	M 1900-2050	
Math/Computer Science	MATH226	Theory of Numbers	Smith	TBA	TBA	Yes
Math/Computer Science	MATH279	Graph Theory	Jones	TBA	TBA	Yes

Figure 15-9. Catalog Displayed using XSL If Operator.

Listing 15-7. Using XSL if Operator (university3.xsl).

```
<?xml version="1.0"?>
<xsl:stylesheet xmlns:xsl="http://www.w3.org/TR/WD-xsl" version="2.0">
<xsl:template match="/">
  <html>
  <title>University Course Catalog</title>
  <body>
    University Course Catalog
    <table border="2" >
      <tr>
        <th bgcolor="yellow">Department</th>
        <th bgcolor="yellow">Course Code</th>
        <th bgcolor="yellow">Course Title</th>
        <th bgcolor="yellow">Professor</th>
        <th bgcolor="yellow">Room</th>
        <th bgcolor="yellow">Time</th>
        <th bgcolor="yellow">See Instructor</th>
      </tr>
      <xsl:for-each select="CATALOG/COURSE">
      <tr>
        <td>
         <xsl:value-of select="DEPARTMENT"/></td>
        <td><xsl:value-of select="COURSE/CODE"/></td>
        <td><xsl:value-of select="COURSE/TITLE"/></td>
        <td><xsl:value-of select="PROFESSOR"/></td>
        <td><xsl:value-of select="ROOM"/></td>
        <td><xsl:value-of select="TIME"/></td>
      <xsl:if match=".[TIME='TBA']">
         <td>Yes</td>
```

```
        </xsl:if>
      </tr>
      </xsl:for-each>
    </table>
  </body>
  </html>
</xsl:template>
</xsl:stylesheet>
```

XSL Conditional Choice

The conditional choice construct is used when you want to translate content one way if a certain match is made, and another way if not. The conditional choice is similar to the if..then..else construct present in most programming languages.

To apply a conditional choice test on the contents of an XML document, we supply three XSL elements in our stylesheet:

- **xsl:choose** is the bounding element that denotes the limits of our conditional choice.
- **xsl:when** is the element that contains the matching rule. The value of the xsl:when element comprises the content that is used when the match is successful.
- **xsl:otherwise** is the element whose value contains the content that is used when the match is unsuccessful.

For example, the code snippet shown below could be used to match all elements relative to the current element that contains a sub-element named "PROFESSOR" with the value "Jones":

```
<xsl:choose>
<xsl:when match=".[PROFESSOR='Jones']">
... some code ...
</xsl:when>
<xsl:otherwise>
... some code ....
</xsl:otherwise>
</xsl:choose>
```

To illustrate this concept, let's modify our stylesheet to shade the background of the table cell dark if the course is an Engineering course, and to shade it light otherwise.

Figure 15-10 shows the output generated from the stylesheet shown in Listing 15-8.

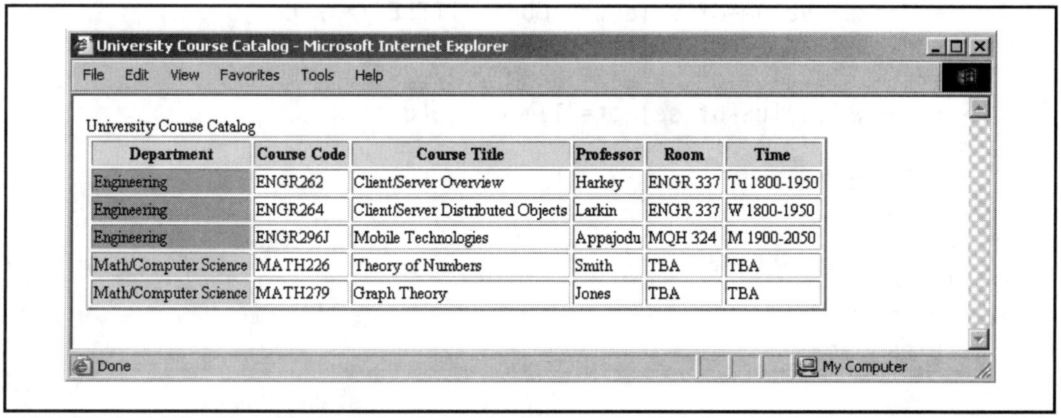

Figure 15-10. Catalog displayed using the XSL Choose Operator.

Listing 15-8. Using XSL Conditional Choice Operator (university4.xsl).

```
<?xml version="1.0"?>
<xsl:stylesheet xmlns:xsl="http://www.w3.org/TR/WD-xsl" version="2.0">
<xsl:template match="/">
  <html>
  <title>University Course Catalog</title>
  <body>
    University Course Catalog
    <table border="2" >
      <tr>
        <th bgcolor="yellow">Department</th>
        <th bgcolor="yellow">Course Code</th>
        <th bgcolor="yellow">Course Title</th>
        <th bgcolor="yellow">Professor</th>
        <th bgcolor="yellow">Room</th>
        <th bgcolor="yellow">Time</th>
      </tr>
      <xsl:for-each select="CATALOG/COURSE">
      <tr>
        <xsl:choose>
         <xsl:when match=".[DEPARTMENT='Engineering']">
          <td bgcolor="#AAAAAA"><xsl:value-of select="DEPARTMENT"/></td>
         </xsl:when>
         <xsl:otherwise>
          <td bgcolor="#DDDDDD"><xsl:value-of select="DEPARTMENT"/></td>
         </xsl:otherwise>
```

```
        </xsl:choose>
        <td><xsl:value-of select="COURSE/CODE"/></td>
        <td><xsl:value-of select="COURSE/TITLE"/></td>
        <td><xsl:value-of select="PROFESSOR"/></td>
        <td><xsl:value-of select="ROOM"/></td>
        <td><xsl:value-of select="TIME"/></td>
      </tr>
      </xsl:for-each>
    </table>
  </body>
  </html>
</xsl:template>
</xsl:stylesheet>
```

Transcoding Engines—an End-to-End example

We've covered a lot of material in this chapter so far. We've learned how devices exchange capabilities via CC/PP and RDF. We've seen how XSL stylesheets can be used to convert XML content from one form to another. We've seen several examples of how XSL stylesheets are constructed internally. We're now ready to put all this knowledge to work and see how a document would actually be transcoded when delivered to three different types of user devices. There's just one last piece we haven't talked about yet—the actual transcoding engine. In the previous examples, we've assumed that "something" has applied the stylesheet to the source XML document, but we haven't said exactly *what* is responsible for this action.

The work of receiving and interpreting client device profiles, applying stylesheets, and so forth, is performed by an application known as a transcoding engine. Many transcoding engines exist. We chose the Apache group's Cocoon transcoding engine for our Campus Portal project, and this is the transcoding engine we will focus our attention on in this section.

Cocoon

Cocoon, an open-source transcoding engine, is a Servlet-based publishing system that provides a framework for developing and deploying content transformers based on the client device. Cocoon also allows dynamic content generation and transformations. It provides a clear demarcation of three different development tasks: content, style, and logic. Although JSPs try to decouple content and rendering, they have not succeeded completely, because the developer must mingle device-rendering code with the business logic. Cocoon tries to maintain the distinction between logic and device rendering by utilizing processing and format instructions in XML documents.

Figure 15-11 illustrates the components of Cocoon. Cocoon supports content generation in XML format (similar to how JavaServer Pages typically generate HTML output), transcoding a document in one XML-based markup language to other XML-based markup languages, and dynamic XML content generation. Cocoon delivers support for dynamic XML content through *eXtensible Server Pages (XSP)*. XSP allows you to embed logic in an XML page in any language, in the same way JSP allows you to embed Java-based logic in an HTML page.

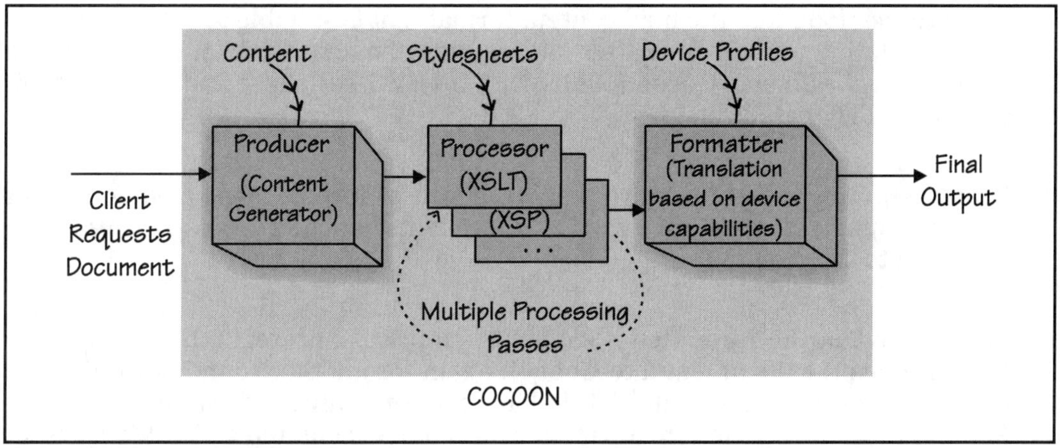

Figure 15-11. Cocoon Processing Flow.

Referring to Figure 15-11, the composite parts of the Cocoon transcoding framework are:

- ***Request:*** This is a wrapper that includes the client's request and contains any ancillary information needed by the processing engine. The request contains information about the client, the URI requested, and the producer that has to handle the request.

- ***Producer:*** Producers handle the client requests, producing XML documents as their output. Producers behave like Servlets and could be comprised of JSP pages or other XML documents. The XML document output by the producer could be either static or dynamically generated. Producers in Cocoon are pluggable, which means end users can implement their own producers. The output XML document from the producer is then fed into the reactor for further processing.

- ***Reactor and Processor:*** The reactor interprets processing instructions contained in the XML documents and routes the document to the processor that should work on the document during the current pass. The processors process the XML content, possibly inserting new processing instructions into the content. The new XML content is then fed back into the reactor for the next pass. When no further processing instructions are encountered, the final content exits the reactor/processor loop. The fact that a processor can re-insert new processing instructions back into the content makes the reactor design pattern different from a processing pipeline since it allows the processing path to be dynamically configurable. Cocoon ships with several processors—these include a XSLT (XSL Stylesheet) processor, a SQL processor, and an LDAP processor.

- ***Formatter:*** Formatters transform the XML document generated by the reactor/producer loop into the final content format that is suitable for the client. The implementation of the formatter determines the output MIME type of the transcoded document. Cocoon ships with an SVG serializer and a Formatting Objects to PDF serializer.

- ***Response:*** This is the response that wraps the tailored, transcoded document with the appropriate MIME type for the client device. The response is then delivered to the client.

- ***Loader:*** When the formatted document comprises an executable format, the Loader loads the document into Cocoon again for another round of processing before delivering to the client. This behavior is generally used with pre-compiled XSP pages (XSPs). When the formatter output is executable code, it is not sent back to the client directly (since the client would have no way of understanding it); instead it is loaded and executed as a document producer. This provides for easier producer development, because precompiled producers can be chained together and loaded by Cocoon as they are required.

Cocoon Processing Instructions

Cocoon's reactor module uses processing instructions embedded in the XML content when determining where to route the content. The Cocoon reactor understands two styles of processing instructions ("PI" for short):

- ***Processing PIs***, which take the form `<?Cocoon-process type="xxx"?>`, identify the processor that will be invoked to handle the content contained in the document. These processors typically comprise Java classes that implement a particular processing behavior. For example, the XSLT processor knows how to process XSL stylesheets, and the SQL processor would know how to access various relational databases.

- ***Formatting PIs***, which take the form `<?Cocoon-format type="yyy"?>`, instruct the Cocoon formatter to encode the output while adhering to a particular MIME type.

The various PIs indicate the processing and formatting path that the document should follow when being transformed and rendered. Listing 15-9 shows how you would instruct Cocoon to route the document through the XSLT processor.

Listing 15-9. How Style Sheets are Specified in XML Documents Using Cocoon.

```
<?xml version="1.0"?>
<?xml-stylesheet href="first.xsl" type="text/xsl"?>
<?Cocoon-process type="xslt"?>
<page>
  <title>Hello World!</title>
  <content>
    <paragraph>Hello World! Welcome to Cocoon!!</paragraph>
  </content>
</page>
```

USING COCOON TO TRANSCODE FOR VARIOUS CLIENT DEVICES

We're finally ready to put together all the information we've learned so far. We've chosen a simple example application that outputs "Hello. This is my first Cocoon page!"

In this example, we will show how the same original source content is transcoded for three different client devices:

- HTML (Regular Web browser)
- VoiceXML
- WML (Suitable for a WAP-enabled cell phone)

We will start by looking at our source XML document, and then see what the output looks like on each type of device as different stylesheets are applied.

Listing 15-10 shows our source XML document.

Listing 15-10. The Hello Cocoon Page (hello-page.xml).

```
<?xml version="1.0"?>
<?xml-stylesheet href="hello-page-html.xsl" type="text/xsl"?>
<?xml-stylesheet  href="hello-page-vxml.xsl"  type="text/xsl"
media="vxml"?>
```

```
<?xml-stylesheet href="hello-page-wml.xsl" type="text/xsl" media="wap"?>

<?Cocoon-process type="xslt"?>

<page>
 <title>Hello</title>
 <content>
  <paragraph>This is my first Cocoon page!
</paragraph>
 </content>
</page>
```

We have provided stylesheets for all three of our client devices, and a processing instruction that will choose the appropriate one, based upon our client's media type. (Cocoon determines this automatically for us based on the headers supplied by our client—remember CC/PP?)

Let's use Cocoon to transcode the above document into HTML. Figure 15-12 shows the HTML output generated when using the stylesheet in Listing 15-11.

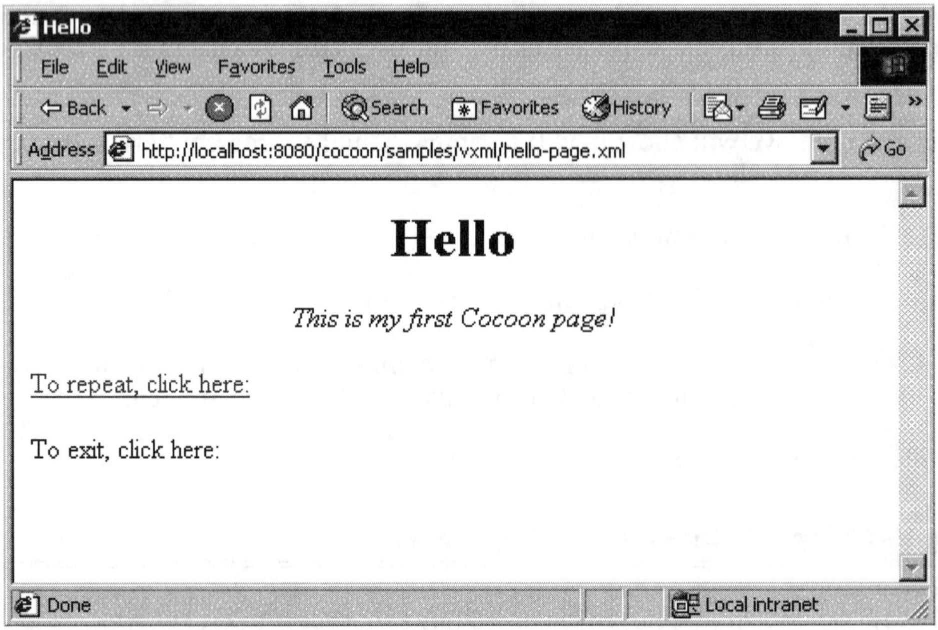

Figure 15-12. Accessing hello-page.xml from a Web Browser.

Listing 15-11. Stylesheet for Rendering HTML (hello-page-html.xsl).

```
<?xml version="1.0"?>

<xsl:stylesheet version="1.0"
xmlns:xsl="http://www.w3.org/1999/XSL/Transform">

  <xsl:template match="page">
   <xsl:processing-instruction
     name="Cocoon-format">type="text/html"</xsl:processing-instruction>
    <html>
     <head>
      <title>
       <xsl:value-of select="title"/>
      </title>
     </head>
     <body bgcolor="#ffffff">
      <xsl:apply-templates/>
     </body>
     <a href="hello-page.xml">To repeat, click here:</a>
     <p/>
     To exit, click here:
    </html>
   </xsl:template>

   <xsl:template match="title">
    <h1 align="center">
     <xsl:apply-templates/>
    </h1>
   </xsl:template>

   <xsl:template match="paragraph">
    <p align="center">
     <i>
      <xsl:apply-templates/>
     </i>
    </p>
   </xsl:template>
</xsl:stylesheet>
```

Now let's look at how Cocoon transcodes the XML document into VoiceXML. Assuming we use a suitable Voice Browser, the output using this stylesheet will be similar to the one below:

```
Hello. This is my first Cocoon page!
To repeat press or say 1
To exit press or say 2
```

Listing 15-12 contains the stylesheet to render first-page.xml in VoiceXML.

Listing 15-12. Stylesheet for Rendering in VoiceXML (hello-page-vxml.xsl).

```
<?xml version="1.0"?>

<xsl:stylesheet
    version="1.0" xmlns:xsl="http://www.w3.org/1999/XSL/Transform">

<xsl:template match="page">
  <xsl:processing-instruction
  name="Cocoon-format">type="text/xml"</xsl:processing-instruction>
 <vxml version="1.0">
  <menu>
   <prompt>
    <ibmvoice gender="female" age="adult">
     <xsl:apply-templates/>
    </ibmvoice>
   </prompt>

   <prompt>
    To repeat, press or say 1:
   </prompt>
   <choice next="hello-page.xml">
    <grammar> 1 </grammar>
    <dtmf>1</dtmf>
   </choice>

   <prompt>
    To exit, press or say 2:
   </prompt>
   <choice event="exit">
    <grammar> 2 </grammar>
    <dtmf>2</dtmf>
   </choice>
```

```
</menu>
 </vxml>

</xsl:template>

<xsl:template match="title">
 <!-- ignore -->
</xsl:template>

<xsl:template match="paragraph">
 <xsl:apply-templates/>
</xsl:template>
</xsl:stylesheet>
```

Finally, let's take a look at how Cocoon would transcode the content into WML. Figure 15-13 shows a screen shot of the output when accessed from a WML browser.

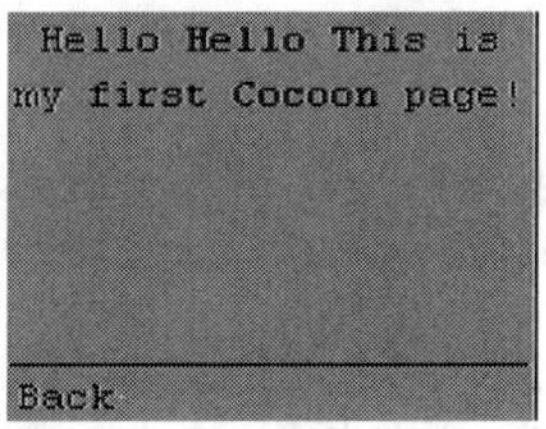

Figure 15-13. Accessing hello-page.xml from a WML Browser.

Listing 15-13 contains the stylesheet used to render first-page.xml in WML.

Listing 15-13. Stylesheet for Rendering in WML (hello-page-wml.xsl).

```
<?xml version="1.0"?>

<xsl:stylesheet version="1.0"
   xmlns:xsl="http://www.w3.org/1999/XSL/Transform">

<xsl:template match="page">
  <xsl:processing-instruction
  name="Cocoon-format">type="text/wml"</xsl:processing-instruction>

  <wml>
   <card id="x" title="xyz">
    <p align="center">
```

```
      <xsl:value-of select="title"/>
      <xsl:apply-templates/>
    </p>
   </card>
  </wml>
 </xsl:template>

 <xsl:template match="paragraph">
  <xsl:apply-templates/>
 </xsl:template>
</xsl:stylesheet>
```

In the previous examples, we have shown how to use the specific markup in the individual stylesheets to take advantage of the device capabilities.

PERSONALIZATION

Personalization is the real-time tailoring of services and/or service interactions to suit each and every end-user. Information about the end-user—commonly known as a customer profile, can be obtained from previous end-user activity, or derived based on the current service used. This profile is utilized to provide a richer end user experience by anticipating and fulfilling the user's needs. The profile could be used to determine automatically which language the user prefers, what color scheme is preferred, and so forth. The profile could also be used to store information about the user, such as their address or account number. This information could be retrieved later to provide automatic form filling and other types of automatic completion. This is especially important on devices like cell phones as their input methods may make it difficult for an end user to repeatedly enter common information.

Content can be personalized based on several different attributes:

- ***Based on the user's preferences:*** The user's preferred content formatting style can be used to transcode the content to a form desired by the user.
- ***Based on the user's device:*** The characteristics of the user's device can be used to personalize the content appropriately for the user.
- ***Based on the user's location:*** The content can be tailored for a particular location or region where the user currently is located.

FYI

Location-based Personalization

Briefing

In order for content to be personalized based on a user's location, there needs to be a way for the enterprise application to determine where the user is. This implies the user's device has to be able to locate itself, or the server has to be able to determine the user's location based on information about the user's connection.

There are several promising technologies that could be used to solve this "Where am I?" problem.

- ***GPS:*** Sitting in orbit around the Earth are many Global Positioning Satellites that a device could use to locate itself. The device could then report this information to the server.
- ***Triangulation:*** This mode applies to wireless devices. As a device enters a transmitter's coverage area, the angle of entry and speed could be used to extrapolate the approximate position of the device. This information could be passed along to the server.
- ***Network Address:*** In some cases, it may be possible to determine where a device is located based upon its network address. This is not always a reliable method.

Once the server had information about where a user was located, the amount of targeted personalization that could be performed is staggering. Targeted advertisement of products or businesses located near the user is just one possibility. Customized reports, news, or any other location-dependant data could be provided by applications that provide location-based personalization. ❑

Personalization can be performed at the service or proxy, at the content publishing framework, or on the client device.

- ***Support for personalization at the client:*** In this model, a software module or application that would be responsible for providing personalization is installed directly on the client device. These kinds of services would only be available if the client device has enough resources (such as memory or processing power).
- ***Support for personalization at the publishing framework:*** You could use Servlets and JSPs to provide automatic input completion (for example, form

filling). We have taken this approach in the Campus Portal, which we will go into more detail in Part 4. If you are using Servlets that conform to Servlet specification 2.3, then you can use *filters* to massage HTTP requests and responses to provide personalized end user experience. Filters can perform a variety of tasks, some of which were discussed in Chapter 13.

- ***Support for personalization at the server:*** In this model, the service is built in such a manner as to be intelligent enough to recognize the end user and provide personalization data for that user. Another way this can be achieved is to provide within each service a method of saving and retrieving a set of personalization data for each user. Whenever a user accesses a service, that particular user's default inputs will be fetched and used to provide the personalized experience.

Personalization—How to Personalize?

Personalization is about providing the right information to the right person at the right time. Figure 15-14 illustrates this concept.

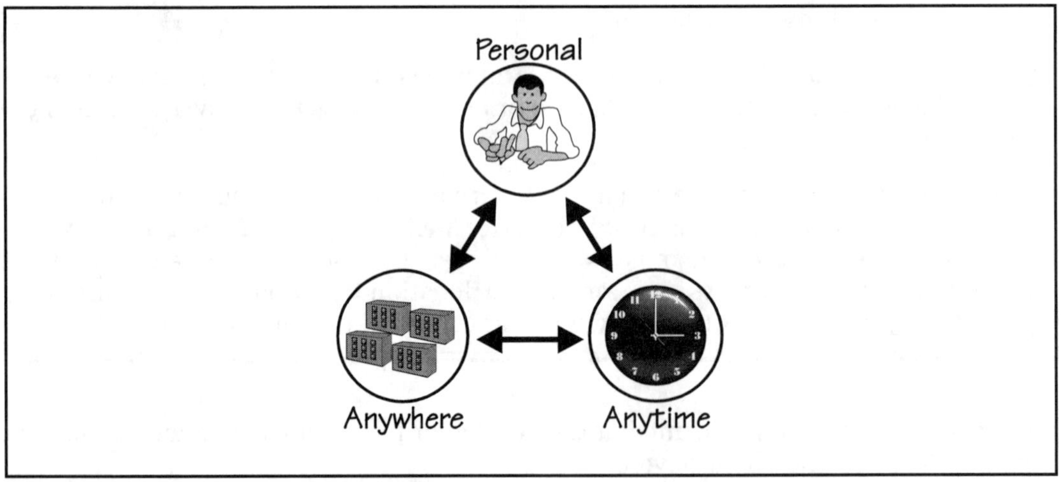

Figure 15-14. Personalized Information Delivery.

An application that contains robust personalization features imparts a richer user experience. There are several approaches that can be used to capture or glean the information needed to construct a user's profile. Of course, the user profile will vary from application to application. One application may need to keep close track of user interface preferences like font sizes and colors. Other applications may not care about these types of preferences. When constructing a user's profile for providing personalization services, the following approaches can be used:

- Creating the user's profile through active user involvement. The user picks and chooses his or her preferences from some menu or list of alternatives.
- Capturing the sites/links user chooses to navigate and analyze this data to try and determine the user's interest. (If this process is used, it is wise to eliminate any personal information obtained in order to protect the user's privacy.)
- Using collaborative profiles. In this approach several individual users' interests are grouped together to come up with a profile that is representative of the entire group.

Personalization has the potential to become a huge market; companies like Vignette are developing personalization frameworks in an attempt to capitalize on this burgeoning market.

FYI Briefing

Specialized Personalization Markup Languages

UIML (User Interface Markup Language) is an XML application that has been designed to describe user interface information in a device-independent manner. UIML can be transformed into any other markup language using XSL/XSLTs or by any program.

UIML considers UI to be made up of user interface elements known as parts. These user interface elements can be text, audio, or video. Users may provide their inputs using user interface widgets like pull-down lists. Since these widgets are different for each device and browser, the mapping between the user interface parts and widgets is done using UIML *style* elements.

UIML also defines a *behavior* element that defines the actions that could occur when a user interacts with UI. The *behavior* element has a set of rules. Each rule has a condition and a sequence of actions. Whenever a user interacts with the UI, an event is generated by the UI. Based on the event, the associated actions are executed.

The following are three types of actions:

- A property of some part in the UI changes.
- A function in a scripting language is invoked.
- A function or method in the application logic is invoked.

In cases of the property change and invoking a function in script language, UIML provides the syntax for describing the calling convention, but does not specify an implementation of how the call is performed (e.g., via RPC, RMI, or any other means).

A UIML document contains two parts. The first part contains the XML version, location of the UIML DTD, and the encoding used. The second part contains the actual meta data. This is made up of the mandatory root tag—*uiml* tag—optional head, interface, peers and template tags. Listing 15-14 is the skeleton of a UIML document:

Listing 15-14. Structure of a UIML Document.

```
<?xml version="1.0"?>
<!DOCTYPE uiml PUBLIC
  "-//Harmonica//DTD UIML 2.0a Draft//EN"
  "http://www.uiml.org/dtds/UIML2_0a.dtd">

<!--The first part ends here and the second part begins below -->
<uiml xmlns='http://uiml.org/dtds/UIML2_0a.dtd'>
<!-- the optional head tag for meta data about the document -- >
<head> ... </head>

<!-- the optional interface tag contains the parts of the UI-- >
<interface> ... </interface>

<!-- the optional peers tag shows the mapping -- >
<!-- between events and properties -- >
<peers> ... </peers>

<!-- the optional template tag for reuse of UIML -- >
<template> ... </template>
</uiml>
```

XForms

The XForms (*www.w3.org/TR/xforms/*) standard is an effort to improve form capabilities and provide flexibility in form pre-filling and auto-filling. The approach taken by XForms is to decouple presentation information from the purpose of the form.

The purpose of the form deals with what the form does while the presentation dictates how the form should be displayed. This separation is particularly useful if the data generated for the form has to be processed in a software application. Hence, XForms do not require hard coding of the user interface information.

XForms allow form pre-filling and auto-filling. XForms track the state of partially filled forms through instance data. These features make it possible to suspend and resume completion of forms.

XForms do not impose a limit on the number of forms in a containing document. The forms are associated with a unique ID and can be referenced using this ID from within the document. The first form in the document can ignore the ID, and any reference references the first form by default. The XForms processor can submit completed forms as XML data.

The XForms user interface consists of a set of XHTML modularization elements. XHTML forms use HTTP to send instance data as name/value pairs. ❑

SUMMARY

In this chapter we defined transcoding and personalization. We then explained the various alternatives available to perform transcoding and explained in detail what a transcoding engine does. We discussed XSL stylesheet transformation and how to construct your own stylesheets. We then explained how to use Cocoon to render data on multiple devices using XSL stylesheets. We also provided a quick overview of XSL and XSLTs.

Personalization was briefly discussed. We will return to personalization in Part 4 when we discuss how we implemented personalization in our Campus Portal project.

Chapter 16

XML Processing Using MIDP

Defining the semantics or meaning of data is one of the most important tasks in designing any application. When an application has to share or exchange data with another application, the rules defining these semantics are of paramount importance. XML, the eXtensible Markup Language, has been widely adopted as a meta-language to address these concerns by providing a rich framework for defining these semantic rules. The adoption of XML by the industry has been accelerating at such a pace that many applications, programming languages, and design frameworks now support processing XML in one form or another. Throughout the previous sections, we have seen several applications of XML. Some of the areas where XML plays a central role are:

- ***Content presentation:*** XHTML, SVG, and WML
- ***Data warehousing:*** XML Databases
- ***Enterprise Application Integration:*** ebXML
- ***Distributed computing:*** SOAP, Web Services

So why has XML become the darling child of data exchange? The rapid adoption can be attributed in part to the following advantages XML has over competing technologies:

- ***XML is language independent.*** As stated earlier, virtually all programming languages support XML processing, usually in the form of a library or add-on API.
- ***XML is network protocol independent.*** Since XML is expressed as a flat text file, it can be transferred successfully over any network link.[1]
- ***Standardization and availability of parsers.*** The APIs for XML parsers have been standardized so that programs written using a specific parser can easily be migrated to another. Further, for the major programming languages in use, there are already many parsers available, which means the application developer does not need to write his or her own parsers. These standardizations reduce the cost to develop the application using the parser.

Further, since XML has become so popular, no major player wants to be "left out." Hence, the industry leaders have all thrown their support behind XML, committing to support the standard.

Unfortunately for us, most existing J2SE XML processors are too large to be of much use in a J2ME MIDP environment. Can we still process XML in our MIDlets? Yes, and the following sections detail the pitfalls you're likely to encounter when processing XML in the limited resource environment provided by most MIDP implementations. Now that we know it can be done, let's look at the various options we have and the trade-offs we need to make when choosing the appropriate MIDP parser.

WHY XML IN MIDP?

As we have seen in previous chapters, MIDP along with CLDC forms a complete J2ME platform that provides networking and storage capabilities. Hence, the applications that run using the MIDP profile can interact with other networked applications on a server or other device. This interaction typically involves sharing and transfer of data. When the applications on both ends of the interaction are controlled by the MIDlet developer, it is easy for them to define the data formats in a format they like. However, in most cases, the data formats used by the server components have already been defined previously. If the server side components use XML as a data exchange language, the MIDlet developer needs a way to interact with the server side component using XML. Hence, the MIDlet needs to process the XML data directly or convert the XML document on the server side to a format that MIDlet understands. The second option is not always possible when the server side component is not easily modified. So when we want to use the first option (or when the first option is the only option available), we need to be able to process XML documents in our MIDP applications.

---

[1] Most XML files are encoded using UTF-8, but XML provides means to encode the file using other Unicode standards.

Being able to parse XML documents in our MIDlets allows us to interact with many applications that are "off limits" to non-XML enabled MIDlets. Such applications could include:

- ***Application Integration/Web Services:*** Being able to construct and understand SOAP messages (which are XML documents) allows our MIDlets to participate in the Web Services world.
- ***Synchronization:*** Synchronization of data from desktops to handheld devices using a common data exchange format such as SyncML.
- ***Data Storage:*** XML has become a popular native format for storing data.

However, the ability to process XML documents comes at a price. First and foremost, XML processing is a "heavy" application, sometimes requiring a large amount of memory and CPU processing power. The limited resources of a handheld device may not be sufficient to allow full-blown XML processing while still retaining enough resources for the rest of the application. We have to be careful in the way we process documents so as not to exhaust the limited resources of our device.

HOW TO PARSE XML IN MIDP?

The current MIDP specification does not mandate any specific XML parser. Hence, several parsers have been written that can be used in the MIDP environment. While these parsers may have been written from scratch or ported down from an existing J2SE parser, each one has its unique strengths and weaknesses. We'll look at a few specific MIDP XML parsers later in the chapter, but in order to make an educated comparison, we first need to understand what features an XML parser may provide.

PARSER TYPES AND FEATURES

All XML parsers, whether they are written for the J2ME platform, the J2SE platform, or any other platform, exhibit certain characteristics.

Tree-Based vs. Event-Based

All parsers are either *tree-based* parsers or *event-based* parsers. The first type of parser, the tree-based parser, reads the entire document into memory and constructs a tree representation of the document's contents. This allows an application to "walk" the in-memory tree and take actions based upon the elements encountered. The second type of parser, the event-based parser, reads the document element-by-element, invoking callback methods in your application based upon which elements are encountered.

So which type of parser is better for your MIDP environment? Well, as usual, it depends on the application. However, unless you know in advance that your documents are going to be small, it generally makes more sense to use an event-based parser in a low-memory environment, which is a typical characteristic of MIDP devices. When using a tree-based parser, there may be no limit to the amount of memory a document may consume, leading to MIDlet instability or crashes.

Validating vs. Non-Validating

In addition to being either event-based or tree-based, parsers are also either *validating parsers* or *non-validating parsers*. A validating parser contains the logic necessary to verify that a given XML document conforms to the rules set forth in some DTD (Document Type Definition) or XML Schema. DTDs and XML Schemas are documents that provide rules detailing how XML documents may be constructed. Such rules could include:

- Element nesting rules to define the hierarchical composition of the document.
- Occurrence limitations (e.g., how many instances of a particular element can the document contain).
- Complex type construction rules (e.g., sequences or collections of elements).

All parsers have code to determine if an XML document is well-formed, meaning that it adheres to the basic rules of the XML specification (elements formed correctly, matching start and end tags, etc). However, there is a key distinction between determining if a document is valid and determining if it is well-formed. An XML document can be well-formed but invalid, if the document does not adhere to the rules specified in its DTD or XML Schema.

Validating parsers generally have a larger memory footprint and higher resource requirements, especially if validating against a large or complex DTD or XML Schema. Consequently, most MIDP parsers are of the non-validating variety. Although this may sound like a serious limitation, as we will see later, there are ways to circumvent this shortcoming.

Support for Mixed Content

Some XML parsers have the ability to handle *mixed content*. Mixed content refers to elements in the document that contain both character data as well as child elements. Listing 16-1 illustrates an element that contains mixed content.

Listing 16-1. Mixed Content In XML Elements.

```
<MyElement>
 This is character data
 <MyChildElement>
    . . .
 </MyChildElement>
</MyElement>
```

HTML is a good example of a mixed content model. Aside from having tags in your HTML documents, you can embed character data in the document basically wherever you please.

The debate as to whether the mixed content model is a good model is a heated one. Those who dislike the mixed content model are quick to point out that the XML 1.0 specification severely limits how your document can be organized if you choose to use mixed content. Proponents of the mixed content model argue that the flexibility gained outweighs the drawbacks. Choose your MIDP XML parser carefully if you think you'll need to handle mixed content; not all parsers support the mixed content model.

Internal vs. External Entities

A parser that validates documents against a given DTD may encounter entity definitions in that DTD that refer to entities outside, or external to, the DTD. This approach is commonly used to associate an entity with an external file or resource. In Listing 16-2, we define two entities. The entity named `local` is an internal entity, defined within the DTD. The entity named *foreign* is an external entity which references the external file "external.xml".

Listing 16-2. Internal and External Entities.

```
<!ENTITY local "This is a local entity">
. . .
<!ENTITY foreign SYSTEM "external.xml">
```

Keep in mind that some MIDP XML parsers cannot handle external entities. If you need to create code that will run correctly using multiple parsers, you should avoid using external entities.

Namespace Support

An XML Namespace is a distinguishing moniker that makes it possible for XML documents generated by different applications to use the same element names in a non-conflicting way. For example, if two online book vendors used the tag <book> in different ways, it wouldn't be possible for an application to determine which type of <book> element it was processing. However, if the book vendors namespace-qualify their book elements, the application can make the determination which vendor's XML file was in use. Vendor A could use the namespace "vendora" and vendor B could similarly use "vendorb." Their <book> elements would then be called <vendora:book> and <vendorb:book>. Their DTDs or schemas would remain the same, but the namespaces would become the distinguishing feature that separated one <book> element from another.

In addition to preventing conflicts between element names, namespaces also provide the benefit of logically grouping elements from one application, making elements from that application easier to recognize. If vendor A from our bookstore example above used the same namespace name consistently across applications, it would be easy to pick out which XML documents belonged to that vendor.

CHOOSING THE RIGHT PARSER

There are many XML parsers available for the MIDP platform. Now that you are familiar with the characteristics that distinguish one parser from another, you can make a more educated choice as to which parser is best for a given situation. In the following sections, we examine a few of the more popular parsers and discuss their particular strengths and weaknesses.

NanoXML

NanoXML is a small, open source (licensed under the zlib/libpng license), non-validating parser for Java. Originally written for J2SE, this parser has been subsequently ported to the KVM, which means we can use it in our MIDlets. The KVM version of NanoXML is based on an earlier version of the NanoXML source code, which means it may not have all the features of its J2SE cousin.

Currently, NanoXML/KVM version 1.6.4 is available. This version is based on earlier NanoXML code; NanoXML has since morphed to become NanoXML2, which provides a much larger feature set. If you need to make use of some of the new features in NanoXML2 (such as support for external entities and other DTD-related enhancements), you may need to wait until the code has been ported to the KVM.

NanoXML is available for download at *http://nanoxml.sourceforge.net/kvm.html*. The page gives instructions on how to download and configure the parser for your MIDlets.

TinyXML

TinyXML is a non-validating parser that was originally written for J2SE. Modifications to TinyXML have since been made that allow it to be used in a J2ME environment.

TinyXML was one of the first lightweight XML parsers available. It provides a proprietary tree-based API that is used to model the document being parsed, but the API is not 100% DOM-compliant.

TinyXML is very small; the entire parser can be packaged in less than 10 KB.

KXML

The kXML project provides an *XML pull parser* and *writer* suitable for all Java platforms including J2ME (CLDC/MIDP/CDC). Because of its small footprint (both JAR size and memory usage), it is especially suited for Applets or Java applications running on handheld devices.

The kXML parser provides:

- XML Namespace support.
- A "relaxed" mode for parsing HTML or other SGML formats (since these documents may not adhere to strict XML syntax).
- A low overhead (small memory footprint) environment.
- Support for writing (e.g., output) XML documents, including support for namespace handling during writing.
- Optional kDOM support (provides DOM APIs for your MIDlets).
- Optional WAP support (WBXML/WML).

The newest version of kXML is kXML2, which provides a pull-based parser. A pull-based parser is a specialized form of the event-based parser, which we discussed earlier. When using a standard event-based parser, the parser emits events as it progresses through your document. Your application catches these events as the parser walks through the document, taking appropriate action based on the events received. In this scenario, the parser "pushes" events to your application. A pull parser, on the other hand, asks the parser for events as needed. The application pulls the events from the parser, asking for each event as it is needed.

Why Use a Pull-based Parser? Why Not SAX?

Details

The SAX parser implementation follows a push model. Depending on the content of the document and the internal state of the parser (e.g., where the parser is currently pointing in the document), events such as "start of element" or "text data" are pushed to a document processor; typically the document processor is your application. From an application standpoint, the push model means extra work for the application developer. First, in the event-handling method, the document processor needs to be able to determine its own internal processing state before it can do anything with the event. This restriction typically results in extra local state variables being introduced inside the application to keep track of what elements have been seen—the parser knows where it is pointing inside the document, but the application also needs to keep track of this information for itself.

In contrast, when the parser is pull-based, the application is in control since it asks for events to be delivered as needed. There is a distinct advantage in using this approach; the processing state can be implemented in a much more natural way in the application or document processor. In addition, an instance variable of a pull-based parser can be passed between different child processors (each processor being responsible for handling a portion of the document), thus supporting simpler modularization and code that is more easily maintainable. ❑

The kXML parser does not support external entities. In addition, the parser does not parse the `<DOCTYPE>` declaration. To compensate for this, the parser emits a "legacy event" when a `<DOCTYPE>` declaration is encountered, so application programmers can catch the event and handle the declaration in their own fashion.

UXML

UXML is a SAX-compliant parser written by John Gash, a graduate student at SJSU. UXML is the parser used by the MIDP components of our Campus Portal. UXML strives to remain SAX compliant while optimizing the size of the parser footprint and parsing speed.

According to the documentation, UXML is nearly SAX 2.0 compliant. It is also one of the faster parsers available, beating out NanoXML, TinyXML, and kXML in our parser speed comparisons (see Table 16-1).

Which Parser to Choose?

The decision as to which parser to choose depends on the requirements of your particular MIDlet. In Table 16-1 we compare features of the parsers we discussed above and show the results of parsing a 3 KB XML document using a few of the parsers. Xerces, the standard heavyweight J2SE parser, is included for comparison purposes, although its sheer size prohibits its use in a MIDlet. The test was not performed on the NanoXML and TinyXML parsers due to the proprietary nature of their APIs.

Table 16-1. A Comparison of MIDP-based Parsers.

	NanoXML	TinyXML	kXML	uXML	Xerces
Parser Type	Proprietary tree-based, SAX with optional Nano-XML/SAX	Proprietary tree-based	DOM (tree-based) with optional kDOM	SAX (event-based)	DOM and SAX
Parser Size (KB)	21	10	20	14	1,809
Memory used to parse 3 KB XML document	n/a	n/a	97 KB	91 KB	134 KB
Time taken to parse 3 KB XML document	n/a	n/a	185.5 ms	60 ms	148.25 ms

What does the comparison shown in Table 16-1 illustrate? The comparison illustrates how you can make your decision as to which parser to use in a MIDlet based on several constraints. Decisions can be based on:

- ***Package Size:*** If MIDlet size is your paramount concern, then you would probably be better off choosing one of the smaller parsers, such as TinyXML or uXML.

- ***Required Functionality:*** If your application requires SAX or DOM support, you may need to choose one of the parsers that support the required API.

- ***Memory Utilization:*** If you need to limit the amount of memory used by your MIDlet, you might want to choose uXML, as it offers the lowest memory requirements. If memory utilization is critical, certainly stay away from DOM or tree-based parsers unless you're certain that the documents you'll be processing will be small. Recall that with a tree-based parser, the entire document is read into an in-memory tree during parsing. If you unintentionally encounter a large XML document, you may exhaust your MIDlet's available memory.

- ***Performance:*** If your application requires the fastest MIDlet parsing speed, you may again want to choose uXML, as it offers the fastest performance of the parsers discussed so far.

There are ways to squeeze the best performance out of any of the MIDP XML parsers described so far. In the next chapter, we'll look at some of these techniques, along with general MIDP programming best practices.

XML Parsing Using MIDP—An Example

Now that we've examined several MIDP XML parsers, let's look at an example program that illustrates how you can integrate XML parsing functions into your own MIDlets.

Our parsing example will consist of two parts:

- A server program which delivers an XML document to our MIDlet.
- The actual MIDlet which performs the parsing.

For our server program, we will create a simple Servlet that could be used to check course registration at a university. Our Servlet will accept a single parameter, called *name*, which the Servlet can use to determine which courses the student was enrolled in. The Servlet will return an XML stream to the caller, who can then parse the stream to determine what courses were returned as results. For this simple example, our Servlet will deliver some "canned" or pre-calculated XML streams. Of course, in a fully functional version, the Servlet could consult some database or other service to gather the real information.

Our Servlet will be configured to return data in the form shown in Listing 16-3. As you can see, we have decided to make our XML stream very simple.

Listing 16-3. Sample XML Stream Returned from CourseServlet.

```
<?xml version="1.0"?>
<courses>
 <course>ENGR264</course>
 <course>ENGR268</course>
 <course>ENGR295A</course>
</course>
```

Let's take a look at our example Servlet code, shown in Listing 16-4. You'll quickly notice that the Servlet code is for simple demonstration purposes (as it only contains data for three people). Further, in a production system, you obviously wouldn't have the courses and student names hardcoded in the source code—we've purposely simplified the Servlet code here in order to concentrate more closely on the MIDlet which will parse the string returned from the Servlet.

Listing 16-4. The Course Servlet (CourseServlet.java).

```
import java.io.*;
import javax.servlet.*;
import javax.servlet.http.*;

public class CourseServlet extends HttpServlet
{
   public void doGet(HttpServletRequest req, HttpServletResponse response)
    {
      try
           {
             String name=req.getParameter("name");
             PrintStream out = new
                          PrintStream(response.getOutputStream());

             out.println("<?xml version=\"1.0\"?>");
             out.println("<courses>");

             if(name!=null)
                 {
                    if(name.equals("Dan"))
                        {
                          out.println("<course>ENGR268</course>");
                          out.println("<course>ENGR296J</course>");
                          out.println("<course>ENGR295A</course>");
                        }
                    else if(name.equals("Shan"))
                        {
                          out.println("<course>ENGR295A</course>");
                        }
                    else if(name.equals("Mike"))
                        {
                          out.println("<course>ENGR264</course>");
                          out.println("<course>ENGR295B</course>");
                        }
                    else
                        {
                          out.println("<course>Unknown</course>");
                        }
                 }
```

```
            out.println("</courses>");
        }
    catch(IOException ioe)
        {
        }
    }
}
```

You will need to compile, package, and deploy this Servlet into your application server's Web container. If you've been following the examples illustrated in previous chapters, you'll already have experience deploying Servlets, so we won't go into any further details now.

Let's now shift our attention to our MIDlet. We've chosen kXML as our MIDlet parser for this example. Although the MIDlet shown below is written using kXML, you should be able to apply the concepts learned in this example when using any of the parsers we looked at earlier.

Our MIDlet's screen provides a TextField for the user to enter the name of the student whose course information is to be retrieved, along with a submit button. We'll keep the user interface simple, as our main focus in this example will be on the XML parsing capabilities of the MIDlet.

Figure 16-1 shows our CourseMIDlet in action. In this case, the MIDlet submitted the name "Mike" to the Servlet, which returned the XML stream shown in Listing 16-5. A quick review of our Servlet code shown above should reinforce this claim.

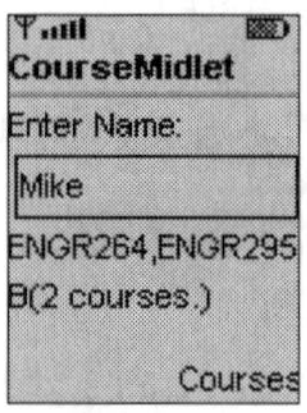

Figure 16-1. The XML Parsing CourseMIDlet.

Listing 16-5. XML Stream Returned When Generating Output Shown in Figure 16-1.

```
<courses>
 <course>ENGR264</course>
 <course>ENGR295B</course>
</courses>
```

Our CourseMIDlet, shown in Listing 16-6, generated the output shown in Figure 16-1.

Listing 16-6. CourseMIDlet.java.

```
import javax.microedition.midlet.*;
import javax.microedition.lcdui.*;
import java.io.*;
import javax.microedition.io.*;
import org.kxml.*;
import org.kxml.parser.*;

public class CourseMIDlet extends MIDlet implements CommandListener
{   private Form mainForm = new Form ("CourseMidlet");
    private StringItem resultItem = new StringItem ("", "");
    private TextField NameField =
            new TextField("Enter Name:", "", 10, TextField.ANY);
    private Command getCourses =
            new Command ("Courses", Command.SCREEN, 1);
    private static String
            SERVLET_URL="http://localhost:8080/CourseServlet";

    public CourseMIDlet ()
    {  mainForm.append (NameField);
       mainForm.append (resultItem);
       mainForm.addCommand (getCourses);
       mainForm.setCommandListener (this);
    }

    public void startApp ()
    {  Display.getDisplay (this).setCurrent (mainForm);
    }

    public void pauseApp ()
    { }

    public void destroyApp (boolean unconditional)
    { }

    public String getName()
    {  return NameField.getString();
    }

    public void commandAction (Command c, Displayable d)
    {  XmlParser parser=null;
```

```
ParseEvent parseEvent=null;
HttpConnection httpConnection=null;
InputStream inputStream=null;
boolean moreData=true;
String s=null;
int numCourses=0;

try
    {
    // open connection to servlet
    httpConnection =
   (HttpConnection)Connector.open(SERVLET_URL+"?name="+getName());
    inputStream = httpConnection.openInputStream();

    parser = new XmlParser(new InputStreamReader(inputStream));

    while(moreData)
       { parseEvent = parser.read();

         switch(parseEvent.getType())
              {
              case Xml.START_TAG:
                   if(parseEvent.getName().equals("course"))
                      numCourses++;
                   break;
              case Xml.END_DOCUMENT:
                   resultItem.setText(resultItem.getText()+
                               "("+numCourses+" courses.)");
                   moreData=false; break;
              case Xml.TEXT:
                   s = resultItem.getText();
                   if(s.length()>0)

                    resultItem.setText(s+","+parseEvent.getText());
                   else
                      resultItem.setText(parseEvent.getText());
                   break;
              }
       }

    if(inputStream!=null)
       inputStream.close();
    if(httpConnection!=null)
       httpConnection.close();
```

```
            }
        catch (Exception e)
            {
               e.printStackTrace ();
               resultItem.setLabel ("Error:");
               System.out.println(e);
            }
    }
}
```

By this point, you're MIDP user interface experts, so we won't belabor the GUI parts of the code. The MIDlet accepts a name (via the NameField **TextField**), and waits for the user to press the "courses" command button. At this point, control is transferred to the **CommandListener** shown near the end of the listing above.

The **CommandListener** formulates a URL based upon the Servlet's base URL (you can change the SERVLET_URL variable to indicate where your Servlet is deployed), plus the name variable that the Servlet is expecting to receive. Note that for this simple example, we aren't properly encoding the name variable for URL inclusion. Hence, if the user enters a name with special (non-alphanumeric) characters, the Servlet may not receive the entire name.

The MIDlet then establishes an HTTP connection to the constructed URL. The input stream of this HTTP connection is opened, and passed to the constructor of our XML parser. The code then enters a loop, continually asking the XML parser for events, until an "Xml.END_DOCUMENT" event is received, signaling the end of the XML document.

Our kXML parser will emit an "Xml.START_TAG" event each time a new element is encountered in the XML stream. Each time an "Xml.START_TAG" event is received, we check to see if the element name matches "course", and if so, we increment a counter to store the number of courses the student is enrolled in. This counter will be displayed in the GUI after the parsing is complete.

When the parser emits an "Xml.TEXT" event, indicating that character data was encountered in the XML stream, we add the text encountered to our GUI output field, after properly adjusting the current text with commas (if data had previously been written into the GUI).

When our parser emits the final "Xml.END_DOCUMENT" event, we display the number of courses read, and exit our main loop. The **CommandListener** finishes up by closing the **HttpConnection** and its associated stream.

You can experiment with this MIDlet by entering names that don't match any of the names the Servlet is expecting. Does the output of the MIDlet match what you expect?

OTHER APPLICATIONS OF MIDP XML PARSING

MIDP 2.0

The next version of the MIDP profile, MIDP 2.0, may have an XML parser included in the profile as a standard package. This will simplify most XML parsing MIDlets, as the developer does not have to package the parser of their choice with their MIDlets. Further, a standard API will provide easier maintenance for these MIDlets as there will only be one parser API that the developer needs to learn.

At the time of this writing, exactly which parser is going to be included is still being debated. It is not clear whether the standard will choose to use an existing parser, or whether one will be developed from scratch.

SOAP

In our Web Services chapter, we examined a MIDlet that used SOAP to invoke a Web Service. Although the Web Service MIDlet did not explicitly make use of the XML parser API, you can be assured that the parser was used behind the scenes when constructing the outgoing SOAP message and parsing the incoming SOAP result. Could Web Services be the XML "killer app" for MIDlets? We'll wait for the jury to come back on that question.

SUMMARY

In this chapter, we examined several XML parsers that are available on the MIDP profile. We discussed reasons to support XML parsing in your MIDlets, and when you would need to. We constructed an example that illustrates how you can use XML parsers in your own MIDlets—our example connected to a Servlet to retrieve an XML stream, but could have retrieved an XML stream from any suitable source.

In the next chapter, we'll close out our discussion of programming technologies by looking at some of the "best practices" you should use when designing your MIDlets.

Chapter 17

Programming Strategies for MIDP

We conclude Part 3 with a look at some of the recommended strategies you might want to use when writing your MIDlets. We've compiled these recommendations from our own experience, as well as from suggestions made from device manufacturers. Although the suggestions covered may seem to be eclectic, as a whole they will help you write MIDlets that are secure, have high performance, and a higher level of usability.

You are all MIDlet experts by now, having learned:

- MIDP programming for the handheld device.
- Network programming using MIDP.
- How to integrate your MIDlets with your J2EE enterprise application.
- How to invoke Web Services from your MIDlets.
- How your MIDlet can be configured to parse and manipulate XML documents.

This chapter will fill in the gaps with strategies you can use to squeeze the most out of the limited environment present in most MIDs.

The examples shown in previous chapters have been very simple. Typically, the example MIDlets we've seen consist of just a few screens with minimal user interaction. However, your own MIDlets may turn out to be much larger in size and complexity. Indeed, if you're not careful with your programming style and technique, your MIDlet may slow to a crawl, become too large to load on the MID, or may even be subverted by hackers!

But there should be no reason to be unduly alarmed now that you've seen the "big picture" of MIDP design from start to finish. What's more, we've chosen to place all the strategies for robust MIDP design into one chapter. The suggestions shown in the following sections are intended to give you some ideas as to how the problems listed above can be avoided. The suggestions and "best practices" detailed in this chapter fall into a few categories:

- Graphics and UI recommendations, including end-user usability factors
- Device hardware considerations
- Suggestions for reducing MIDlet size
- Obfuscation
- Network programming suggestions
- Efficiently handling MIDlet images (PNG files)
- Vendor libraries
- XML parsing recommendations
- Miscellaneous suggestions
- We'll look at each of these categories in order. Let's get started!

GRAPHICS AND USER INTERFACE RECOMMENDATIONS

In Chapter 9, we mentioned some suggestions that you might want to follow when designing the user interface for your MIDlets. Let's revisit the suggestions again, this time adding some further ideas.

- ***Use the high-level API whenever possible.*** Aside from remaining portable across MIDs, by using the high-level API, your application's widgets will more often resemble the native applications that ship with the MID. If your application resembles the native applications on the device, the user is more likely to understand how to interact with your application without extensive training.

- ***Provide easy navigation buttons.*** If your MIDlet involves navigating through nested menus or screens, try to provide a navigation shortcut menu, or Next/Back buttons, as appropriate. You might consider having a "Menu" or "Help" key that takes the user to a navigation screen where they can jump around between screens in your MIDlet. If you add this type of functionality, you should probably also add titles to each screen so the user knows where they want to navigate.

- ***Provide appropriate choices for prompts.*** For example, if your MIDlet requires a YES/NO answer to a prompt, don't require the user to type in "YES" or "NO"—provide a single keypress option for YES and another for NO.

- ***Don't assume anything about the MID's display capabilities.*** Although some MIDs (Pocket PCs or Palm devices) may support color, don't force your code to use certain colors. Instead, detect whether or not the MID has color support, and handle the graphics appropriately.

By following the strategies above (as well as general common-sense UI design), you will ensure your users will have an easy time using your application.

MID HARDWARE CONSIDERATIONS

Different MIDs have different hardware, and your code should not break or behave strangely depending on the hardware encountered. If you want your user to experience the best that your particular MID has to offer, you may want to query your MIDs hardware support in the constructor of your MIDlet. That way, your code can quickly check whether or not a given feature is present.

The hardware features you should check for include:

- ***Screen resolution:*** Saving the screen resolution is useful if you have position-dependent display elements that you want to have scale properly on MIDs that have larger (or smaller) screens.

- ***Screen color depth:*** Saving the screen color depth is useful if you are going to be making use of colorful user interfaces. Further, saving the color depth allows you to load particular images stored in your MIDlet's JAR file, depending on their color depth. We'll talk more about this strategy when we discuss images in a later section.

- ***Existence of directional and other special keys:*** Your MIDlet should not depend on a four-way directional key, present on most cellular phones. Further, your code should also not depend on the presence of a pen or stylus, present on most Palm and Pocket PC devices.

An astute reader will notice that you can avoid the pitfalls described above by sticking with the high-level APIs. However, as noted in earlier chapters, there are times when you need to query the MID's hardware and make use of the low-level APIs (if you were making a game MIDlet, for example). If your MIDlet is going to fall into this category, make sure you don't make any assumptions about the MID onto which your application will be deployed.

REDUCING MIDLET SIZE

Strive to have your MIDlet JAR consume as little space as possible. MIDs vary greatly in the amount of RAM they contain to house MIDlets. If your MIDlet's JAR file exceeds 20-30k, you might experience problems loading the application into some MIDs.

To complicate matters further, some MIDs download the JAR file into RAM, then copy or install the MIDlets in the suite from RAM to another location in memory. If the other location happens to be the same place as where the JAR file was downloaded, the deployment process will effectively take *double* the amount of memory as expected!

There is also an economic factor involved if your MIDlets are too large. If the user of your MIDlet suite downloads the suite to their MID via OTA, they will be charged a fee by their wireless provider. This fee, typically charged per KB, can become an issue if your JAR file is large. A 30 KB JAR file, downloaded at $0.10/KB can result in a $3.00 charge being added to the user's bill. This $3.00 is on top of the purchase price charged by the OTA vending machine server. Although $3.00 may seem like a small amount, some users may not appreciate this "MIDlet tax" appearing on their bill.

So, what can one do to reduce the size of the MIDlet suite? There are a few strategies you can use:

- ***Use an obfuscator.*** An obfuscator, which we'll discuss next, has the useful bonus of reducing your code size. A good obfuscator will perform a number of optimizations, including removing unreferenced variables and methods, and stripping out debugging information like line numbers and symbol tables. The obfuscator also replaces your method names with single or double character names. *MyReallyLongMethodName(..)* could be renamed to *a(..)*, a savings of 21 bytes each time the method name appears in your code! The obfuscator handles searching through your code and changing all the references—you don't have to do anything more.

- ***Move resources to Web servers.*** If your MIDlet will be running on MIDs that are permanently connected to a wireless network, you can move your image files and other resources to a Web server. The MID can download the images as needed, thereby eliminating the need to store the image files locally inside the JAR file. If the MIDlet tried to download the image from the server and couldn't do so for some reason (network outage, etc.), the MIDlet could replace the image on the screen with an alternate "dummy" image or text string. The promotion of images to the Web server can save a tremendous amount of space if your MIDlet is image-intensive.

There is another consideration that needs to be made when considering a MIDlet's "size." Up until now, we have equated the MIDlet's "size" with the size of its JAR file. Although the size of the JAR file is important and should be managed carefully, the memory required by our MIDlet while running is equally important. There are a number of techniques you can use to reduce the memory requirements of your MIDlet as it runs:

- ***Move set-up code to another MIDlet***. If your MIDlet needs to populate a data file (typically a **RecordStore**), move the code which does this population to another MIDlet in the same MIDlet suite (recall that MIDlets from the same suite can access each other's **RecordStore**). The user can then invoke the "set-up" or "preparation" MIDlet first, before they run the real MIDlet. Since the code that sets up the **RecordStore** is not contained within the same MIDlet, you will have more memory available to the real application when needed.

- ***Allocate objects in the constructor***. Try to allocate as many objects as you can in the MIDlet constructor, if your application design supports this behavior. If a memory allocation error is going to occur, it's better to be able to find out about the problem early, rather than letting the user get halfway through your application, only to have it die unexpectedly.

- ***Reuse as many objects as possible***. This is not a MIDP-specific strategy, and you should already be doing this with your other Java programs. Garbage collection is expensive, even more so on a MID. Don't force the Garbage Collector (GC) to run by creating hundreds of objects when a handful could have sufficed.

- ***Be careful with the "static" modifier***. If you're only going to be accessing the value within a large static object a few times, it makes more sense to leave the static modifier out. This way, you can set the object reference to null when you're finished with the object and the GC can reclaim the memory for you. Of course, if you need the static modifier because you need to reference the object from within a static context, then you may have no choice.

MIDP memory tuning is a black art—there's no one set of suggestions that is appropriate for all cases. The strategies listed above, when used in the right combination for your application, should help you optimize your MIDlet's memory usage, while running or while stored.

CODE OBFUSCATION

In addition to providing the useful side-effect of truncating your method names to single or double characters, a code obfuscator provides another important function—it hides your code from prying eyes.

Some MIDs provide the capability to extract, or download, the MIDlet suites installed in their RAM to a host PC or other system. This fact means that you should never assume that your MIDlet can't be copied. While there are good reasons for copying MIDlets (backup, archive, etc.), there will always be hackers out there that delight in stealing your code, possibly modifying it for their own nefarious purposes. While an obfuscator can't prevent the unauthorized copying of your MIDlet, it can help in preventing unauthorized modifications.

Since Java bytecodes can be disassembled back into Java source code, use of an obfuscator is critical if you want your code to be as safe as possible from modification.

Details

Obfuscation—How Much is Too Much?

Some obfuscators will more aggressively hide your code, including translating all string literals into unintelligible garbage. Further, some obfuscators will inject invalid data into the *.class* file, which will cause decompilers to go crazy trying to figure out how to represent the code. However, the invalid data will cause the MIDP preverify step to fail, and the resulting *.class* file won't load on your MID!

Hence, there is a point at which too much obfuscation can hurt you. As with other security concerns, you'll have to weigh how much obfuscation you really need in your project. ❑

There are many commercial obfuscators available on the market. A Web search for 'Java obfuscator' will yield the most popular titles.

Details

The DashO Obfuscator

PreEmptive Solutions' DashO obfuscator is the obfuscator tool we used when designing our MIDlet components for our campus portal. We found that DashO significantly reduced the size of our MIDlets, which was our primary reason for using an obfuscator in the first place. Since our campus portal MIDlets were created in an academic environment, we were not explicitly concerned with hiding our MIDlet source code from anyone. ❑

NETWORK PROGRAMMING

Strategies for efficient network programming using MIDP were discussed in Chapter 10. Expanding on these ideas, we offer the following strategies when your MIDlets need to communicate over the network:

- ***Use HTTP for MIDlet-to-server communication.*** As stated in Chapter 10, MIDP implementations are required to support the HTTP protocol for data exchange. Using HTTP, as opposed to some proprietary socket communication scheme, ensures that your MIDlets will be portable across implementations.
- ***Keep communication exchanges short.*** Avoid the "MIDlet tax" described above by continually striving to keep the amount of network traffic to an absolute minimum. If you need to exchange HTTP headers, use short strings (as few characters as possible) for the header names, and keep their associated header values short.
- ***Make the most out of each network connection.*** Since HTTP is a request/response protocol, you can only send one message and receive one answer. If possible, try to aggregate multiple messages that are being sent to the server into one communication, instead of opening a new connection for each message being sent. This will make the application run faster as less time is spent building up and tearing down network connections.

EFFICIENT IMAGE (PNG) MANIPULATION

If you are making use of PNG images in your MIDlet, you need to balance the size of the image file (which directly contributes to the size of your MIDlet's JAR file) with the application's needs. Fortunately, the PNG standard provides good image quality, even at low color depths.

Consider the set of images shown in Figure 17-1, which shows the same image rendered at different color depths. The original image was reduced in depth at each successive stage by 50%, and then saved as a PNG file suitable for inclusion in a MIDlet's JAR file. This particular image doesn't start to degrade appreciably until the color depth reaches 16 colors. Ask yourself the question whether or not your users will be able to appreciate (or even notice) the difference in color depth between a 256-color PNG file and a 16-color PNG file. As shown in Figure 17-1, the difference in size is threefold—using the 16-color version would allow you to store three images in the space occupied by the 256-color version.

Further, if you are certain that your audience will own MIDs with grayscale or black-and-white displays, there is no reason to even include color PNG files. In fact, you probably don't want to include color PNGs unless absolutely necessary as there is an overhead involved in processing color images.

Figure 17-1. A Comparison of PNG Images.

USING VENDOR-SUPPLIED OR THIRD-PARTY MIDP LIBRARIES

Many MIDP libraries have been developed to help smooth over the deficiencies of the current MIDP specification. You can find libraries that:

- Manage **Canvas** and **Display** objects.
- Provide new and improved user interface elements.
- Manage your MIDlet's **RecordStore** objects.
- Persist data across the network.

If you feel like the MIDP graphics classes did not provide a rich enough set of functionality, you aren't alone. J2SE programmers used to Swing and AWT programming tend to scratch their heads a bit when confronted with the dearth of widget and layout classes. Compared to AWT and Swing, programming a robust GUI using MIDP is difficult. Several companies and individuals have answered the call, providing more complex UI elements that sit on top of MIDP (see the following briefing box).

The same can be said for the base MIDP RMS architecture. Try storing a complex data structure like a hash table or linked list using RMS and you'll quickly see that a higher level of abstraction is needed. RMS is at *too* low of a level to be easy to use—you have to concern yourself with bits and bytes and offsets inside records. Finding the desired record inside the **RecordStore** can also be a challenge, as you saw in our example in Chapter 10. Again, third parties have stepped up to the plate and written layers that sit on top of RMS and handle the nitty-gritty work for us.

Libraries written by third parties or MID vendors can generally be used by developers when designing new MIDlets. Some have stricter licensing requirements than others, so be sure you check the library's license agreement before including it with any MIDlet you deploy.

FYI Briefing

Third-Party MIDP Libraries

Here are some of the popular MIDP third-party libraries:

- **kAWT**. The kAWT library (available from *www.kawt.de*) provides an AWT-like framework for MIDP. This library allows the developer who is familiar with AWT programming to leverage his or her skills in the MIDP environment. The kAWT package provides AWT-like layouts and supports many of the same types of widgets as AWT. Figure 17-2 shows an illustration of a sample kAWT application.

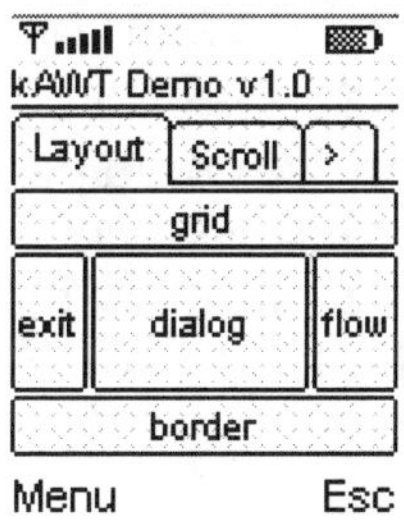

Figure 17-2. A kAWT Application.

- **Nextel Open Source J2ME Toolkits.** There are actually three Nextel toolkits (available from *nextel.sourceforge.net*) that provide UI, RMS, and networking enhancements. The Nextel UI toolkit provides widgets for scrollable lists, 3-dimensional buttons, layout management, grid displays (tables), and specialized text input fields for money values and dates. The Nextel RMS toolkit provides a mapping for persistent arrays and vectors on to the low-level RMS classes. This means that you can use the Nextel toolkit's **OPersistentArray** and **OPersistentVector** classes to automatically persist the contents of an array or vector into your MIDlet's **RecordStore**. Without such classes, you'd have to write the code to extract the contents of the array or vector, locate the appropriate record(s) in the **RecordStore**, handle how to resize the records, and so forth. The designers of the Nextel RMS toolkit have already undertaken this task.

- Finally, the Nextel net toolkit provides a persistence mechanism to persist your data not to a **RecordStore**, but rather to a server configured to receive persistent storage requests. This allows MIDlets running on different MIDs to share data seamlessly. At the time of this writing, the Nextel net toolkit was still in development (e.g., beta release).
- **Nokia UI API**. The Nokia UI API (available from *forum.nokia.com/java-forum*), like the Nextel UI toolkit, provides extra widgets and layout facilities for MIDP devices. ❑

There are two distinct drawbacks to using any vendor or third-party toolkit. First, the toolkit may not be free, meaning you may have to pay a licensing fee to the toolkit vendor for each copy of your MIDlet that is deployed. Second, even if the toolkit is free, you'll still need to package the toolkit code with your MIDlet, which could increase the size of your JAR file. This increase in size translates into less space available for your application.

EFFICIENT XML PARSING

The previous chapter illustrated how we could perform XML parsing in our MIDlets. However, we did not extensively discuss the best way in which to encode our documents for MIDP use. As with the strategy discussed earlier dealing with HTTP connections, a good strategy to follow with MIDP XML documents is to keep the tag names as short as possible.

However, this poses a problem if the servlet/content generating application is already deployed and outputs documents that are lengthy or use tags with long names. For example, the content generator may output an XML stream similar to the one shown in Listing 17-1, but what we'd really like to receive is the document shown in Listing 17-2. If you look closely, you'll notice that each document conveys the same information, but the document shown in Listing 17-2 does so at a much lower cost. We could certainly parse the document shown in either listing with our MIDP XML parser, but to reduce our memory usage and speed up the application, we'd like to parse as few tags as possible, and the tags we do process, we'd like them to have short (ideally single-character) names.

So, how can we take the document shown in Listing 17-1 and convert it to the document shown in Listing 17-2? The answer is simple. We use an XSL stylesheet to translate the document shown in Listing 17-1, and create the document shown in Listing 17-2. Since you are all XSL experts by now, we'll leave the actual implementation to you, but you should be able to see how it can be easily done.

Listing 17-1 . XML Document with Extra Tags.

```
<meeting>
  <topic>Planning for FY2002 budget</topic>
  <location>
    <building>130</building>
    <roomNumber>1135</roomNumber>
  </location>
  <scheduledOn>
    <time>3:00pm PST</time>
    <day>November 21, 2001</day>
  </scheduledOn>
</meeting>
```

Listing 17-2. XML Document Condensed for MIDlet Consumption.

```
<m>Planning for FY2002 budget</m>
<l>B/130 R/1135</l>
<d>11/21/01 3pm PST</d>
```

If you look closely at the two documents above, you'll see that not only did we reduce the number of tags and the size of each tag's name; we also flattened the hierarchical structure of the document. We no longer have any nested tags. This was intentional, as the resulting flattened document consumes less memory than a document with structure. The reason this is true is due to the way most XML parsers are implemented. Nesting tags leads to recursion inside the parser, which is not a frugal operation. Nesting tags too deeply can cause more memory usage than we would have encountered had the document been flattened.

You need to balance the structure of your documents with the memory needs of your application. Too flat, and your document loses the benefits of being an XML document altogether—you might as well have used some proprietary format to express the same data. Too nested, and your memory requirements skyrocket. Like all the other strategies in this chapter, it's an exercise in tuning.

MISCELLANEOUS RECOMMENDATIONS

We'll close with some strategies that didn't fit in any of the previous sections.

- ***Threads:*** In general, the places where you should use threads are exactly the same places you would likely use them in any other J2SE program. These areas generally include I/O, network, and UI handling (including animation). A hung user interface does not make for a happy end user. You should put your I/O and network interactions in a separate thread if you expect them to take longer than a second or so. Unfortunately, the same rigors of thread synchronization also apply in the MIDP environment, so you have to ensure access to your shared variables is properly synchronized, just as in J2SE. Threads are particularly good at displaying progress indicators when paired with the **Gauge** class.

- ***Error Recovery:*** If you are going to be implementing your own RMS management system for storing data into the MIDlet's **RecordStore**, you may want to consider writing a separate MIDlet that performs data recovery on a corrupted/damaged **Record**. How this is done is implementation-dependent—perhaps a backup record could be written or some other parity/checksum could be performed. If you can't supply a recovery system, you might want to perform extra testing to ensure that a corrupt record can't be created. Try stopping your MIDlet during a write operation—is the data still intact?

SUMMARY

The strategies detailed above are guidelines you can use when designing your MIDlets. They are not written in stone and need not be followed for each and every MIDlet. However, if you employ as many of the strategies as is possible in a given situation, or modify the strategies to fit your application, you'll find that your MIDlets are more robust, secure, and perform better than if you hadn't.

This concludes Part 3, our journey into the server side of our wireless application environment. We've discussed everything from Servlets to JSPs, Web Services to transcoding, and finished up with some good ideas you might want to follow when designing your MIDlets. You now have the background needed to understand our campus portal, with its rich framework for hosting wireless enterprise applications.

Part 4
Grand Finale: The Campus Portal For Wireless Devices

An Introduction to Part 4

With Part 4 we begin to tie everything together that we've covered so far. We do this using a prototype application called the *Campus Portal for Wireless Devices* or *Campus Portal* for short. This is a platform for deploying applications for a university environment that are accessible from a number of different clients including handheld devices and browsers. Some possible applications that you can imagine are for finding a classroom location, retrieving a grade for a class that was just completed (always a tense moment), sending messages from one student to another, and might even include games and an application to locate entertainment events. The possibilities are endless. The platform can support a number of handheld devices and more can be added as they appear on the market.

More importantly, the platform integrates all of the technologies we have described so far—including J2EE, Web Services, transcoding, and personalization. This is not to mention the technologies and best practices for delivering client applications using WML, J2ME, and VoiceXML.

Finally, we wrap up the part with a look to the future for wireless enterprise applications. Here we step "out of the box" for a moment and try to predict what is arriving in the future and how it will impact our lives as developers of wireless enterprise applications (and perhaps as consumers also). We also take a moment to look at some of the commercial applications that try to provide similar function to the Campus Portal.

Here's a brief description of the chapters in Part 4:

- ***Chapter 18*** gives a high-level overview of the Campus Portal from a user's and administrator's perspective.
- ***Chapter 19*** takes a peek at the details of the implementation of the Campus Portal.
- ***Chapter 20*** let's us provide some final words of wisdom about the future of wireless enterprise applications development.

It's been a long trip but we're close to the finish line. A quick trip to the coffee bar and off we go!

Chapter 18

Campus Portal for Wireless Devices

We've been enjoying a nice journey through a number of client, server, middleware, and networking technologies. We've seen client technologies that present content on a handheld device, a programming environment for handheld devices using the Java platform, and voice-driven technologies. On the server side, we've presented technologies that are based on J2EE, XML, and Web Services for providing application services, transforming application output for multiple devices, providing personalization, and extending the reach of our applications.

It's time to put all of the technologies together in a way that makes it easy to create powerful applications that provide personal communication and data access, anytime and anyplace. The client technologies that we have examined, and the handheld devices they run on, provide the personal aspect of our applications. They are assumed to always be within arm's reach. Our client technologies—WML, J2ME, and VoiceXML—allow us to tailor client applications for these devices.

That is the good news: the connectivity of handheld devices gives us a way to extend our applications to the enterprise infrastructure. The bad news—you've heard it many times before—is that handheld devices provide a limited environment for delivering and executing applications. Server-side technologies are a solution to this problem of limitation. The server-side infrastructure also solves other problems that handheld devices introduce. With a plethora of devices in the market, the architecture chosen has to be able to handle many types of devices and have the capacity to easily introduce new devices.

The server-side architecture also should be able to easily introduce new applications targeted for handheld devices. Applications for the wireless world will be small but numerous. The sheer number of applications that may be deployed requires that the deployment structure be simple and flexible so that new applications can be introduced with minimal effort.

Finally, the architecture should also be able to compensate for some of the limitations of handheld devices. For instance, given that the need for personalization is very high for wireless applications it becomes necessary to minimize the amount of data a user has to enter and to minimize what is transmitted to a handheld device.

We will address these issues with a platform that takes advantage of an architecture we proposed in Chapter 1. We call the platform the *Campus Portal for Wireless Devices*.

CAMPUS PORTAL FOR WIRELESS DEVICES

The *Campus Portal for Wireless Devices*, or Campus Portal for short, was a project conceived at San Jose State University (SJSU) to study and implement an architecture suitable for delivering services to university students via handheld devices. We'll enumerate these services later but you can imagine that services that allow a student to locate the classroom for an upcoming class or contact classmates while enjoying a cup of coffee at the local campus coffee shop would provide a new category of student services only possible with handheld devices.

The Campus Portal project was the effort of several students completing their master's projects during the Summer and the Fall semester of 2001 at SJSU. The student projects explored architecture issues, implemented a substantial portion of the architecture, or implemented domain applications using the Campus Portal platform. At the completion of the projects, the efforts of these individuals were integrated to form a prototype Campus Portal.

The Campus Portal was developed using a university campus as the setting and the domain for applications developed for the portal. However, the "campus" setting applies equally well to a corporate campus. From an application delivery perspective, a student rushing to class and using a handheld device to locate the building and classroom is no different than the corporate employee rushing to a meeting and trying to locate the building and conference room.

A number of services can be imagined that have a very similar structure between a university campus and a corporate campus. These services can support information retrieval, calendaring and scheduling, expense tracking, and messaging. You can also imagine more esoteric services that provide for sharing information using Peer-to-Peer (P2P) technologies and accessing maps or weather information using location-based information from the handheld device.

Of course, there will also be services that are more appropriate for a university campus than a corporate campus. These might include multi-user game services, dating services, and auction services to sell course textbooks the moment a course is finished.

Humor aside, the point is the services provide new value because they provide new types of communication between students (or employees) and new types of information access based on the premise that it can be accessed anywhere and anytime via a mobile device. The services provided might be tailored to a specific domain—for example, a university campus—but there are many domains that can be supported by the Campus Portal architecture.

Campus Portal: Features, Functions, Benefits

Let's take a deeper look into the functions provided by the Campus Portal. We'll look at the Campus Portal in this section from two perspectives:

- ***From a services perspective:*** A few application services that are specific to a university environment (Grade Finder, Class Finder, Faculty Finder, and Messaging) will be described from a user's viewpoint.
- ***From an architecture and platform perspective:*** We'll begin to provide the next level of detail for the architecture that was proposed in Chapter 1.

In this chapter we'll stay at a high-level view. We'll leave the implementation details for the next chapter.

THE CAMPUS PORTAL APPLICATIONS

The Campus Portal is a framework of various J2EE components that enables the deployment of services for handheld devices. The services can be implemented using a variety of technologies and then deployed as a Campus Portal Web Service. In the Campus Portal a number of applications were proposed and were divided into the following broad categories:

- Administration Services
- System Services

- Information Services
- Messaging Services
- Entertainment Services
- Calendaring and Scheduling Services
- Peer-to-Peer Services
- Miscellaneous Services

We'll take a look at the applications in each of the categories in the following sections. Some of the services enumerated above were implemented so we'll give you a few screen shots to illustrate those applications. Other applications were left for future students (or readers of this book) for implementation. So, in those cases we'll give a quick description.

Administration Service

A simple browser-based administration service is provided to administer users, groups of users, and services. Users can be authorized to use the system using the administration service. Groups of users can also be created using the same service. Services are registered with the *Services Framework* (we'll see more about the Services Framework in the next chapter) and customized using personalization pages provided by the Campus Portal Administrative Service. Users can also personalize their preferences for each of their devices.

Figure 18-1 shows the home page for the Campus Portal. Users or administrators can select the appropriate login function (using either a simple password or a Java Card) from this page or they can select the "personalize" option to customize their user experience for each service deployed in the portal.

Figure 18-2 shows the login screen for the Campus Portal. Both the user login and admin login selections, when chosen, result in the screen in Figure 18-2 being shown. Upon clicking the "go" link, the portal will display one of the following screens depending on the style of login used:

- ***User Login:*** The portal displays a list of available services, as shown in Figure 18-9 on page 611.
- ***Admin Login:*** The portal displays a screen with options for adding and deleting users, and managing groups (not shown).

Also available in the left side navigation bar is an option the user can use to personalize the services housed in the portal. Upon selecting the "personalize" option, the portal displays a personalization Web page, as shown in Figure 18-3.

Figure 18-3 shows the service personalization page that could be used to personalize the services by the *service provider* or the *service user*. In the service provider

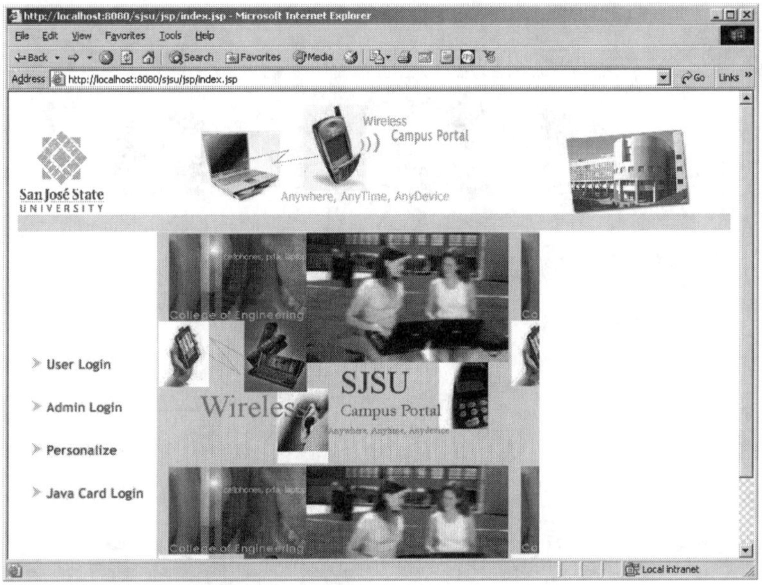

Figure 18-1. The Campus Portal Home Page.

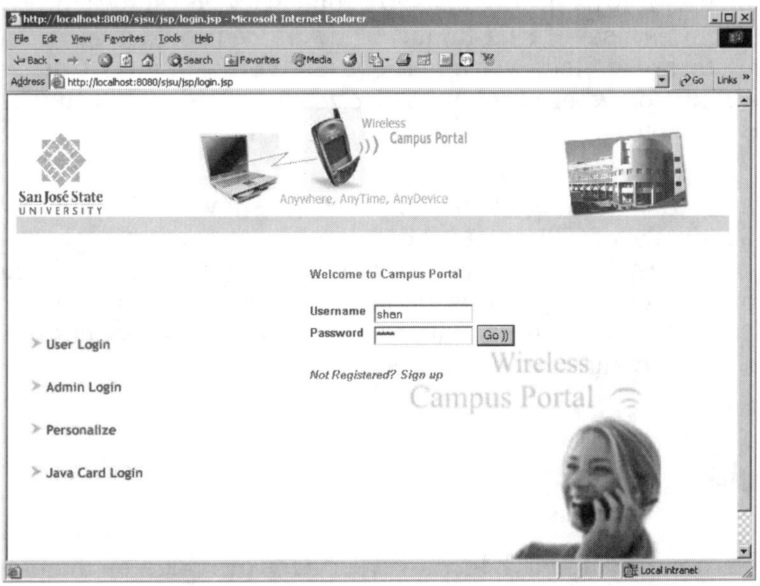

Figure 18-2. The Campus Portal Login Screen.

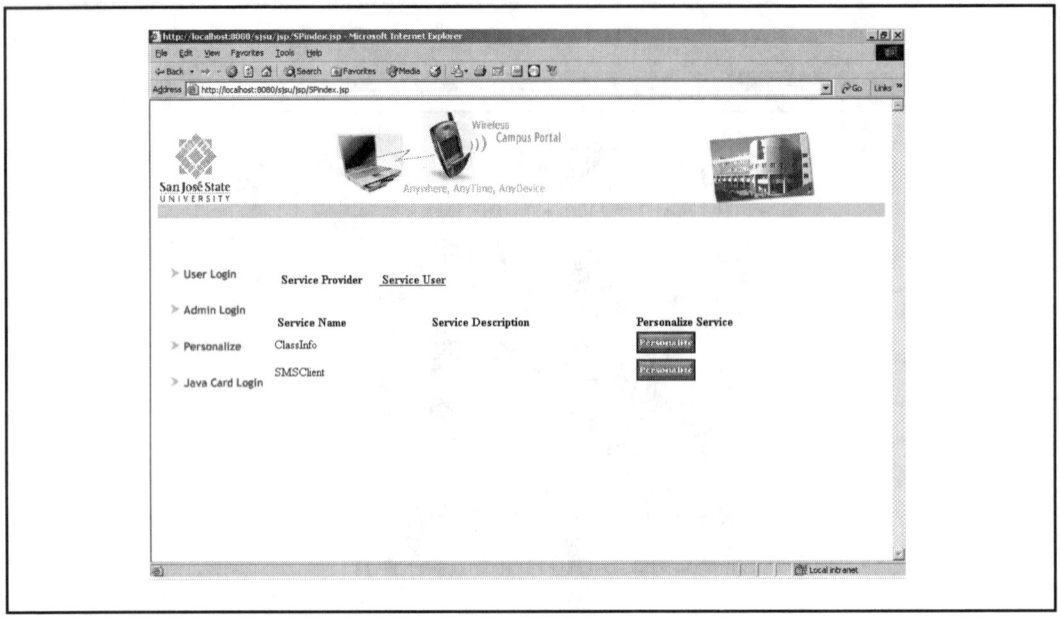

Figure 18-3. The Campus Portal Service Personalization Web Page.

screen, upon selection of one of the Personalize buttons, the portal displays a specific personalization page for that particular service (shown in Figure 18-4). Each service requires a set of user inputs. The service provider can personalize the way these inputs are collected from the user. For example, *Course Name* is collected in a Text Box whereas *Course Section* is a Radio Button with Section A and Section B as choices.

When the *Service User* link is clicked on the page shown in Figure 18-3, the page in Figure 18-5 is displayed. This page allows the user to customize the default values for inputs to a service. For example, a student can set his student ID here and the Personalization engine would fill this value as default in any form that requires his student ID. This details page is shown in Figure 18-6. This reduces the number of inputs that user has to enter on the device.

System Services

System services are fundamental, low-level services provided by the Campus Portal System for use by other higher-level services. The system services are available to any service deployed in the portal, and include such services as Single Sign-On, the Services Framework, and the Transcoding Engine. We'll see more on the Services Framework and Transcoding Engine in the next chapter.

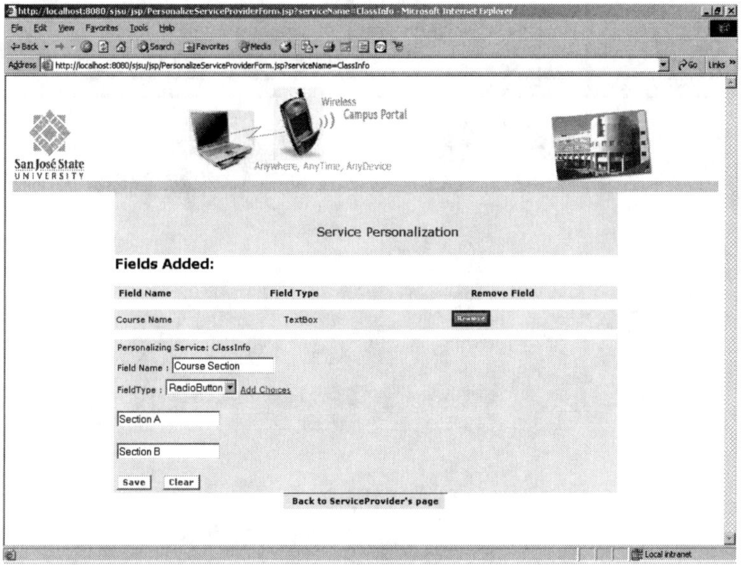

Figure 18-4. Entering the Campus Service Personalization Details.

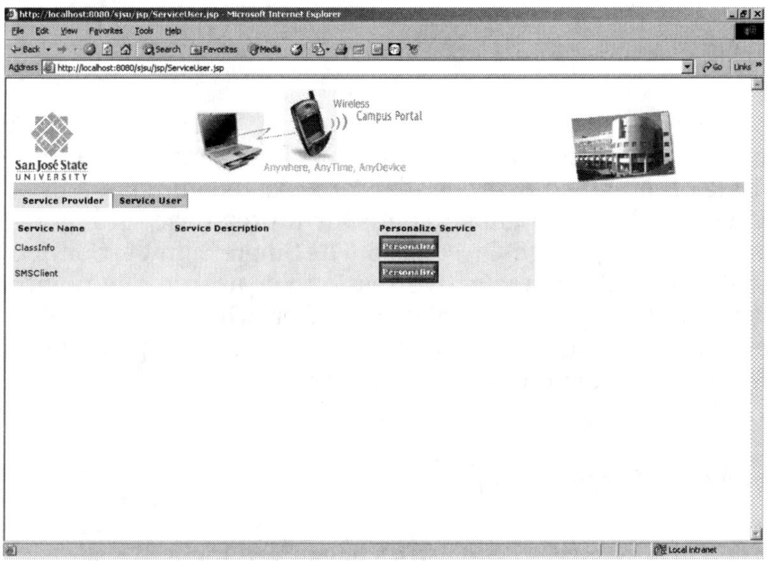

Figure 18-5. The Service User Personalization Web Page.

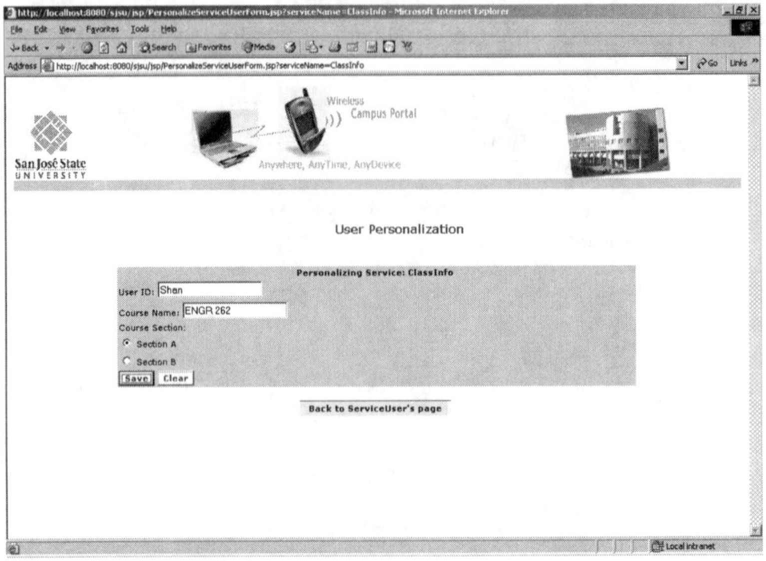

Figure 18-6. The User Personalization Details Web Page.

Single Sign-On

The Single Sign-On service is a service that allows users to log into the Campus Portal and access all services without having to log in for each additional service used. At the time of initial login, credentials are created and passed to each service as it is being accessed. The Login screen, shown previously in Figure 18-2, is the starting point for the Single Sign-On service. The Single Sign-On Service keeps track of the user logins and allows one active session per user at any point in time. The credentials are valid for a period of time, after which they expire. When the credentials expire, a user has to reenter their user name and password. Figure 18-7 shows that the user's credentials expired and prompts the user to re-authenticate.

Single Sign-On with Java Card

The Single Sign-On service also supports a Java Card login. A user can log into the portal by just inserting a Java Card into the card reader and entering his PIN number. The Java Card login module is made available as an applet (shown in Figure 18-8) and also as an application (not shown).

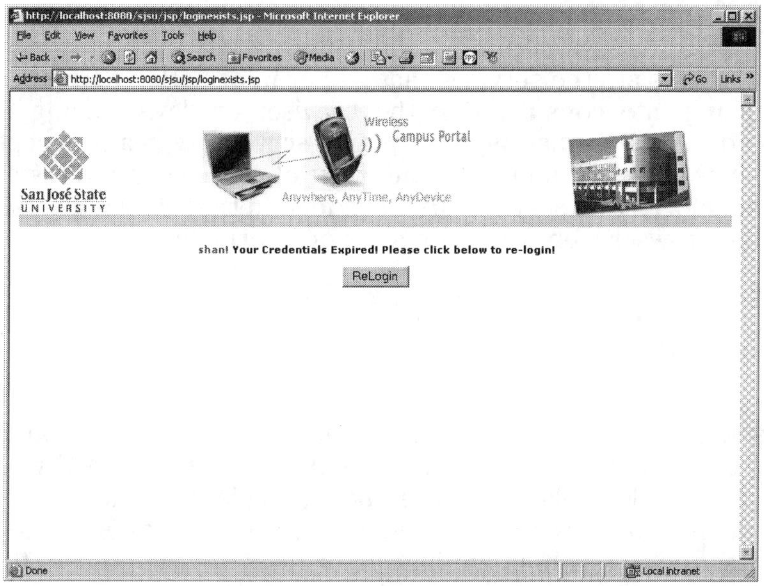

Figure 18-7. The Campus Portal Expired Credentials Web Page.

Figure 18-8. Campus Portal Login using Java Card.

Information Services

Information Services are the services that gather information and present it to the user while taking into consideration the browser or device being used. When gathering the desired information, information services typically accept or require input from the user to determine what information the user is interested in seeing. Some possible information services are described in the following sections. Figure 18-9 shows the browser Web page with a list of Information Services.

Class Information

This service provides basic catalog and schedule search function based on a keyword. For example, a search for courses with the keyword "XML" would result in those courses with XML-related lectures being displayed, while also providing a brief course description for each matched course. The service is also able to obtain the course description (such as the instructor teaching the course, the schedule, and the classroom location) by entering the course ID.

When using a browser, all the course details are displayed (as shown in Figure 18-10). For a handheld device, the service only displays the instructor's name and concise details based on the device's capabilities. Figure 18-11 displays a Class Info search that would be displayed if the user used a WML-enabled phone. In this example the default information is already provided at the invocation of the service and the forms are filled with appropriate customized values. Figure 18-12 shows the VoiceXML interactions for the same service when dialed from any telephone.

Registration

This registration service allows a student to enroll in a course. Here are the steps that would be involved in using the registration service:

- The user obtains his username and password from the university.
- The user logs into the system with his username and password.
- The user clicks on the registration service.
- The service prompts for required information such as course ID and addcode (a code that is used by instructors to grant enrollment in a particular course).

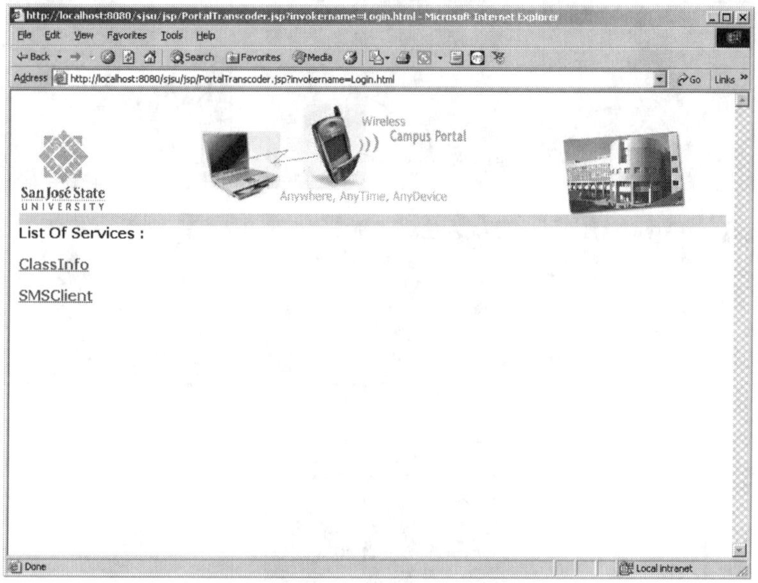

Figure 18-9. The Campus Portal List of Services.

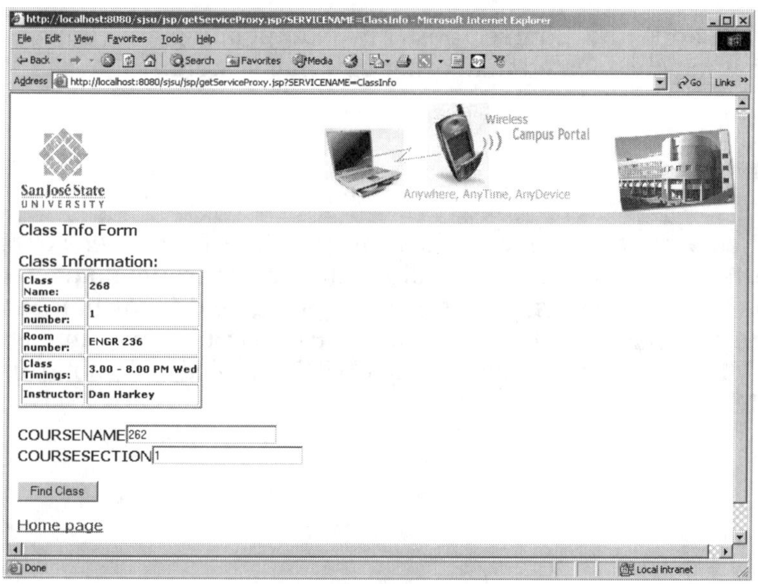

Figure 18-10. The Campus Portal Class Info Service in a Browser.

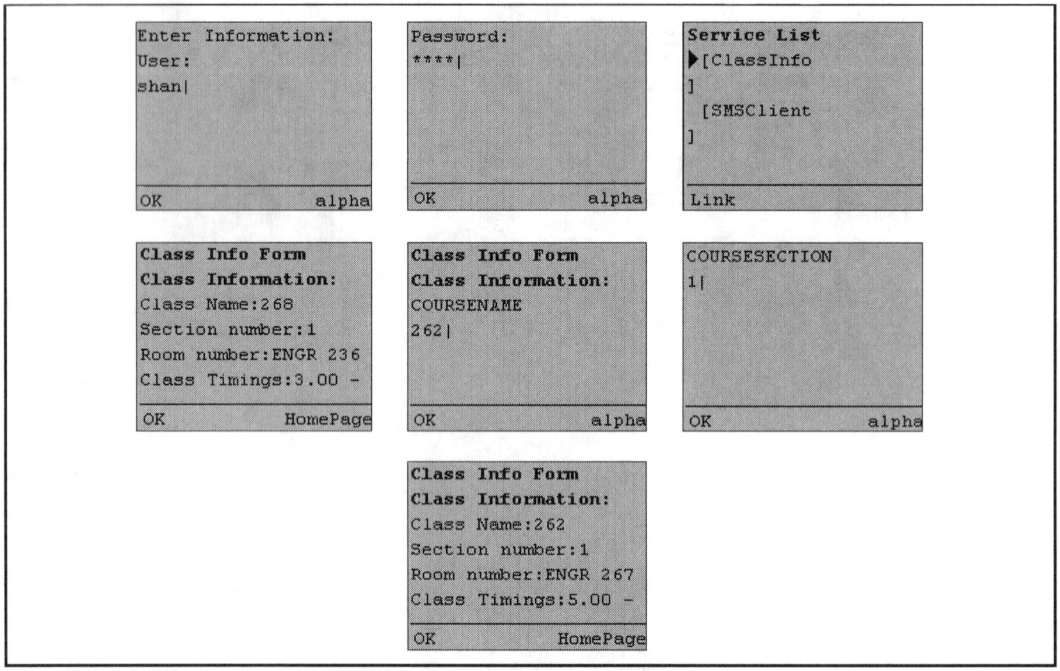

Figure 18-11. The Campus Portal Class Info Service in a WML Simulator.

- The service adds the course, possibly manipulating the university records system that may be located in another external system.

Class Finder

This service helps students to locate the room numbers for a given course. Students would enter the course ID and it would illustrate the classroom number and possible directions to the building where the class was being held (shown in Figure 18-13). As part of the Campus Portal, an interactive campus map was implemented as a MIDlet. This MIDlet can be downloaded to a MIDP-enabled cell phone or PDA device (see Figure 18-14).

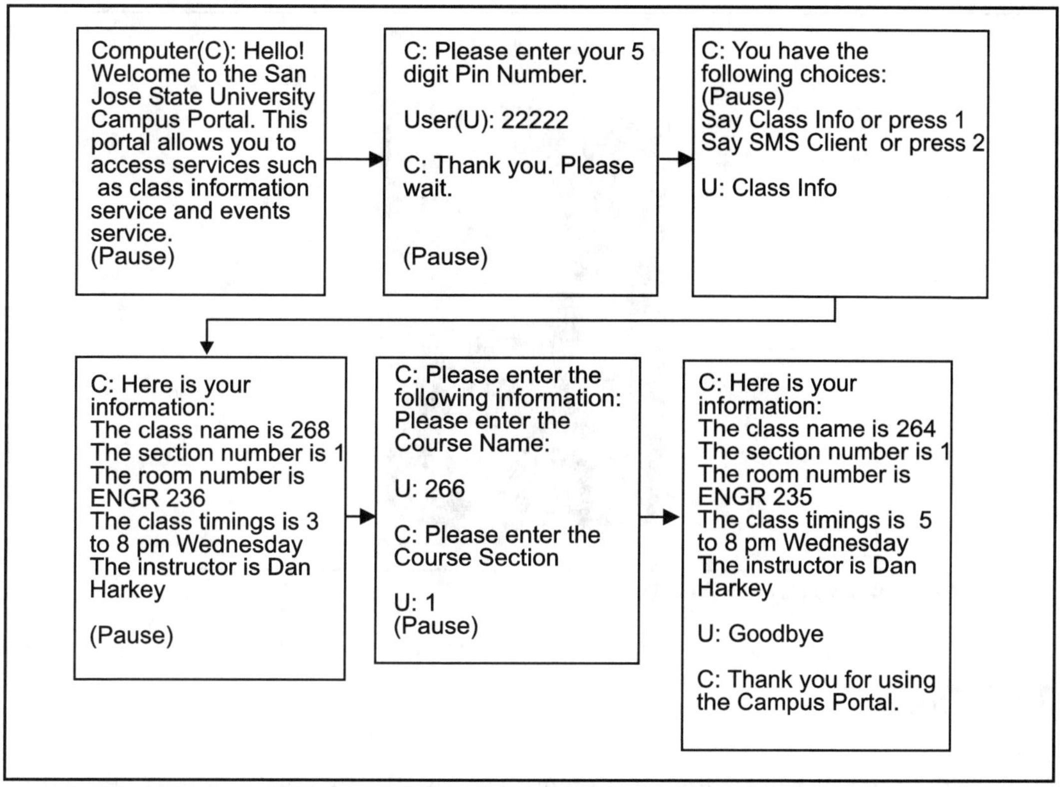

Figure 18-12. A Simulation of the Campus Portal Class Info Service using VoiceXML.

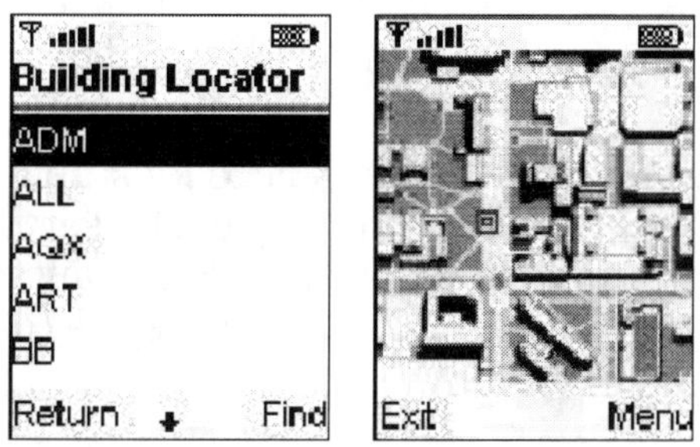

Figure 18-13. Campus Map on a MIDP Simulator.

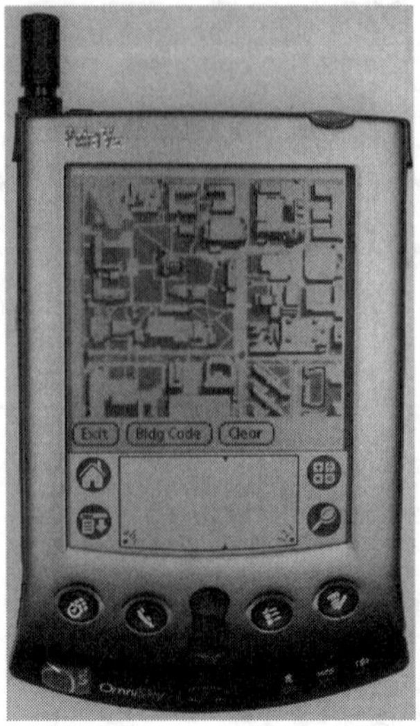

Figure 18-14. The Campus Map on a Palm Device.

Grade Finder

The Grade Finder service allows a student to access the final grade for a course they have just completed.

Classmate Finder

The Classmate Finder service allows a student to look up another student's contact information similar to the telephone company's 411 service. Students are able to search on a classmate's first and last names in a given course. Once the user is found this service could interact with the Messaging Services (discussed next) to send a short message.

Messaging Services

The Messaging Services provide simple communication between students using short messages (SMS), broadcast messages, chat, and email. These are types of communication we are dependent upon with our browser-based Internet access. The benefit provided by the Campus Portal Messaging Services is that a user can participate in such communication using a handheld device. The different messaging services are described in the following sections.

Short Messaging Service (SMS)

With the SMS Messaging Service, a user can send and receive a short message using a handheld device. Figure 18-15 illustrates sending an SMS message when using a WML interface.

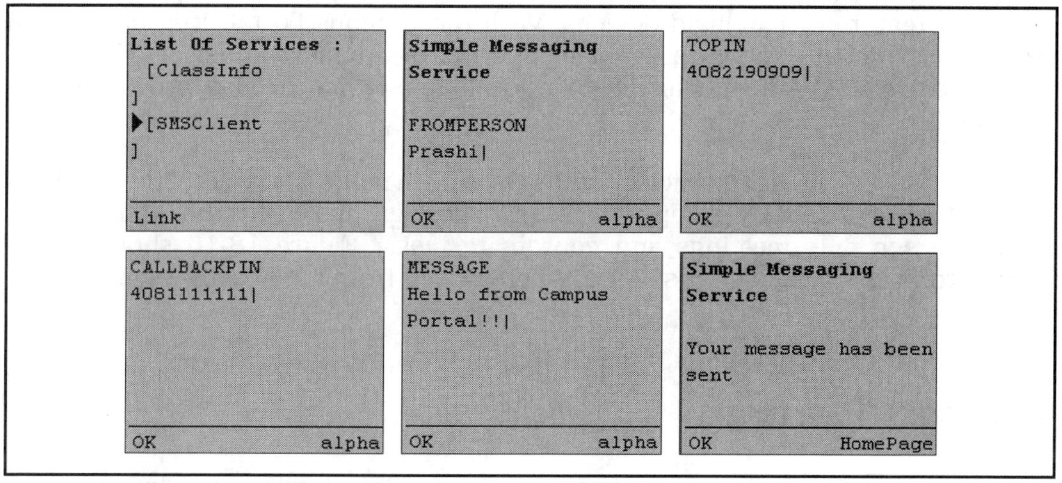

Figure 18-15. The Messaging SMS Service via WML.

Broadcast Messaging Service

The Broadcast Messaging Service is similar to the SMS service, except it has multiple recipients. For example, a professor might want to send a broadcast message to the entire class with an important, time-sensitive announcement.

Chat Messaging Service

The Chat Messaging Service provides real time and continuous message exchange between a group of users participating in a chat session. Chat sessions are already very popular on the Internet. Delivering Chat capabilities to wireless devices would allow a person to chat anywhere and anytime.

Email Messaging Service

The Email Messaging Service allows a user to send or receive an email message using the Campus Portal. The service allows the origination of messages from browsers as well as handheld devices. Given the limited capabilities of handheld devices, attachments are not supported on some devices.

Calendaring and Scheduling Services

The ability to manage tasks using to do lists and manage a calendar is always an important feature of handheld devices. With the Campus Portal, it is possible to provide this function in a consistent manner across numerous devices. The connected nature of the handheld device allows the service to also provide group collaboration.

By using the Calendaring service, students can manage class activities such as project discussions, study groups, and so on. These groups can use the Calendaring service to schedule meetings and coordinate tasks. Figure 18-16 shows a few screens from the Calendaring service, when viewed from a WML browser.

Entertainment Services

A student's life is often very intense with heavy class loads and other stressful elements. Entertainment is a way to relieve these stresses. The Entertainment Services strive to help students enjoy themselves and take their minds off their studies from time to time. With a handheld device, entertainment possibilities can be available immediately, anytime, anywhere. The following sections describe some of the possible entertainment services that could be deployed in the Campus Portal.

Games Service

Handheld devices have always been a platform for games. However, with connected devices, multi-user games become a possibility. Most likely, these games will take advantage of the advanced programming capabilities of handheld devices.

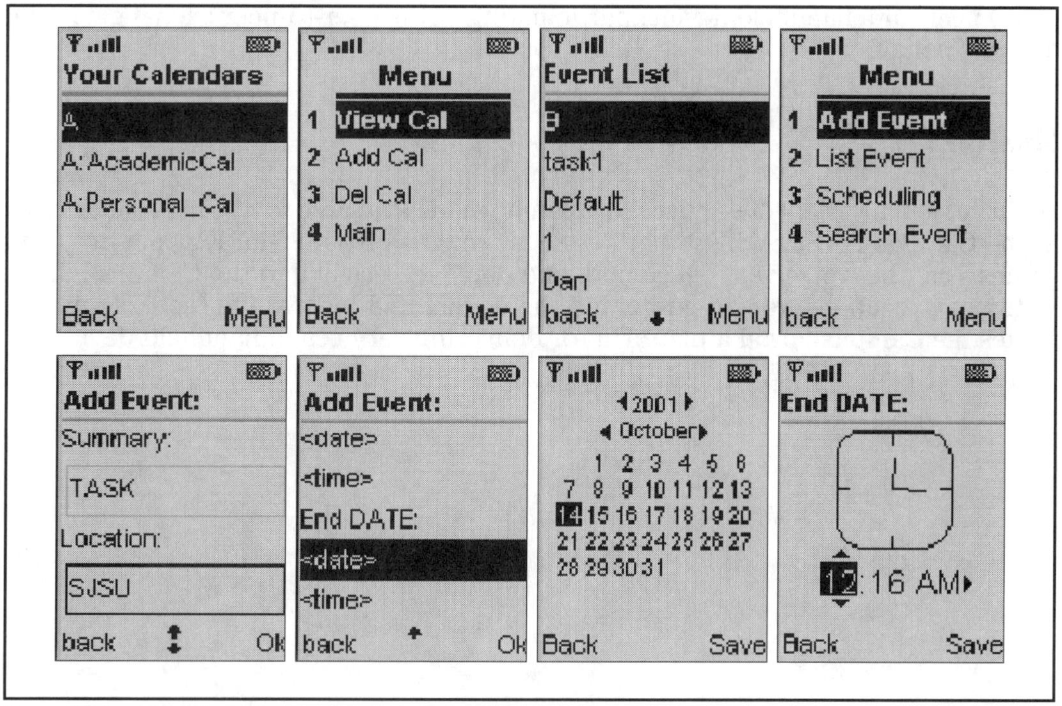

Figure 18-16. The Calendaring Service Shown on a MIDP Simulator.

Event Finder Service

Playing games on a handheld device is one, but not the only source of entertainment in a student's life. After all, there are movies, concerts, sporting events, and travel (at least during spring break). Finding and planning these events—both on and off campus—are important. This service provides event search and details about events. Given its connected nature, it can draw on other resources available on the Internet such as ticket sellers or travel reservation services and can even collaborate with the calendaring service to add the event to a student's schedule.

Peer-To-Peer Services

Peer-to-peer applications on the Internet have become popular with the advent of systems like Napster and Gnutella that allow file sharing. With the more capable handheld devices, there is the possibility of being a direct participant in a peer-to-peer network. For less capable devices, the Campus Portal can act as a proxy peer for a handheld device, participating in the peer-to-peer exchange on the device's behalf.

SUMMARY

We have seen a variety of services that can be made available to a number of devices in the Campus Portal. Many other services—such as forums, an FAQ service, and games—can be conceived, developed, and deployed easily into the Campus Portal. In the next chapter, we will go behind the covers and look at the technology that makes the Campus Portal a platform for delivering services to handheld devices.

Chapter 19

Campus Portal: A Look Under the Covers

This chapter reviews the architecture and describes implementation details of the Campus Portal for Wireless Devices. The architecture can be viewed as both a portal and a framework. It is a portal in the sense that a variety of applications can be delivered and interaction with the applications can be personalized. A traditional portal provides a flexible user interface that can be personalized with a variety of applications. Given the limited display capability of handheld devices, this level of personalization is not a feasible function for the Campus Portal. On the other hand, the Campus Portal needs to be flexible in support of many devices (including handheld devices), so the portal function appears in different form.

The Campus Portal can also be viewed as a framework—that is, a system that provides a set of system services and a platform for delivering applications in a piecemeal way. Here are some of the goals of the Campus Portal framework:

- Provide multiple client support and be able to easily integrate new devices.
- Provide single sign-on capabilities to the portal framework and its services.
- Provide an easy process for deploying new services to the framework.
- Provide personalization at various levels such as user, service, and form.

Before you proceed to the remainder of the chapter, you may want to review the enterprise architecture for wireless devices we proposed in Chapter 1 including

Figure 1-2. Figure 19-1 shows some of the subsystems defined in the Campus Portal that were used to implement the architecture and achieve the goals previously stated. The diagram also shows the major flow of information between the various subsystems and where data is stored to capture information about users, Web Services, personalization, and application data.

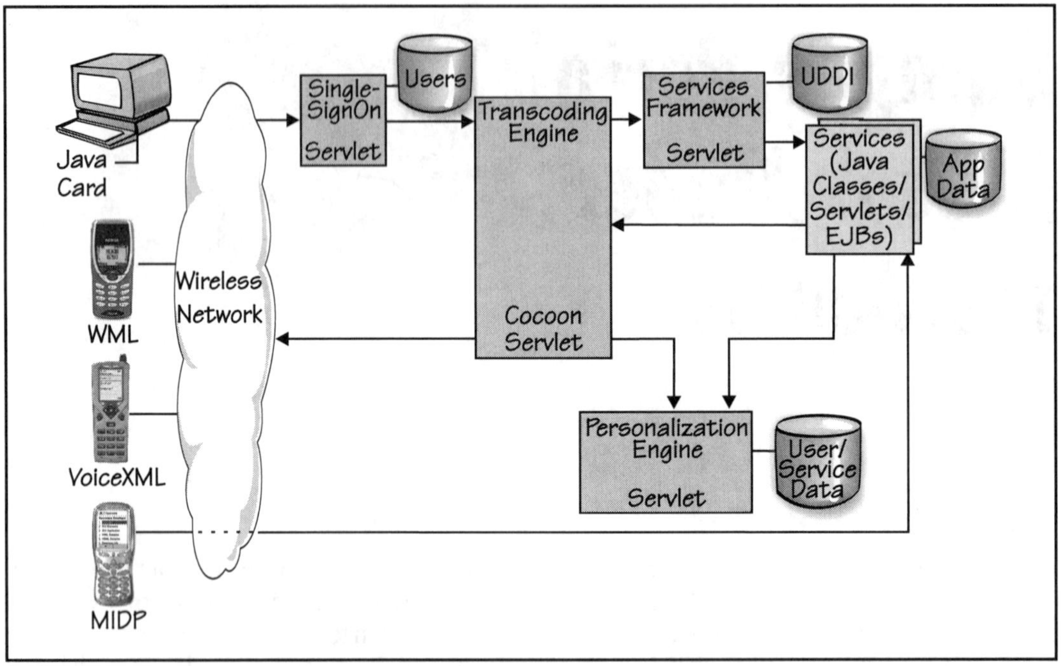

Figure 19-1. The Campus Portal Implementation Details.

In the remainder of the chapter we will explore the next layer of implementation details for the Campus Portal. For the most part, you will find descriptions of the subsystems and components and the technologies that were used for implementation. There are also interaction diagrams that will guide you through the steps performed between the different subsystems. You will find the descriptions to be mostly devoid of coding examples, as they can be quite lengthy. However, don't despair, as the code examples are available on the Web site that accompanies this book. You can visit it at *http://wirelessjava.engr.sjsu.edu.*

CAMPUS PORTAL CLIENT TECHNOLOGIES

The Campus Portal supports a variety of devices and is flexible enough to add new devices as they become available. This can include PDAs, WAP/J2ME-enabled cell phones, VoiceXML, and set-top boxes. Since these devices do not have the same rendering capabilities as desktop computers, it is necessary for Web content to be adapted, or *transcoded*, for proper presentation on the client device. This introduces a need for Web content transcoding, which seamlessly and transparently translates existing content from XML, HTML, or non-SGML sources to mobile devices to support a multitude of operating systems, markup languages, micro-browsers, form factors, and protocols. Table 19-1 gives an overview of some of the device characteristics that need to be supported.

Table 19-1. An Overview of Device Characteristics.

Characteristic	Internet Appliance				
	PC	Handheld PC	Palm	Cell Phone	Voice Phone
Can Download Applications	Yes	Yes	Yes	Yes	Yes
Output Devices	>= 800x600 color + speaker	240x480, mono, gray, or color	Portrait orientation, gray	3 or 4 line display	Speaker
Input Devices	Keyboard, pointing device, Microphone	Keyboard, stylus, touch screen	Keypad, stylus, touch screen	Keypad, microphone	Keypad, microphone
UI Metaphor	GUI, multiple windows	GUI, single window + message box	GUI, single window + message box, graffiti	Cards	Speech synthesis, voice recognition, keypad input
UI Development Tools	DHTML, XwingML, XUL	OS specific toolkit, Java	OS specific toolkit	WML	VoiceXML
Graphics	Yes	Yes	Yes	No	No
Primary Memory (MB)	4–256	4–64	1–8	<1	None
Secondary Memory (MB)	Gigabytes	none	None	None	None
Processor Speed MHz)	233–1000+	75–190	16–75	<=50	None

We discussed various ways of rendering content to different kinds of devices in Chapter 15. For the Campus Portal, we chose to implement transcoding as part of the Campus Portal platform. So instead of the client directly accessing the service, a middle-tier subsystem that acts as a proxy between a client and a service was implemented. In this approach, the proxy or transcoding engine acts as a virtual client to the service. The service consumes and produces content in XML.

Campus Portal Server Technologies

The middle tier is composed of the Transcoding Engine, the Services Framework, and the individual services registered with the Services Framework. All interactions between the client and the Campus Portal are mediated through the Transcoding Engine. Thus, the Transcoding Engine has two main functions:

- Control of the workflow for receiving a client request and processing it.
- Display of information to the end user in a recognizable format specific to a device.

Workflow is established and maintained through the use of HTTP POST and/or GET requests to JSPs deployed on a Web server. Information is transcoded by rendering XML to different formats through the use of *eXtensible Stylesheet Language Transformation (XSLT)*. An XML format specific to the Campus Portal called *PortalXML* is used to return the reply to the Transcoding Engine for transformation to a device-specific format. PortalXML will be described in more detail in upcoming sections.

However, prior to rendering the data, the information requested by the end user has to be retrieved by an individual service from one or more back-end data sources. The Services Framework, in conjunction with the individual service, returns data that represents the output of a service call. J2EE-based services (such as the Information Services) retrieve data by the use of Java applications that perform Java Naming and Directory Interface (JNDI) lookups for one or more Enterprise JavaBeans (EJBs) deployed in a Java 2 Enterprise Edition (J2EE) server.

Non-J2EE-based services (such as the messaging and calendaring services) are implemented as Java applications that use other specialized technologies (such as SMS, Java Mail, and LDAP) to retrieve data. In all cases, these Java applications are converted into Web Services and published with the Services Framework and deployed in a *Simple Object Access Protocol (SOAP)* server. Once this is done, JSPs in the Campus Portal invoke the operations exposed by a Web Service by using a SOAP message. The Campus Portal uses a number of technologies—such as XML, XSL, WSDL, UDDI, and SOAP—to provide a middle-tier platform for delivering application function to a variety of clients. See Figure 19-2 for a high-level overview of the technologies used and the flow from client to service and back.

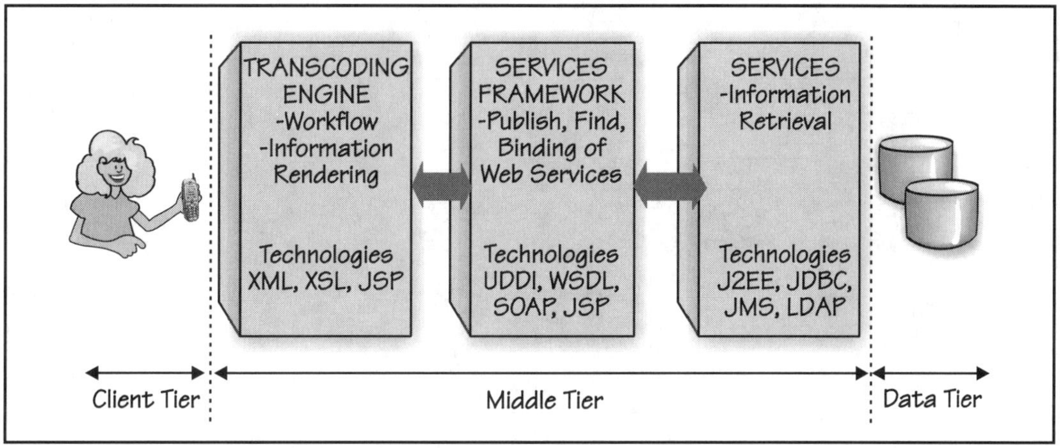

Figure 19-2. The Campus Portal Middle-Tier Technologies.

An End-to-End Scenario

Figure 19-3 shows an end-to-end scenario wherein the Campus Portal presents a list of services to a client, the client chooses a service, then the specific service is invoked. The following provides a step-by-step description of that exchange.

1. A client logs into the Campus Portal and requests a listing of all available services.
2. The Transcoding Engine forwards the request to the Services Framework.
3. The Services Framework queries the UDDI registry for all published services and returns a list of these using PortalXML.
4. The Transcoding Engine renders the PortalXML document to the client in appropriate device format.
5. The end user chooses a service of interest and submits a request to connect to the specified service.
6. The Services Framework fetches the Web Services proxy from the session.
7. The Services Framework forwards this request to the selected service.
8. The individual service evaluates the information that needs to be entered by the client to invoke a specific operation. It then returns a PortalXML document to the Transcoding Engine. The Transcoding Engine uses the information in the PortalXML document to create a form that contains the list of all the form elements that are required to collect the necessary information from the end user. The Transcoding Engine returns the form to the client after rendering it to the appropriate client device format.

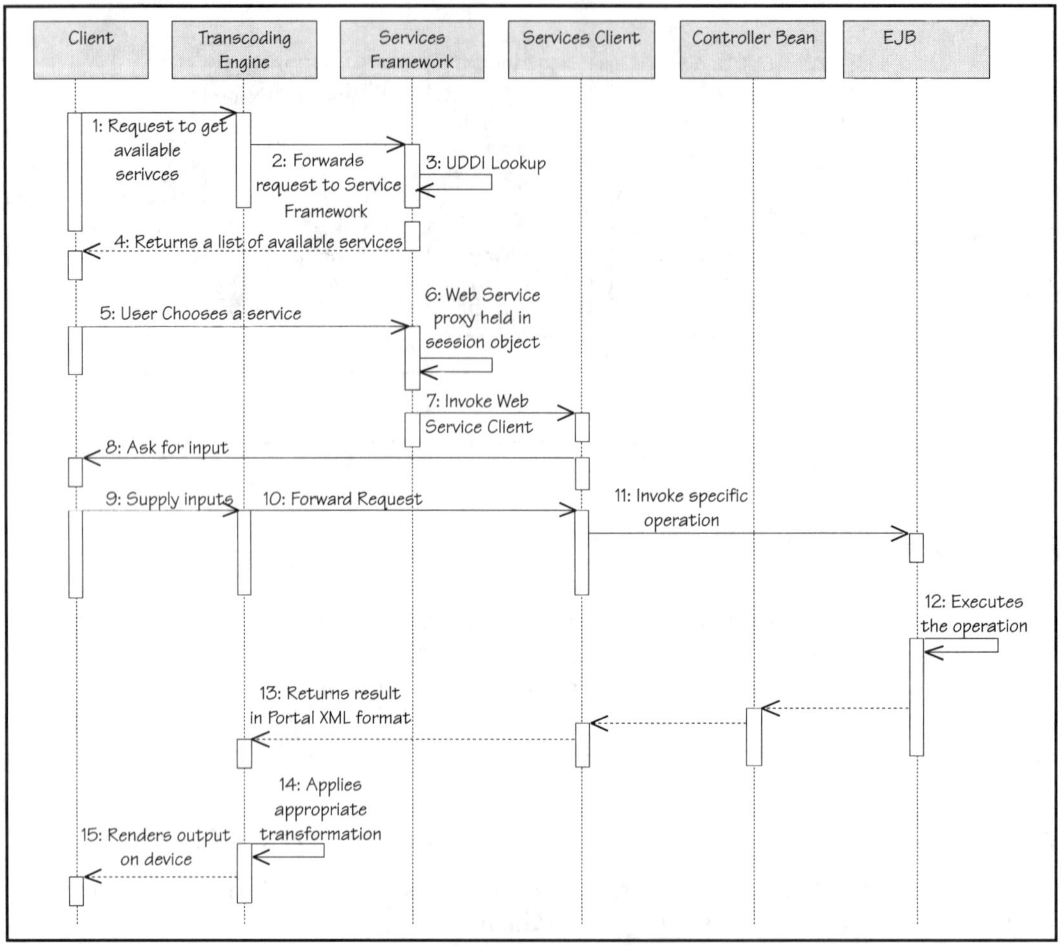

Figure 19-3. The Campus Portal Middleware End-to-End Scenario.

9. The client inputs all of the necessary data in the form and submits the request for processing.
10. The Transcoding Engine again forwards this request to the selected service.
11. The Services Framework uses the service proxy to invoke a specific operation on the service.
12. The service retrieves data from a back-end data source.
13. The service returns the result to the Transcoding Engine as PortalXML.
14. The Transcoding Engine applies the appropriate transformations to the PortalXML.
15. The Transcoding Engine renders the PortalXML for final display to the client-specific device.

A more detailed discussion of the subsystems mentioned above follows in the next sections.

PortalXML

Any multi-tier architecture is comprised of a number of components and is often organized as layers of functionality. These components and layers need to exchange and share data in order to perform their functions. Therefore it is necessary to define the semantics of this data to make the components modular. A complex architecture might have a number of such data formats. For simplicity, we chose to have only one data format in the Campus Portal and it is used by all components. We call it PortalXML. APIs were also designed to construct PortalXML using the DOM model. We will describe how components such as the Transcoding Engine, Services Framework, and application services consume and produce PortalXML.

PortalXML follows the natural top-down structure of an XML document. It is modeled after User Interface elements rather than the data representation of the service. The reasons for this are:

1. It makes the architecture more service-agnostic, as we do not know the nature of the service and its data representation to model the data format.
2. Portals are normally used to present information to the user in many customizable formats. For the Campus Portal, there is a need to present information to different device types. Most User Interface (UI) elements supported by devices are similar, though the details of their implementation may be different. For example, a textbox UI element is present in HTML, WML, and MIDP but each is rendered differently. So, it is possible to define a generic textbox using PortalXML, and then render the representation to a specific target device.
3. It is also easy to achieve device personalization and form filling with this approach (return to Chapter 15 for a review of these topics).
4. It is easy to extend this format to accommodate new User Interface elements as required by the addition of new devices.

Figure 19-4 shows the DTD that defines PortalXML.

As you can see the design is based on actions that are normally performed by the user from a client. These actions include:

- Accessing the Campus Portal to get a list of available services
- Invoking a specific service by filling in form data
- Getting the results from the service

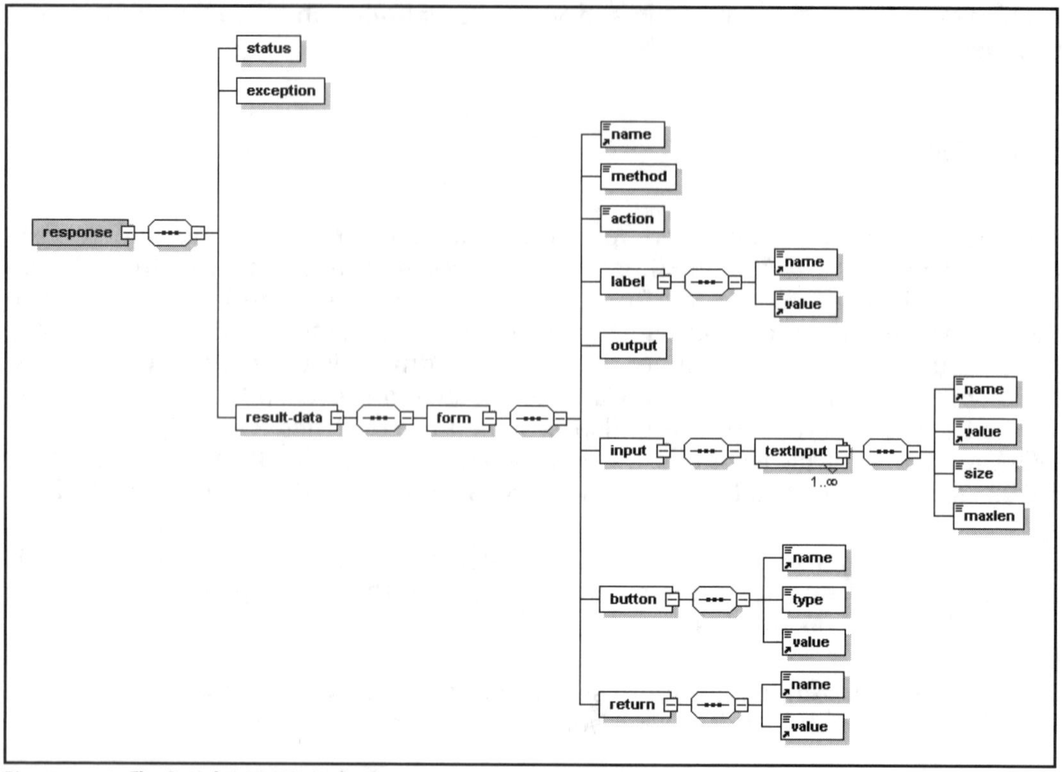

Figure 19-4. The Portal XML DTD Definition.

Transcoding Engine

The Transcoding Engine is the bridge between clients and services. In the Campus Portal architecture, the Transcoding Engine also performs functions that enable personalization.

The Transcoding Engine receives requests from devices via standard protocols. The transcoder component of the Transcoding Engine identifies the device, collects data for personalization and forwards the request to the service. The service returns the output, which again is intercepted by the transcoder. The transcoder renders the content of the response depending on the device type, and forwards it to the client. The entire system architecture aims at a simple design, which can be easily extended to different clients such as VoiceXML, J2ME, and WAP devices.

Personalization for the service user and provider is also handled by the Transcoding/Personalization subsystem. The response from the service is received as a PortalXML document. Stylesheets transform the PortalXML documents into either

WML, HTML or XHTML formats depending on the device type identified. The architecture of the transcoding engine is shown in Figure 19-5.

The Transcoding Engine has several components:

- Session handler
- Device Profiler
- Device Identification
- Controller (Workflow Manager)
- XML/XSL Repository
- Transcoder
- Personalization Engine

Figure 19-5 shows the interactions between components of the Transcoding Engine.

The actual steps in a service request are as follows:

1. The client makes a request to access a Campus Portal Service.
2. The Single Sign-On module determines if the client has unexpired credentials with the authority to access the service. If this check succeeds, the Single Sign-On module allows the user to access the service. If not, it will prompt the

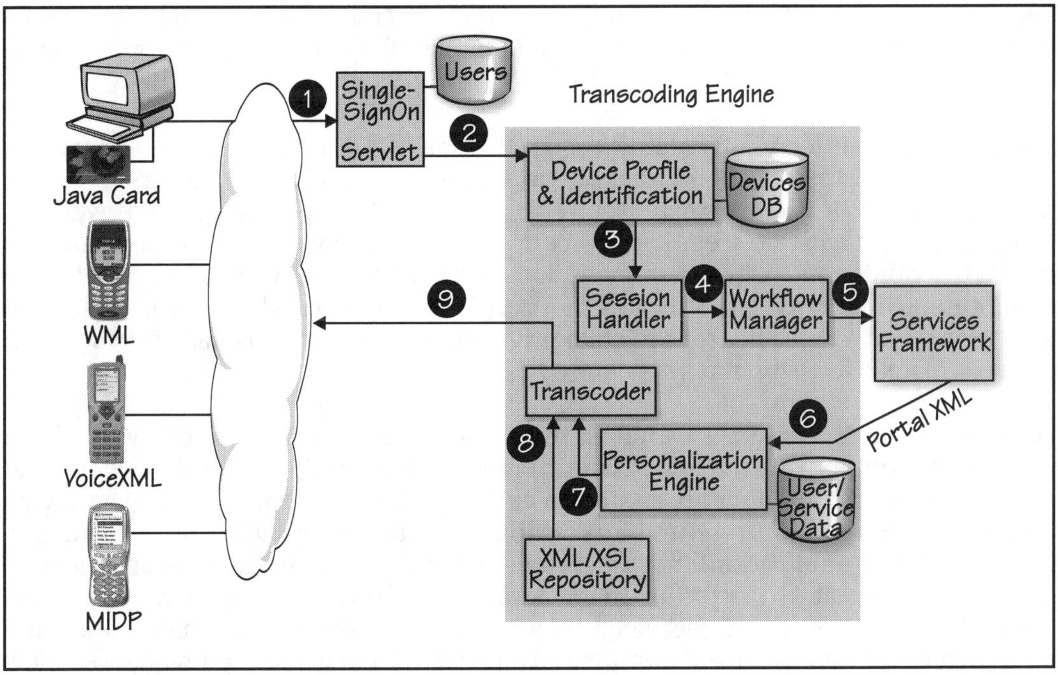

Figure 19-5. A Transcoding Scenario.

user to enter his user ID and password, and new credentials are created. This module keeps track of the user credentials.

3. The device is identified and matched with the devices in the device profile.
4. The Session Handler tracks the device session.
5. The Workflow Manager then forwards the request to the appropriate service.
6. The service returns the result as a PortalXML object.
7. The PortalXML object is filled with any personalized information by using form filling.
8. The appropriate stylesheet is selected from the XML/XSL repository and transformations are applied.
9. The result is rendered to the device.

The following sections describe the details of each of these components.

Single Sign-On (SSO)

The Campus Portal architecture is comprised of several subsystems, each with their own security mechanisms. Security is loosely divided into three categories: authentication, authorization, and information protection (including encryption). Transport level security is provided by SSL, and almost all devices and server protocols support it.

A popular authentication model is referred to as a shared secret (or password). Authentication is established under the assumption that if a user knows a private phrase it must be the correct user. This scheme does not work very well for validating users and services that are part of electronic systems. The problem lies in that the validation is only one way; it does not verify the service to the client. This is important in light of *Man in the Middle (MTM)* attacks where a rogue service impersonates a service in order to steal private user data (such as passwords, credit cards, or social security numbers).

To minimize MTM threats, a mutual authentication process is required, where both parties must verify to each other who they are. Once authenticated, a client and server establish a time-limited session to exchange information. This project investigated the framework required to establish a trusted relationship between a wireless device application and a service. When establishing a session between wireless devices, it is important to be aware of a device's lack of a continuous network connection to the service. To address this concern, we have taken the approach of passing the user's credentials as part of each service invocation. This is similar to URL rewriting.

Each service available on the portal might use its own authentication mechanisms and security models. The Single Sign-On component allows the user to access all the services provided by the system by providing his or her credentials once and without having to present them for each and every resource that is accessed. A Single Sign-On system is needed for the following reasons:

- Wireless devices are resource constrained and it is cumbersome for a user to enter a username/password every time a service is accessed. Single Sign-On allows the user to enter his or her username/password once to login into the system and use the services for a given period of time before his or her credentials expire.

- Different services might use different types of authentication and authorization mechanisms. Each service might be using a different security model on the back end. Single Sign-On provides a single interface to hide the complexities.

- The connection from the wireless device to the server might be intermittent. Therefore maintaining a session needs transparent reauthentication with the device.

Implementation Details

The Single Sign-On (SSO) component is implemented using JavaServer Pages and JavaBeans. The JSP pages are used primarily to communicate between the User Interface and the middle tier, which are the JavaBeans, that communicate with the database. The following three beans are used:

- **SSOHandlerBean** handles the core functionality of SSO. When user information is successfully verified, an entry is made in a temporary session table with a timestamp. This bean is also used to delete timed out user entries in the session table.

- **SSOValidator** is used to verify the active logon status of a user. This bean is called by services of the portal service to verify a valid sign in into the campus portal system.

- **SSOPassport** provides the logon verification service API for other services of the campus portal.

The beans are used as session scope variables in JavaServer Pages. Using this design approach, the connection is established once during the life of a user session and the same connection is used subsequently to minimize database connection time.

Special Cases

For HTML, WML, and other markup-enabled devices, the username and password is entered by the user using a form and the session handler on the server side takes care of maintaining the credentials of that user for a given time.

With MIDP, the credentials can be persisted in the local record store and then they can be used to reauthenticate with the server. This is possible because of the more sophisticated functions such as record store available on MIDP devices. It is also possible for different MIDlets to share the same credentials as long as they are in the same MIDlet Suite. This way user need not reenter credentials for each MIDlet.

Device Profiling and Identification

The Transcoding Engine maintains a database of supported devices and their specifications, such as format, protocol, screen size, and configurations. W3C has a proposed standard to extract this information by a Composite Capabilities/Preference Profile (CC/PP) feature. An application can query the device to get this information.

Unfortunately, the CC/PP standard is not ratified yet. We chose to work around this issue by maintaining the information in a local database. We also chose to maintain what application or service is available for a given device. This is very important because the Transcoding Engine can give an instant reply back to the device if the particular service is not available on that device.

The Transcoding Engine identifies the device by using the User-agent configuration in the HTTP request header. It tries to do a fuzzy match to the closest possible device if the device cannot be identified.

XML/XSL Repository

PortalXML documents are received as responses from the Services Framework after the transcoder sends a request to a Web Service. These documents are rendered to devices using stylesheets from the XML/XSL repository in the Transcoding Engine. PortalXML defines the vocabulary for the content in the Campus Portal. The Transcoding Engine uses one or more XSL documents for each device type. If there are new devices added, the repository needs to be updated with the appropriate XSL. The section below describes the XSLs for some of the device types currently supported by the Campus Portal.

We will be applying various stylesheets for the following PortalXML document and examine how the output is rendered on the various devices.

Listing 19-1. PortalXML for List of Services.

```
<?xml version="1.0"?>
<?xml-stylesheet  href="portal-html.xsl"  type="text/xsl"
media="explorer"?>
<?xml-stylesheet href="portal-vxml.xsl" type="text/xsl" ?>
<?xml-stylesheet href="portal-wml.xsl" type="text/xsl" media="wap"?>
<?cocoon-process type="xslt"?>
<response>
  <status>
  </status>
  <exception>
  </exception>
  <result-data>
    <form>
      <name>Service List</name>
      <method></method>
      <action></action>
      <list>
        <link>
          <name>ClassInfo</name>
          <value>getServiceProxy.jsp?SERVICENAME=ClassInfo</value>
        </link>
        <link>
          <name>SMSClient</name>
          <value>getServiceProxy.jsp?SERVICENAME=SMSClient</value>
        </link>
      </list>
    </form>
  </result-data>
</response>
```

Rendering to a Browser

When the stylesheet in Listing 19-2 is applied to the PortalXML document in Listing 19-1, the output generated looks like Figure 19-6.

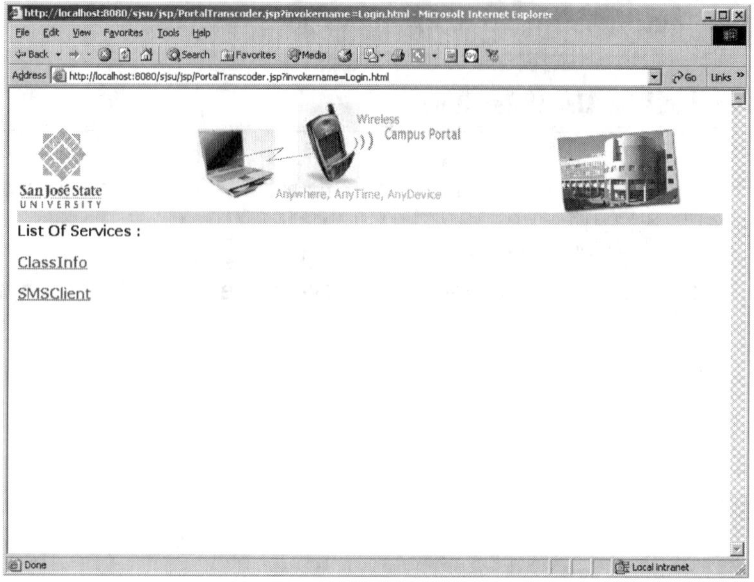

Figure 19-6. XML Browser Output.

Listing 19-2. Stylesheet for HTML

```
<?xml version="1.0"?>
<xsl:stylesheet xmlns:xsl=
   "http://www.w3.org/1999/XSL/Transform" version="1.0">

  <!-- template for document root -->
  <!-- template for list of services-->

  <xsl:template match="response/">
    <xsl:processing-instruction
        name="cocoon-format">type="text/html"
    </xsl:processing-instruction>
    <html>
      <body >
        <xsl:apply-templates select = "status"/>
        <xsl:apply-templates select = "exception"/>
        <xsl:apply-templates select = "result-data"/>
      </body>
    </html>
  </xsl:template>
```

```
<xsl:template match="status">
  <xsl:value-of select= "."/>
</xsl:template>

<xsl:template match="exception">
  <xsl:value-of select= "."/>
</xsl:template>

<xsl:template match="result-data">
  <xsl:apply-templates select = "form"/>
</xsl:template>

<!-- template for title -->
<xsl:template match="form">
  <xsl:element name="form">
    <xsl:attribute name="method">
      <xsl:value-of select="./method"/>
    </xsl:attribute>
    <xsl:attribute name="action">
      <xsl:value-of select="./action"/>
    </xsl:attribute>
    <html>
    <link rel="Stylesheet" href="PortalStyleSheet.css"
          type="text/css"/>
    <body class="gray" align="center">
    <table width="100%" >
      <td>
      <b>
      <font>
        <xsl:apply-templates select = "*"/>
      </font>
      </b>
      </td>
    </table>
    </body>
    </html>
    <br/>
  </xsl:element>
</xsl:template>

<xsl:template match = "label">
  <xsl:value-of select= "./name"/>
  <br/>
</xsl:template>
```

```
<xsl:template match = "method">
  <xsl:variable name="dummy" select="a"/>
</xsl:template >

<xsl:template match = "name">
<xsl:variable name="dummy" select="a"/>
</xsl:template>

<xsl:template match = "action">
<xsl:variable name="dummy" select="a"/>
</xsl:template>

<xsl:template match="list">
  <xsl:apply-templates select = "link"/>
</xsl:template>

<xsl:template match = "link">
  <a><xsl:attribute name="href">
   <xsl:value-of select="./value"/>
  </xsl:attribute>
  <xsl:value-of select="./name"/><br/></a>
  <br/>
</xsl:template>

<xsl:template match="input">
  <xsl:apply-templates select = "textInput"/>
</xsl:template>

<xsl:template match = "textInput">
  <xsl:value-of select="./name"/>
    <xsl:variable name="textName" select="./name"/>
    <xsl:variable name="textValue" select="./value"/>
    <xsl:variable name="textSize" select="./size"/>
    <xsl:variable name="textMaxLen" select="./maxlen"/>
    <xsl:element name="input">

    <xsl:attribute name="type">
      <xsl:value-of select="text"/>
    </xsl:attribute>

    <xsl:attribute name="name">
      <xsl:value-of select='$textName'/>
    </xsl:attribute>
```

```
    <xsl:attribute name="value">
      <xsl:value-of select='$textValue'/>
    </xsl:attribute>

    <xsl:attribute name="size">
      <xsl:value-of select='$textSize'/>
    </xsl:attribute>

    <xsl:attribute name="maxlength">
      <xsl:value-of select='$textMaxLen'/>
    </xsl:attribute>

   </xsl:element>
   <br/>
</xsl:template>

<xsl:template match = "button">
  <xsl:variable name="buttonName" select="./name"/>
  <xsl:variable name="buttonType" select="./type"/>
  <xsl:variable name="buttonValue" select="./value"/>
  <xsl:element name="input">
    <xsl:attribute name="type">
      <xsl:value-of select='$buttonType'/>
    </xsl:attribute>
    <xsl:attribute name="name">
      <xsl:value-of select='$buttonName'/>
    </xsl:attribute>
    <xsl:attribute name="value">
      <xsl:value-of select='$buttonValue'/>
    </xsl:attribute>
  </xsl:element>
</xsl:template>

<xsl:template match="output">
  <xsl:apply-templates select = "resultOutput"/>
</xsl:template>
<xsl:template match = "resultOutput">
  <xsl:value-of select="./name"/>
  <xsl:value-of select="./value"/>
  <br/>
</xsl:template>

<xsl:template match="return">
  <a><xsl:attribute name="href"> <xsl:value-of select="./value"/>
    </xsl:attribute>
```

```
    <xsl:value-of select="./name"/><br/></a>
    <br/>
  </xsl:template>
</xsl:stylesheet>
```

Rendering to a WML Browser

When the WML Stylesheet shown in Listing 19-3 is applied to the PortalXML document for the List of Services in a WML browser, the result would look like Figure 19-7.

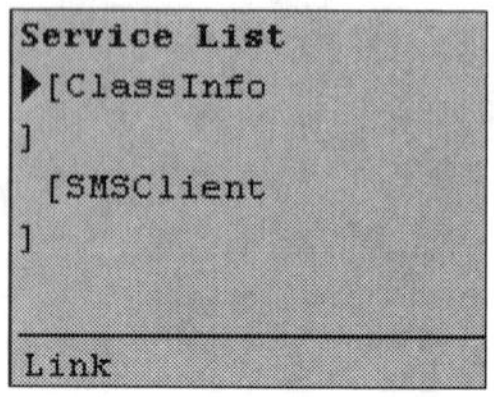

Figure 19-7. The "List of Services" Presented on a WML Device.

Listing 19-3. Stylesheet for WML.

```
<?xml version="1.0"?>
<xsl:stylesheet  xmlns:xsl="http://www.w3.org/1999/XSL/Transform"
version="1.0">

  <!-- template for document root -->
  <xsl:template match="response">
  <xsl:processing-instruction name="cocoon-format">
    type="text/wml"
  </xsl:processing-instruction>
  <wml>
    <xsl:for-each select="status">
      <xsl:apply-templates/>
    </xsl:for-each>
    <xsl:for-each select="exception">
      <xsl:apply-templates/>
    </xsl:for-each>
    <xsl:for-each select="result-data">
      <xsl:apply-templates select = "form"/>
    </xsl:for-each>
```

```
</wml>
</xsl:template>

<xsl:template match="form">
  <xsl:apply-templates select="output"/>
  <xsl:apply-templates select="input"/>
  <xsl:apply-templates select="list"/>
</xsl:template>

<xsl:template match="list">
  <card id="serv1" title="services" newcontext="true">
    <p>
     <b><xsl:value-of select="//name"/></b><br/>
       <xsl:for-each select="link">
         <a>
         <xsl:attribute name="href">
           <xsl:value-of select="./value"/>
         </xsl:attribute>
         <xsl:value-of select="./name"/><br/>
         </a>
       </xsl:for-each>
    </p>
    <do type="prev" label="Back">
      <prev/>
      </do>
   </card>
</xsl:template>

<xsl:template match="output">
  <card id="serv2" title="userinfo" newcontext="true">
    <do type="accept">
      <go href="#serv3"/>
    </do>
    <p>
      <b><xsl:value-of select="//name"/><br/>
      <xsl:value-of select="//label/name"/><br/></b>
      <xsl:for-each select="resultOutput">
        <xsl:value-of select="name"/>
        <xsl:value-of select="value"/><br/>
      </xsl:for-each>
    </p>
    <do type="accept" label="HomePage">
      <xsl:element name="go">
        <xsl:attribute name="href">
          <xsl:value-of select="//return/value"/>
```

```
        </xsl:attribute>
      </xsl:element>
    </do>
  </card>
</xsl:template>

<xsl:template match="input">
  <card id="serv3" title="userfill" newcontext="true">
    <p>
    <b><xsl:value-of select="//name"/><br/>
      <xsl:value-of select="//label/name"/><br/></b>
      <xsl:for-each select="textInput">
        <xsl:value-of select="name"/>
        <xsl:variable name="textname" select="name"/>
        <xsl:variable name="textvalue" select="value"/>
        <xsl:variable name="textsize" select="size"/>
        <xsl:variable name="textmaxlen" select="maxlen"/>
        <xsl:element name="input">
        <xsl:attribute name="name">
          <xsl:value-of select='$textname'/>
        </xsl:attribute>
        <xsl:attribute name="value">
          <xsl:value-of select='$textvalue'/>
        </xsl:attribute>
        <xsl:attribute name="size">
          <xsl:value-of select='$textsize'/>
        </xsl:attribute>
          <xsl:attribute name="maxlength">
          <xsl:value-of select='$textmaxlen'/>
        </xsl:attribute>
      </xsl:element>
    </xsl:for-each>
    </p>
    <do type="accept">
     <xsl:element name="go">
       <xsl:attribute name="href">
         <xsl:value-of select="//action"/>
       </xsl:attribute>
       <xsl:attribute name="method">
         <xsl:value-of select="//method"/>
       </xsl:attribute>
       <xsl:for-each select="textInput">
         <xsl:variable name="textname" select="name"/>
         <xsl:element name="postfield">
           <xsl:attribute name="name">
```

```
              <xsl:value-of select='$textname'/>
            </xsl:attribute>
            <xsl:attribute name="value">
              <xsl:text>$</xsl:text>
              <xsl:value-of select='$textname'/>
            </xsl:attribute>
          </xsl:element>
        </xsl:for-each>
      </xsl:element>
      </do>
      <do type="prev" label="Back"><prev/></do>
    </card>
  </xsl:template>
</xsl:stylesheet>
```

Rendering to J2ME

J2ME devices are not content browsers. Given that (as we have seen in Chapter 16) the J2ME platform is capable of processing XML, the PortalXML is therefore directly processed by J2ME. However, some applications may require preprocessing of PortalXML in which case a stylesheet can be added that is specific to the J2ME device.

Rendering VoiceXML

When the stylesheet in Listing 19-4 is applied on the List of Services PortalXML document, the voice browser renders the output as shown in Figure 19-8.

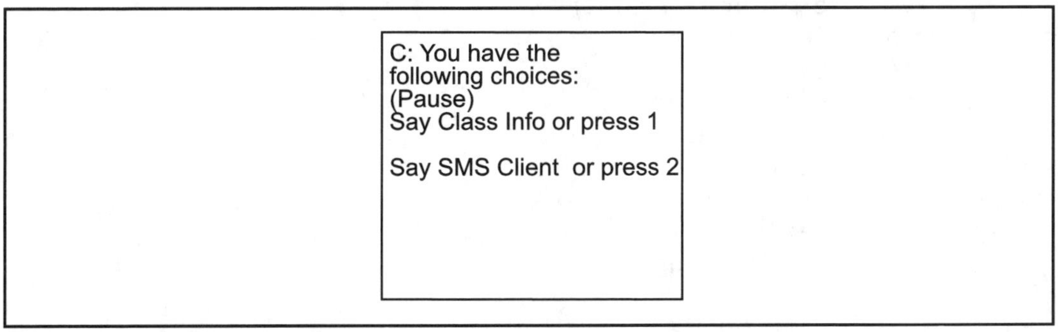

Figure 19-8. Output from VoiceXML.

Listing 19-4 shows the Stylesheet for VoiceXML.

Listing 19-4. Stylesheet for VoiceXML.

```
<?xml version="1.0" encoding="UTF-8"?>
<xsl:stylesheet version="1.0"
xmlns:xsl="http://www.w3.org/1999/XSL/Transform"
xmlns:fo="http://www.w3.org/1999/XSL/Format">
<xsl:template match="/">
<xsl:processing-instruction
name="cocoon-format">type="text/xml"</xsl:processing-instruction>
<vxml version="1.0">
    <xsl:for-each select="response/status">
        <xsl:apply-templates/>
    </xsl:for-each>
    <xsl:for-each select="response/exception">
    <xsl:apply-templates/>
    </xsl:for-each>
    <xsl:for-each select="response/result-data">
        <xsl:apply-templates select = "form"/>
    </xsl:for-each>
    <link event="exit">
    <grammar type="application/x-jsgf"> [Good] bye</grammar>
    </link>
    <catch event ="exit">
    Thank you for using the campus portal
    <exit/>
    </catch>
</vxml>
</xsl:template>
    <xsl:template match="status">
        <xsl:value-of select= "."/>
    </xsl:template>
    <xsl:template match="exception">
<xsl:value-of select= "."/>
    </xsl:template>
    <xsl:template match="form">
<xsl:apply-templates select = "output"/>
<xsl:apply-templates select = "input"/>
<xsl:apply-templates select = "list"/>
    </xsl:template>
    <xsl:template match="input">
<form id="new">
  <xsl:variable name="url">
   <xsl:value-of select="//action"/>
```

```
  </xsl:variable>
  <block>Please enter the following information</block>
  <xsl:for-each select="textInput">
   <field ><xsl:attribute name="name">
           <xsl:value-of select="./name"/></xsl:attribute>
           <xsl:attribute name="type">digits</xsl:attribute>
     <prompt> Please Enter the <xsl:value-of select="./name"/></prompt>
   </field>
  </xsl:for-each>
  <filled>
     <submit>  <xsl:attribute  name="next"><xsl:value-of  select='$url'/>
</xsl:attribute></submit>
  </filled>
</form>
     </xsl:template>
     <xsl:template match="output">
<form id="result">
  <block>Here is the information <break size="small"/>
  <xsl:for-each select="resultOutput">
   The <xsl:value-of select="./name"/>is <xsl:value-of select="./value"/>
   <break size="medium"/>
  </xsl:for-each>
   <goto next ="#new"/>
  </block>
</form>
     </xsl:template>
     <xsl:template match="list">
<menu>
    <prompt><ibmvoice  gender="female"  age="middle_adult">You  have  the
following choices <break size="medium"/>
          <enumerate>
          Say <value> <xsl:attribute name="expr"> _prompt</xsl:attribute>
</value>
        or press <value> <xsl:attribute name="expr"> _dtmf</xsl:attribute>
</value>
   </enumerate>
   </ibmvoice>
  </prompt>
  <xsl:for-each select="link">
  <choice>
   <xsl:attribute name="dtmf">
        <xsl:number count="*" />
   </xsl:attribute>
           <xsl:attribute    name="next">    <xsl:value-of
select="./value"/></xsl:attribute>
```

```
    <xsl:value-of select="./name"/>
  </choice>
    <break size="medium"/>
  </xsl:for-each>
</menu>
    </xsl:template>
</xsl:stylesheet>
```

Transcoder

Applying transformations to PortalXML and rendering the output to devices is key to the Campus Portal architecture. Performance is also very important, as applying various transformations is a process-intensive task. We chose to use Cocoon as our transformation engine. Cocoon is an Open Source Project from the Apache Software Foundation. It is a complete content publishing framework with many other capabilities. For the Campus Portal, we are using it as a transcoder for the following reasons:

- Cocoon is stable and well-proven technology to perform XSL transformations.
- It is a Java Servlet technology-based engine. So, we could easily integrate it with our framework.
- It is available as Open Source.

You may want to review the details of Cocoon by revisiting Chapter 15.

Transcoding Engine Implementation Details

Cocoon is a Servlet-based engine. This requires that the following entry be made in the Servlet engine's configuration (in our case Tomcat).

```
<servlet>
  <servlet-name>org.apache.cocoon.Cocoon</servlet-name>
  <servlet-class>org.apache.cocoon.Cocoon</servlet-class>
  <init-param>
    <param-name>properties</param-name>
    <param-value>WEB-INF/cocoon.properties</param-value>
  </init-param>
</servlet>
<servlet-mapping>
```

```
  <servlet-name>org.apache.cocoon.Cocoon</servlet-name>
  <url-pattern>*.xml</url-pattern>
</servlet-mapping>
```

This configures the Servlet engine to invoke the Cocoon Servlet when an XML document is requested. However, adding to the configuration of Cocoon will not make Cocoon apply any transformations. The Cocoon reactor also requires processing instructions to forward the document to the right processor or formatter.

These processing instructions are:

- For processing:

```
<?cocoon-process type="xxx"?>
```

- For formatting:

```
<?cocoon-format type="yyy"?>
```

These processing instructions are used to indicate the processing and formatting path that the document should follow. Despite using these, Cocoon would still not know which stylesheet to use when transcoding the XML document unless the XML document specifies the stylesheet.

The formatting processing instruction is used to indicate what formatter should be used to transform the document tree into a suitable form for the requesting client.

```
<?xml version="1.0"?>
<?xml-stylesheet href="portal-html.xsl" type="text/xsl"
media="explorer"?>
<?xml-stylesheet href="portal-vxml.xsl" type="text/xsl" ?>
<?xml-stylesheet href="portal-wml.xsl" type="text/xsl"
media="wap"?>
<?cocoon-process type="xslt"?>
```

So based on the media type, the appropriate stylesheet is applied to the XML. Here are some media types (se Table 19-2).

Table 19-2. Media Type Mapping.

Browser	Media
Explorer	explorer
VoiceXML	vxml

Table 19-2. Media Type Mapping. (Continued)

Browser	Media
MIDP devices	MIDP-1.0
WAP/WML	wap
Netscape/Mozilla	mozilla5/mozilla6

Personalization Engine

Personalization is the use of technology and end-user information to tailor interactions between a service and each individual user. Using information either previously obtained or provided in real-time about the user, the exchange between the parties is altered to fit that user's stated needs as well as needs perceived by the service based on the available user information. The purpose of this technology is to:

- Better serve the user by anticipating their needs.
- Make the interaction efficient and satisfying for both parties.
- Build a relationship that encourages the user to return for subsequent transactions.

Personalization allows content to be customized for an individual user's preferences, type of device, and time and location where the device is used. It is a key element when it comes to mobile devices by making interactions faster and easier. For example, given that it is highly inconvenient for a user using 15 keys of a cell phone to punch in data, user preferences are recorded and customized whenever a service is accessed.

The subsystems described in the following sections form the Personalization Engine.

User Profiling and Form Filling

Users are profiled in a relational database. This module stores the form information that the user has previously entered and uses that information to auto-fill the forms when reused. XForms are used to achieve this. This can be very useful on devices with limited input capabilities. XForms, the device-independent XML form definition model, provides the capability for forms to collect data as XML instance data.

Through the use of this instance data, XForms-aware applications can auto-fill and pre-fill forms. The user interface provides a standard set of controls that are directly usable in XML tag sets. When a user fills in a form, the inputs are stored and each form is cached in the database. So when the same form is presented the next time, the form is automatically populated with the input instance data from the cache. Users can choose to use the defaults based on their previous entry or choose to clear and fill in the form with new data. This user profiling and form filling is the basis for our implementation of personalization features in Campus Portal. The PortalXML designed for the Campus Portal is based on the XForms concept.

Service Provider/User Personalization

In addition to auto-form filling, there is also a need to personalize service content for the users. The two main types of users are identified as the Service Provider and the Service User. The Service Providers log into the system and personalize any fields for the services they provide. Service users view these personalized services and personalize content for ease of use. When the user visits the system the next time, the data already personalized by the user appears as default values. This aids the user by eliminating the need to re-enter form data each time the same information is requested.

Services Framework

While the transcoding engine provides a proxy between the clients and the services, the important component that makes the services available to users of the Campus Portal is the Services Framework.

We have chosen a service-oriented architecture over other distributed component models for the following reasons:

- Service oriented architectures are more loosely coupled than other distributed component models.
- The Web Services model is transport protocol-independent and only requires the ubiquitous HTTP as the minimum transport protocol, while traditional CORBA and EJB requires RMI/IIOP.
- Using SOAP as an RPC model enables us to create modular and loosely coupled services.
- Tunneling SOAP over HTTP promises more interoperable services than binary protocols such as IIOP.

- It is easy to wrap existing services and distributed components as Web Services.

The Services Framework is a simplified layer that utilizes the standard Web Services technologies to provide a runtime environment to perform the following functions:

- Dynamic deployment of Web Services
- Registration of services
- Lookup of the services
- Proxy creation for the services
- Invocation of the services

Let's look at how the Services Framework provides these functions in detail. We used IBM's Web Services Tool Kit (WSTK version 2.3) for our implementation.

The WSTK provides the following components that are used in our implementation:

- A private UDDI registry
- An embedded Apache SOAP server
- A Web server

Warning: Evolving Standards Ahead

Warning

The Web Services ToolKit (WSTK) from IBM at the time of the writing of this book was a beta product and was undergoing significant changes. It also uses the Axis runtime for SOAP implementation, and it was also changing rapidly. However, the concepts we present will remain the same regardless of which SOAP implementation used. We could have used Axis for our Campus Portal, but it was available only after we had made signifigant progress during the development of the Campus Portal.

One of the issues we would have needed to address is the difference between how Apache Soap and Axis handle custom SOAP serializers. Both products support custom SOAP serializers, but do so in slightly different ways. Apache Soap uses a **SOAPMappingRegistry** to manage the catalog of defined custom serializers. Axis, as we saw in Chapter 14, uses an XML file to maintain this catalog. ❑

All services in the Campus Portal are published into a private UDDI registry and deployed into the SOAP server. Service discovery and invocation was done from JavaServer Pages (JSPs) deployed on to a Web server (Jakarta Tomcat).

The UDDI4J API provides access to published WSDL service files (and thus all the necessary information needed to invoke a service) to a client given a unique service name. Client applications can easily generate instances of the client stub (also known as a service proxy) and perform service invocations. The services framework model uses these procedures in all the Java clients that invoked Web Services.

WSTK version 2.3 supported conversion of EJBs deployed on IBM's WebSphere application server only. It was undesirable to tie service providers to specific J2EE implementations. Hence, the alternative conversion of a JSP into a Web Service was considered briefly. However, since JSPs could be directly accessed via HTTP Post/Get methods it was thought that no special advantage would be gained by the merging of the SOAP-based Web Services and JSP technologies. Moreover, JSPs are of special merit when business logic needs to be embedded into presentation code. Since all of the data display is handled by the transcoder component and all of the business logic is handled by the service component, the use of JSPs is not an attractive option. The final model adopted is the use of a simple Java client that uses JNDI lookups to locate one or more EJBs (see Figure 19-9).

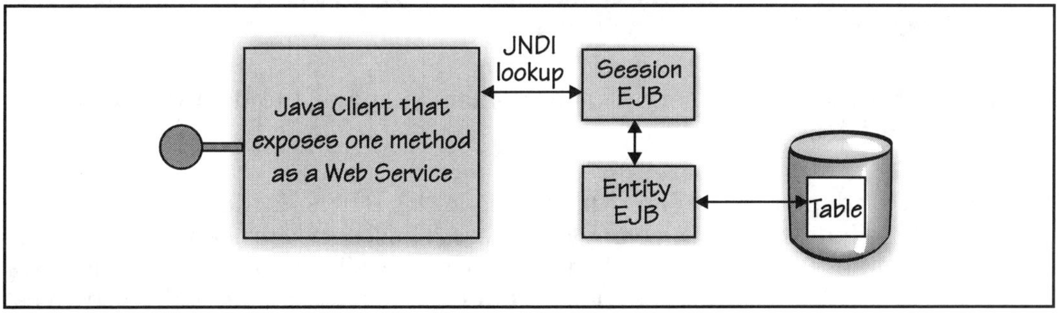

Figure 19-9. Java Client as Web Service.

This model is simple, application server-independent, and easy to implement and test in a standalone mode. The implementation process, however, emphasized the need to further standardize the interface between the system components by the use of:

- Standard function signature formats for all exposed methods of a Web Service
- A standard, pre-agreed upon format for returning requested information, i.e., PortalXML

The first issue is addressed by the specification that all exposed methods accept String or PortalXML as arguments. The return type is always a PortalXML object. A related issue addressed is whether a service should be standalone or linked to other Web Services. The latter approach requires a means of specifying interactions between Web Services. The IBM Web Services initiative defines such a language: the Web Service Flow Language (WSFL). This language provides specifications for:

- The usage pattern of a collection of Web Services, in such a way that the resulting composition describes how to achieve a particular business goal

- The interaction pattern of a collection of Web Services in order to obtain a description of the overall partner interactions

However at the time this book was going to press, this WSFL specification is incomplete. Thus, a design decision was made to use standalone Web Services with one exposed method per Web Service. Some workflow and interactions are provided by the controller component of the Transcoding Engine.

Service Invocation Functionality

Two common tasks are required to invoke any service in our Campus Portal:

- Procurement of a callable service instance or proxy to the Web Service from the UDDI registry

- Actual invocation of the service and un-marshalling of the data received from the Web Service

The approach adopted is to hide the SOAP and UDDI specific implementation details from the individual service implementers. Thus, service-independent code is factored out into separate JSPs. These JSPs are easily included into the JSPs implemented by the individual service through the use of JSP include directives. Furthermore, it was decided that a service proxy should be instantiated behind the covers and added to the session to establish and maintain a session once a user logged into the Portal. Thus, to invoke a service, individual services only need to retrieve the proxy from the shared session, then provide basic information (such as specifying the values for the arguments passed in while invoking a service).

SOAP Serialization of User-defined Data Types

The SOAP API provides transparent support for the serialization and de-serialization of basic Java data types and other standard data types such as arrays and vectors. However, specialized SOAP mappings must be provided for user-defined data types.

We mentioned that all the Web Services deployed on the Portal consume and produce PortalXML objects. As you have noticed in our PortalXML discussion the PortalXML object comprises a lot of nodes and structures.

Thus, it is necessary to implement the creation of the appropriate SOAP Mapping Registry (SMR) for user-defined object instances that are passed back and forth between the various Campus Portal components. In order to hide these SOAP-specific implementation details from individual service implementers, this functionality is factored out into a base class (called **PortalService**) and all individual services are derived from this base class. As additional services are added to the Campus Portal, the use of an inheritance model ensures that implementation changes are restricted to a single base class. This simplifies the integration process.

Services Implementation

Using the Campus Portal, we can develop and deploy a variety of services. We have described several such services in Chapter 18. Let's take a look at some of the more interesting implementation details for some of those services.

Information Services

Information Services deal mostly with getting the data from a data source based on user input. Typically these services are developed using the J2EE EJB model. We can expose the Session bean as a Web Service as shown in Figure 19-10.

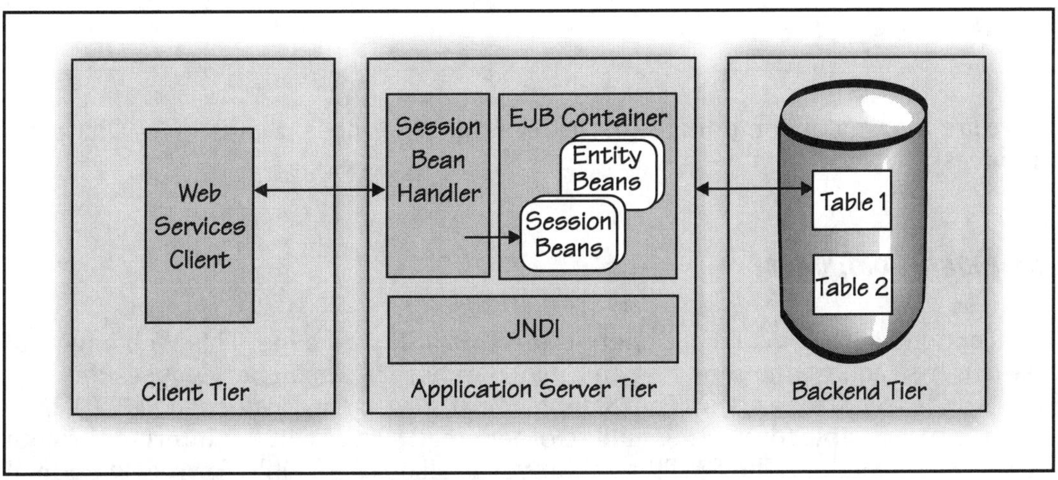

Figure 19-10. Session Bean Exposed as a Web Service.

As we have stated earlier, all the Campus Portal services are implemented as Web Services and adhere to a standard calling convention. Our Service would contain the code needed to invoke methods on the Session bean, whose methods would accept the arguments passed to the Web Service from the Services Framework, which in turn received the arguments from the user. The exposed Web Service would then create the PortalXML document that represents the service's content.

Messaging Services

The Messaging Services use a variety of diverse technologies to perform their functions (see Figure 19-11).

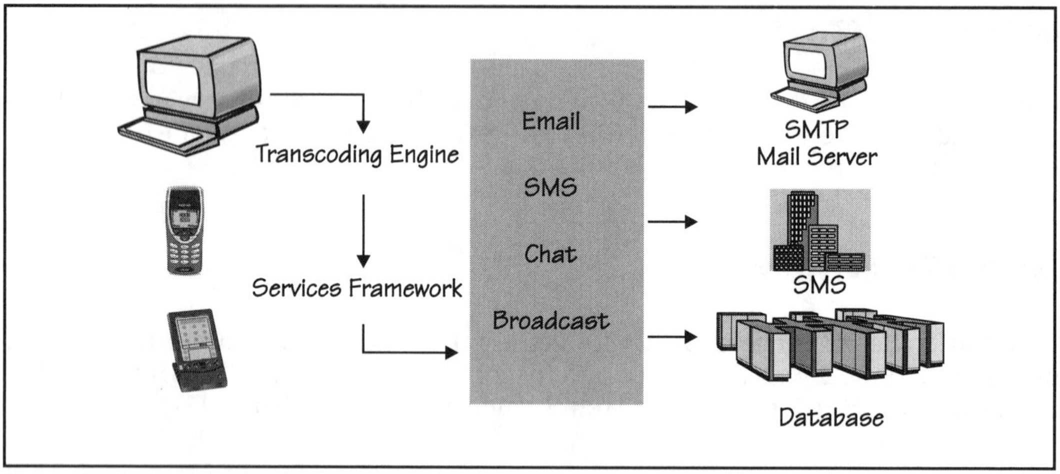

Figure 19-11. Messaging Services.

There are four major components in the messaging services: Email, SMS, Chat, and Broadcast.

The Email Component

The front end of the email component is a HTML/JSP Web page. This Web page will take a user's request and send it to an Email Servlet. The Email Servlet uses the Java Mail API to send messages. The *Simple Mail Transfer Protocol (SMTP)* mail server is responsible for routing mail to other email servers on the Internet through an SMTP connection. The SMTP mail Server is also responsible for receiving mail from other email servers on the Internet through a POP3 connection (See Figure 19-12).

This email component is exposed as a Service in our Campus Portal. Wireless devices will be able to connect to the service framework through the Transcoding Engine. Thus, wireless devices will be able to use this Email Web Service.

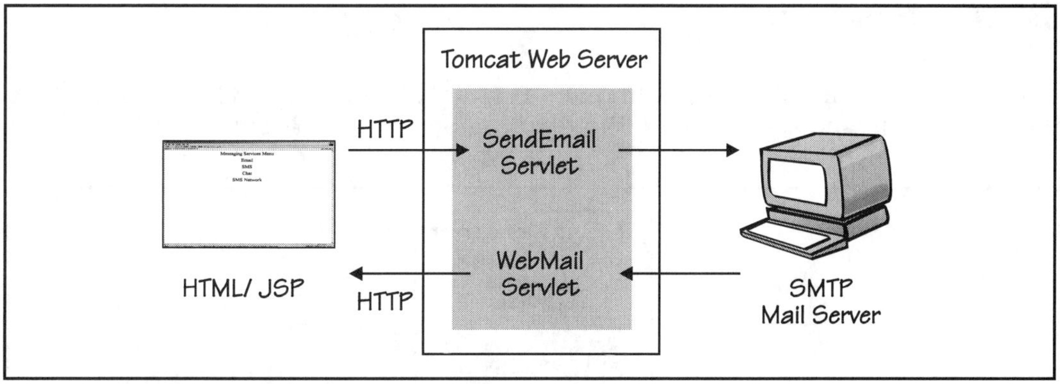

Figure 19-12. The Messaging Service Email Component.

Simple Messaging Service (SMS) Component

The simple messaging service component consists of three major components. The front end is a HTML/JSP Web page that will take the user's input data and send the input to an SMS Servlet for processing. This Servlet uses Simplewire's Java SMS toolkit to compile the data and send it to the server on the Simplewire.com network. The SMS server will forward the message to an appropriate phone carrier (See Figure 19-13). We used Simplewire for our SMS service.

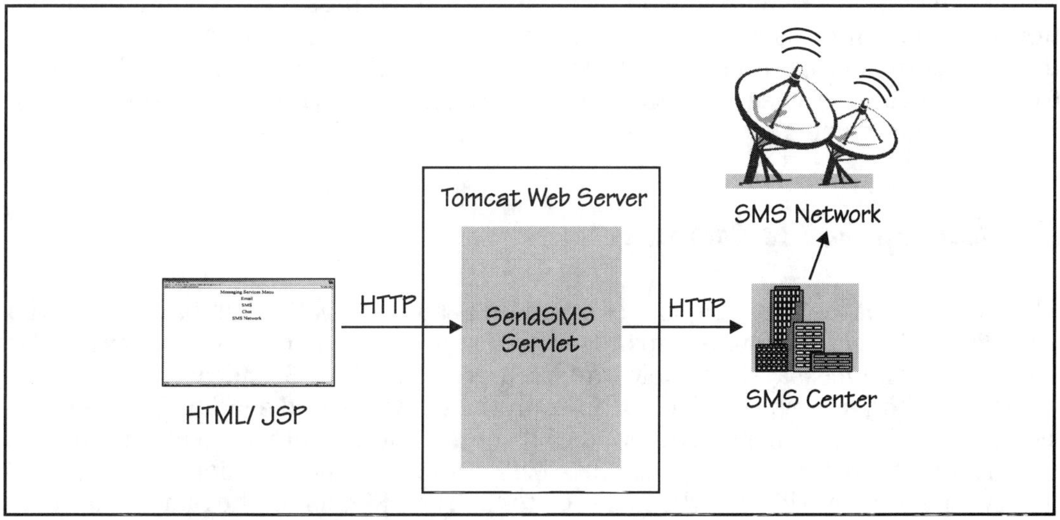

Figure 19-13. The Messaging Service SMS Component.

Chatting Component

The front end of the Chatting component is an HTML/JSP page. This page will take the user's input and send them to the Chatting Servlet (see Figure 19-14).

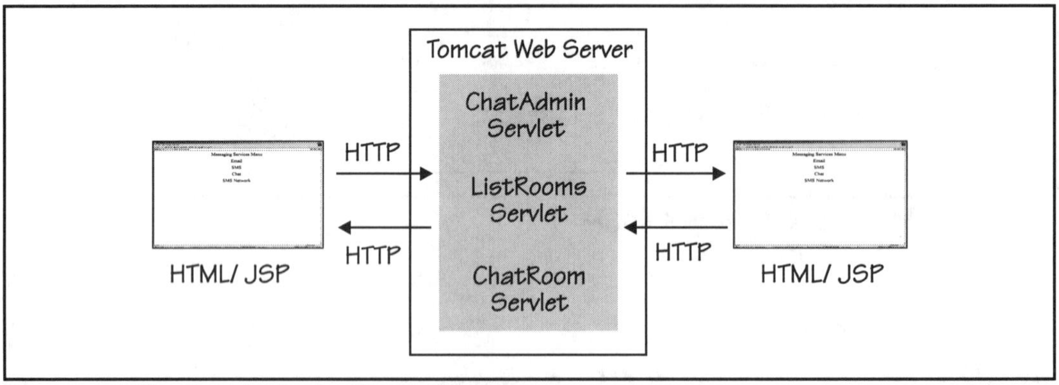

Figure 19-14. The Messaging Service Chatting Component.

The first requirement for Chatting to work on wireless devices is to have a browser-enabled device, like a smart phone or PDA with Internet capability. The second requirement is to sign up with a Chat Service Provider such as *Jabber.org*. The next thing is to set up a server to store the necessary information such as user instant message preferences, user profiles, buddy list, groups, and devices. Any user can access their instant message data through any Web-enabled mobile device and send an instant message to any other Web-enabled wireless device. While information is stored on the server, users don't need to download additional software to their wireless devices to use the messaging service. Technically, users will be able to communicate with users of other IM clients such as AOL Instant Messenger, ICQ, MSN Messenger and Yahoo! Messenger.

Broadcasting Service Component

The front end of the broadcasting component is an HTML/JSP Web page. This Web page lists different mailing groups. Users can subscribe/unsubscribe to a particular group and send messages to their preferred group(s). The Broadcasting Servlet is responsible for retrieving information from the database and sending the message according to the recipient's preference. If the recipient prefers email format, this Servlet will send the message to the SMTP server. If the recipient prefers SMS format, the Servlet will send the message to the recipient using the Simplewire.com network (see Figure 19-15).

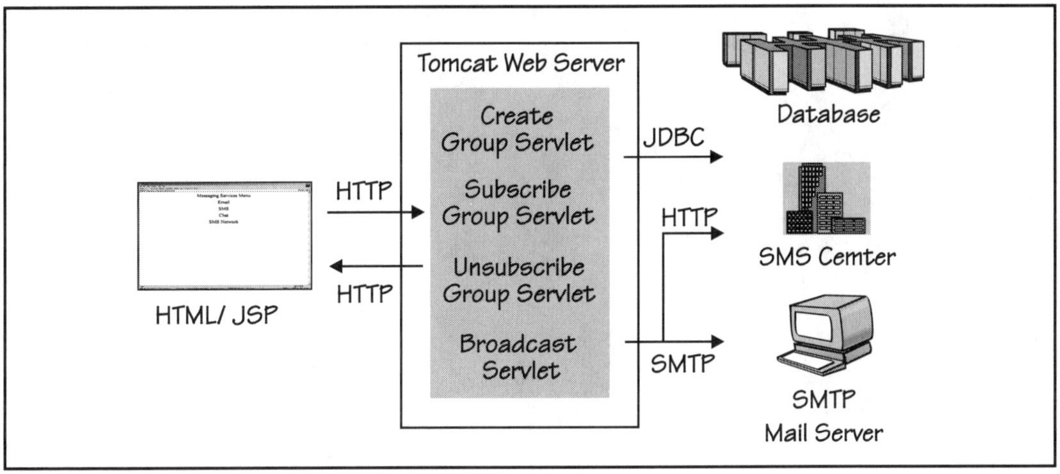

Figure 19-15. Broadcast Component Architecture.

All the messaging components are registered as a service on the Campus Portal service framework. Wireless devices will be able to connect to the service framework through the transcoding engine. Thus, wireless devices will be able to use this broadcasting Web Service.

SUMMARY

In this chapter we discussed the implementation details of the Campus Portal Framework. We discussed the design decisions we had to make due to technology constraints and the lack of well-established standards. This discussion has been a quick overview of the implementation. In order to further understand the details, visit the code repository on the accompanying Web site at *http://wireless-java.engr.sjsu.edu*. We will conclude our journey in the next chapter by reviewing the current state of wireless technology that we introduced in Part 1, and take a look at the exciting possibilities that lie ahead in mobile and wireless application development.

Chapter 20

The Enterprise Goes Wireless

We finish our long journey with a look ahead at the future of enterprise wireless application technology. This look ahead promises to tempt us with some exciting new ideas and application development possibilities. These exciting new technologies will take the form of new handheld devices with ever-increasing processing power and network connectivity, new 3G and 4G wireless networks capable of pushing gigabytes of data around effortlessly, and new application design tools and platforms to ease our development burden. These technologies, when taken together, will result in an entirely new breed of enterprise wireless application. At first glimpse, the future looks bright indeed!

Wireless networks, handheld devices, and the applications that utilize them are all currently in their infancy. As we progress toward the wireless utopia envisioned in Part 1 (and reiterated here), there will be growing pains. The wireless application "holy grail" environment is still down the road a ways. We'll get there, but it may take some time.

In this final chapter, we're going to reflect on some of these new technologies that will make all our lives as application developers much easier, while keeping our feet on the ground. We will discuss what we can expect from device manufacturers, tool vendors, and wireless providers, and when we can realistically expect to see them. We hope that by the end of this chapter you'll be eagerly awaiting the new technologies lying just over the horizon.

THE HANDHELD DEVICE OF THE FUTURE

Enterprise wireless handheld devices, whether they are highly functional PDAs or more limited J2ME- or WAP-enabled cell phones, are constantly being enhanced with new features. Within a short time, we'll see PDAs that rival the computational power of today's desktop PCs. As the processors in the PDAs of tomorrow increase in speed and processing power, the ability to provide many new functions becomes available:

- Rich multimedia features
- Voice Recognition and Synthesis
- Improved handwriting recognition

This progression is shown in Figure 20-1.

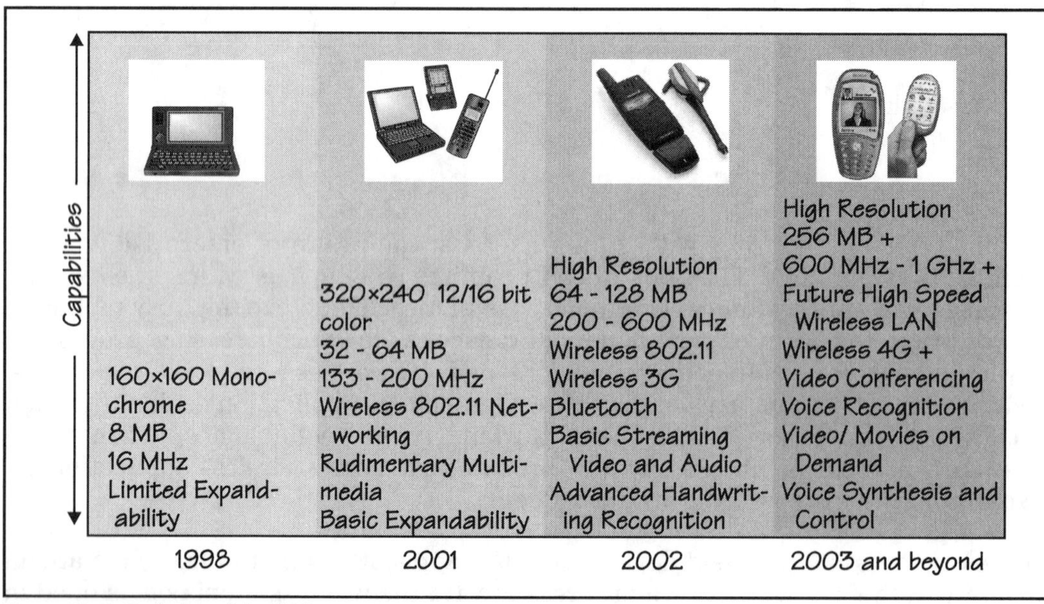

Figure 20-1. Evolution of the PDA.

The PDA is not the only handheld device that is being improved. Cellular phones are continually undergoing enhancement, with new features added to each successive model. A new phone with the latest and greatest technology seems to appear every few months. There's no reason to assume this trend won't continue into the foreseeable future.

Ultimately, the simple mobile phones of today will evolve into true handheld wireless communicators. The communicators of tomorrow will have:

- Advanced, high-resolution color screens
- Multimedia capabilities (both audio and video)
- Location awareness via GPS receivers or other means
- A faster processor with more memory for tomorrow's demanding applications

However, without a fast wireless network to support these new devices, we won't be much better off than we are today. Fortunately, the major wireless providers are already hard at work designing the next generation of wireless network infrastructure.

FYI

Communicators of Tomorrow

Briefing

We discussed communicators in Chapter 2. Communicators are devices with substantially more computing resources than the typical mobile phone or PDA; typically including a keyboard, large screen, and fast processor.

- We've just discussed the PDA and mobile phone of tomorrow, and what new features they're likely to contain. So, what features will the communicator of tomorrow provide? We'll likely see more cell phone/PDA combinations (devices like the Nokia 9290 discussed in Chapter 2), with more devices based on Microsoft's PocketPC platform.

- However, the trend seems to be toward merging the small notebook PC platform with the PDA. These ultra-light, ultra-portable, high-performance devices will begin to appear in late 2002 or early 2003.

- Platforms such as Microsoft's TabletPC will deliver desktop-like performance on a handheld device, slightly larger than today's handheld PCs, but smaller than today's PDAs.

Looking ahead even further, new technologies like flexible LCD displays (such as LCD displays that can be rolled up like a newspaper) may make it possible for us to carry our communicators around in a tube not much larger than an oversized ink pen. Combined with the latest high-speed wireless networking (by that time, 4G or 5G), the combination of these two technologies would lead to truly pervasive computing. The "computing anytime, anywhere" mantra would truly be accurate! ❑

3G...4G... And Subsequent Networks

The next generation of wireless network infrastructure promises greatly improved bandwidth that our handheld devices can truly take advantage of. We're just beginning to see the 3G networks rolling out to a few select customers. The initial results of the first 3G deployments, although offering higher performance than our current 2G and 2.5G networks, don't seem to be fulfilling the true promise of what a 3G network can provide. Bandwidth measurements seem to be indicating that we aren't seeing the transfer rates touted by the wireless providers.

Why is this the case? If you recall from our Chapter 3 discussion about 3G networks, the 3G deployments that have already been made are still in "beta mode." In other words, current 3G networks aren't quite production-ready. Wireless network providers are still at work addressing bugs and other glitches in their systems.

Things are bound to change once these bugs are shaken out. Once the promise of 3G's drastically improved bandwidth becomes a reality, the hardware capabilities of tomorrow's handheld devices can really be put to use.

Unfortunately, if Moore's Law holds and computing power increases at its current rate, even the vast bandwidth promised by 3G networks may be insufficient for the enterprise application of the future. Real-time video and audio, along with robust email retrieval can saturate even the fastest network, and for this reason, wireless network providers are starting to look at 4G, 5G, and beyond.

Figure 20-2 shows what types of applications each generation of wireless network supports. Today, we're stuck in the realm of 2G and 2.5G, which provide barely enough bandwidth for simple email retrieval. Future 3G and 4G networks promise enough bandwidth for sophisticated applications including advanced multimedia, voice over IP, and fast data transfer.

Alas, at the end of the day, neither an advanced handheld device nor a fast wireless network is of much use without applications to take advantage of their resources. How will tomorrow's enterprise wireless application take advantage of the seemingly immense (by today's standards) computational and network resources available to it?

Current IP based networks (of which the Internet is a good example), are based on an old addressing scheme called IPv4. IPv4 was a protocol that served its time well, but the need has clearly arisen for a newer standard. There are several areas where IPv4 is deficient—IPv6 attempts to solve them.

IPv4, initially developed over 20 years ago, provides 32-bits worth of addressing, which in theory could yield about 4 billion addresses. Due to some restrictions on addressing, the number is substantially less than that, and is expected to be

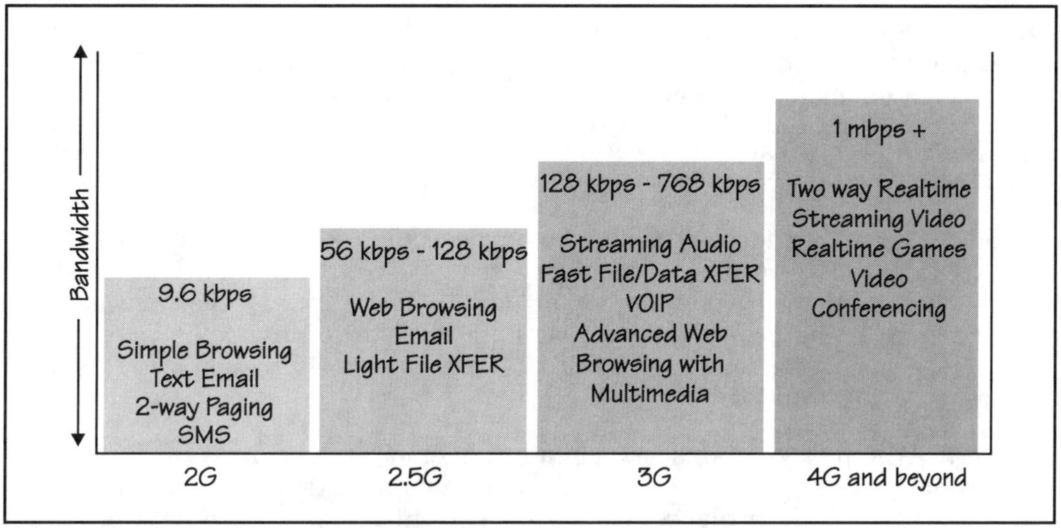

Figure 20-2. Evolution of the Wireless Network.

completely exhausted within the next few years. By increasing the number of usable addresses to 10^{38}. This means that there would be enough usable addresses to give each person on the planet several thousand (if not several million) addresses all to themselves. This amount should be plenty for the foreseeable future.

FYI

IPv6

Briefing

Current IP based networks (of which the Internet is a good example), are based on an old addressing scheme called IPv4. IPv4 was a protocol that served its time well, but the need has clearly arisen for a newer standard. There are several areas where IPv4 is deficient—IPv6 attempts to solve them.

IPv4, initially developed over 20 years ago, provides 32-bits worth of addressing, which in theory could yield about 4 billion addresses. Due to some restrictions on addressing, the number is substantially less than that, and is expected to be completely exhausted within the next few years. By increasing the number of usable addresses to 10^{38}. This means that there would be enough usable addresses to give each person on the planet several thousand (if not several million) addresses all to themselves. This amount should be plenty for the foreseeable future.

There are other problems with IPv4 besides its limited addressing:

- ***Security:*** IPv4 security is grafted on to the protocol at higher layers. Security was not built into the protocol from the beginning. IPv6 has integrated, mandatory security defined as part of the standard.
- ***QoS:*** IPv4 has had varying levels of support for Quality of Service. IPv6 formalizes the level of QoS required for implementers.
- ***NAT:*** Network Address Translation, or the "hiding" of one network behind a single address, became a necessary crutch due to the limited addresses present in IPv4. NAT systems can oftentimes block or otherwise garble incoming connection requests, making it difficult to provide services from behind the NAT gateway. IPv6, with its vastly larger address space, does not require NAT solutions.

Early IPv6 implementations are already underway—most routers, PCs, and operating systems support IPv6 already. To switch the entire Internet to IPv6 will take years, so it's clear that we will need to live in a mixed IPv4/IPv6 world for a while. IPv6 is capable of living on top of the existing IPv4 network as a temporary solution while the Internet is converted en masse. ❑

Tomorrow's Enterprise Wireless Application

With powerful new handheld devices and ultra-fast wireless networks on which to run, what can we expect from tomorrow's enterprise wireless applications? With the improvements discussed in earlier sections, an entirely new class of applications becomes possible. These new applications, crafted explicitly to take advantage of the enhanced power and speed of tomorrow's enterprise wireless environment, will usher in a new era in wireless computing.

Users of these applications will see the most benefits, from applications with enhanced multimedia capabilities to applications utilizing the vast network resources available to the device.

Other new types of applications could include:

- ***Location-aware applications:*** Future handheld devices will have the capability to intrinsically know where they are located, or will have the capability to determine their locations from GPS signals (or by other means). With this capability, wireless applications will be able to provide custom content based upon the location of the device. Sales and advertising are just a few of the areas where this capability could be exploited. For example, as one strolls through a shopping mall, sales and other advertisements can be displayed on the screen of the wireless device, based on nearby stores.

- ***Personalization:*** Future wireless applications will be highly personalized, providing custom content based on the user's identity (and location, as described above).

- ***M-Commerce:*** Mobile commerce, or the ability to conduct business transactions anytime, anywhere, and from any device, will provide an entirely new way for retailers to do business. M-Commerce could include anything from using your mobile browser to navigate to a vendor's Website and purchase goods, to using location-based technologies to pay for goods and services in a store without waiting in a check-out line.

- ***Streaming video:*** Online on-demand movies, realtime video conferencing, and media file sharing become possible with the next generation of wireless applications.

The wireless device's user is not the only person who will benefit from these new improvements. As application developers, we too will benefit from the merging of existing technologies and the emergence of new technologies that are just beginning to be seen today.

FYI Briefing

Java Card Revisited

One interesting area where the Java Card standard is receiving a lot of attention is in the GSM arena. GSM chips, if you recall from Part 1, are chips embedded in certain mobile phones that store the identity of the phone and its owner.

Several mobile phone vendors are looking into enhancing the GSM chip with Java Card capabilities. This would open up a whole new area for personalization and security on MIDlets running on the mobile phone, as the benefits of the Java Card API (described in Chapter 12) would be available to MIDlets. A user's personal preferences could be stored in the Java Card, and the MIDlets running on the phone could use the information stored inside when displaying content to the phone's owner. Further, if the owner switched to another GSM phone, the GSM chip could easily be moved to the new device, and the owner would not have to re-enter their preferences again. ❑

So what benefits are we, as application developers, likely to see in the coming years? In the Java space, we're likely to see a further integration of the various Java2 editions. This integration translates directly into more robust wireless applications, as our wireless devices can more seamlessly browse and manipulate content from our J2EE servers. If you recall from our discussion in Chapter 13 about J2EE servers, currently the trend is to push as much processing as possible to the server, leaving our client applications "lean and mean." As devices and networks become more powerful, perhaps some of the work can be shared between the server and the client. The potential homogeneity of future Java2 editions means our roles as application developers become simplified.

For example, with a truly integrated environment, our mobile clients can take advantage of J2EE applications without the need for an intermediary Servlet or JSP. As the computing potential and processing power of our devices grow, we'll soon reach the time in which our mobile clients can utilize native J2EE application client containers when transacting with J2EE-housed applications. This simplifies the development of our applications significantly, as we no longer have to design, test, and maintain the middle layer of logic (in this case, the Servlet or JSP).

Another area where the integration of Java2 "families" may play an important role is in transcoding: the tailoring of content for a given device. Transcoding, discussed at length in Chapter 15, currently requires substantial server-side processing resources to tailor content for the myriad of devices in use. If some of this processing could be pushed back to the mobile device (when the devices attain sufficient computing power to handle the task), it would certainly simplify our server-side coding and greatly reduce the class of server side hardware needed.

So, where does this leave J2ME? J2ME will still be a viable programming environment for the foreseeable future, thanks to its ability to accommodate new devices through different profiles and configurations. J2ME profiles like the Foundation, Game, and RMI profiles may be supported by the next generation of devices, while newer profiles we haven't seen yet will be created to support devices that are still under wraps. It's clear that J2ME is being accepted by the industry, with most mobile device manufacturers including support for J2ME in existing or planned products.

What's more, the gravitation towards J2ME by device manufacturers might change the landscape of potential application development environments. Of all the application platforms we have discussed until now (WML, VoiceXML, and J2ME), J2ME offers the highest quality of user experience when paired with a proficient application developer. This claim is backed up by J2ME's advanced graphics, user interface elements, networking capabilities, and persistence mechanisms. Figure 20-3 compares the application development platforms we've discussed thus far from this perspective.

Of course, both WML and VoiceXML still play an integral role in today's marketplace—there are millions of phones deployed with WML browsers in use today. There are even more phones deployed that have no WML browser, necessitating the development of VoiceXML interfaces to our applications. Neglecting these potential users would be a foolish move. However, over the next few years, the role of WML and VoiceXML may diminish as more types of devices become able to house a MIDP or future J2ME implementation.

Clearly, we all desire the highest quality of user experience, both as users and developers of applications. As shown in Figure 20-3, MIDP 2.0 will bring us the highest quality of user experience through its new APIs (which we discussed in Chapter 9).

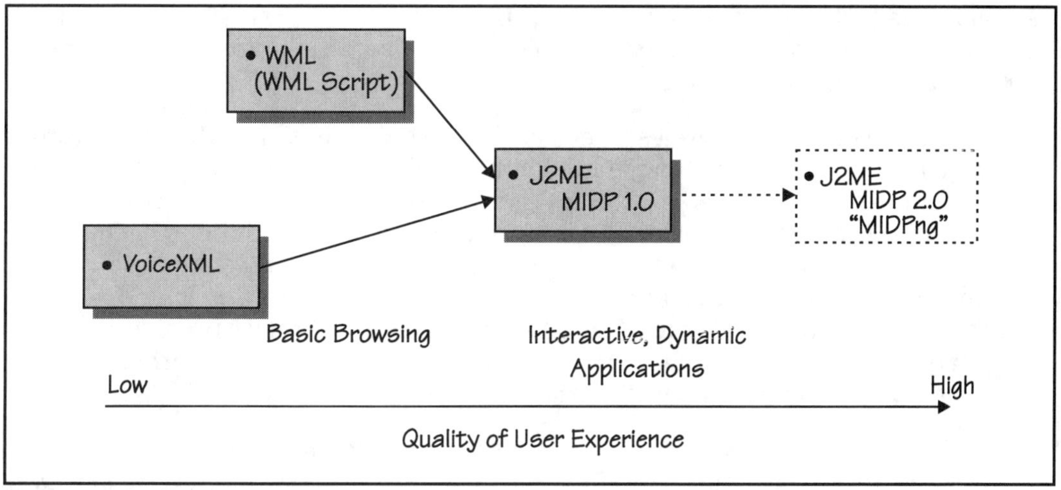

Figure 20-3. Quality of User Experience.

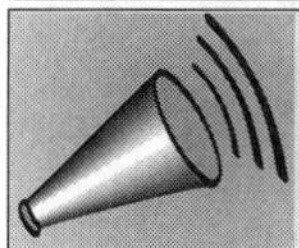

Soapbox

Wireless Technology's Killer App

There are several reasons why wireless applications have not found immediate, widespread acceptance in the U.S. First, consumers in the U.S. tend to view their handheld devices in a more limited fashion than do users in overseas markets like Europe or Japan. European and Japanese consumers tend to view their phones as extremely personal objects, like a wallet or purse, whereas U.S. consumers treat their phones as simple gadgets. Why does such a disparity exist? Does the U.S. consumer see something lacking in today's wireless offering? Or does the U.S. consumer simply desire different functions from what obviously is so popular elsewhere?

It could be a little of both. What's clearly missing from the U.S. consumer's viewpoint, however, is a killer app—some application that the consumer just can't live without. The explosion of the Internet in the early 1990s was fueled by two killer apps—the Web and email. Japanese customers have found their killer apps in i-Mode's games and video conferencing. European customers have also found applications they rely upon as well.

So, what would it take to design a wireless application that the typical U.S. consumer couldn't live without? Do we have one yet? Is one being developed? As far as the U.S. consumer is concerned, they have not seen such an application yet. However, several areas of wireless technology have the potential to generate a killer app in coming months:

- ***Games:*** Games on wireless devices have the potential to be immensely popular. Games using Japan's i-Mode network are already enjoyed daily by millions of subscribers, and there's no reason to believe U.S. consumers would enjoy them any less. From simple board games and card games to other multiplayer experiences, games have the potential to supply a killer app. Simple games that take a few minutes to play could be a welcome respite from waiting in lines or riding on the bus or subway. Wireless providers could charge users based on the number of plays, or packets transferred. Game developers could charge per download, or per vend.

- ***Corporate VPNs:*** Being able to access corporate data from your wireless device could also comprise a killer app. VPNs, or *virtual private networks*, allow your handheld device to become an extension of the corporate LAN. This means you have full access to corporate applications from wherever you are. Imagine meeting with a customer and being able to retrieve data from your company's database via your wireless device to answer the customer's questions right there at the meeting. Being able to access product data, drawings, costs, etc. without having to promise to email the data to the customer later, could be a deal-winner. If corporations buy into this type of application, the wireless VPN has the potential to be another killer app.

- ***Multimedia:*** We've already seen the explosion of MP3 and video sharing over the Internet. While some avenues of media exchange might have questionable legal issues, a secure and robust media exchange system may prove to be yet another wireless killer app. Imagine being able to retrieve any music file, anywhere, from any artist, any time you wanted to hear it. This idea could also be extended for video and movie files. Video conferencing is another possibility that could stem from these same ideas.

The underlying reality is we can't deliver the potential killer apps described above with today's technology. For example, access to corporate data through wireless VPNs requires a fast 3G or 4G network for reasonable success. Similarly, games and multimedia applications require the same sort of networks, as well as advanced high-resolution color displays and sound features. Perhaps we'll have to wait for the next generation of devices and networks before we see an application we just can't live without. ❑

The Future of Web Services

Finally, we'll hopefully see the various Web Services camps settle their differences and agree on the more esoteric Web Services extensions. Everyone seems to have settled on SOAP, WSDL, and UDDI so far, but the industry still needs to settle on standards for transactions, security, and service invocation flow. We're beginning to see the dust clearing in the security realm—IBM, Microsoft, and VeriSign have submitted their WS-Security specification to OASIS for standardization, leaving only WS-Security and SAML as realistic competitors in this area.

Transaction and conversation flow standards are still up in the air—only time will tell what standards will win out.

But standards are not everything. While it is certainly a feather in any company's cap to have their standards adopted by the industry at large, the painful truth is that we do live in a world of mixed platforms and mixed standards. Web Services are intended to solve the panacea of problems encountered when integrating diverse applications. Although this is the goal, implementers often add platform-specific extensions to their implementations, leading to subtle incompatibilities.

That's the bad news. The good news is that vendors are noticing these incompatibilities and introducing modifications to handle each other's idiosyncrasies. These modifications are resulting in better, more portable and more interoperable code. Further, the desire to not lag behind the curve is leading to implementations that support as many new standards and features as possible.

Another notable area where the Web Services computing paradigm is gaining strength is in the area of application development tools and toolkits. Early Web Service adopters had to work with obtuse, cumbersome tools, if any were available at all. Services were frequently coded and implemented using a variety of text editors, XML parsers, and command-line compilers. Now, we're starting to see visual composition editors for Web Services, making it easy to generate service stubs and invocations with just a few mouse clicks. It won't be too long before we will be completely isolated from coding Web Services by hand. New tools will make it possible to assemble services on the screen, much like assembling a jigsaw puzzle.

In order for these visions to be realized fully, the industry needs to settle down and adopt a few key standards quickly. The Web Services environment is currently too fluid, changing all the time. Few developers will want to base an application on standards they're not sure will be around in a month, let alone a few years.

FYI Briefing

The Open Mobile Alliance

The Open Mobile Alliance (OMA) is a consortium of wireless network providers, device manufacturers, tool vendors, and Internet service providers. The OMA's goal is to bring the industry together, in an attempt to adopt or draft standards that will hopefully reduce the current incompatibilities between existing proposed standards.

The OMA hopes to formalize standards that will encompass the needs of the wireless community as a whole. These needs, which include wireless transmission standards (CDMA, TDMA, EDGE, etc), content formats (WML, etc.), security (SAML, WS-Security, etc.), and application development platforms (J2ME, etc.), are currently quite diverse, as you have undoubtedly realized through the course of this book.

The OMA was originally formed by combining the Open Mobile Architecture initiative and the WAP Forum. However, the OMA will now also benefit from the work of the Location Interoperability Forum (the LIF), the SyncML initiative, the MMS Interoperability Group (MMS-IOP), and Wireless Village; each of these groups focus on mobile service enablement. And all of them have agreed to eventually be folded into the OMA.

Why go to all this effort to form such an organization? What was wrong with having the individual groups handle their own standards? The answer lies in the fact that an all-encompassing standards body such as what the OMA hopes to become is capable of managing the "big picture," drawing from each area the important pieces needed to solve tomorrow's wireless problems. By absorbing each member group, the OMA can reconcile any problem areas that lie outside the scope of the separate member groups, while reducing duplicated effort that occurred when the groups were autonomous. ❑

Campus Portals for Everyone?

In Parts 1–3, we laid the foundation for the technologies used in our Campus Portal project. In Chapters 18 and 19, we applied the technologies and described the implementation of our Campus Portal project. In an academic environment such as ours, the design and lessons learned along the way are the primary goals. These in turn are entirely applicable in a business environment, where a portal system like our Campus Portal could be quite useful. However, in a business environment, it would be undesirable to reinvent the entire system as we did. So, short of writing one's own Campus Portal, what options are available to address this situation?

Fortunately, some solutions exist. Several companies have created portal products that you can purchase and deploy with minimal effort. The best news is that if you use one of these pre-packaged portal products, you only need to code your service providers—the services framework, transcoding engine, and all the other pieces are provided as part of the portal package.

Major portal software foundations that support wireless devices include:

- ***Oracle 9iAS Portal:*** Provides a rich environment for deploying portals in the enterprise, including support for J2EE, Web Services, and wireless devices (when paired with Oracle's 9*i*AS Wireless product), as shown in Figure 20-4. However, portlets (see next briefing box) used by wireless clients can interface with Oracle's Portal product when creating content for the user.

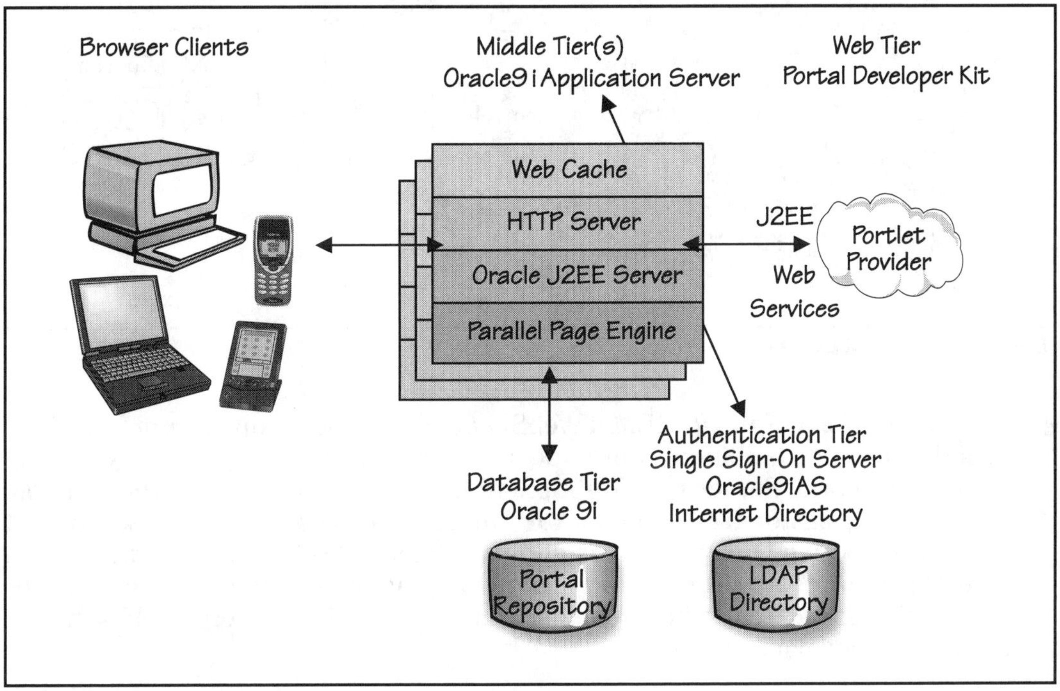

Figure 20-4. Oracle 9iAS Portal.

- ***BEA WebLogic Portal:*** BEA's WebLogic Portal delivers a complete portal framework for wireless devices. WebLogic Portal is part of the overall WebLogic Enterprise platform. It also contains intelligent administration services, personalization and interaction management, and allows custom tailoring for specific devices (see Figure 20-5).

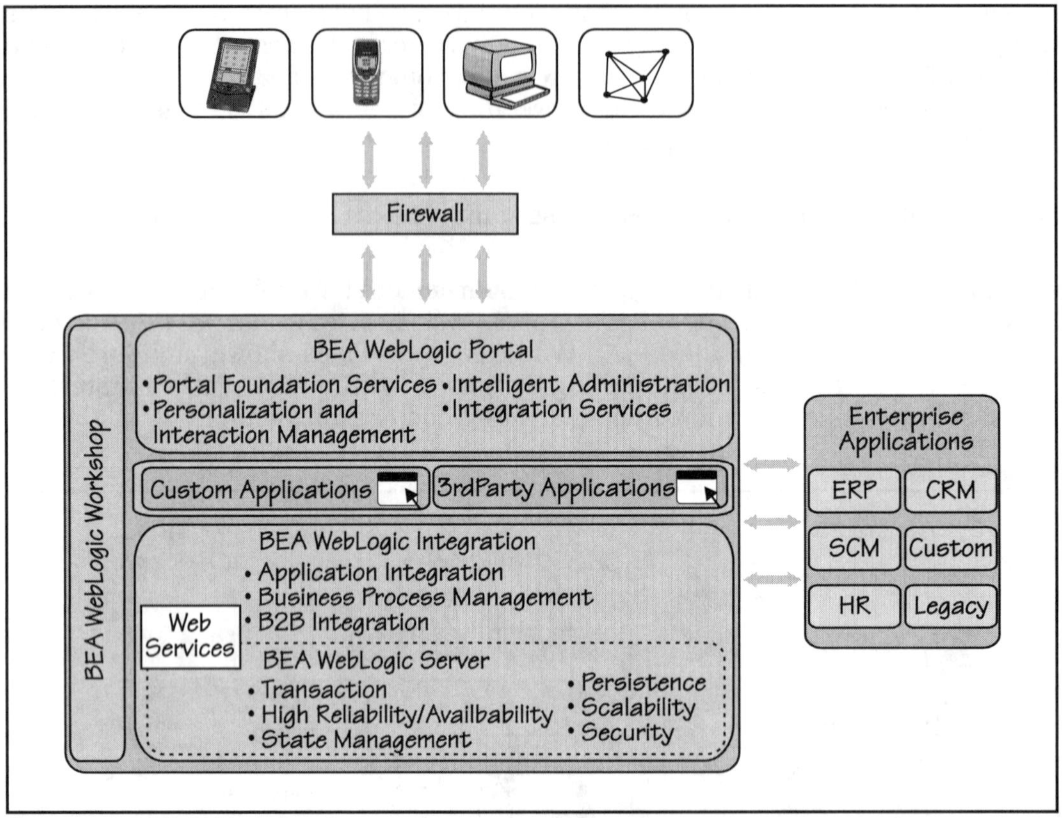

Figure 20-5. BEA's WebLogic Portal.

- ***IBM WebSphere Family:*** IBM's WebSphere family of products contains offerings that can be combined to provide a Campus Portal-like system. By pairing WebSphere Portal Server with WebSphere Everyplace Server, a system capable of delivering customized content for wireless devices from diverse content generators can be constructed, as shown in Figure 20-6. Other features, such as IBM's SecureWay LDAP server can also be added to increase your portal's utility. Your portal's content can be generated from J2EE servers, Web Services, or other legacy data sources.

- ***Apache Jetspeed:*** Apache Jetspeed is an open-source (free) portal server from the same organization that gave us the Apache Web server and Apache Axis. You can construct a robust portal server from Apache components, but there's a bit more work involved than with the other commercial implementations. Jetspeed can provide similar functionality to the other implementations, at a fraction of the price, when paired with Cocoon (for transcoding content to mobile devices), Apache Axis (for Web Services support), Tomcat (for content delivery), and a suitable EJB container.

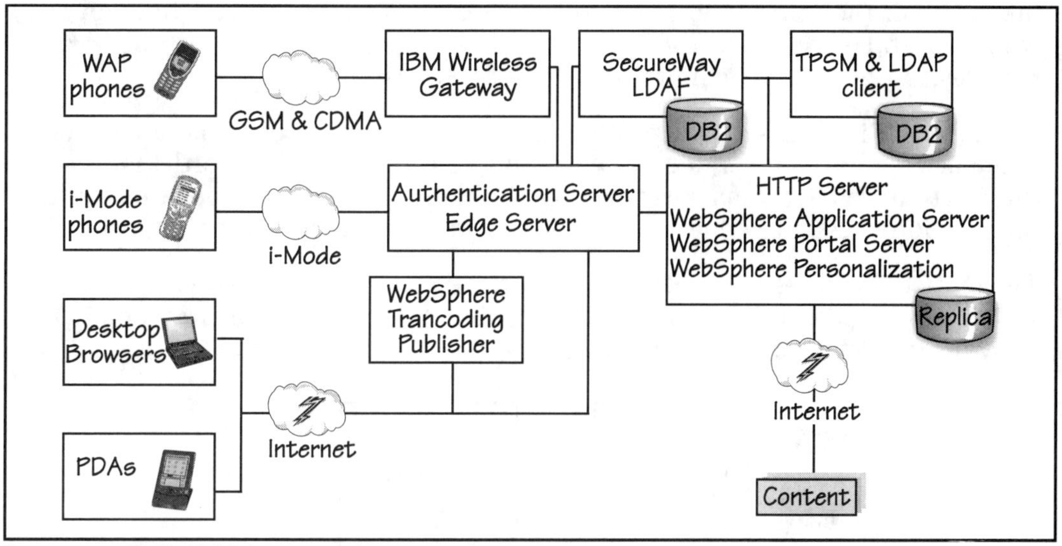

Figure 20-6. IBM's WebSphere Everyplace.

FYI

Briefing

Portals and Portlets

So what are portlets, anyway? A portal server, like the ones we mentioned above, provide a portal experience for users. Not unlike our Campus portal project, the general idea behind a Web portal is to provide personalized content to each and every user, based on that user's requirements and desires.

A portal server accomplishes this by providing a selection of portlets to each user. We've all seen portlets before on our favorite Web launch pages like AOL, MSN, or Yahoo. Portlets provide the content in the tiny configurable side boxes on these Web portals, and may even provide the main page content. Portlets are typically used to deliver customized content like news, traffic reports, stock quotes, and the like.

Each portal server has a portlet catalog from which users can select the services they are interested in viewing. Our Campus Portal had the same idea in its use of individual services.

Whether or not you want to choose one of the commercial implementations, assemble a portal server from open source components, or write your own portal server from scratch like we did is up to you. The factors at work in your particular business case will dictate the appropriate course of action. After taking in all the information you've been exposed to, we're sure to soon see robust portal servers, festooned with all the latest and greatest bells and whistles, supporting all manner of different devices!

As shown in Figure 20-7, each portlet has an associated content generator, which could be an EJB, Web Service, or something else entirely. Each portlet is passed information about the user requesting the content, including personalization information needed to generate the correct content for that particular user.

The portlet generates the content (in a portlet-specific way since each is unique), which is then passed through the transcoding engine and formatted appropriately for the device making the request.

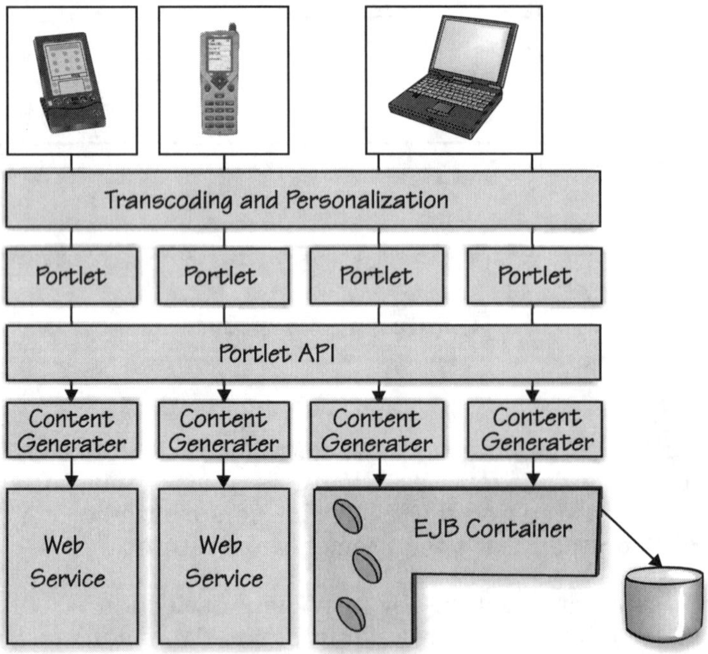

Figure 20-7. An Overview of Portlets.

Each user can have multiple portlets chosen for each portal page, and with some portal servers can even choose different portlets to be active depending on the device they are currently using. ❑

CONCLUSION

With all this material under your belt, you're ready to begin your own journey into the world of wireless enterprise application development. There's little doubt that the wireless wave has arrived. Whether you choose to develop your soon-to-be famous applications using WML, VoiceXML, J2ME, Bluetooth, or all of the above, we expect you'll have great success.

As you move forward with your wireless development, remember to keep your skills honed! Although we've introduced you to the latest and greatest technologies at the time of this writing, there will undoubtedly be new standards, platforms, and technologies created at every turn. Remember that technology marches steadily onward and those who are complacent and sit idly by may end up going the way of the dinosaur.

Where to Go for More Information

We compiled the following list of resources to help you find more information on the topics we covered in this book.

Devices

http://www.nokia.com
http://www.ericsson.com
http://www.motorola.com
http://www.palm.com

Technologies

WAP

http://www.openwave.com
http://www.openmobilealliance.org
http://www.nokia.com/wap

i-mode

http://www.nttdocomo.com

Compact HTML

http://www.w3.org/TR/1998/NOTE-compactHTML-19980209/

XHTML

http://www.w3.org/TR/xhtml1/

Syncml

http://www.syncml.org

Web Clipping for Palm OS

http://www.palmos.com/dev/tech/webclipping/

GSM

http://www.gsmworld.com

CDMA

http://www.cdg.org

GPRS

http://www.mobilegprs.com
http://www.mobilegprs.com/

SMS

http://www.simplewire.com

802.11

http://standards.ieee.org/getieee802

VOIP

http://www.protocols.com/voip.htm

J2ME

http://wireless.java.sun.com

MIDP OTA

http://java.sun.com/products/midp/OTAProvisioning-1.0.pdf

Development Kits

http://www.borland.com/jbuilder/mobileset/index.html
www.zucotto.com
http://e-www.motorola.com/webapp/sps/site/homepage.jsp?nodeId=03M0ym4sDZx

Java Card

http://wireless.java.sun.com/javacard
http://www1.slb.com/smartcards
http://www.gemplus.com
http://www.ibutton.com
http://www.cyberflex.slb.com

Java Community Process

http://www.jcp.org

Java Server Technologies

http://java.sun.com/j2ee/
http://www.borland.com
http://www.bea.com
http://www-3.ibm.com/software/info1/websphere/index.jsp

Portals

http://www.oraclemobile.com
http://www.bea.com/products/weblogic/portal/index.shtml

Bluetooth

www.bluetooth.com
www.bluetooth.org
http://www.ericsson.com/bluetooth
http://www.zucotto.com
http://www.rococosoft.com

Opensource Web Sites

Cocoon: *http://xml.apache.org/cocoon*
Axis: *http://xml.apache.org/axis*
Apache SOAP: *http://xml.apache.org/soap*
Batik: *http://xml.apache.org/batik*
Tomcat: *http://jakarta.apache.org*

XML Parsers and Technologies

KXML: *http://www.kxml.org*
KSOAP: *http://www.ksoap.org*
NanoXML: *http://nanoxml.sourceforge.net*
TinyXML: *http://www.gibaradunn.srac.org/tiny/index.shtml*

Code Related

DashO obfusicator: http://www.preemptive.com/tools/index.html

Web Services

Axis: *http://xml.apache.org/axis*
UDDI: *http://www.uddi.org*
WSDL: *http://www.w3.org/TR/wsdl*
SOAP: *http://www.w3.org/TR/SOAP*
HP espeak: *http://www.e-speak.net/*
IBM Web Services Tool Kit: *http://www.alphaworks.ibm.com/tech/webservicestoolkit*

Books, Articles, and Reports

Durlacher's m-commerce report: *http://www.durlacher.com/downloads/mcomreport.pdf*
How to Write Java Card Applets : A Developer Guide by Zhiqun Chen (Addison-Wesley, 2000)
http://www.javaworld.com/javaworld/jw-07-1999/jw-07-javacard.html

Mobile Information Portals

http://www.wirelessdevnet.com
http://www.microjava.com

Index

Numerics

A

B

C

D

E

F

K

L

M

N

P

R

S

T

U

V

W

X